THE ROUTLEDGE COMPANION TO THE SOUND OF SPACE

This companion explores a range of conceptual and practical relationships between sound and space across various disciplines, providing insights from technical, creative, cultural, political, philosophical, psychological, and physiological perspectives. The content spans a wide range of spatial typologies, from large reverberant buildings to modest and intimate ones, from external public squares to domestic interiors, and from naturally formed environments to highly engineered spaces. These compiled insights and observations explore the vast diversity of ways in which sonic and spatial realms interact.

This publication therefore forms important bridges between the intricate and diverse topics of technology, philosophy, composition, performance, and spatial design, to contemplate the potential of sound and space as tools for creative expression and communication, as well as for technical innovation. It is hoped that by sharing these insights, this book will inspire practitioners, scholars, and enthusiasts to incorporate new perspectives and methodologies into their own work.

Through a rich blend of theory, practice, and critical reflection, this volume serves as a valuable resource for anyone interested in exploring the intricacy of relationships between space and sound, whether they are students, professionals, or simply curious. Our companion provides a cross-section through shared territories between sonic and spatial disciplines from architecture, engineering, sound design, music composition and performance, urban design, product design, and much more.

Emma-Kate Matthews is an architect, composer, musician, and researcher. Her work explores the creative intersections between sonic and spatial practices through the production of site-responsive and spatialised audiovisual projects.

Jane Burry is an architect and Head of the School of Architecture and Civil Engineering at the University of Adelaide. Her research leverages digital fabrication with simulation and feedback to create better, more sensitive, human-centric spaces through linking to environmental and perceptual experiences including the auditory.

Mark Burry AO is an architect and the Founding Director of the Smart Cities Research Institute (SCRI) at Swinburne University of Technology. His role is to lead the development of a whole-of-university research approach to 'urban futures', helping ensure that our future cities anticipate and meet the needs of all – smart citizens participating in the development of smart cities.

In memory of Professor Mark Taylor (1955–2024)

THE ROUTLEDGE COMPANION TO THE SOUND OF SPACE

*Edited by Emma-Kate Matthews,
Jane Burry and Mark Burry*

LONDON AND NEW YORK

Designed cover image: Dr. Finneas Catling

First published 2025
by Routledge
4 Park Square, Milton Park, Abingdon, Oxon, OX14 4RN

and by Routledge
605 Third Avenue, New York, NY 10158

Routledge is an imprint of the Taylor & Francis Group, an informa business

© 2025 selection and editorial matter, Emma-Kate Matthews, Jane Burry and Mark Burry; individual chapters, the contributors

The right of Emma-Kate Matthews, Jane Burry and Mark Burry to be identified as the authors of the editorial material, and of the authors for their individual chapters, has been asserted in accordance with sections 77 and 78 of the Copyright, Designs and Patents Act 1988.

With the exception of Chapter 17, no part of this book may be reprinted or reproduced or utilised in any form or by any electronic, mechanical, or other means, now known or hereafter invented, including photocopying and recording, or in any information storage or retrieval system, without permission in writing from the publishers.

Chapter 17 of this book is freely available as a downloadable Open Access PDF at http://www.taylorfrancis.com under a Creative Commons Attribution-Non Commercial-No Derivatives (CC-BY-NC-ND) 4.0 license.

Any third party material in this book is not included in the OA Creative Commons license, unless indicated otherwise in a credit line to the material. Please direct any permissions enquiries to the original rightsholder.

Trademark notice: Product or corporate names may be trademarks or registered trademarks, and are used only for identification and explanation without intent to infringe.

British Library Cataloguing-in-Publication Data
A catalogue record for this book is available from the British Library

Library of Congress Cataloging-in-Publication Data
Names: Matthews, Emma-Kate, editor. | Burry, Jane, editor. | Burry, Mark, editor.
Title: The Routledge companion to the sound of space /
edited by Emma-Kate Matthews, Jane Burry and Mark Burry.
Description: Abingdon, Oxon: Routledge, 2025. |
Series: Routledge international handbooks |
Includes bibliographical references and index.
Identifiers: LCCN 2024028999 (print) | LCCN 2024029000 (ebook) |
ISBN 9781032388540 (hardback) | ISBN 9781032388557 (paperback) |
ISBN 9781003347149 (ebook)
Subjects: LCSH: Sound—Social aspects.
Classification: LCC QC247 .R68 2024 (print) | LCC QC247 (ebook) |
DDC 534—dc23/eng/20240809 LC record available at https://lccn.loc.gov/2024028999
LC ebook record available at https://lccn.loc.gov/2024029000

ISBN: 978-1-032-38854-0 (hbk)
ISBN: 978-1-032-38855-7 (pbk)
ISBN: 978-1-003-34714-9 (ebk)

DOI: 10.4324/9781003347149

Typeset in Times New Roman
by codeMantra

Access the Support Material: https://www.routledge.com/The-Routledge-Companion-to-the-Sound-of-Space/Matthews-Burry-Burry/p/book/9781032388540

CONTENTS

List of figures ix
List of tables xv
List of contributors xvi

   Introduction 1
   Emma-Kate Matthews and Jane Burry

PART I
Architectural acoustics **15**
Jane Burry

1 Designed extremes 19
   Shane Myrbeck

2 Introducing acoustic shadows 28
   Pantea Alambeigi and Jane Burry

3 Transferring the impression of real and imaginary spaces 38
   Philip J B Jackson and Philip Coleman

4 Beyond the sweet spot: Sound, space and emotion 52
   Raj Patel, Iain Forsyth & Jane Pollard and Gerrie van Noord

5 Architectural acoustics of the Sagrada Família Basílica 68
   Sipei Zhao and Mark Burry

6	Intimate acoustic environments on record *Emil Kraugerud*	87
7	Embracing subtlety: Reflections on an acoustic surface of glass *Zackery Belanger, Catie Newell and Wes McGee*	98

PART II
The psychology and physiology
Jane Burry
107

8	Immersive Ambisonic spatial audio design for extreme environments *Stuart Favilla*	110
9	In an open (music) field. Space and time notation for representing landscape *David Buck and Carla Molinari*	126
10	Lend me your ears *Michael Fowler*	139
11	Imagining together *Nina Garthwaite*	153
12	Aural diverse spatial perception: From paracusis to panacusis loci *John Levack Drever*	164
13	On sonic growth and form; biometric evolution of sound and space *Paul Bavister*	180
14	Infrastructures of inaudibility: The spatial politics of assistive listening *Jonathan Tyrrell*	197

PART III
Philosophy and politics
Mark Burry
209

15	Reading aloud: The vocalisation of living space *Paul Carter*	213
16	From affordances to value chains: Probing the system of sound, space, and public *Sven Anderson*	224

17	On vibrational architectures *Gascia Ouzounian (text) and Jan St. Werner (images)*	238
18	House of silence, of stillness, of solitude *Mark Taylor*	255
19	Dimensionless space (with serrated edges and sucking noises): Intimacy, ASMR, micro-magic, sensory scholarship, and other taboos *David Toop*	264
20	Posthuman listening to the more-than-human soundscape *Jordan Lacey*	274
21	Toward a topology of music *Ildar Khannanov*	285
22	Sound's spatial-material circuitry *Raviv Ganchrow*	298
23	Place, sound, and architecture *Jeff Malpas*	317
24	Shaping sounds of future environments *Eleni-Ira Panourgia*	329

PART IV
Sound art and music — **343**
Emma-Kate Matthews

25	Spatiosonic dialogues: Exploring architecture's role in music composition and performance *Emma-Kate Matthews*	346
26	In praise of emptiness: A future for performance venues *Fabricio Mattos*	360
27	Opera in the bathhouse: Exploring an acoustically led approach to dramaturgy and scenography *Rosalind Parker and Pedro Novo*	373
28	Sound, space and the IKO loudspeaker – the apparent paradox of diversity with unity *Angela McArthur and Emma Margetson*	385

29 Intimate sound: Making known, curating and composing for small spaces *Lawrence Harvey*	399
30 Listening with, listening toward: Proposing graphic transcription as a means of (re)hearing space *Ben McDonnell*	412
31 Site-oriented music curation. Contouring the listening spaces *Sasha Elina*	426
32 Notes from the far field *Philip Samartzis*	437
33 Fluid architectures and aural sculpturality – towards an aesthetic of sonic spatio-temporal environments *Gerriet Krishna Sharma*	452
34 Acoustic Atlas – an orchestra of echoes *Cobi van Tonder*	471
Index	*483*

FIGURES

1.1	Inside the Sound Column (Shane Myrbeck)	20
1.2	Impulse response analysis of the flutter echo phenomenon at SFMOMA (Arup)	21
1.3	Dirigible hangar (Shane Myrbeck)	23
1.4	Orbit Pavilion (Dan Goods)	26
2.1	FabPod 1	32
2.2	Acoustic shadows in plan and section	34
3.1	Idealised schematic of a room impulse response with direct sound, early reflections and late reverberation developing up to the mixing time and then decaying	40
3.2	Consistent distance SRIR and BRIR measurements in Lecture Theatre D	42
3.3	RSAO parameter encoding of SRIR into early reflection (upper, solid) and late reverberation (lower, dashed) parameters	42
3.4	Illustration of visual characterisation of a meeting room for generating immersive sound with six degree-of-freedom interaction within a VR experience: (top, left) a perspective of the estimated geometry of cuboid objects with their colour-coded classification described by the key in the centre; (top, right) the same perspective overlaid with the visual appearance of the objects mapped from the camera images used for the classification; (bottom, left) the 3D environment and material properties translated into a games engine with a sound source (clarinet with loudspeaker symbol) and observer (grey manikin in a T-pose with a video camera symbol); (bottom, right) user wearing VR headset with headphones and hand-held controllers alongside a video display of the user's perspective	46
4.1	Arup SoundLab, 1998, conceptual rendering, listeners at the centre of an ambisonic sphere of loudspeakers, able to experience measured sound from a space, or sound created from a virtual model	54
4.2	Silent Sound, 2006, live 40-minute performance at St. George's Hall, Liverpool, with limited-edition CD and ambisonic installation at the A Foundation, Liverpool; Score: J. Spaceman (Spiritualized); Team: Raj Patel and Alban Bassuet at Arup Acoustics, Dr. Ciáran O'Keeffe, Charles Poulet,	

Figures

	Andrew Bolton, Noah Rose, Jhon Bell at Metametric, and Dan Howard-Birt. A second live iteration with recording and production of a CD took place at Middlesbrough Town Hall in 2010 as part of the AV Festival	56
4.3	Radio Mania, 2009, 29-minute two-channel stereoscopic video installation with ambisonic sound, presented at the BFI Gallery, London. Script: Kirk Lake; Music: Nick Plytas	58
4.4	The Death of Bunny Munro, 2009, spatialised 3D audiobook, 8 hours. 15 minutes. Music by Nick Cave and Warren Ellis, published by Canongate in multiple formats, including a 7-CD version with a bonus DVD, and downloadable as an enhanced iPhone app. Recordings mastered by Will Quinell at Sterling Sound, NYC. Sound spatialised by Terence Caulkins, Anne Guthrie, Raj Patel and Ryan Biziorek	59
4.5	Soon, 2011, 12-hour durational sound and light installation, commissioned for Nuit Blanche in Toronto. Curator: Nicholas Brown; Score: Scanner; Team: Raj Patel, Ryan Biziorek, Brian Stacy and Star Davis at Arup	61
4.6	Bish Bosch: Ambisymphonic (with Scott Walker), 2013, ambisonic sound installation, 25 minutes, commissioned and realised at the Sydney Opera House for the Vivid Festival. Team: Charles and Cathy Negus-Fancey, Pete Walsh, Raj Patel, Terence Caulkins, Fergus Linehan and Ben Marshall	63
4.7	Diagram of vocal movement positions in Bish Bosch	65
5.1	The Sagrada Família Basilica interior in 2019	75
5.2	An exemplary room impulse response in a small meeting room	76
5.3	Diagram of a RIR measurement system	78
5.4	The sound source positions (points indexed with a starting letter S followed by a number) and the measurement locations (points indexed with a number)	82
5.5	The RIRs measured in the left ear at (a) M1 and (b) M13 corresponding to the sound source position S1	83
5.6	The average reverberation time (T_{20}) and early decay time (EDT) in the Sagrada Família Basilica	83
5.7	The measured middle frequency (500–2,000 Hz) (a) clarity (C80) and (b) binaural quality index ($1 - IACC_E$) at each measurement location corresponding to the sound source positions S1, S2, and S3. The red and green dashed lines indicate the optimal range for concert halls as a reference only	84
7.1	A portion of Long Range at Riverbank Acoustical Laboratories	100
7.2	Simulation stills of sound in a 2D corridor made of rigid boundaries that progressively increase in curvature and perforated openings. The top still references an earlier time step. From left to right the energy is coherent, then dispersed, then reduced, indicating continuity between acoustic reflection, diffusion, and absorption	101
7.3	The two-layered wall system of Long Range with shifting variations in curvature, openings, and direction of curvature	102
8.1	'The astronauts', c. 1,000 BC	111
8.2	2D virtual speaker mappings for first-order microphone soundfield warping and beamforming as viewed from above. The outer circle represents a boost of +96 dBV while the inner circle is zero volume, courtesy of the author	116
8.3	Normal speech at a distance of 10 m, Ambisonic beamformer off (full noise) and on, revealing speech and formant patterns	119

x

Figures

9.1	Two outstanding products have been created at the Scriptorium: St. Gallen neumatic notation, one of the most advanced forms of open-field notation for Gregorian chant discussed earlier, and the St. Gallen plan	129
9.2	A distorted perspective, intentionally wrong, which exaggerates the urban presence of the subject, eliminating any form of symmetry of the oval and trying to escape the finiteness of the volume. Piranesi portrayed a complex landscape, which could not be defined by only one point of view and is subject to multiple times, in a similar way to the multiple scenes painted by the Picturesque artists of the period	131
9.3	These auditory symbols were located within the space and time of the visual sequences of Kent's design	133
10.1	Notation of the overtone system after Schenker (Harmony, 6)	140
10.2	Part three of Peter Ablinger's Listening Piece in four parts (2001). Parking lot at 4th Street/Merrick Street, Los Angeles	143
10.3	Auditory map of Skruv	147
13.1	A biometrically evolved volume overlaid over a digital model of the Musikvereinssaal	181
13.2	Evolutionary programming	185
13.3	GSR data from a response to Mozart with waveform and spectrograph analysis	187
13.4	Spaces used in the evolutionary soundscape experiment, from top left; Entrance to Barbican Cold Stores, Charterhouse Cloisters, Fabric Nightclub and the Barbican Foyer (all images Solen Fluzin)	188
13.5	Variable reflectors (left) are independently addressable to enable a modulated envelope (right) to be evolved from biometric responses to input stimuli	189
13.6	Raytracing analysis of a hall evolved to suit Beethoven's 7th Symphony	189
13.7	Beethoven Hall viewed from the Top with IRIS plot showing acoustic reflection data	190
13.8	Deconvolved Electrodermal Activity shown below analysis of the source stimuli, in this case, a section of Beethoven's 7th Symphony. The red lines on the electrodermal response (EDA) graph represent recorded emotional change. This represents a positive valence, which was triangulated with the post-test questionnaire	190
13.9	Beethoven Hall overlaid on a digital model of the Musikvereinssaal – note the exaggerated width	193
14.1	Three types of induction loops: Left: TFL Ticketing Window, Centre: Help Point (note the double-negation of the non-functional telecoil symbol), Right: Listening Point	201
14.2	Left: Central induction loop at Euston station device occluded from view and blocked from access by a marketing display. Centre: Large Listening Point detail (also obstructed by a construction barrier). Right: Accompanying symbolic denotation mounted further up the column	204
14.3	Left: Consulting The Oracle, John William Waterhouse, 1884 (Wikimedia). The oracular voice is embodied in the 'Teraph', a mummified head towards which the priestess strains to listen. Right: Author conducting listening to custom-built induction loop relaying signals from a geophone	205
16.1	Continuous Drift. A public artwork that blurs the boundaries between public sound installation, architectural intervention, and curatorial framework.	

	The project allows the public to use their mobile devices to trigger sounds by different artists that play back from eight loudspeakers integrated in the architecture of Meeting House Square in Dublin, Ireland. Installed in 2015, additional works were added in 2016 and 2017. Developed within the context of the public art commission The Manual for Acoustic Planning and Urban Sound Design, the project is still active in 2024	226
16.2	Proposal for the UK Holocaust Memorial International Design Competition. Honourable Mention. 2016. London (UK). Design team: Heneghan Peng Architects, Sven Anderson, Gustafson Porter + Bowman, Bruce Mao Design, Arup, Event, Bartenbach, BuroHappold, Mamou-Mani, Turner & Townsend, PFB, Andrew Ingham & Associates and LMNB. Competition organised by Malcolm Reading Consultants. Image courtesy of Heneghan Peng Architects	228
16.3	The Sound-Frameworks design tool has four core indicators: 'Value', 'Narrative', 'Legibility' and 'Integration'. Users input information concerning the significance and rating of each indicator to develop a visualisation of these parameters in each project. These four indicators receive prominent attention to place emphasis on the importance of communicating about the value of working with sound in specific contexts. In this case, the user has set their 'significance' at 3, 3, 2, and 3 (on a scale of 1–3) and their 'rating' at 1, 2, 3, and 3 (on a scale of 1–3), respectively	232
16.4	The second diagram in the Sound-Frameworks design tool allows users to select and configure up to eight custom indicators selected from a list, including Accessibility, Acoustics, Activism, Ambiance, Architecture, Atmosphere, Biodiversity, Building Codes, Climate Change, Commercial Space, Community, Conflict, Crisis, Culture, Density, Ecology, Education, Environmental Acoustics, Equity, Events, Exclusion, Gentrification, Heritage, information and communication technologies (ICTs), Inclusion, Industrial Space, Landscape, Lighting, Masterplanning, Mobility, Morphology, Music, Nature, Night-Time Use, Noise, Noise Complaints, Noise Regulations, Non-Human Life, Pedestrian Space, Personal Space, Physical Plant, Public Art, Residential Space, Resilience, Redevelopment, Safety, Shelter, Silence, Soundscape, Sports, Sprawl, Suburbs, Sustainability, Tranquillity, Transportation, Vibrance, Waste Management, Water Features, Weather, Zoning. In this case, the user has selected Shelter, Night-Time Use, Inclusivity, and Silence, and set their 'significance' at 3, 3, 2, and 2 (on a scale of 1–3) and their 'rating' at 1, 1, 2, and 2 (on a scale of 1–3), respectively	233
16.5	The third diagram in the Sound-Frameworks design tool maps the linkages between core indicators and custom indicators, showing how specific aspects of the project contribute to its value, narrative, legibility, and integration. The weights of the connections demonstrate how much each custom indicator contributes to each core indicator. The colours are derived from the settings used in the core indicator diagram and custom indicator diagram. This diagram provides a means of identifying which dimensions of a project require attention to ensure they can progress	234
17.1	Vibraceptional architecture sequence 1 (2023) by Jan St. Werner	246
17.2	Vibraceptional architecture sequence 2 (2023) by Jan St. Werner	247
17.3	Vibraceptional architecture sequence 3 (2023) by Jan St. Werner	248

Figures

17.4	Vibraceptional architecture sequence 4 (2023) by Jan St. Werner	249
17.5	Vibraceptional architecture sequence 5 (2023) by Jan St. Werner	250
21.1	Beethoven, Piano Sonata op.57, Appassionata, Slow movement	292
21.2	Chopin, Mazurka op. 68 no. 4, F minor	294
21.3	Chopin, Scherzo op. 38, B minor	294
22.1	Device for measuring radio wavelengths with a copper coil tapped at various points in the windings. Invented by Adolf Slaby developed at Telefunken, Berlin c. 1902–1904	303
22.2	Pressure–wave interactions from spark-gap discharges in various reflection/refraction configurations. Images were produced by Schlieren photography, a method developed by the German physicist August Toepler to visualise fluid dynamics in mediums such as air. Sequence from A. Toepler's book Beobachtungen bach der schlierenmethode (1906). Ostwalds Klassiker Nr. 158. P117	304
22.3	Waveforms plotting the frequency of 'perception' versus 'sensation' (ngrams) on Google's corpus of English language books (1750–2019). Graph illustrating the frequency of occurrence over time (x-axis publication years and y-axis frequency of occurrence)	306
22.4	Fossilised whale inner ear (tympanic bulla), c. 18 million years, South Carolina	308
22.5	Exxon/Vail curve. Waveform charting 540 million years of sea-level fluctuations was initially published in 1977 by geologists from the petrochemical industry	311
22.6	Left: Coccolithophores (phytoplankton) capable of forming calcium carbonate (calcite) structures called coccoliths in the process of biomineralisation. Right: Compound eyes of a trilobite (genus Coltraneia), made of the rigid mineral calcite lenses. Early Devonian trilobite from Hamar Laghdad, Morocco	312
22.7	Map of cotidal lines along which tides meet, constituting still nodes of oscillation convergence, at each given hour of diurnal oceanic rhythms	313
24.1	Landscape in Nuthepark, Potsdam	334
24.2	Recording underwater and surface vibrations on a tree bark in Nuthepark	335
25.1	BSF performance layout diagrams (Emma-Kate Matthews, 2017)	352
25.2	Left to right: Dr. Jim Barbour with ambisonic microphone array'; view of microphone array from balcony; microphone against the altar (Emma-Kate Matthews & Finneas Catling, 2017)	353
26.1	On the left, Angela Burgess Recital Hall, at the Royal Academy of Music; on the right, the Main Hall at Sands End Arts and Community Centre, London, both in London	368
26.2	National Sawdust, in New York	368
27.1	Gala Pool showing the location of the ambisonic microphone, and loudspeaker measurement locations (1–4)	376
27.2	Measurement 1 – Upper Graph – Early impulse response, W component. Lower graph – direction of incidence of 3 sets of strong reflections identified on the upper graph (CATT Acoustic software)	378
27.3	Measurement 3 – Upper Graph – Early impulse response, W component. Lower graph – direction of incidence of 2 sets of strong reflections identified on the upper graph (CATT Acoustic software)	378
28.1	IKO loudspeaker	386
28.2	Inside-out system of the IKO loudspeaker (illustrating direct sound only)	390

28.3	Outside-in system of a traditional loudspeaker orchestra array with listeners	391
28.4	Map of relationality between the IKO, place, sound and listener	392
28.5	A listener affects a sound beam by casting an acoustic shadow	393
29.1	3D model and image of Speaker Orchestra installed in the Primrose Potter Salon, Melbourne Recital Centre, 2009. The performance shown is for La lontananza nostalgica utopica futura by Luigi Nono, where the violinist journeys between music stand 'stations' positioned through the space	406
29.2	SITE and SOUND: Sonic Art as ecology practice. Diagram produced for early planning and image of final installed Speaker Orchestra. Note the elevated CODA Tube speakers and audience in three sectors with side speakers	407
29.3	The Planting, Castlemaine State Festival, 2023. The diagram on the left shows two wide seating banks. In the final version pictured, the seating banks were narrowed and made into three rows. However, the gently curving layout and hemispheres can be seen in the image	408
29.4	Sound Bites City. Early render of the space and image of final installation. Due to deadlines for exhibition opening, some loudspeakers were positioned outside and not on the structure: Image Nick Williams	408
29.5	SITE and SOUND 3D model and exhibition image. The diagram indicates the speakers sharing different spatialisations, intended to operate as zones or pools of sound: Image Ross Mcleod	409
30.1	'Listen' (2023) graphic score	413
30.2	'Crit Transcription, Design School' (2023) graphic transcription	414
30.3	Detail 'Crit Transcription, Design School' (2023) graphic transcription	415
30.4	Crit Transcription, Art/Education School' (2023) graphic transcription	416
30.5	Detail 'Crit Transcription, Art/Education School' (2023) graphic transcription	417
32.1	Wilkes Station	440
32.2	Casey Station	440
32.3	Newcomb Bay	442
32.4	Mount Whaleback Mine	444
32.5	Fortescue River	445
32.6	Bogong High Plains	447
32.7	Pretty Valley	448
32.8	Kiewa Valley	448
33.1	IKO speaker	454
33.2	IKO beamforming graph	455
33.3	Odio-App (screenshots)	458
33.4	Odio-App (graphical scheme)	459
33.5	SPS (schematic illustration)	463
34.1	When convolving the incoming dry signal with the impulse response (IR) the end product is an 'Auralisation' – which sounds as if the dry signal is now in the virtual acoustic space	473
34.2	Globe UI for Acoustic Atlas	477
34.3	Some of the sites in the current archive	479
34.4	Artwork for the album Echoes and Reflections, created by nine artists together with the author	480

TABLES

8.1	Summary of spatial auditory cues	113
27.1	Acoustic parameters used to characterise performance spaces and their objective and subjective meanings	377
27.2	Acoustic measurements results at four locations	377
29.1	The 22 key teaching activities described in the next section. In practice, most are not stand-alone but combined with two or more others	400

CONTRIBUTORS

Pantea Alambeigi is a lecturer in architecture at the Swinburne University of Technology. Her research focuses on 'Architectural Acoustics' and the mutual interaction of human perception of sound and the built environment. She touches on architecture in a multisensory approach and studies the mutual impact of human perception of a built environment and design performance. Her interdisciplinary research explores and extends the concept of 'Seamless Architecture' through computational data-driven design and integrates the architectural emerging technologies and environmental parameters with human behaviour and spatial perception to inform and transform the strategies of a performance-driven design.

Sven Anderson is an artist, educator, and architectural consultant based in Dublin. He is the Director of the Masters in Digital Arts and Intermedia Practices and an Assistant Professor in the School of Creative Arts at Trinity College Dublin. In 2021, he was awarded a Marie Skłodowska-Curie Actions Individual Fellowship to develop Sound-Frameworks with Theatrum Mundi. As a member of Annex, he co-curated Entanglement, Ireland's pavilion for the 17th International Architecture Exhibition of La Biennale di Venezia. With Heneghan Peng Architects and Gustafson Porter + Bowman, Anderson developed proposals for the Pulse National Memorial and Museum (Orlando) and the UK Holocaust Memorial (London). He co-edited the books *States of Entanglement: Data in the Irish Landscape* (Actar) and *Signal Spectre System: A Late Evening in the Future* (Verlag Für Moderne Künste). As an artist and architectural consultant, Anderson leads projects that explore sonic practice in the context of architecture and urbanism. His work ranges from urban sound installations and multichannel video systems to self-initiated artist placements and public services. His work has been exhibited at venues from the Douglas Hyde Gallery (Dublin) to Secession (Vienna) and is featured in the permanent collections of the Arts Council of Ireland and the Office of Public Works. Anderson holds a PhD from the Graduate School of Creative Arts and Media (GradCAM) and a BA from Cornell University.

Paul Bavister is an architect, researcher, and academic, serving as a Project Director at Flanagan Lawrence and an Associate Professor (Architecture) at the Bartlett School of Architecture, UCL. He specialises in designing for performance and interaction, drawing from his extensive experience, with a recent PhD from UCL focusing on biometric evolutionary practice with sound and space. At Flanagan Lawrence, Paul helps develop performance projects with clients such as the BBC, the San Diego Symphony Orchestra, and the Royal Welsh College of Music & Drama. Outside of architectural practice, Paul regularly contributes to academic and professional institutions as a guest lecturer, and regularly judges at the World Architecture Festival. Paul has had research exhibited in the UK at the Tate Modern, the Science Museum, RIBA, the Barbican, and internationally in Ars Electronica Linz, IRCAM France, Austria, Finland and Japan. Paul has had research published at UCL Publications, Forum Acousticum, and the Institute of Acoustics.

Zackery Belanger is an expert on the intrinsic acoustic influence of shape and material in architecture. He is educated in physics and architecture and has practised on the design teams of projects such as Grimshaw's Experimental Media and Performing Arts Center (EMPAC) at Rensselaer. He was the inaugural Researcher-In-Residence at EMPAC and is the author of the essay Acoustic Ornament, which argues for the intentional acoustic use of non-acoustic surfaces and objects. In 2017, he founded the Detroit-based studio Arcgeometer. Belanger was deeply involved in the recent merging of acoustics and lighting and is a prolific speaker on acoustics.

David Buck is a landscape architect and educator with a special interest in integrating theoretical and speculative approaches to examining the temporality of landscape. His Doctorate by Design from The Bartlett, UCL, was nominated for the RIBA's President Award for Research and was published by Routledge as a sole-authored monograph *A Musicology for Landscape*. He has also appeared on Radio 4's flagship programme Music Matters. His research examines alternatives to the static picture plane of the perspective, investigating how the temporal qualities of landscape materiality and space can be better drawn. He uses notation, defined by Ferruccio Bussoni as 'the transcription of an abstract idea', to develop drawing tools for investigation and reflection and to document research processes and outcomes. His research offers insights into both music and landscape architecture through notation and affords a focused and sensitive exploration of temporality and sound in both fields.

Jane Burry is an architect and Head of the University of Adelaide's School of Architecture and Civil Engineering. Jane's research leverages digital fabrication with simulation and feedback to create better, more sensitive, human-centric spaces through linking to environmental and perceptual experiences including the auditory. Jane is the lead author of *The New Mathematics of Architecture*, T&H, 2010; editor of *Designing the Dynamic*, Melbourne Books, 2013; co-author of *Prototyping for Architects*, T&H 2016; coeditor of *Making Resilient Architecture*, 2020 UCL Press; guest coeditor of *Urban Dystopias*, 2023, Architectural Design 93 (1); and has over 120 other publications on architecture and computational design topics, acoustics among them. Jane's involvement as an architect and researcher at Antoni Gaudí's Sagrada Família church in Barcelona stimulated research into the impact of Gaudí's use of geometry on sound.

Mark Burry AO is a registered architect and the Founding Director of SCRI at the Swinburne University of Technology. His role is to lead the development of a whole-of-university research approach to 'urban futures', helping ensure that our future cities anticipate and meet the needs of all – smart citizens participating in the development of smart cities. He has published internationally on two main themes: putting theory into practice with regard to procuring 'challenging' architecture, and the life, work and theories of the architect Antoni Gaudí. He has been a Senior

Architect at the Sagrada Família Basilica Foundation since 1979, pioneering distant collaboration with his colleagues based on-site in Barcelona concluding in late 2016.

Paul Carter is an interdisciplinary scholar and artist. His publication *Amplifications: Poetic Migration, Auditory Memory* (New York: Bloomsbury, 2018) surveys and contextualises his sound art practice and is supported by the publication of his radiophonic scripts (*Absolute Rhythm: Works for Minor Radio*, Aberystwyth, UK: Performance Research Publications, 2020), together with online access to their first productions (https://performance-research.org/absolute-rhythm.html). Recent multichannel sound installations in Melbourne include Cooee Song (2021) and No Man's Land (2022). His drawing practice is showcased in *Return of the Centaurs: A Field Guide* (Naples: Artem, 2024) and (forthcoming) *Neglected Dimensions: Rough Sketches for Public Space* (Barcelona/New York: Actar, 2025). Paul is co-director of the Aboriginal cultural heritage and co-design agency, Nyungar Birdiyia, and Professor of Design (Urbanism), School of Architecture and Urban Design, RMIT University, Melbourne.

Philip Coleman (PhD, MAES) is a Senior Immersive Audio Research Engineer within the Immersive Audio team at L-Acoustics, researching tools and technologies for acoustic enhancement and large-scale immersive sound reinforcement. Previously, he was a Senior Lecturer in Audio at the Institute of Sound Recording, University of Surrey. He received his PhD degree from the Centre for Vision, Speech and Signal Processing, University of Surrey, in 2014.

John Levack Drever operates at the intersection of acoustics, audiology, environmental studies and sound art. He has a special interest in soundscape methods, in particular field recording and soundwalking. In 1998 he co-founded the UK and Ireland Soundscape Community (a regional affiliate of the World Forum for Acoustic Ecology). He was awarded a PhD from Dartington College of Arts in 2001. Drever is an avid collaborator and a member of Blind Ditch. He has devised sound in many different configurations and contexts. Commissions range from the Groupe de Recherches Musicales, France (1999) to Shiga National Museum, Japan (2012). Most recently, he collaborated with Rural recreation on the School of Insects (2021–2022) at Trumpington Park Primary School, Cambridge. Drever is a Professor of Acoustic Ecology and Sound Art at Goldsmiths, University of London, where he co-leads Sound Practice Research (SPR). He is co-director of LAURA, Leverhulme Trust Aural Diversity Doctoral Research Hub.

Sasha Elina is a curator, musician and researcher based in London. Her focus lies in experimental and contemporary classical music, sound art, as well as broader interdisciplinary practices. Sasha is the founder and artistic director of the international project Music Space Architecture and the curator of the Eternal series of music events in London. As a vocalist and flautist, Sasha performs composed and improvised music both solo and in various collectives. She was a founding member of The Same Ensemble (2013–2023), performed with the Moscow Contemporary Music Ensemble, the Moscow Scratch Orchestra, and many others. She received her Music BA from Moscow State Tchaikovsky Conservatory (2017) and Music MA in Research from the University of Huddersfield (2023).

Stuart Favilla's diverse research spans Defence Systems, Medical Technologies, Wildlife Conservation, and STEM education for the visually impaired. Collaborating with Sonja Pedell at Swinburne University, he has pioneered tablet apps and social music interactions for dementia patients. Head of the Swinburne Sonic Research group, Favilla is an internationally recognised audio designer with a three-decade career. He has won prestigious awards like the Karl Szucka Preis and has been a finalist for the Bourges Electroacoustique Prize. Favilla's work in computer music and instrument design has been featured in various media outlets and academic publications. He has performed at prominent European festivals and is skilled in software development and audio-video signal processing, earning him recognition at the 2020 International Design Awards for his generative life video wall projections.

Iain Forsyth & Jane Pollard are artists and BAFTA-nominated film directors working with documentary, 3D sound, installation and drama. They've worked exclusively as a collaboration since meeting at Goldsmiths 30 years ago. Central to all their work is the transformative power of music. Their debut feature film about Nick Cave, 20,000 Days on Earth, premiered at Sundance and won several awards including Best Debut Director at the British Independent Film Awards. They collaborate regularly and astutely with a wide spectrum of musicians including Scott Walker, Gil Scott-Heron, Marianne Faithfull, Jarvis Cocker and Spiritualized. In 2023 they curated The Horror Show, the landmark exhibition at Somerset House in London that presented a twisted tale of 5 decades of British creative culture in three acts – Monster, Ghost and Witch. Their art work has been exhibited around the world and is collected by museums and institutions including Tate at the British Government Art Collection.

Michael Fowler is an independent researcher whose work is interdisciplinary. He is primarily interested in techniques for the mapping, analysis and design of exemplary sound spaces. He has a DMA in Piano (CCM) and pursued an international concert career as a pianist and sound artist before pursuing research. His postdoctoral work has concentrated on various topics such as the auditory semiotics of Japanese garden design, concepts of sound and listening in architecture and landscape architecture, the music of Karlheinz Stockhausen, and the utilisation of techniques from mathematics and AI for representing indeterminate musical spaces in the scores of John Cage. His artistic practice has also included the presentation and design of immersive sound installations that have been presented in Australia, China and Japan. He is an alumnus of the Alexander von Humboldt Stiftung research fellowship programme.

Raviv Ganchrow researches the interdependencies of sound, locale and hearing through installations, writing, and the development of transduction technologies. His sound works attend to spatial-material manifestations of oscillations in conditions such as environmental infrasound, telluric currents, long-range radio, ocean acoustics, and anechoic chambers. Recent installations employ in-situ circuits patched directly into locales, making tangible relays of contextual dynamics. He publishes, workshops, and lectures broadly on auditory contexts and the spatial-material agency of sound and is currently a faculty member at the Institute of Sonology, University of the Arts, The Hague.

Nina Garthwaite is the founder of In The Dark. Since 2010, In The Dark has been holding gathered listening events – a little like a film screening but for sound – in the UK and beyond. During this time, In The Dark has held over 200 events in venues ranging from rooms above pubs to art galleries, cinemas, theatres, fields, churches, boats, cemeteries, and more. Nina is also an interdisciplinary artist and teaches creative audio through In The Dark.

Lawrence Harvey is a composer, sound designer and director of SIAL Sound Studios, School of Design, RMIT University. He co-leads the Design and Sonic Practice Research Group, collaborated on founding the RMIT Sonic Arts Collection and directs public concerts and exhibitions on SIAL Sound Studios speaker orchestra. Harvey has curated over 30 performances and exhibitions and led teaching and research with musicians and artists, interior, digital and industrial designers, and architects. He has directed large research projects funded by the Australian Research Council, and industry-funded projects in urban soundscape research and spatial sound performance. In addition to electroacoustic compositions, he has produced gallery and urban sound installations, spatial sound designs for VR and theatre, and performed in Melbourne, Seoul, Huddersfield, The Hague and Vienna. His writing, teaching and postgraduate supervisions explore spatial sound composition and performance, soundscape studies and sound as a model for other cultural practices. For further details <https://sialsound.studio/>.

Philip Jackson (BA Cambridge, 1993; PhD Southampton, 2000) is a Professor of Machine Audition at the Centre for Vision, Speech & Signal Processing at the University of Surrey and Fellow of the Surrey Institute for People-Centred Artificial Intelligence. An expert in acoustical array processing, spatial audio and machine listening, directs research in object-based media, audio AI and audiovisual AI towards personalisation for media producers to create exciting, high-quality user experiences. Funded by £92M awards, 225 co-authored academic publications with 3,493 citations (h-index 29), including 36 journal papers and 5 patent filings. His research on next-generation flexible media technology changes how we experience sound, via collaboration with entertainment giants BBC and Bang & Olufsen. Founded CVSSP's machine listening group in 2004. Led Surrey's research stream in the highly successful S3A programme with BBC R&D (2013–2019) on next-generation audio. Currently research stream lead, ethics lead and co-architect alongside PI Hilton of AI4ME, Surrey's £15M BBC-EPSRC prosperity partnership on the use of object-based media to achieve personalisation for mass audiences. He is a member of University Senate (2021–2024).

Ildar Khannanov, Ph.D. (University of California, Santa Barbara 2003), ABD (Moscow Conservatory, 1993), study with Jacques Derrida (UC Irvine 1998–2001). Currently Associate Professor of Music Theory, Peabody Institute, Johns Hopkins University. Publications: 'Line, Surface and Speed: Nomadic Aspects of Melody' in *Sounding the Virtual: Gilles Deleuze and the Philosophy and Theory of Music* (Ashgate, 2011), 'Extension and Directionality. A Sketch for Musical Topology', in *Music and Space* (SANA, Belgrade 2021), as well as 'Rameau and the Sciences: The Impact of Scientific Discoveries of the Lumières on Rameau's Theory of Harmony', in *Proceedings of the Worldwide Music Conference* (Springer, 2021). Dr. Khannanov is a member of the Scientific Committee of EUROMAC.

Emil Kraugerud is a postdoctoral research fellow in the Department of Musicology at the University of Oslo, Norway. His work takes a musicological approach to the study of record production, including the technological and aesthetic affordances of music production technologies, with a particular focus on the production and meaning of acousmatic intimacy in recorded popular music. Kraugerud is currently part of the research project 'The Platformization of Music Production' – funded by The Research Council of Norway – which is concerned with the development and use of music production technology in the online environment. He also has practical experience from working as a musician and recording engineer.

Jordan Lacey is a sonic thinker/practitioner, and senior lecturer in the School of Design at RMIT University. Much of his work is concerned with binding sonic practices with vitalist flat ontologies. His two books *Sonic Rupture (SR)* and *Urban Roar (UR)* rethink acoustic ecology with affect theory and sound art installation practices (SR) and apply psychophysical approaches to the reimagining of relationships between artistic bodies, expressive environments, and the generative waves of the psyche (UR). He is associate editor, and lead book reviewer, of the Journal of Sonic Studies. A comprehensive list of his theoretical and practical contributions to sound studies and soundscape design practices can be found at: jordan-lacey.com.

Jeff Malpas is an Emeritus Distinguished Professor at the University of Tasmania, an Honorary Professor at Latrobe University and the University of Queensland, and a Fellow of the Australian Academy of Humanities. He has published on a wide range of topics in philosophy, as well as other fields, including art, architecture, and cultural geography. His most recent books are *The Fundamental Field* (with the poet Kenneth White – 2021), *Rethinking Dwelling* (2021), and *In the Brightness of Place* (2022).

Emma Margetson is an acousmatic composer and sound artist. Her research interests include sound diffusion and spatialisation practices; site-specific works, sound walks and installations; audience development and engagement; and community music practice. She has received a variety of awards and special mentions for her work, including first prize in the prestigious L'Espace du Son International Spatialisation Competition by INFLUX (Musiques & Recherches), klingt gut! Young Artist Award in 2018 and Ars Electronica Forum Wallis 2019. She is a Senior Lecturer in Music and Sound at the University of Greenwich and co-director of the Loudspeaker Orchestra Concert Series. http://emmamargetson.co.uk/.

Emma-Kate Matthews is an architect, composer, musician, and digital artist. Her work explores the creative intersections between sonic and spatial practices through the production of site-specific and spatialised audiovisual projects. In addition to composing on the London Symphony Orchestra's Panufnik scheme, her work has been performed internationally at acoustically distinctive sites such as the Sagrada Familia, the Southbank Centre, the Barbican and Brighton Festival. She has released solo electronic-classical works on labels including Accidental Records, Algebra Records, and NMC Records. In addition to making music, she also designs and makes her own instruments which she calls 'resonant bodies'. In 2022 she was nominated for the Lumen Prize and the Aesthetica Art Prize and her work has been 'highly commended' in categories for the 'Sound of the Year Awards' for the past two years. She hosts an eclectic radio show on RTM.FM called Hunter Gatherer. See https://www.ekmworks.com/ for more details.

Fabricio Mattos is a musician deeply committed to finding new paths for music performance in the 21st century. Founder of the innovative WGC – Worldwide Guitar Connections, he has been performing regularly worldwide, having premiered dozens of works for guitar and toured in five continents over the last years. Mattos is also a multiple award-winner, including the prestigious 'Julian Bream Prize' awarded by the legendary British guitarist Julian Bream. In 2023 Fabricio Mattos became the first Latin-American musician to receive a PhD from the Royal Academy of Music, researching the interconnections between space, dynamics, layouts and terminology in music performance. Mattos is the co-founder of the New Stages Creations and works regularly as creative director of live performances and bespoke video projects commissioned by a wide variety of ensembles and institutions.

Angela McArthur is an artist, academic and interdisciplinary advocate. She leads a master's programme in spatial sound in the Department of Anthropology at University College London.

She has undertaken many artist residencies and interdisciplinary collaborations, including a five-month residency working with the IKO loudspeaker at the Institut für Elektronische Musik (IEM) in Graz. She initiated the first UK tour of IKO works in 2019. Her work centres around the practice and theorisation of aesthetics in sound, underrepresented (including other-than-human) onto-epistemologies, and ocean environments. She champions diversity in access and representation. She's worked in studio, live and location environments and founded Soundstack, an annual series of workshops, masterclasses and concerts about spatial sound aesthetics.

Ben McDonnell is an artist based in south London. He works with expanded lens-based practices and sound and is drawn to temporary structures that support, pedagogy and the relationship these activities have with the spaces they occupy. Ben graduated with a master's degree in photography from the Royal College of Art in 2017, following his undergraduate studies in jazz composition and performance at Leeds College of Music. Currently residing and working in southeast London, UK, his artistic practice incorporates photography, sound, and sculpture to explore themes of stability, support, and structure, drawing inspiration from sound, music, and the built environment. In 2019, he was selected for the Chisenhale Studio Residency and received a residency in Vancouver from the Outset Contemporary Art Fund, supported by the British Council and Vancouver Biennale. Additionally, he completed a studio residency and exhibition with PADA in Lisbon, Portugal, in 2018 and was commissioned to create new work for Concealer, part of the Peckham 24 photography festival. His work has been exhibited internationally and is held in both public and private collections. He also serves as a Senior Lecturer on the Fine Art BA course at Norwich University of the Arts.

Wes McGee is an Associate Professor in Architecture and the Director of the Fabrication and Robotics Lab at the University of Michigan Taubman College of Architecture and Urban Planning and a principal at Matter Design. His research involves the interrogation of design and material production in architecture, with the goal of developing new connections between design, engineering, materials, and manufacturing processes as they relate to the built environment.

Carla Molinari is a Senior Lecturer in Architecture and BA Course Leader at Anglia Ruskin University. She teaches architectural history and theory, and Design Studio. Carla has a PhD in Theory and Criticism of Architecture and has published on cinema and architecture, on the conception of architectural space, and on cultural regeneration. Before joining ARU in 2022, she taught at Leeds Beckett University, University of Gloucestershire, University of Liverpool, and University Sapienza of Rome. In 2020 she was awarded a Paul Mellon Research Grant for her archival research on Gordon Cullen and in 2016, she was awarded a British Academy Fellowship by the Accademia Nazionale dei Lincei for her research on Peter Greenaway and Sergei Eisenstein. Carla's research engages with architecture and media, innovative interpretations of montage and cinematic design methods, theory and history of space, and urban narrative strategies.

Shane Myrbeck, a Los Angeles-based sound artist, composer, and acoustician, is renowned for his work that explores the immersive power of sound. He uses spatial audio systems and architectural forms to shape and enhance experiences, often focusing on onsite-specific multichannel audio installations. His work has earned worldwide recognition and has been displayed in prestigious venues globally. Myrbeck has had residencies at institutions like the Montalvo Arts Center, San Francisco's Exploratorium, and NASA's Jet Propulsion Laboratory. He also serves as the Associate Principal at Arup, a leading architectural engineering firm. Here, he leads the Acoustics,

Audiovisual, Theatre, and Experience Design team in Los Angeles and has worked on projects involving notable figures such as Björk, SFMOMA, and Lou Reed. Additionally, Myrbeck manages Arup's SoundLab, where he conducts acoustic simulations, creates sonic VR experiences, and composes innovative pieces. These activities further highlight his expertise in blending art and technology within architectural environments.

Catie Newell develops material and optical assemblies that amplify our connection to a living and spinning earth. As the founding principal of the architecture and research practice Alibi Studio and a Professor of Architecture at the University of Michigan, Newell's research and creative practice has been widely recognised for exploring design construction and materiality in relationship to location and geography. The work ranges in scale from buildings to products and explores the world most deeply with material systems and optical captures. Alibi Studio deploys material explorations, illumination and darkness, and novel modes of occupation, all created through rigorous prototyping and custom fabrication tools. The process of fabrication is a vital act in the work, often amplifying material effects and situational influences, intertwining the processes of making and design across the entire project.

Pedro Novo utilises his expertise in acoustic engineering to improve the well-being of people in the buildings he designs. With a record of over 50 building projects, Pedro has had the opportunity to collaborate with renowned architects and institutions, including several collaborations with Zaha Hadid Architects, respected museums such as the Victoria & Albert and the Science Museum and universities like Oxford and Cambridge. His work includes the application of auralisation techniques, using both Ambisonics and Binaural reproduction, to demonstrate the potential acoustic effects of design alternatives. His approach extends to artistic installations where he has used active reverberation systems and conducted detailed 3D impulse response measurements to ensure spaces are acoustically optimised. Pedro's approach blends technical precision with creative innovation to subtly influence our spatial experience through sound.

Gascia Ouzounian is a sonic theorist whose work explores sound in relation to space, architecture, urbanism, and violence. She is an Associate Professor of music at the University of Oxford, where she leads the project Sonorous Cities: Towards as Sonic Urbanism (soncities.org). Ouzounian is the author of *Stereophonica: Sound and Space in Science, Technology, and the Arts* (MIT Press 2021), and she has contributed articles to leading journals of music, visual art, and architecture. Recent projects include Scoring the City, which takes inspiration from the graphic score and other unconventional notations in experimental music to develop new modes of 'urban scoring' (scoring.city); and Acoustic Cities: London & Beirut, which brought together ten artists in creating works that explore the sonic, social, and spatial conditions of two cities.

Eleni-Ira Panourgia is a sound and visual artist, and researcher. Her work focuses on the development of new forms of expression and creative methods that combine sound, objects, spaces and environments. She explores the potential of such complex morphologies within artistic, design, social and ecological processes. Eleni-Ira's work has been presented internationally in museums, galleries, festivals, exhibitions, radio shows, academic journals, edited volumes and conferences. Eleni-Ira completed a PhD in Art at the University of Edinburgh as a Scholar of the Onassis Foundation. She is co-founder and managing editor of the journal Airea: Arts and Interdisciplinary Research. Eleni-Ira is currently a Teaching and Research Fellow at Gustave Eiffel University. https://eleniirapanourgia.com/.

Contributors

Rosalind Parker is an opera director and co-founder of underscore, a new international partnership that works with an interdisciplinary research-praxis hermeneutic, engaging creatives and specialists to realise site-specific works which expand the way that opera can be performed and experienced. Using classic text, they explore disrupting traditional audience–performer relationships and enable new ways of physically and acoustically engaging with opera.

Raj Patel has worked at the intersection of sound, music, art, technology, design, and engineering for over 30 years and is noted for his artist collaborations and international portfolio of projects in the built environment focusing on sensory-driven solutions. Pioneering the use of spatial audio as a design and artistic medium, he has co-created ground-breaking immersive tools for listening to, feeling, and visualising the built environment. This knowledge has also been applied to conceive and create spaces specifically for the development and presentation of immersive spatial and 3D audio/video artwork multimedia installations and exhibitions. His collaborative works with a range of artists including Ai Weiwei, Phillip Glass, Lou Reed, Scott Walker, Bjork, Nick Cave, Doug Wheeler, Iain Forsyth and Jane Pollard have been presented globally including the Tate Modern, The Whitney, MoMA (NY), The Sydney Opera House, and The British Film Institute. He is a Fellow at Arup.

Philip Samartzis is a sound artist, scholar, and curator who specialises in the social and environmental conditions that shape remote wilderness regions and their communities. His artistic practice involves extensive fieldwork, using advanced sound recording technology to document natural, anthropogenic, and geophysical forces. These recorded sounds are integrated into exhibitions, performances, and publications to illustrate the transformative power of sound within a contemporary art and music context. He is particularly interested in exploring concepts of perception, immersion, and embodiment to offer audiences nuanced experiences of space and place. Philip is a professor at RMIT School of Art and the artistic director of the Bogong Centre for Sound Culture.

Gerriet K. Sharma is a composer, sound artist, and artistic researcher in spatial practices. Within the past 20 years, he was deeply involved in the spatialisation of electroacoustic and instrumental compositions in Ambisonics and Wave-Field Synthesis. Furthermore, he was extensively concerned with textural transformation processes into 3D-sound sculptures and the development of the IKO and 393 beam-forming loudspeakers in cooperation with the IEM Graz. 2017/2018 he was appointed Edgard Varèse professor at the electronic studio of the TU Berlin. His works were presented at international festivals, conferences and symposia. Sculptural works for the icosahedral loudspeaker (IKO) and loudspeaker hemisphere were presented amongst others at Darmstädter Summer Courses, Music Biennale Zagreb, Kontakte Festival Berlin, Wiener Festwochen, and Atonal Festival Berlin. He received numerous awards and scholarships. Publications in international journals and books on spatial practices and sound. Initiator of the Special Interest Group 'Spatial Aesthetics and Artificial Environments' within the Society for Artistic Research (SAR) and co-founder of the 'Lab for Spatial Aesthetics in Sound' (spaes) at Funkhaus Berlin in 2020. He lives in Berlin.

Mark Taylor is a Professor of Architecture at Swinburne University, Australia. His primary research focus is the history and theory of the modern architectural interior with an emphasis on cultural and social issues. He has held two competitive Australian Research Council (ARC) grants and is currently a CI for the ARC Training Centre for Next-Gen Architectural Manufacturing. Mark has published widely through journals and book chapters and has authored and edited several books, including *Interior Design and Architecture: Critical and Primary Sources* (Bloomsbury 2013), *Flow: Interiors, Landscapes and Architecture in the Era of Liquid Modernity* (Bloomsbury 2018) and *Domesticity under Siege Threatened Spaces of the Modern Home* (Bloomsbury 2023).

David Toop has been developing a collaborative practice that crosses boundaries of sound, listening, music and materials since 1970, encompassing improvised music performance (solo and in groups), writing, electronic sound, field recording, exhibition curating, sound art installations and opera. It includes eight acclaimed books, including *Rap Attack* (1984), *Ocean of Sound* (1995), *Sinister Resonance* (2010), *Into the Maelstrom* (2016), *Flutter Echo and Inflamed Invisible* (2019), along with catalogue essays, theoretical papers and music criticism. His solo records include New and Rediscovered Musical Instruments on Brian Eno's Obscure label (1975), Sound Body on David Sylvian's Samadhisound label (2006), Entities Inertias Faint Beings (2016) and Apparition Paintings (2020). His 1978 Amazonas field recordings of Yanomami shamanism and ritual were released on Sub Rosa as Lost Shadows (2016). He is an Emeritus Professor at London College of Communication.

Jonathan Tyrrell is an Associate Professor at the Bartlett School of Architecture where he teaches on the Design for Performance and Interaction, and Design for Manufacture postgraduate programmes. He joined UCL in 2021 after teaching for seven years at the University of Waterloo School of Architecture in Canada. He has over a decade of experience in award-winning design research practices including Dereck Revington Studio (Toronto), where he led a series of large-scale public art projects and memorials; and with experimental architect Philip Beesley (Toronto) leading complex installations at major international art festivals. Jonathan is currently completing his PhD in Architectural Design at the Bartlett School of Architecture where his UCL-funded research explores relationships between sound, matter, and the politics of listening. His spatial-sonic installations have been exhibited in Toronto, London, Berlin, and most recently in Salzkammergut (Austria) for the European Capital of Culture 2024.

Gerrie van Noord teaches the MA Curating Contemporary Art at the Royal College of Art, London, and works as a freelance editor/curator of publications. She is particularly interested in the discourse around collaborative ways of working and the role of publishing in the perception of creative practices. Most recently, she has co-edited and contributed to *Not Going it Alone: Collective Curatorial Curating* (apexart, 2024), *Curious* (Open Editions, 2024) and *Kathrin Böhm: Art on the Scale of Life* (Sternberg Press, 2023). Alongside working on publishing projects with individual artists, she was managing editor of the critical anthologies *Between the Material and the Possible* (Sternberg Press, 2022), *Curating after the Global* (MIT, 2019), *How Institutions Think* (MIT, 2017) and *The Curatorial Conundrum* (MIT, 2016). Since 2016, she also works for open-access peer-reviewed PARSE Journal.

Cobi van Tonder is a practice-led researcher and interdisciplinary artist. Cobi has been awarded a Marie Skłodowska-Curie Fellowship with her project ACOUSTIC ATLAS – Cultivating the Capacity to Listen and is currently a research fellow at the University of Bologna. Her work and research focus on the art of listening in the context of acoustics, and audio-physical experience of sound and composition. She completed a PhD in Music Composition at the Digital Arts & Humanities Program of Trinity College, Dublin, an MFA Art Practice degree at Stanford, USA; and a BHons in Music in History and Society (Musicology) at the University of the Witwatersrand, Johannesburg, South Africa.

Contributors

Jan St. Werner is co-founder of the experimental music group Mouse on Mars and releases music and artwork under his own name via the Edition Fiepblatter Catalogue, distributed by Thrill Jockey and Sonig. In the mid-1990s, he was part of the Cologne sound collective A-Musik, worked together with Markus Popp (Oval) as Microstoria, and developed music for installations and films by visual artist Rosa Barba. In the 2000s, Werner was artistic director of STEIM, the Dutch studio for electro-instrumental music. Werner presented works at Ural Biennale, documenta 14, Lenbachhaus Munich, ICA London, Staatliche Kunsthalle Baden-Baden, MSU Zagreb a.o. In 2024, he was one of six artists representing Germany at the Venice Biennale. Werner taught at the MIT Program in Art, Culture and Technology and was Professor for Dynamic Acoustic Research at the Academy of Fine Arts Nuremberg.

Sipei Zhao earned his PhD in Electrical Engineering from RMIT University, Australia, in 2018, following a Bachelor's degree in Electronics Engineering and a Master's degree in Acoustics from Nanjing University, China, in 2012 and 2015, respectively. He served as a postdoctoral research fellow at the Centre for Audio, Acoustics, and Vibration of UTS from July 2018 to December 2020, and since January 2021, he has been a Lecturer at the School of Mechanical and Mechatronics Engineering at UTS. Zhao's exceptional contributions have been recognised with awards such as the Best Student Paper at ICSV22 in 2015, the Best Paper Prize at CAADRA2016, and the Best Paper Award at ICA2019. He also received the Young Professionals Grant from I-INCE and the Ford Publication Commendation Prize from RMIT University in 2017.

INTRODUCTION

Emma-Kate Matthews and Jane Burry

Sound is inherently spatial. Whether it is the effects of wind currents and ice movements captured in the remote polar regions of the globe, or familiar voices broadcast over the radio into the intimacy of our homes, sound is inextricably linked to the space it inhabits. We all access and experience these phenomena differently, through the way we hear and feel the vibrations that inhabit and define spaces, both natural and built. 'The Routledge Companion to the Sound of Space' charts a journey through design, creativity, technology, philosophy, and beyond, providing a panoramic view of the many ways in which space and sound interact and ways to design and create for this.

Between spatial and sonic fields

This Companion explores the intersection of spatial and sonic practices across various disciplines, examining the current 'state of the art'. Each chapter offers insights, observations, and ideas based on the technical, creative, cultural, political, philosophical, psychological, and physiological contexts of working with space and sound. This content covers a wide range of spatial typologies, from the extremely large and reverberant to the narrow and intimate, from the public and external to the domestic and interior, and from the naturally formed to the highly engineered. The insights and observations compiled within these pages explore the incredibly diverse range of interactions and relationships between space and sound. These perspectives come from researchers and practitioners spanning various disciplinary areas, which, despite their diversity, share a high level of compatibility. From contributors engaged in the design and composition of space and sound, including architects, sound artists, acousticians, and cultural theorists, to those investigating the sensory and physiological dimensions of spatial perception in our encounters with sonic phenomena, as well as those exploring the political and philosophical implications of spatial and sonic practices. Through examining the connections between space and sound from the perspectives of both creators who shape our experiences and analysts who study our perceptions, across both physical and virtual realms, we have the opportunity to transcend disciplinary limitations that might otherwise restrict our comprehension and detailed understanding of these subjects.

This exploration is not just about observing and understanding these topics, but also about active participation and experimentation. This book invites readers to contemplate the potential of space and sound as tools for creative expression and communication, as well as for technical innovation. The various case studies presented offer a number of practical insights that complement the theoretical content also discussed throughout the book. By examining real-world examples, readers gain a deeper understanding of the relevance and adaptability of spatial and sonic practices across diverse applications and contexts.

This publication therefore forms an important bridge between the intricate and diverse realms of technology, philosophy, composition, performance, and spatial design. These pages discuss many pertinent topics in spatial and sonic practices including, but not limited to, acoustic design, music composition and performance, sound production, digital simulation, and event curation. Additionally, they examine the impact of technological advancements, including recent developments in digital fabrication techniques and spatial audio tools, on professional practice. Specifically, the ways in which these technologies affect the design and construction of spaces and surfaces, and the resulting impact on acoustic characteristics and other sonic behaviours. Despite the inherent complexity of these topics, the chapters presented cover a broad range of disciplinary perspectives, from highly technical to the poetic.

Through such a rich blend of theory, practice, and critical reflection, the volume serves as a valuable resource for anyone interested in exploring the intricacy of relationships between space and sound, whether they are scholars, researchers, practitioners, or simply curious.

Disciplinary resonance and 'togethering'

The idea of 'disciplinary resonance' lies at the very heart of this work providing resources, examples, and frameworks for sharing knowledge and ideas across specialist fields. It aims to combat the 'siloing' of information, a common issue among practices with unique vocabularies, notational conventions, and areas of expertise. The acoustic metaphor of 'resonance' is used to discuss a collaborative state where compatibilities and overlaps between these fields are explored and developed in unison. The word 'resonance' typically describes a condition in which two or more distinct elements align, leading to a mutual amplification of their principles, mechanisms, and outcomes. This book highlights examples of projects where knowledge is not ring-fenced or disconnected from its neighbours, as is sometimes observed in collaborations between highly specialised fields such as architecture and music. Rather than keeping skills and expertise compartmentalised, these projects demonstrate a true integration of diverse disciplines, fostering mutual development and collaboration across boundaries.

The word 'resonance' has always had a particular connection to sonic phenomena, with its etymological roots in the Latin resonare, where the prefix 're' means 'again' or 'back' and 'sonare' is a verb meaning 'to sound', related to 'sonus' simply meaning 'sound'. There are many different ways in which disciplines might resonate. Theoretical physicist Basarab Nicolescu notably advocated for 'resonance' between the sciences and humanities with his concept of 'transdisciplinarity'. This approach, as the term implies, transcends disciplinary boundaries, encouraging not only the creation and exchange of knowledge within disciplines but also across 'and beyond all discipline'.[1] The concept of transdisciplinarity intends to cultivate open dialogue and encourage a holistic approach to collaboration among wildly diverse practices. Nicolescu's efforts to dissolve boundaries between the sciences and humanities align with the combination of empirical and scientific methodologies necessary for the exploration of complex interactions between space and sound. Architect and acoustician Michael Forsyth echoes this sentiment with regard to acoustics, stating that acoustic design is 'a matter of musical judgement [as much as it is] a scientific process'.[2]

Nicolescu's *Manifesto of Transdisciplinarity* explicitly challenges the perceived barriers between different practices and advocates for the integration of knowledge and methodologies across fields that have disparate ways of working and thinking. He describes disciplinary fragmentation as a situation where 'each discipline … [becomes] more and more specific'[3] and suggests that technology contributes to this fragmentation. In his view, the increased complexity of communication and thought in the modern world enhances disciplinary specificity, in contrast to the 'simple, one-dimensional reality of classical thought' Whilst this complexity benefits the development of each discipline, it also makes it more challenging to foster the unity that inter- and transdisciplinary modes of practice seem to offer. At this point, it is useful to clarify key differences between multi-, inter-, and transdisciplinary modes of practice. According to Nicolescu, 'Multidisciplinarity concerns studying a research topic not in just one discipline, but in several at the same time'.[4] This creates the potential for conversations and overlaps across domains, but it also results in each discipline maintaining a highly individual and self-contained body of knowledge. Interdisciplinarity takes collaboration a step further by enabling 'the transfer of methods from one discipline to another'.[5] Interdisciplinarity also offers opportunities for the cross-pollination of ideas and methodologies but remains 'within the framework of disciplinary research'.[6] Both of these practice structures have unique advantages, with many examples of interdisciplinary and transdisciplinary projects in this book. However, transdisciplinarity is particularly suggestive of a mode of practice where individual contributions within a collaborative output become inseparable. Collaborators are not simply intersecting, but developing interdependencies and corresponding in ways that contribute to shared vocabularies, knowledge, and methodologies. This way of working is analogous to anthropologist Tim Ingold's notion of 'togethering', as examined in his 2017 publication on the topic of 'Correspondences'.[7] The concept of togethering provides a useful theoretical framework for understanding differences between projects where collaborators 'interact' but keep within the typical extents of their disciplinary fields, and those where ideas, skills, knowledge, and apparatus are openly shared and in constant correspondence. As Ingold States: 'If interaction is about othering, then correspondence is about togethering'.[8] Architecture and music stand as two particularly useful examples of fields that are used to 'togethering'. Both architectural and musical outputs are the product of collaborative efforts that involve multiple contributors with different. As architect Yeoryia Manolopolou notes, 'Architecture is rarely the manifestation of a single author's efforts, despite how much that author might seek to represent it as such'.[9] Similarly, the composition and performance of music, particularly orchestral music, is a collective effort involving multiple musicians, conductors, curators, and audiences.

Many of the correspondences that are discussed in this book relate to things or events which often are not physically tangible, clearly visible, or easy to measure or quantify. Philosopher Timothy Morton invented the word 'hyperobject'[10] as a means of describing 'all kinds of things that you can study and think about and compute, but that are not so easy to see directly'.[11] Morton states that hyperobjects, with their lack of clearly defined boundaries and enormously complex constitutions, cannot ever be seen in their entirety: 'one only sees pieces of a hyperobject at any one moment'.[12] However, they can be 'thought and computed'.[13] In the context of the contents of this Companion, we encounter similarly large and evasive entities in things such as the 'creative reciprocity' that exists between the already vast, multifarious fields of architecture and sonic practice. We cannot examine transdisciplinarity in its entirety, we can only see fragments of it evidenced in the scores, drawings, conversations, performances, recordings, and reviews that are made during its creation, execution, and recollection. Similarly, we cannot define the edges of the correspondences that characterise transdisciplinarity and its far-reaching outputs, nor can we even begin to identify every single one of its fundamental parts.

Whilst the disciplinary areas identified here already naturally intersect, they still need some assistance to understand and leverage each other's expertise more fully. This book advocates for the progression of such intersections beyond multidisciplinary 'othering' and towards a transdisciplinary 'togethering'. Such a shift can help to erode the disciplinary boundaries that are unhelpful and encourage a less restricted sharing and overlapping of skills, knowledge, techniques, and tools – or what we might call disciplinary resonance.

Space, sound and technology

To nurture potential for such 'disciplinary resonance' *The Routledge Companion to the Sound of Space* serves as a reference for those seeking an enhanced and detailed knowledge of the myriad ways in which space and sound interact. This Companion is aimed at all spatial and sonic practitioners, researchers, and scholars. This includes, but is certainly not limited to, architects, interior designers, musicians, sound artists, acoustic and auditory engineers. It is also written to capture the imagination and attention of readers who are not specialists in any of these areas but are curious to understand more about the creative synergies that continue to emerge and develop between spatial and sonic disciplinary realms.

In this era of rapid technological advancement, our relationship with space and sound is transforming. Over the past century, improvements in audio and spatial recording and processing tools have significantly changed our experience of sound in 'virtual' spaces. Given this context, the importance of our book is particularly highlighted, as it provides insights into the evolving dynamics of technology-mediated spatial and auditory experiences. In the early part of the 20th century, sound recording technology was not sophisticated enough to capture both instrumental sound and the acoustic response of the room in which it was situated. This in turn generated a desire to simulate spatial effects such as reverberation to 'create the illusion of space in sound recording'.[14] Since these early developments, we now have a huge range of recording, production and listening tools at our disposal for creating and augmenting ideas of space in audible media, typically with the ambition to make the experience of playback appear more spatially immersive. Stereo and surround-sound methods of playback have been widely available for several decades. However, their three-dimensional capabilities are now considered rather limited compared to the spherical flexibility of comparably new ambisonic recording, simulation and playback technologies. Recorded or exported ambisonic spatial audio encodes sound sources with details about their direction and distance, which enables precise positioning and movement within a 3D audible space during playback. This technology has significantly improved the spatial realism and immersive quality of audial experiences in multimedia applications such as virtual reality, gaming, music production, and film.

In parallel to this, continued improvement to the portability of recording and listening devices is changing the way that we understand and experience relationships between space and sound. Recorded content in music and sound art is now very often dissociated from the physical context of its origins in that such content is often captured in spaces distinct from where we end up listening to it. As a result, a large proportion of the music that we listen to, unknowingly embodies varying degrees of spatial information, and often lacks any deliberate interaction with its spatial situation. Despite the obvious advantages to such technological developments across both spatial and sonic disciplines, the resulting lack of planned interaction between space and sound appears to forget the historically productive symbiosis between the exemplary practices of architecture and music. As a result of this trend, the autonomy from physical sites that manifests is contrarily accompanied by a desire to access and replicate the acoustic character of other spaces, as evidenced in the routine application of electroacoustic reverberation, echo and delay effects to recorded (and sometimes performed) sound. Typically, these effects give a generic illusion of an acoustically reflective space, with the exception of convolution reverb which has the capacity to correspond to particular spaces, though is rarely used with the space in which the original impulse response was captured, in mind. The conceptual and practical disconnect that has become common between physical space and sound is also evidenced in the way that spaces for music are typically designed. Architects and engineers often do not question the design paradigms that fundamentally influence the way we experience indoor musical performances. Design decisions relating to aspects such as audience/performer relationships and acoustic propagation are becoming increasingly standardised, albeit informally, thus fostering a disjointed dynamic between the realms of architecture and music in which the discussion of each other's disciplinary zeitgeist is neither explicitly acknowledged nor sought out. This continued divergence is surprising, considering the wealth of conceptual and practical overlap between these two worlds.

Despite these problems, technology is also helping to democratise access to knowledge and tools that may have previously been 'out of reach' to non-specialists. An example of this is in the way that 3D spatial modelling tools and acoustic simulation tools have developed in recent years. Digital sound simulation is migrating from the strict domain of the expert acoustic or audio engineer to the working palette of the architectural designer, musician and sound artist. It is now simpler than ever to analyse and auralise the behaviour of sound in spaces that either haven't been built yet or are physically inaccessible for some reason. Numerous tools for convolution reverb are now widely available and able to easily utilise impulse responses that have either been captured in real spaces or simulated in ray-traced models. Such high-fidelity spatial tools are used by both spatial designers and sonic composers alike. These simulation tools not only facilitate the creation and execution of projects but also provide opportunities to construct their own uncanny and virtual realities. These realities are 'real' enough to define their own existence: as sociologist Sherry Turkle writes, on simulation 'Screen versions of reality will always leave something out, yet screen versions of reality may come to seem like reality itself'.[15] The projects discussed in this book highlight both the challenges and opportunities associated with spatial audio technologies, providing invaluable insights for researchers and practitioners working in such a rapidly evolving landscape. Several chapters examine the current state of 'spatial audio' and speculate on future developments and uses of associated technologies and tools within various disciplinary contexts.

Developing dialogues

Despite the concurrent evolution of spatial and sonic practices, there remains a notable scarcity of publications and formal collections of texts that aim to consolidate a unified body of information and knowledge into one comprehensive resource. Numerous discipline-specific handbooks have already been written, yet those addressing the coevolution of spatial and sonic practices tend to concentrate on the historic narratives of the symbiotic relationship between architecture and music, rather than the discussion of the contemporary landscape and future possibilities.

This book serves to fill this gap by offering a comprehensive overview of the current state of research, scholarship, and practice concerning the many and intricate relationships between space and sound.

The genesis of this volume traces back to the third and most significant in a series of symposia that emerged from our exploration of the highly reverberant interior of the Basílica de la Sagrada Família. This project, titled 'Architecture for Improved Auditory Performance in the Age of Digital Manufacturing', spanned three years, and received public funding from the Australian Research Council to investigate the dual topics of acoustic privacy and acoustic intelligibility through a highly architectural lens. As our project concluded, we identified a shared interest in the further exploration of interactions between space and sound. To unpack this exploration further, we issued an international call for proposals in the summer of 2019. This call invited potential contributors to propose presentations, performances, or demonstrations centred around the theme of 'making spaces for sound and music and making sound for interesting or unusual spaces'. We disseminated this call to a diverse array of practices, institutions, and groups, extending far beyond our individual networks. The response exceeded our expectations, as we received a wide range of proposals for talks and performances from an even broader spectrum of disciplinary backgrounds than we initially anticipated.

The two-day 'Sound of Space Symposium' took place at the Bartlett School of Architecture's Here East building, University College London, on 11 December 2019. It served as a platform for discussing the challenges, creative opportunities, and specific technical considerations in working with extreme and unique spatial conditions. These discussions spanned the related disciplines of architecture, engineering, music composition, and performance. The event featured talks, presentations, performances, and demonstrations by a diverse group of academics and industry professionals. On the first day, our symposium featured presentations by esteemed acoustic experts including Prof. Trevor Cox from the University of Salford and Adam Foxwell from Arup alongside installations and performances from musician and artist Ben McDonnell, and sound artists and researchers Gerriet. K. Sharma and Angela McArthur. We addressed several key topics and questions, including the advantages and limitations of working with acoustic simulation tools, ways of measuring and representing invisible phenomena in architectural space, and comparisons between empirical and computational/mathematical methods in space and sound research. We also discussed approaches in responding to and designing for extreme spatial and acoustic characteristics using case studies to understand real-world scenarios. The second day followed a more informal seminar format, with sessions on each of the disciplinary areas now forming the main sections of this book. Focusing on the disciplinary areas that now constitute the main sections of this comprehensive volume. Initially, our intention was to publish the symposium papers in a condensed pamphlet format. However, our post-symposium reflections revealed a resounding call for a more expansive publication. Attendees, as well as individuals within our extended networks, expressed a clear desire for a more comprehensive resource documenting the breadth of expertise, knowledge, and insights shared across academic and practice-based projects and research endeavours, serving as an interdisciplinary companion.

Following the endorsement of our shared vision to create a more extensive curated volume, we invited our symposium presenters to expand their papers into more comprehensive chapters. This initiative aimed to provide readers with a more in-depth exploration of the topics across each subject area. Simultaneously, we issued another global call for proposals to engage those who might have initially missed the symposium. Once again, we were humbled by the enthusiastic responses and support that we received in return. Through our outreach efforts, we were able to compile an impressive list of authors who collectively contribute an exceptional depth and breadth of knowledge. As we embarked upon the writing process, we identified both overlaps and divergences in thoughts, approaches and opinions across chapters, presenting opportunities for interdisciplinary dialogue and collaboration among chapter authors. This led us to organise a second mini-symposium in a hybrid format, accommodating both in-person and online attendance. Many of the discussions throughout the event were characterised by their complexity and nuance, highlighting the depth of expertise and insight among the participants. Importantly, these discussions led to beneficial introductions between contributors which sparked several new interdisciplinary projects and collaborations. This demonstrates the power of our initiative, to foster meaningful connections and the exchange of ideas and knowledge. Even before its publication, the book showed its potential to bridge gaps between disciplines and act as a catalyst for innovation and collaboration across academic and professional spheres.

As the editors, we are primarily united by our backgrounds in architectural practice, as well as our varying interests and first-hand experiences across the various technical and creative fields addressed here. Therefore, this publication uses architecture as a starting point for opening up conversations relating space to sound both practically and conceptually. Architecture is inherently multidisciplinary, requiring a synthesis of knowledge, skills, and tools from numerous disciplines in the act of space-making. The synthesis of diverse content witnessed in architectural practice provides a conceptual framework for the organisation of content as rich and varied as that presented here. Within each section, the content of each chapter is loosely organised around spatial typologies, maintaining a focus on relationships between space and sound, and enabling contributors to discuss their topics with great depth and specificity. This is something which doesn't often happen outside of the bounds of individual disciplines.

Despite its all-encompassing multidisciplinary nature, architecture often neglects things that cannot be drawn, quantified, or visualised. Architectural design is primarily ocularcentric[16] and aspects related to the non-visual senses are frequently overlooked in the design and discussion of architectural spaces. Contemporary philosophers like David Michael Levin have noted that 'Western culture has been dominated by an ocularcentric paradigm, a vision-generated, vision-centred interpretation of knowledge, truth and reality'.[17] They also question what a future 'beyond the governance of ocularcentrism' might look like.[18] This is a particularly relevant question in practices such as architectural design, where working tools and methods are predicated on visual representation and visual thinking. Buildings are primarily designed for visual appeal, with decisions centred around the appearance of their spatial organisation, massing, materiality, and so forth. Architectural drawing and modelling, as visual mediums, are the main methods of designing and representing buildings. However, our interaction with architecture is multisensory. We use all our senses to grasp and appreciate the unique characteristics of the built environment. Despite our individual psychological and physiological differences, this multisensory interaction gives us a comprehensive understanding of the dimensions, metrics, and qualities of the spaces that we inhabit. Although vision is often regarded as the 'dominant sense'[19] in architecture and spatial design, there is a distinct synergy between space and sound. This is not only evident in the phenomenological interaction between space and sound but also in the parallel evolution of spatial design and sonic expression over centuries. This symbiotic history has been well documented through the centuries,[20] from the speculative work of Vitruvius and his 'Sounding Vessels in the Theatre'[21] to the more recent and high-tech design of concert halls with adjustable interiors[22] and installation of active reverberation systems.[23] Despite the rich history and the availability of numerous tools and techniques for calculating and analysing acoustic behaviours like reverberation, some less tangible and nuanced aspects of space and sound often get overlooked. Architectural briefs rarely focus on shaping sonic experiences beyond the basic acoustic requirements for speech or music intelligibility. Exceptions are sometimes made for spaces dedicated to music performance. However, even in these cases, designers often adhere to long-established 'rules of thumb' for defining geometry and dimensions. And although we have a long history of observing acoustic behaviours, the discipline of acoustics is still relatively young. It faces the challenge of fostering productive interactions between spatial and sonic disciplines instead of alienating them. This book therefore focuses on sound, aiming to highlight the intangible or invisible elements in architectural discourse from various disciplinary perspectives, highlighting key examples where interdisciplinary collaboration has been crucial to a project or practice's development.

Sound is not only a crucial factor in building design, but it also plays a key role in our perception of space. It provides insights into the physical characteristics of a place, such as its size, material, and surface qualities. It also informs us about the way space is occupied, reflecting the movement and directionality of dynamic events occurring within it. At lower frequencies (around 20 Hz for humans), sound transitions from being audible to being tactile. Through sound alone, we can understand the spatial complexities and nuances of our surroundings. Sound can also be manipulated as a material to produce, simulate, and conduct spatial experiences. Sound art practices, music composition, and performance can deliberately exploit and interact with acoustic conditions as well as express spatial conditions and concepts in relation to both virtual and real spaces. This book comprises chapters that discuss and expand upon various completed and theoretical projects, documenting methods and tools that consider space in the organisation of sonic content and, conversely, regard sound as a crucial factor in the construction and organisation of space. This applies whether in a literal acoustic or physical sense or in conveying more abstract and metaphorical content.

In public spaces, sound often serves as a medium for both artistic and socio-political expression, underlining a deep relationship between sonic intervention and collective experience. These shared public spaces provide a window into our behaviours and lifestyles, using sound as a lens. Our urban centres are becoming increasingly populated, leading to higher levels of background or ambient noise due to human activity.[24] This was particularly noticeable during the recent global lockdown when our usual outdoor activities were temporarily paused. The sudden 'quiet' allowed us to observe phenomena usually drowned out and rendered undetectable by our daily movements.[25] The growing popularity of noise-cancelling headphones also reflects our growing desire for personalised and portable acoustic environments in an increasingly noisy world. As such technologies become more prevalent in our everyday lives. They are also causing a disconnect between the world we hear and the world we inhabit. This disconnection is perhaps evident in the noise complaints from those who move to non-urban areas expecting the same undiscriminating quiet provided by personal audio technology. Instead, they encounter a different type of 'quietness' characterised by non-human sounds. In some cases, such sounds have come under threat from such complaints and laws have been introduced to protect them in cases where they are integral to the identity and cultural heritage of a place that.[26] The complex and ever-evolving 'acoustic politics of space' is examined in detail by artists and writers such as Brandon LaBelle.[27] In conjunction with these publications, our book addresses some of these ongoing debates, providing a starting point for further conversation and observation.

This book is published at a time when rapid technological development is one of the major factors responsible for shaping dialogues between spatial and sonic realms in practice, research, and daily life. This collection of thoughtful and insightful chapters provides an opportunity to ensure an ongoing interconnectedness of space-making and sound-making practices. It curates a wealth of insights, observations, speculations, and knowledge from a variety of perspectives at the intersection of multiple disciplines, all with a shared interest in shaping and understanding the interaction between spatial and sonic realms. As the history of development between these realms is already well documented, our publication places a focus on the future, providing a snapshot of the current 'state of the art' and offering a platform for speculation about its future.

Scope and structure

The contents of our Companion are organised around a multi-dimensional matrix of topics and disciplinary focii. In the interest of fostering interdisciplinary collaboration and discussion, the chapters are organised around the disciplinary themes of architectural acoustics, psychology and physiology, philosophy and politics, and music and sound art. Interwoven within each of these sections, a range of spatially focussed subcategories provides a framework for discussing a range of particularities in relationships between space and sound. These subcategories are inspired by Edward Hall's categorisation of space in 'The Hidden Dimension'.[28] In Hall's book, he identifies space as being either intimate, personal, social, or public. We have expanded this taxonomy of space – designed and experienced through sound – to include a few more categories, ranging from the intimate to the intangible.

The concept of intimacy is examined through the acoustic and spatial conditions of **interior** and **domestic spaces**, as well as through acts of shared listening experiences across remote locations. Chapters that discuss the importance of curation and collaboration in community-based, sound-led projects, particularly those with an educational focus, explore the topic of **socio-political** space. With some conceptual overlap, the chapters that examine the design and use of **performance** spaces and other **public** settings specifically look at sonic acts in shared spaces and buildings or constructs dedicated to music performance. **Exterior** spaces are discussed through urban sound art projects and a number of practical and conceptual relationships between landscape and sound. The concept of **extreme** space is not distinct from those already listed, but specifically deals with environments that are particularly unusual in their acoustic characteristics or spatial layout. Finally, both **technological** and **virtual** spaces, with their capacity to transcend physical scale and tangible intimacy, are discussed in the context of spatial and sonic interactions within computational or cognitive contexts. This list of disciplinary areas and spatial typologies is of course not exhaustive. Also, due to the interdisciplinary nature of this book, many chapters, although categorised under a specific section, cover multiple topics and overlap with various categories simultaneously. Despite these caveats, this organisational framework captures a broad cross-section of this complex area of practice and study, providing ample opportunity for exploring the many intricate relationships between space and sound.

To briefly summarise each of the disciplinary themes addressed in each section:

The section titled **Architectural acoustics** discusses the design and experience of the built environment in relation to sound. It explores how different material compositions and geometric configurations influence the behaviour of sound across a variety of different physical and virtual spatial contexts, with a particular focus on reverberation. From extremely reflective to completely absorbent spaces, the challenges, and advantages of each are discussed in detail. Consideration is given to issues of intelligibility in speech and music, while also exploring intentional methods for shaping and articulating sound within environments with varying acoustic characteristics. The book also includes examples where sound is used for understanding architectural character, through both scientific analysis and intuitive exploration. The book emphasises the importance of considering acoustics in spatial design. It demonstrates how the careful manipulation of sound in space can enhance our experiences of architectural spaces, from improving speech intelligibility in learning environments to creating immersive experiences in performance spaces.

The way our bodies and brains respond to sonic stimulus and its spatial contexts is covered in the section titled **Psychology and physiology**. Chapters within this section place a particular focus on the way that sound is used to navigate and comprehend different spaces from the urban realm to uninhabited landscapes and virtual environments. This section also highlights the diversity of our audial physiology, acknowledging that we each perceive and interpret sound and its interaction with space in our own unique ways. This difference in audial perception can be influenced by several factors, including our physical health, mental state, cultural background, and personal experiences. Understanding how we perceive and interpret space through sound is complex, deeply personal, and subjective. This diversity is reflected in how sound is both felt and heard, and how our sensitivity to this cross-sensory stimuli varies greatly between individuals, spaces, and cultural contexts. The chapters in this section also contemplate how our individual ways of hearing might provide insights into the way that spaces for music are designed in the future and the necessary tools for their realisation. This is especially relevant when considering potential virtual rather than physical environments. Future-focused approaches that use artificial intelligence and machine learning to create tools that adapt to our emotional responses in real time are explored, introducing a new level of personalisation to our music experiences.

Philosophy and politics are presented as two interconnected realms where sound plays a crucial role in shaping and discussing both societal structures and abstract concepts. From a political standpoint, the utilisation of sound in public spaces can either amplify or suppress the voices of certain communities, serving as a tool for empowerment or control. Correspondingly, philosophy has the capacity to explore the nature and impact of sound on our perception of reality and our comprehension of spatial and sonic constructs. Together, these disciplines offer a critical lens through which we can analyse our sonic environment and its influence on shaping our social and spatial experiences and interactions. This publication aims to provide an overview of the complex and rapidly evolving political dimension of space and sound. It touches upon various political themes and emphasises the role of sound in socio-political discourse across different spatial contexts. It pays special attention to the navigation and design of urban environments, particularly for those who rely on assistive listening devices. And to the value of sound in opening cross-disciplinary conversations in the management and production of urban space. Similarly, the philosophical exploration of space and sound is vast, and this volume can only scratch the surface by presenting a handful of examples. Nonetheless, these examples are crucial in architectural discourse, where the intangible aspects of sonic phenomena are often overlooked. Discussing these phenomena provides essential intellectual and conceptual frameworks for spatial and sonic practices, especially in an era when technology is able to blur the boundaries between represented and real spaces.

Like architectural acoustics, **Music and sound art** has its own dedicated section in the book, as representing a particularly fertile ground for exploration, reflection, and speculation regarding the interplay between space and sound. Advancements in spatial sound composition and reproduction have made it increasingly accessible and immersive, offering composers and musicians the ability to experiment with spatialisation and acoustic response using tools traditionally associated with architecture and engineering. This enables them to refine their ideas before performances, enriching the compositional process. Virtual audio technologies provide opportunities to explore sound in fantastical or fictional spaces, allowing for real-time design and interaction. This section explores projects where space plays a pivotal role in the construction of sonic art or musical composition, and vice versa, where music and sound are considered integral to an architectural concept or experience. The interplay between physical and digital spaces in music and sound art continues to expand, leading to the creation of hybrid compositions that can be experienced in both traditional settings like concert halls and innovative contexts such as virtual or mixed reality environments. The projects explored in this section showcase how space and sound might develop interdependencies, each influencing and pushing the other to make the outputs of music and sound art more spatially immersive, multisensory and interactive. Through this integration of space and sound, new dimensions of artistic expression can emerge, simultaneously pushing and dissolving the boundaries of all involved disciplines and offering audiences unique and enriching experiences.

We acknowledge that the list of disciplinary areas covered in this publication is not exhaustive, nor does it claim to capture every detail of these vast and complex fields. However, the samples presented here offer a diverse cross-section of starting points for readers to explore the broader associated territory. By providing an expanded introduction to these areas, our aim is to facilitate a deeper understanding of these disciplines and encourage readers to embark on further exploration and study.

Introduction

A multimodal resource

The content compiled in this publication represents a wealth of knowledge spanning a wide range of disciplinary fields related to the creation and experience of space and sound. Our intention is for this publication to serve as a catalyst for further cross-disciplinary collaboration and conversation, fostering the development of a shared vocabulary. This vocabulary encompasses linguistic terms, conceptual frameworks for understanding the more nuanced interplays between spatial and sonic practices, and visual notational methods used in both sonic and architectural realms for the rehearsal and execution of works. Such specialise vocabularies often remain confined within disciplinary boundaries, and even reinforce the exclusivity of those fields. However, this Companion aims to increase accessibility of this information by consolidating it into one publication for a broad readership. Through these efforts, we aim to broaden perspectives and promote inclusivity in discussions about space and sound.

The integration of diverse modes of communication, as observed across various practices, offers numerous benefits and reflects the core ethos of the work presented here. Embracing diverse voices and perspectives, whether through drawings, transcripts, scores, or sound, promotes collaboration and combines different viewpoints and tools. This approach is akin to multidisciplinary and transdisciplinary approaches. The structure of this volume mirrors the inclusive nature of such collaboration. We encourage the unique voices of the individual contributors within each mode. We deliberately did not standardise the presentation of content as we recognise that different disciplines have different ways of presenting information and using language. This variety of tones is part of creating a shared vocabulary. Some chapters are presented as first-person transcripts of informal conversations, whilst others are written in the third person. Similarly, the inclusion of images and sound alongside text allows for a more comprehensive exploration of the topics discussed, appealing to different learning styles and sensory experiences. The method of curation and presentation of content here highlights the importance of respecting and appreciating the individual voices of each contributor. Just as cross-disciplinary teams recognise the expertise and perspectives of individuals from various fields, this book acknowledges and celebrates the diverse talents and viewpoints of its authors. In short, by embracing cross-disciplinarity in both content and presentation, this collection exemplifies the power of collaboration and integration in fostering a deeper understanding and appreciation of a complex range of topics.

Aside from the written text, this publication is as multimodal as possible within a book format, presenting a selection of supporting visual and sonic information. We urge readers to interact fully with this content, and to expand their knowledge by further researching the work of our contributors and those in their networks. Our goal is to cultivate an inclusive, cross-disciplinary dialogue that bridges different fields and promotes collaboration across creative and technical fields. We hope that by sharing these insights, we can inspire practitioners, scholars, and enthusiasts to incorporate new perspectives and methodologies into their own work.

Outlooks

In conclusion, this comprehensive volume offers a diverse exploration of the intersection of space and sound across various disciplines. We embarked on this editorial project to explore the complex interplay between space and sound through the various disciplinary perspectives represented in these chapters. Each contribution provides a unique viewpoint on this dynamic relationship. Thanks to the collective efforts of scholars, researchers, and practitioners from various fields, we have gained substantial insights that we're excited to share with the global community. Our sincere gratitude goes to everyone who has contributed their expertise and passion to this project. We intend for this volume to stand as a testament to our continuous fascination with the evolving interdisciplinary inquiry between spatial and sonic realms. Identifying and categorising this content is merely a small part of developing dialogues between space and sound. Therefore, this book should not be understood as a 'how to' manual prescribing methods for such work, but more as a guide to facilitate productive conversations across diverse disciplines. We hope that the information within these pages inspires further discussion, sparks innovative inquiries, and continues to promote a deeper understanding of space and sound, from the poetic to the technical.

Notes

1 Nicolescu, *Manifesto of Transdisciplinarity*. P44.
2 Forsyth, *Buildings for Music*. P16.
3 Nicolescu, *Manifesto of Transdisciplinarity*. P34.
4 Nicolescu, *Manifesto of Transdisciplinarity*. P42.
5 Nicolescu, *Manifesto of Transdisciplinarity*. P43.
6 Nicolescu, *Manifesto of Transdisciplinarity*. P43.
7 Ingold, *Knowing from the Inside: Correspondences*.
8 Ingold, *Knowing from the Inside: Correspondences* P41.
9 Manolopoulou, 'Open Score Architecture'. P233.
10 Morton, 'Forward Thinking'. Pp130–135.
11 Morton, 'Introducing the Idea of "Hyperobjects"'.
12 Morton, *Hyperobjects: Philosophy and Ecology After the End of the World*. P4.
13 Morton, *Hyperobjects: Philosophy and Ecology After the End of the World*. P3.
14 Horning, 'Spatial Effects: Sound, Space, and Technology in Twentieth Century Popular Music'. P138.
15 Turkle, *Simulation and Its Discontents*. P17.
16 Pallasmaa, *The Eyes of the Skin*. P21.
17 Levin, *Modernity and the Hegemony of Vision*. P2.
18 Levin, *Modernity and the Hegemony of Vision*. P2..
19 Pallasmaa, *The Eyes of the Skin*. P12.
20 Forsyth, *Buildings for Music*.
21 Vitruvius, 'Book V'.
22 Kongresszentrum, 'Acoustics'.
23 Dahlstedt, 'Electronic Reverberation Equipment in the Stockholm Concert Hall'. P627.
24 Ritchie, Samborska, and Roser, 'Urbanization'.
25 Ball, 'Why Lockdown Silence Was Golden for Science'.
26 *BBC News*, 'French Rooster Maurice Wins Battle over Noise with Neighbours'.
27 LaBelle, *Acoustic Territories: Sound Culture and Everyday Life*.
28 Hall, *The Hidden Dimension: An Anthropologist Examines Man's Use of Space in Public and in Private*.

Bibliography

Ball, Philip. 'Why Lockdown Silence Was Golden for Science'. *The Guardian*, 20 June 2020. https://www.theguardian.com/science/2020/jun/20/why-lockdown-silence-was-golden-for-science.

BBC News. 'French Rooster Maurice Wins Battle over Noise with Neighbours'. *BBC*, 5 September 2019. https://www.bbc.co.uk/news/world-europe-49593954.

Dahlstedt, Stellan. 'Electronic Reverberation Equipment in the Stockholm Concert Hall'. *Journal of the Audio Engineering Society* 22, no. 8 (1 October 1974): 627–631.

Forsyth, Michael. *Buildings for Music: The Architect, the Musician, the Listener from the Seventeenth Century to the Present Day*. CUP Archive, 1985.

Hall, Edward Twitchell. *The Hidden Dimension: An Anthropologist Examines Man's Use of Space in Public and in Private*. Anchor Books/Doubleday, 1969.

Horning, Susan Schmidt. 'Spatial Effects: Sound, Space, and Technology in Twentieth Century Popular Music'. In *Kompositionen Für Hörbaren Raum / Compositions for Audible Space*, edited by Martha Brech and Ralph Paland, 137–149. transcript Verlag, 2015.

Ingold, Tim. *Knowing from the Inside: Correspondences*. The University of Aberdeen, 2017.

Kongresszentrum, Kultur-Und. 'Acoustics'. *KKL Luzern*. Accessed 18 May 2022. https://www.kkl-luzern.ch/en/about-us/architecture-acoustics

LaBelle, Brandon. *Acoustic Territories: Sound Culture and Everyday Life*. A&C Black, 2010.

Levin, David Michael. *Modernity and the Hegemony of Vision*. University of California Press, 1993.

Manolopoulou, Yeoryia. 'Open Score Architecture'. Edited by Matthew Butcher and O'Shea. *Expanding Fields of Architectural Discourse and Practice. Curated Works from the P.E.A.R. Journal*, 27 November 2020. https://www.uclpress.co.uk/products/126106.

Morton, Timothy. 'Forward Thinking'. In *The Ecological Thought*, edited by Timothy Morton, 98–136. Harvard University Press, 2010.

Morton, Timothy. *Hyperobjects: Philosophy and Ecology After the End of the World*. University of Minnesota Press, 2013.

Morton, Timothy. 'Introducing the Idea of "Hyperobjects"'. Accessed 11 April 2021. https://www.hcn.org/issues/47.1/introducing-the-idea-of-hyperobjects.

Nicolescu, Basarab. *Manifesto of Transdisciplinarity*. SUNY Press, 2002.

Pallasmaa, Juhani. *The Eyes of the Skin: Architecture and the Senses*. 2nd edition. Wiley, 2005.

Ritchie, Hannah, Veronika Samborska, and Max Roser. 'Urbanization'. *Our World in Data*, 23 February 2024. https://ourworldindata.org/urbanization.

Turkle, Sherry, ed. *Simulation and Its Discontents*. Simplicity. The MIT Press, 2009.

Vitruvius. 'Book V'. In *Ten Books on Architecture*, edited by Vitruvius. 129–162. Harvard University Press, 1914.

PART I

Architectural acoustics

Jane Burry

In this first section of the book, we open with Architectural Acoustics, the physics of the interaction of sound and the material world. Conceptually it is the contextual side of the sound of space equation. The chapters that follow explore the crafting of space in ways which shape sound, whether through physical analogue interventions, enclosure, surfaces, texture and materiality, or virtual manipulation to produce a particular spatial experience in the way the sound is manifest or reproduced. The spatial or acoustic extremities, in this sense, are sound in the free field, in other words, in an open-air environment in which sound is produced and its energy disperses with minimal barriers or surfaces to reflect it, a situation in which there is very little sounding of the space itself and what is heard is the direct sound from the source. At another extreme, sound produced in an enclosed chamber consisting of completely sound-absorbing surfaces (generally the contemporary starting point for the recording studio) similarly removes the reflections and spatial information. It also reduces the sound experience to something close to the direct sound from the source. Both these instances can be considered as boundless, dimensionless spaces in which to hear. But in most situations, many characteristics of the space can be read through hearing sound in it: its size and surfaces, geometry, materiality, level of occupancy are all encoded in the way that it sounds. Much is gleaned consciously and subconsciously even without complementary visual appraisal of the space, although for most people, the two interact in profoundly influential ways.[1] While the focus is on interiors, urban interiors, streets, laneways, urban plazas and even forests and corn fields all have characteristic acoustics. Living within earshot of rock concerts and football matches, fireworks and crowd events, demonstrates that the whole city displays acoustic resonances. Powerful reflections which can often 'throw' the sound source and provide a completely inverse direction for the source.

The time of human interaction with acoustics is both very long and relatively short. Examples such as inhabitation and ritual use of caves and other prehistorical ceremonial spaces, the shape and sounding jars of ancient Greek theatre, the writing of Vitruvius, and gothic churches all demonstrate an ancient empirical understanding of how to exploit, manage and design with the affective characteristics of sound in occupied space. Yet the history of representing the physics of sound and space mathematically for explanatory and predictive purposes is relatively brief.

Wallace Clement Sabine, a Harvard Professor of Physics (and incidentally bachelor's and master's graduate of Arts) is credited with the birth of acoustic science at the turn of the 19th and 20th centuries. Tasked with improving the acoustic of the Fogg Lecture Hall, he and his students studied the differences between this space and comparable ones, experimenting empirically with cushions and bodies by night, establishing that the acoustic was a function of room size and the amount of sound absorbing material. Through measurement, a person in the audience was found to be worth about six cushions. He established the concept of reverberation time (time taken for the sound to reduce by 60 DB), and through continued empirical experiment and measurement, established the Sabine equation. He discovered that successful concert halls had a reverberation time of approximately 2–2.25 seconds. In the 20th century the number of acoustic measures and parameters has increased exponentially, and the science has flourished, particularly in response to notable acoustic failures that required extensive ameliorative measures. Objective measures of the behaviour of spaces have been complemented with more refined human-centric measures such as Clarity, Speech Transmission Index and A-weighting of sound pressure levels to take account of perceived loudness across the spectrum, rather than just its raw air pressure wave metrics.

The science understandably started in public and performance spaces, large concert halls and theatres, for example, where their size and use make their acoustic behaviour mission critical, whether for music or the drier acoustic needed for the spoken word. But acoustic design and standards have infiltrated 'everyday' spaces, for instance, the impact on learning in school classrooms, the ability to work effectively in open plan offices combining rule of thumb with simulation, with adding and adjusting absorbent material to surfaces remaining the prevalent approach. As we will encounter in this section the consideration of the sound environment of small and even domestic interiors is now a significant area for research and commercial opportunity.

In the first chapter in the Architectural Acoustics section, Shane Myrbeck, Los Angeles (LA)-based sound artist, musician and acoustician, writes about sound installation into extraordinary and extreme spaces, both found and designed. In his role as Associate Principal in the Arup LA's Acoustics, Audiovisual and Theatre consulting team, he reveals something of the process of designing a bespoke Reverb Chamber for Bjork to record not in her huge powerful stage voice, but in her intimate singing-at-home voice. This is an exhilarating and emotionally led journey through spaces so extreme that the acoustic is a main event. The collaboration with Bjork on the chamber for her Cornucopia concert is an unparalleled imaginative and creative adventure. Myrbeck concludes with reflections on right-sizing of spaces for the sound in them giving the value to that sound. This is a veritable tale of extremes with a right-size conclusion.

Jane Burry, one of the editors, writes with Pantea Alambeigi, Architectural Acoustic research academic, on the curious world of acoustic shadows, and how we and all material things cast them. Shadows are one of the ways the environment, and what is happening within it, is understood. Whether it is to hear and detect movement or detect the detail of what lies hidden in the depths of the ocean, or the depths of the body, acoustic shadows are an increasingly significant phenomenon as data gathering and deep learning increase their legibility. This chapter explores how the phenomenon can be exploited in the design and layout of interior spaces to create lagunas of relative quietness and differential sound transmission. Examples of prototypical structures and experiments to test their performance expose the part that shadows play.

Also working across scales in interior spaces, Phil Colman and Philip Jackson extend what is known collectively as Next Generation Audio through encoding the impression of a room using the concept of Reverberant Spatial Audio Objects. These use a combination of direct sound, early reflection, and more diffuse reverberation objects to record and recreate an impression of room acoustics that can be rendered on any speaker system or even headphones. The spatial room impulse responses (SRIR) effectively reconstruct the room, its size, principal reflective surfaces, whether open or closed and absorbent contents. The impulse response is a profile that varies across frequencies. In their work they have collected SRIRs from many and diverse spaces small and large, using a 48-microphone array on two concentric spheres and also drawing the Open-Air dataset for spaces such as York Minster. They have also done binaural recordings with the two-eared dummy. Beyond the opportunity to reproduce different spatial acoustic environments at home or on any speaker set-up, there are also opportunities to creatively edit and manipulate the spatial impression for virtual environments.

Mark Burry and Sipei Zhao write about a project also involving the construction of a virtual acoustic model, in this case of Gaudí's Sagrada Família Basílica.[2] They also use impulse response tests locating the sound source and listening binaural dummy head in a grid of locations throughout the basilica interior to collect the data. Sipei unpacks the acoustics theory and practice in which such a grand endeavour can yield the acoustic twin of the space, not only in terms of reverberation characteristics but other qualities and opportunities of the sound. The data is ultimately used to be able to create an application in which to hear different instruments playing in different locations in the 200,000 cubic metre volume, heard from different listening points. Thus, while the emphasis in the Colman and Jackson's work is to be able to instil the appropriate acoustic space, however impressive and exceptional, for reproduced sound in the most basic of living room interiors, the focus of the work at the Sagrada Família Basílica is to be able to understand the impact of instrumentation in one exceptional and unique space. It resulted in the creation of listening and simulation tools to be able to design both the disposition of the planned pipe organs for the church and to be able to compose space-specific music for the space (refer to Emma Kate Matthew's Chapter in the Music and Sound Art section.) In our editorial taxonomy, we attributed the first to the sound of interior and domestic spaces, and the second to the sound of virtual spaces, which just illustrates what a loose taxonomy this is.

Raj Patel's conversation with artists Iain Forsythe and Jane Pollard creates a break in the general literary pattern of the book. While each contributor in this book has been invited to speak with their own voice, this trio has chosen to emphasise the importance of the collaborative nature of their work across the disciplines of public art and acoustic engineering through a loose dialogue spanning their project work together over seven years from 2006 to 2011. Raj opens by contrasting this work, in which collective creative endeavour is acknowledged and celebrated, with the contemporaneous culture of 'starchitects and celebrity artists' in which the process by which work came into being are assiduously concealed from the public gaze. This conversation should be read with the linked works and sound files themselves. Just as the conversation cross references their expertise, the full dimensionality comes from cross-referencing the other linked media.

Emil Kraugerud introduces some intriguing reflections on spatial intimacy in music recording in Intimate Acoustic Environments on Record. Under the banner for which we coined 'virtual space', he examines both the acoustic intimacy of home environments for better or worse, starting from Gaston Bachelard's very positive emotional associations with the spaces of the childhood home and their connection to aspects of memory, and consciousness through a phenomenological lens.[3] Kraugerud considers the opportunities afforded by 20th- and 21st-century recording technology and specialist studios including the current facilitation of remote online collaboration from intimate and often home-based spaces, and where this may lead in the future. Mediated Distance (the listener's perceived distance from the performers and performance), Portrayed Distance (the perceived distance between performers) and Objective and Subjective Distance (respectively whether from the standpoint of detached listener, or as an apparently active participant within the sound tableau) are all adopted as intimacy parameters from their origins in more multisensory media such as TV and film. Simultaneity is the perception of music with multiple acoustic signatures combined, as though emanating from an immediate and more distant room, illustrated in David Byrne's work. Combined, these characteristics make up the 'acousmatic space' of intimate listening and its emotional resonance.

Finally in this section, we close with a different approach to extremity in acoustics. Zachary Belanger and Catie Newell bring us the acoustic manipulation of a material that arguably poses one of the greatest challenges to the design of every type of space that requires good natural lighting: glass. Glass is a solid form of liquid that when deployed in-built fabric, usually as windows, has an unusually flat, smooth, hard, and flawless surface. Little could be as reflective to sound and, as such, create such, often unwanted, long reverberation, and bright, noisy acoustics. Their installation, *Long Range* demonstrates how acoustic gradients of absorption and diffusion can be accomplished by differentially slumping a single flat hexagonal glass 'blank' with lines scored by a waterjet and combining the resulting custom panels. Similarly to the FabPods and other prototypes referenced in the Burry and Alambeigi chapter, the case is made for the possibility of a more intrinsic shaping of sound by the surface shape and design of the building fabric in architecture, in comparison to the prevalent, after-the-fact applique of acoustic materials to the space. Long Range showcases a fabrication approach that minimises moulds and waste while creating a highly customisable architectural acoustic system.

Notes

1 P. Alambeigi, J. Burry, S. Zhao, E. Cheng (2020) A study of human vocal effort in response to the architectural auditory environment. *Architectural Science Review*. doi: 10.1080/00038628.2019.1708259.
2 P. Alambeigi, J. Burry, M. Burry (2024) Auralising the soundscape of Sagrada Família Basilica: a virtual journey through sound. *Architectural Science Review*, 1–11.
3 G. Bachelard (2014) *The Poetics of Space* (English translation of Gaston Bachelard, La Poétique de l'Espace, 1958, Presses Universitaires de France). London: Penguin Classics.

Bibliography

Alambeigi, P., Burry, J., Burry, M., 2024. Auralising the soundscape of Sagrada Família Basilica: a virtual journey through sound. *Architectural Science Review*, March, 1–11.
Alambeigi, P., Burry, J., Zhao, S., Cheng, E., 2020. A study of human vocal effort in response to the architectural auditory environment. *Architectural Science Review*. doi: 10.1080/00038628.2019.1708259
Bachelard, G. 2014. *The Poetics of Space*. London: Penguin Classics.

1
DESIGNED EXTREMES

Shane Myrbeck

Introduction

An extreme acoustic is an emotion. Wave physics may be objective and unconcerned with our experience, but our perception of sound is anchored in emotional response, often rooted so deep down that it can be difficult to articulate. As designers we aspire to translate those memories into architecture: it's honouring the feeling of when we were kids letting ourselves into the abandoned opera house to play shows, Pauline leading us through the old cistern, the chamber concert in the hardest rain we'd ever felt. Of listening to snow on the beach, calm plush here on shore, out there the roaring grey eternity.

Sound can be perceptually dominant as it can be too subtle to notice. This section will deal mostly with the former, but not always in the sense of overwhelm. Sound's extremities are multi-faceted, and its dynamism could be from loudness, rhythm, timbre, spectral complexity, duration, or most often, subtle aspects of all of these at once.

You could say that a venue is only ever as good as what goes on inside of it, but on occasion, its acoustic can be the point or at least the spark of our emotional response. That's what Björk asked for. She wanted the acoustic design process to transcend its engineering comfort zone and jolt us into feeling. To take an intimate acoustic moment, capture it like a firefly and send it out to more of an audience than could have experienced it before. Let the listener be the explorer, with the design to guide. When we designed the reverberation chamber for her Cornucopia tour, she wanted its wild acoustic to become an element of the tour's landscape.

She is unique in that perspective though, as she is in many things. Most often we don't design the extreme spaces, we find them. An extreme acoustic is often an unbridled one. When we design acoustic spaces, it's more common to want to control them, to quiet or wrangle them to some expected familiarity.

But every space has an acoustic, with or without a designer. Just because we can build without any acoustic intentionality doesn't mean the result won't be dramatic. The extremes of our built sonic environment frequently occur because no one thought of it, didn't care, or decided it wouldn't be important enough to do anything about. In many ways, this is wonderful – that some of the more uncanny acoustic experiences we can have exist just because some air particles are exploring a new space in ways we hadn't considered before. To explore a few examples:

Headlands tunnel

The hike to Battery Townsley in the Marin Headlands is beautiful, and foreboding yet hopeful. We've wrested beauty from brutality, thereby reclaiming war structures intended as the first line of defence from transpacific mid-century foe that never came. The chaparral will eventually conquer the concrete, but there will be a long disintegration – when you build something to gird it against the terror of explosive impact, the erosion by salt and roots might take millennia. In the meantime, its acoustic will linger. Even in July, the Headlands fog makes the massive concrete tunnels damp and freezing. You can hear the cold as the wind howls through the chases, even more so now that the guns and barracks are gone, as the coughs and shivers and scraping of feet echo for impossible lengths. We've extracted the gentlest thing from the most destructive, swallows instead of ordnance, sound art biennials instead of war games.

Sound Column

The Palace of the Fine Arts in San Francisco was built for the 1915 Panama-Pacific International Exposition as a temporary structure.[1] As it aged, decayed, and remained, it became a fondly regarded fixture in the city. Eventually, a campaign to rebuild and preserve the palace as a permanent cultural institution gained momentum. Nonprofits spearheaded the effort to raise funds and lobby for its preservation, and the city agreed to retrofit and restore the dilapidated structure.

One of the major upgrades was the reinforcement of the giant 60-ft tall columns supporting the Rotunda dome, which were lined with new concrete as part of the strengthening. What was not intended was the surreal containment of sound energy inside the pillars. Long after the renovations were complete and San Francisco's Exploratorium moved in, some intrepid experiential science enthusiasts discovered a new alien acoustic behind its locked door. At such a strange geometry, storage wasn't the best use of the space, and luckily, composer and gamelan player Dan Schmidt was invited to activate it (Figure 1.1).[2]

Figure 1.1 Inside the Sound Column (Shane Myrbeck).

Designed extremes

The keys of the large xylophone in the image above are tuned to the resonant 'standing waves' of the column. The mosaic on the wall behind represents how the room shapes those frequencies – the sounds get louder at the wider parts of the wave pattern and quieter at the pinch points. The tuning of the instrument to the room allows users to viscerally experience the long, deep reverberance of the space.

SFMOMA stair

I worked with my colleagues at Arup as part of the acoustics design team for the San Francisco Museum of Modern Art expansion,[3] with the architect Snohetta. Gallery acoustics can be fairly light touch, with most of the work focused on HVAC (Heating, Ventilation, and Air Conditioning) noise control, flexibility, and specialty rooms like theatres, media galleries, or event spaces. Suffice it to say it's not the type of project that you would think much about the acoustics of staircases. Here, however, inside the curved, textured east facade is a communicating stair that winds its way up to each gallery level, connecting them. During design, there had been some discussion of adding sound-absorbing material to this space, but the museum and design team decided to let it run free. As we were commissioning the space shortly before opening, the site architect asked me somewhat sheepishly to investigate something that was happening at the stair. He stood on one of the landings, clapped, and unleashed one of the most glorious flutter echoes I've ever heard. The reflected sound cascaded up and down the unique geometry for an impossibly long time. Typically, we would work to quell this type of acoustic artefact, but to do so in this case would have been costly, wasteful, and ultimately a bit sad. Leaving it untreated gives the stairs their own sound signature to be experienced and activated by guests and artists alike. This was a special opportunity to redefine 'good', by letting the space be what it is, have its acoustics match its form, and let the building clap back (Figure 1.2).

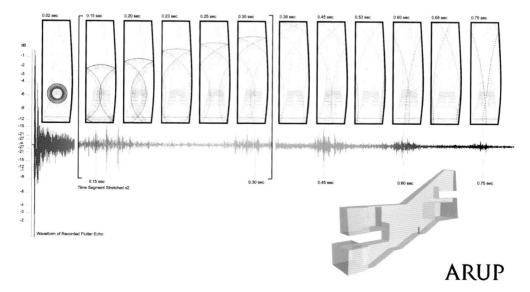

Figure 1.2   Impulse response analysis of the flutter echo phenomenon at SFMOMA (Arup).

Hangar

The buildings where we built airships a century ago are breathtaking. First, there is the sheer size, with some eclipsing 1,000' in length and 200' in height. These colossal structures are among the largest enclosed spaces in the world, the immense size being necessary to accommodate the dimensions and machinations of dirigibles. These are also typically constructed with a clear-span design, meaning they have no internal support columns or beams, allowing for unobstructed floor space. This provides ample room for the airships to manoeuver and for maintenance and repairs. The construction materials used for the hangars have to serve a dual purpose of lightness to contain overall building weight and a strength capable of withstanding high winds, snow loads, and other environmental factors.

Imagine what that does to sound. When we think of large volumes with hard surfaces, the term 'reverberation' often comes to mind, but at these volumes, the cohesion of dense reflections associated with reverberation loosens to something different. Reverberation is often thought of homogeneously, as a characteristic blend of reflected sound that provides a contained signature of a space.

The large, open space of the hangar allows sound waves to bounce and reflect off the walls, creating a canyon-like reflection pattern. But no canyon has exactly parallel concrete wall bases 300' apart, or 500-ton doors 1,000' from each other. Sound travels at roughly 1,000' per second, which in acoustics we usefully break down to the millisecond scale (1'~1 ms). But the hangar acoustic is a different order of magnitude. Rather than any sort of cohesive 'reverberation', these are uncanny discrete echoes in a cruciform cavern (Figure 1.3).

Björk's reverberation chamber

Then there are, of course, times when we do design to extremes. Björk's team approached Arup with a dream she had of being able to create an acoustic chamber that she could sing in on tour, with the intent of bringing a level of intimacy to audiences that she'd previously found elusive.[4]

During early discussions about what this meant to her, the form of the idea came into clearer focus. She described herself as having two voices: one is her on-stage voice, which has strong projection, bravado, and presence. It's active, sometimes aggressive – it's a strong voice. The other is what she calls her 'acoustic' voice, which is when she's singing at home and to herself. She described that she often writes music by simply singing melodies while she's on a walk outside or in her studio at home. There was also a Proustian aspect to this, where she would recall the lifelong understanding of the intimacy of this acoustic voice, starting at an early age. On her albums, you can hear both – those private, quiet moments and the loud, grandiose, full moments.

But on tour, based on the constraints of the sonic environment of the stage, she felt she was mostly able to achieve that bigger, louder moment. This new idea was inspired by the hope that she could create a chamber that would not only create some protection for her but also have a lush natural acoustic on its own where she could sing in a more dynamic, private voice and project that to the audience. This became our design brief. She wanted to straddle a line of the public and intrapersonal. The chamber helped the performances become a blending of the ancient and natural and the very digital, post-human future. As the design progressed, she was using words like 'sanctuary' and 'private contemplative moment'. In the throes of design, you end up with a project vocabulary for things. This quickly became the 'chamber' or 'chapel' based upon those conversations.

There are some precedents for a room like this: a remarkable sound sculpture near a fjord in Iceland called Tvísöngur, which is a series of connected stone huts that visitors can hike to and activate the long resonances of five connected concrete volumes. Another more technical precedent

Designed extremes

Figure 1.3 Dirigible hangar (Shane Myrbeck).

was reverb chambers from old recording studios: before the advent of digital sound processing and the ability to create rooms synthetically, mix engineers would send the recording of a voice or instrument down to a concrete room in the basement and record the sound of that room. This is often the reverb that can be heard on pop records from the 50s and 60s – the chambers at Capitol, Abbey Road, Sunset Sound, Universal Recording and many others were and are famous. The Exploratorium's Sound Column described above was also a point of reference.

With these, we had our key precedents of small rooms sounding like big rooms, and each on a physical scale that could conceivably fit on a stage. There was also the major secondary design challenge of being able to get that sort of massive resonant sound and sense of structure in something that was lightweight and could be moved from venue to venue.

Starting the design process with these shapes meant we had acoustic measurements of real rooms to compare to computer models and recordings, thus giving us confidence in the accuracy of our modelling process for this new purpose. When we asked for visual references, she sent simple sketches of quick one-motion shapes, as if to say it could be anything, and she wanted to explore: if it's shaped like an egg or like a flower or an articulated orchid shape, what do all these things do acoustically? We then took inspiration from those ideas and then built virtual models of them. From the beginning, Björk was very clear about the fact that in terms of our design, sound came first. She was insistent that we as the acoustic designers didn't need to worry about what it looked like. She and her set designers could make anything work, and they were interested in the concept of being visually inspired by a sonic form. She was even hesitant to show us the production storyboards ahead of time so as not to influence the ideas of what it could be because she just wanted it to be a purely sound-derived volume. This entire concept was beautiful and daunting for an acoustic designer to hear.

She then gave us her isolated vocal track from a song on her utopia record. We took all those things together and then built computer models of the different forms. We would send her batches of binaural listening examples that virtually placed her voice within those acoustic shapes to give her a general sense of which typology she preferred. By this process, we were able to get some parameters for how tall and wide the chamber needed to be. Eventually, we convened in Arup's London SoundLab, and were able to be in the room together to experience a real-time version of the sound simulation where she could talk and vocalise in the virtual space to get a sense of which resonances were the most evocative.

This is how we developed the general shape, but then there was the practical matter of how to disassemble it and pack it in the tour trailers. That process was a collaboration with her tour manager and set builders to discuss some of the practical aspects of re-assembly and durability. The set design team gave the chamber additional visual curvature by covering it with a stretched fabric, but the basic acoustic form is eight 10′ plywood sheets in an octagon with a vaulted ceiling. The octagon provides strong parallel reflection in several different directions so it doesn't have the strange ringing that might be familiar from an empty apartment or stairway because there are enough different reflection angles to create a blend and diffuseness. The vaulted ceiling shape helps to further scatter but contains sound. Sound travels upward and is redirected by the curvatures, which creates a lushness by avoiding any focusing effect typical of domed-shaped rooms, for example. We were able to do some expressive articulation with the ceiling based on the decision that the cap didn't have to break down – upon disassembly that piece is craned off and crated as is. And through this process, we ended up with a form that the set design team could dress for the stage. We provided the interior, acoustic heart of it, and they dressed it to fit into the visual world of the Cornucopia tour.

There's a distinct memory I have about the design process that demonstrates how she'll push the envelope of practical ideas to inspire her collaborators to follow their creativity to a more inspired design answer than they may otherwise have. We were in London, all in the SoundLab together, and she was evaluating the various shape options we had modelled. Then she started to do the movement blocking, thinking okay, I walk into it, I sing Claimstaker. And then she says, oh, hang on, could I wear it? And everyone was floored, because we've been assuming the chamber would be concrete, or the heavy gypsum-like resin material it ended up being, or something similarly weighty. The chamber would have to have a lot of mass to achieve the type of acoustic she was looking for. So, the initial reaction from the design team was, well, probably not. But then you think about operas and Broadway productions where actors fly around and massive set pieces drop down effortlessly. So conceivably there is a way with a proper theatrical rigging system that you could have the impression of wearing something of that size. It was just much different than what we'd been talking about. But the fact that everyone thought through it was an inspiring and useful way to explore what this chamber could be. What is the distilled thing that we're after with this design? What about it is the most important? And if it had to float and be lifted up, what about what we've talked about so far could survive? In the end, we decided to leave it on the ground and have it be a chamber that she walked into, but through the process, we learned something about its core purpose. And once you see the larger production that came out of a similar type of ideation and you think, of course, she could wear it.

Many traditional acoustic instruments are shaped the way they are to project sound the way they do based on their interaction with architecture. A violin's shape evolved symbiotically with the rooms it was perceived to sound best in, and if it wasn't meant to be played in a specific type of room, it would be a different sounding and shaped instrument. There's always an interaction between the sound source and what the architecture does, but when the two blend harmoniously, it can be a breathtaking experience. As such, a fundamental design challenge here was that Björk's intimate acoustic voice would need that kind of a call and response with the chamber acoustic. The room needed to respond in a way that sounded good to sing inside, but also that could spur her voice to respond to it. In that way, the chamber is not only its own instrument that blends with Björk's voice, but it is also influencing the way she sings inside, as was the original intent. Her interaction and activation breathe incredible life into its natural static acoustic.

Björk has an impressive amount of technical knowledge, but she's deeply collaborative. She allows people to express their own creativity by giving a prompt and enabling a true design conversation. All of the elements of the Cornucopia tour – the projection mapping, the costumes, and the set design, while part of her world-building, are all from very involved collaborators, people expressing their own selves through their expertise. She has a deep enough knowledge of the various components to be able to know what to want from them, but then she will set up a framework for people to express themselves rather than control. All the while, she's doing so within the terms and constraints of a world tour. She'll know how to push boundaries far and yet not to ask for things that are potentially impossible, to be totally groundbreaking but also to have a tour that can actually run. It was a striking level of sophistication with so many of the different tools she used to build the world in which each element operates. It was only at the opening that I fully understood the gestalt of Cornucopia after having been so focused on my little part of it.

Cornucopia was the first big concert I saw coming out of the pandemic lockdown, and it reminded me of all the magic, hope, and possibilities that were so easy to lose sight of in isolation. After seeing the show for the second time I realised being back in that world reminded me of what is possible when you approach creativity with so much joy, excitement, and confidence that it's going to amaze and inspire. To have a show at that specific point in time about rebirth, uplift, and community was indescribably powerful.

Orbit Pavilion

Another room-size sculpture that I worked on, that used sound to convey a contradictory vastness was Orbit Pavilion.[5]

NASA (The National Aeronautics and Space Administration) uses a fleet of satellites to observe Earth – its weather patterns, atmosphere, sea levels and ocean currents, winds and storm warnings, and freshwater resources. This sculpture is a piece that represents the orbital movement of these spacecraft using an immersive outdoor sound system. Dan Goods and David Delgado, both artists at NASA's Jet Propulsion Laboratory, initially developed the concept and then commissioned me to compose the soundscape and Jason Klimoski of Studio KCA (Studio Klimoski Chang Architects) to design the form. Jason's idea for the massive nautilus shape was inspired by the thought of holding a shell close to the ear and hearing ocean-like sounds inside. In this case, listeners walk into the shell and are immersed in a soundscape played over a hemisphere of loudspeakers.

The piece is in two parts, each with one sound following the path of a satellite. One section demonstrates the movement of the satellites by compressing a day's worth of trajectory data into one minute, so listeners are enveloped by a symphony of 19 sounds swirling around them. The other section represents the real-time position of the spacecraft: each satellite currently above our hemisphere will 'speak' in sequence, and when a sound is playing, if a listener points to the direction of the sound they are pointing to the satellite orbiting hundreds of miles above us. The position is translated to the latitude/longitude of each new location of the exhibit.

I want the composition to evoke something about the satellites, both where they are and what they study. When the sounds are representing their real-time location, I want people to think about the actual spacecraft out there in orbit working away, so I chose electronic and mechanical textures, playing with the types of sounds that hopefully make people think of satellites. When the orbits are sped up, I decided to focus more on the missions, which is a combination of the data being collected by the satellites and the people down here making it useful. These satellites are all part of Earth science missions, studying our atmosphere, oceans, and geology – they are helping us better understand how our planet is changing, and potentially how we can be better stewards of it. In that way, I see them as kind of sentinels or protectors. To evoke this, I created a soundscape that relies on field recordings mixed with musical tones, creating a symphonic ecosystem that is intended to be both enveloping and comforting (Figure 1.4).

Figure 1.4 Orbit Pavilion (Dan Goods).

The next spaces

At a certain scale, a sound exhausts its natural ability to address its audience. It gets too big or small for its room to hold it acoustically. It needs either an amplification that fights against its intimacy or a packaging that belies its size. Our mastery of physics can offer engineering solutions, but only through empathy and close listening can we retain magic through these spatial translations.

If I whisper the right thing in your ear, any room is the right size. If I speak under my breath across the room, it becomes too big. If I shout, most rooms become too small, and when a space is too big even for shouting it can become dangerous.

Take this concept back to familiar music performance scenarios, and we all have fond examples of a good fit, whether a great sound reinforcement system in a great room, or an acoustic venue exactly sized for an ensemble. It's a problem set that's been solved for many scenarios, but even so, there remains extremity in the right-sizing of sound and room, the rarity of a transcendent match, the ephemeral perfection.

As music evolves, as art becomes more multisensory, as our definitions of words like performance, experience, and exhibit blur, this scaling problem reintroduces itself with an entirely new set of parameters. When we have mature technologies like Arup's SoundLab that can translate an experience from a small room on up to the largest immersive venues in the world, what do these concepts of intimacy and scale mean? If we can virtualise impossible scales of vastness or closeness either right at your eyes and ears and at venue scale with the wide array of platforms at our disposal what is the new perceptual and emotional lexicon? We don't yet know the extremes because we're still beginning to explore.

Notes

1 'Palace of Fine Arts Event Venue in San Francisco'.
2 'Dan Schmidt'.
3 'SFMOMA'.
4 'Designing an Intimate Acoustic Chamber for Björk's Cornucopia Tour'.
5 'Orbit Pavilion'.

Bibliography

'Designing an Intimate Acoustic Chamber for Björk's Cornucopia Tour', Accessed 12 April 2024. https://www.arup.com/projects/bjork-reverberation-chamber.
Exploratorium. 'Dan Schmidt', Accessed 14 December 2022. https://www.exploratorium.edu/arts/artists/dan_schmidt.
Palace of Fine Arts. 'Palace of Fine Arts Event Venue in San Francisco', Accessed 20 April 2017. https://palaceoffinearts.com/.
'SFMOMA', Accessed 12 April 2024. https://www.arup.com/projects/sf-moma.
Shane A. Myrbeck. 'Orbit Pavilion', Accessed 20 March 2021. https://shanemyrbeck.com/portfolio/orbit-pavilion/.

2
INTRODUCING ACOUSTIC SHADOWS

Pantea Alambeigi and Jane Burry

A shadowy absence

This chapter takes an acoustic journey into not what is there so much as what is absent, not what is heard so much as what is occluded. We explore the role of acoustic shadowing within experiments carried out primarily to create and understand the opposing phenomena of human speech privacy and human speech intelligibility. The domain of experimentation for this exploration is open-plan knowledge work areas. Our dual foci are first, the constructible possibilities and second, the potential for impact and functionality of deploying semi-open architectural pods distributed within such open workspaces. Before venturing in any detail into that work, we first take a step back and consider, from first principles, the mysterious tenebrous, umbriferous world of acoustic shadows.

The phenomenon

What a delight it is to observe the art of the shadow puppeteer. There is a thrill in the certain knowledge that between the light source and the moving image there are real puppeteers and their carefully crafted silhouette-casting puppets. Yet we are transported and read the narrative through the flat monochrome abstraction of the light that is blocked. The story becomes all the stronger and transporting for dispensing with the extraneous details of the source.

Acoustic shadows are a sound phenomenon highly analogous to shadows cast on surfaces by light when it is differentially blocked. Frank Dufour (2011) notes that acoustic shadow theory introduces a silent player into the understanding of audition, which we otherwise traditionally consider as involving the three constituents of sound source, sound propagation, and sound perception or reception.[1] It is a phenomenon most perceptible in noisy environments with diverse background noise that includes high frequencies in its bandwidth. It is a strong addition to spatial perception through the hearing of not sound but the absence of it. It is a powerful aid to detecting the size, nature, and movement of a sound obstructer through space nearby. One can only imagine the evolutionary opportunities of such sensitivity during dark nights in forest or savannah. Dufour's contemporary example is the detection of movement of other passengers, including their speed, direction, and size, when sitting on a plane at night with one's eyes closed or covered.

The theory is grounded in James J. Gibson's ecological approach to perception.[2] A distinction is made between the perception of silent objects versus the perception of sound shadows and sound holes. Blind people use acoustic shadowing and other auditory phenomena to detect obstacles. Bats deploy echolocation, reflection, and absorption in combination with acoustic shadowing to understand and navigate spaces and even locate their prey through shadowing when they are poorly acoustically camouflaged on leaf surfaces.[3]

> A surface obstructing a sound source may mediate, reflect, refract, and absorb the acoustic signal. The resultant structure originating from the source sound may be better conceptualized as an acoustic shadow than as an occluded sound per se.[4]

Also analogous to visual perception, acoustic shadowing is better perceived when the listener is on the move relative to the perceived sound obstacle or shadow caster. Experiments to detect and collect data to simulate the phenomenon involve moving the silent blocking sound absorber that is creating the acoustic shadow quite quickly past the microphone or receiver.[5]

Lately useful acoustic shadows in mapping, fisheries, and health diagnostics

Acoustic shadow refers to a concept in sonar (Sound Navigation and Ranging), a low-frequency sound technology first developed to detect submerged icebergs.[6] Sonar and Ultrasonic navigation systems use the full spectrum of frequencies from very low, even infrasonic (less than 20 Hz) for long range to high frequencies up to 500 kHz for short range. They encounter shadowing and blocking at all frequencies, as do marine animals using low-frequency sound for sensing and navigation.[7] Acoustic shadows are a now familiar concept from such ultrasonic scanning technologies, from seeking out fish[8] in the ocean to diverse medical applications. Dual frequency sonar Fourier analysis of acoustic shadows has been used to identify and distinguish between specific fish species for more than a decade.[9] More recently, there are examples of the automatic detection of fish stocks in the ocean by use of acoustic cameras and Deep Learning algorithms on a combination of direct sound, acoustic shadows, and the combination of the two.[10] In medical diagnostics, acoustic shadows, often viewed as a nuisance in ultrasonography, can now, through the detection of their overlap with organs, and the application of Machine Learning, be used to get a more precise location of lesions for treatment.[11] Defined as an interface where the acoustic energy is almost completely lost at a tissue-to-tissue, or air-to-tissue interface, acoustic shadows help to detect gallstone, bone, and benign tumours.[12]

Within interior architecture and urban design

The concept of acoustic shadow is important in sound engineering and architectural design, where efforts are usually made to minimise these areas and ensure that sound reaches all intended listeners. However, there are both interior and urban space instances where an acoustic shadow is desirable and advantageous in establishing a quiet area or ensuring greater privacy for speech.

In indoor environments, acoustic shadows can occur when sound waves encounter obstacles such as walls, columns, or furniture. These obstacles can either absorb, reflect, or diffract the sound waves, causing a region of reduced sound intensity or (rarely) complete silence on the other side of the object. This can lead to uneven sound distribution in the room, which impacts speech intelligibility, music quality, and overall acoustic performance.

Acoustic shadows in indoor environments can have a significant effect on the quality of sound in large spaces, such as auditoriums, concert halls, and conference rooms. Diffraction is one way to create different sound levels and acoustic conditions in such spaces. Diffraction is a phenomenon that occurs when sound waves bend around an obstacle, creating a region of sound enhancement behind the object. This can be used to create more even sound distribution in large spaces by strategically placing diffraction elements such as sound diffusers or reflective surfaces in the room. By carefully designing the shape, size, and location of these elements, acoustic shadows can be minimised, and a more uniform sound field can be achieved.

More relevant to our preoccupations, the phenomenon of acoustic shadowing can be exploited synthetically in the auditory spectrum of sound within built interiors to produce a highly differentiated acoustic across large open multipurpose shared spaces. Just as the best architecture subtly exploits gradients of light, we can achieve gradients of sound within designed spaces. Using geometrical and material tropes to manipulate acoustic behaviour and auditory perception, it is possible to produce and calibrate acoustic shadows and differentiated auditory experiences, like stepping in and out of pools of light or moving within the shadows. Within an environment with a perceptible level of background sound or noise, we all cast acoustic shadows. These shadows follow us and mark our passage, much like our visible shadow on a sunny day. The more intense the sunshine, the more densely black and sharp our shadow appears. The greater the level of background sound or noise, the more audibly perceptible our acoustic shadow is to others nearby.[13]

Similarly, architectural form and mass cast acoustic shadows, either directional in relation to a sound source or by selectively blocking the experience of the diffuse ambient sound when we are in proximity to the shadow caster. Creating pockets, or poche, has a long architectural tradition, going back to Neolithic wheelhouses, small spaces carved out of the earth or solid stone walls, subspaces, and nooks that not only define a differentiated use from the main part of the space but provide an acoustic retreat for sleep or small conversations.[14] Even more ancient examples are primeval caves, and interesting geological formations that are renowned for their acoustics. The phenomenon of acoustic shadows plays a significant role in the acoustics of these caves and helps to create singular and fascinating soundscapes. The rough and irregular surfaces of the cave walls and ceilings can create unique reflection patterns and consequently shadows that reduce or eliminate sound in certain areas.

More refined versions of the negative carving out of space (and acoustic) can be witnessed in Hellenic and Baroque architecture, in the side chapels of churches, for example.[15] We are interested in a contemporary reinterpretation of the creation of micro acoustic environments, islands of customised sound effects, carved out of a larger space. We explore not only the impact that the formal geometrical and material design and creation of such semi-open subspaces can have on the auditory experience within them but also the distinct impact of the shadow they cast on the macro interior in which they exist. We explore the exploitation of the difference between the inner and outer architectural shape and surface, not pure poche, but extraction of micro acoustic space and extraction from macro acoustic space: a two-faced architectural creator of shadow.

Health well-being and areas of privacy and respite in open-plan work areas

In contrast to 'found' examples of rich acoustic shadowing in caves and historic architecture, we focus on semi-enclosed spaces or sound barriers within the large open-plan layouts in contemporary workplace architecture. Three overriding challenges of the state-of-the-art open-plan layout that prompt zoning are, the lack of privacy in an immense open interior, the overwhelming perception of emptiness, and the blank architecture with no distinct functional areas. The outcome can result in a large space that appears chaotic, with occupants who are either disturbed or actively searching for a suitable place to rest or communicate. The achievable solution in the market is to have more visually or physically disconnected spaces to offer smaller zoned pockets with more intimacy and privacy. Semi-enclosed interiors in open-plan layouts are becoming increasingly popular, allowing for privacy and seclusion while still encouraging a collaborative atmosphere through visual and actual connectivity. They are valuable in promoting a sense of cohesiveness in the space while dividing an open-plan area into distinct sections for various functions.[16]

Just as cave formations result in micro-auditory space within the larger macro environment of the whole subterranean space, in zoning the contemporary open-plan layout, micro-auditory environments are generated within a macro soundscape.

In designing and creating intimate zones, access to the sunlight and shadow cast on other objects and surfaces in space is typically the first sensory parameter to be investigated. While light's importance on design appearance, colour, comfort, performance, and productivity is undeniable, sound as a main distracting factor in interior spaces should be counted as a critical form determinant. Acoustic shadows are invisible but are creators of important functional micro and macro auditory environments. Acoustic shadows are the results of the interaction of sounds with form and geometry within the space. The reflection and refraction patterns off the surfaces differentiated at various scales of form, texture, and materiality, create acoustic shadows on both sides of any structure or enclosure.

Shells and shadows

The notion of acoustic shadows within extensive open spaces is not a recent subject. Indeed, since the outset of the widespread implementation of this architectural arrangement, particularly in office spaces, the integration of acoustical panels and partitions has been introduced to obstruct or absorb sound waves, consequently producing a quieter zone within the broader space. Various pieces of furniture, including vertical acoustic panel dividers, bookshelves, and other sound-diffracting objects, generate acoustic shadows and alter the soundscape on either side, influenced by the location of the sound source. Despite the availability of numerous solutions, studies consistently demonstrate that occupants of open-plan offices are dissatisfied with their working conditions due to a lack of acoustic and visual privacy. Most common solutions for improving acoustics in interior spaces have focused on the material properties of surfaces for absorbing, reflecting, and diffusing sound. However, the influence of geometry on the overall soundscape has been understudied. As a response and to reconcile the conflict between privacy and the advantages of an open layout, architectural shells were introduced that eliminated doors and ceilings. This approach aimed to maintain the essential features of an open-plan design while still accommodating the need for privacy. The introduction of architectural shells brought to light the subtle interplay between sound and geometry, resulting in the creation of micro-auditory environments. As examples, we can list a few architectural shells, which shared a focus on acoustic performance but featured diverse approaches and workflows to their design and creation, including Distortion I and II,[17] MPAS (Manufacturing Parametric Acoustic Surfaces),[18] Origami partitions[19] such as PleatPod,[20] RMIT (Royal Melbourne Institute of Technology) meeting pavilion,[21] FabPod I,[22] and FabPod II.[23]

The projects

Deploying the possibilities of form, geometry, materiality, and surface articulation, we and others have undertaken experimental work through the design, sound performance simulation, and construction of built prototypes to create areas of reduced background noise, controlled sound transmission and to affect the acoustic of the overall open interiors into which these prototypes have been placed. This has led to quite incidental but significant discovery about the specific effects of acoustic shadowing and its possibilities for sculpting particular types of architectural and social space out of larger interiors. We will next introduce and detail these projects and the ways in which these discoveries were made.

FabPod I

The idea for prototyping FabPod I emerged from the desire to explore the sound scattering properties of hyperbolic form and surface across multiple scales. Doubly curved hyperboloid surfaces are deployed across the Gaudí's Sagrada Família Basílica in both circular and elliptical forms[24] and our interest lay in exploring the geometry of the hyperboloid at a smaller scale of a meeting pod. We discovered that the hyperboloid has a significant impact on sound scattering and has radical potential to shape the soundscape, used in conjunction with reflective and absorbent materials. FabPod I was constructed using a series of extracted hyperboloid cells, distributed on larger scale partial spherical convex intersecting shapes. The overall geometry was formed by intersecting spheres, complementing surface articulations of hyperboloid cells, and, based on simulation, intentional distribution of materials across the surface, ranging from fully absorptive to reflective (Figure 2.1).[25]

Once the semi-enclosed meeting pod was constructed and placed in an open-layout office, we conducted two experiments with different targets to investigate the soundscape within the pod. Our aim was not only to understand the experience of the pod users but also to gain insights into how the pod impacted the office occupants near it.

Figure 2.1 FabPod 1. FabPod acoustic meeting room, Image credits: John Gollings a) exterior creating differentiated external acoustic alcoves, b) open doorway facing long metal grid clad wall, with high quotient of absorbent panels, c) occupied interior, d) interior, facing entrance opening

Sound perception and privacy experiment

For the first experiment,[26] we invited a group of participants to have a meeting inside the FabPod on a topic of their choice, in a natural setting. Following the meeting, we asked them to rate their perception of the speech privacy of the pod. At the same time, another group of office occupants sat arrayed around the pod. Each sat at an identical distance from the sound source and pod. This group rated the privacy of the conversation occurring inside the pod by what they were hearing and what they were able to comprehend while they were attempting to undertake other tasks, such as their email. It was anticipated that the conversation would be heard significantly by the listener sitting at the open entry point. Through objective, quantitative sound measurement, this was known to be the main source of sound leakage. However, the picture painted by the differing perceptions of speech by the listeners sitting at different positions around the FabPod was more complex than anticipated. Their perceptions were intriguing and warranted further detailed exploration.

The outcomes of simultaneous objective sound measurements were found to be consistent with the participants' rankings. Additionally, we conducted a digital simulation of the experimental setup and compared the results of simulated speech privacy with human perception. All three methods confirmed that the FabPod creates multiple distinct micro-auditory environments around its external shell in the open-plan area where sited.

To determine the reason behind the findings, all influential parameters were meticulously examined. The source and receiver distance were consistent in all the locations, and so were the thickness and structural layers of the FabPod. Additionally, the ratio of different surface materials on the wall cells of the pod, ranging from highly sound-reflective anodised aluminium, to formed acrylic, to high-performing absorbent acoustic felt conformed to a fairly similar distribution on all sides, except the entry side. After controlling for various factors, we found that the overall geometry, orientation, and location of the pod in relation to the open-plan office space, walls, windows, and ceiling were the only remaining variables affecting the generation and alteration of micro-auditory environments. As the overall form of the FabPod I's shell is constructed by the intersection of multiple spheres with varying radii, the resulting irregular curved wall intersections in both plan and section, create a 'sound umbrella' and generate acoustic shadows in specific areas around the outside of the pod.

The transmission of sound through space can occur via various channels, with leakage from different openings being a common culprit. In semi-open spaces, the transmission of sound can take place through either vertical (open door) or horizontal (open ceiling) planes, depending on the specific configuration of the meeting space. In the case of FabPod, however, sound leakage occurs through both vertical and horizontal planes. The findings indicate that the conversation inside FabPod was completely intelligible to participants who were situated near the open entry and measurements confirmed the increased intelligibility for individuals in the vicinity of the entrance. Based on our analysis, the open entry serves as the primary source of sound leakage, with the ceiling above the pod, on the horizontal plane, also contributing to the transmission of sound into the open-plan office, despite its textured nature and incorporation of sound absorption.

When sound escapes from FabPod, it disperses into the surrounding environment and spreads through reflections. The role of vertical (walls) and horizontal (ceiling) surfaces of the 'parent' enclosing space, in reflecting and thus propagating the leaked sound within an open-plan office is significant, and it depends on the magnitude of sound leakage from each opening. In the case of the FabPod, it was identified that the open entry was the primary source of sound leakage. As a result, the initial and early reflections from the wall facing the entry (vertical plane) are particularly important.

Consequently, the shadows that arise from the FabPod's irregular plan layout are more significant than those that stem from the bespoke curved vertical section (Figure 2.2).

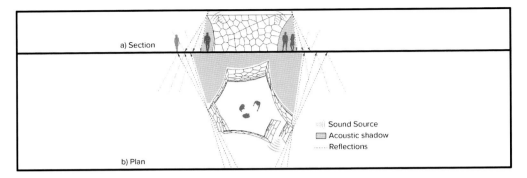

Figure 2.2 Acoustic shadows in plan and section.

The orientation and design of the FabPod both have a significant impact on the acoustic environment of the whole open-plan office, particularly in proximity to the pod. The presence of openings and the geometry of the pod contribute to the way sound is propagated and reflected. By understanding the sources of sound leakage and the significance of reflections from various surfaces, adjustments can be made to the design to improve acoustic performance increasing the protection of the speech privacy of pod inhabitants and the shielding of others working outside the pod from intelligible, distracting and disturbing transmission of speech from within.

In the second iteration of the pod, modifications were made – more limited openness to ceiling and the addition of a door – to reduce the speed of sound propagation, increase the rate of dissipation of sound energy within, and improve the acoustic environment. By limiting the openness to the ceiling, the presence of vertical sound channels was reduced, thereby slowing down the spread of sound. Additionally, the inclusion of a door that followed the pod's overall geometry provided an effective barrier to sound transmission and improved privacy.

FabPod II

When delving into the intricacies of FabPod I, the open entrance emerged as the primary source of sound leakage. This leakage occurred despite the 'wrapping' or overlapping of the walls like a shell, to create a curved passageway into the pod with a predominance of absorbent material on the cellular walls of this 'passage'. The entrance was consciously oriented toward a passageway and a blank wall, and away from work areas to minimise the disturbance and loss of privacy both from people coming and going and from escaping sound. Despite the highly textured nature of this wall opposite the entrance with its sound-scattering metal grill, the reflections of the escaping sound nevertheless impacted other areas around the pod in highly unexpected ways. This discovery emphasised the critical role played by reflections from the wall facing the entry, resulting in the loss of acoustic shadows in some of the concave areas around the outside walls of the pod. With the incorporation of a door in the design of FabPod II, it became evident that the changing vertical profile of the pod's walls also assumed a dominant role in shaping these acoustic shadows. Consequently, this underscored the need to allocate greater attention and focus to the pod's section.

As the project's requirements were scrutinised, an additional vital factor came to light. The pod's shell had to address the seemingly conflicting demands for speech intelligibility and clarity within the meeting room and speech privacy from the open-plan space. Recognising the challenge posed by this dichotomy, a double-skin architecture was introduced, enabling distinct geometrical and textural treatment of the two sides. Through the implementation of a double-layered approach, the pod's acoustic performance became more sophisticated, allowing for effective management of both speech intelligibility and privacy requirements.

In iterating the section of the FabPod II, the performance of flat, concave, and convex walls were analysed. Flat walls allow reflections to build up into standing waves and create stronger sound signals; however, concave and convex walls exhibit distinct characteristics when it comes to sound performance, each offering unique benefits and challenges. A concave wall, with its inward curving surface, has the potential to focus and amplify sound waves, enhancing their projection and resulting in holding the sound within a space for a longer time. On the other hand, convex walls, with their outward-curving surface, tend to disperse sound waves, diffusing them across a wider area. This can contribute to a more even distribution of sound and help reduce the concentration of sound energy in specific spots. While concave walls can enhance sound projection and create a focused listening experience, convex walls facilitate sound dispersion and contribute to a more balanced and even acoustic field.

When it comes to creating acoustic shadows, concave, and convex walls exhibit contrasting behaviours. A concave wall, with its inward curvature, can act as a focal point for sound waves, directing them toward a specific area. This can result in the formation of a more pronounced acoustic shadow behind the concave wall, where sound energy is reduced or even blocked. The curvature of the concave surface can help in capturing and absorbing sound waves, minimising their propagation into the shadowed region.

In contrast, convex walls, with their outward curvature, tend to disperse sound waves rather than focusing them. As a result, the formation of a distinct acoustic shadow behind a convex wall may be less prominent. Sound energy is more likely to spread and diffuse around the convex surface, reducing the intensity of a concentrated acoustic shadow.

The innovative FabPod II double-skin architecture, integrated into a lightweight steel-framed structure, incorporates a biconcave shell design that effectively generates acoustic shadows in the surrounding areas of an open-plan office. This unique feature significantly enhances both speech intelligibility within the space and speech privacy within the open-plan office, surpassing the performance of its competing counterparts. By strategically implementing the biconcave shell, the FabPod II successfully mitigates sound propagation and addresses the challenges of privacy in adjacent areas by creating acoustic shadows and zones of reduced acoustic interference. It minimises the negative impact of sound leakage, effectively preventing distractions and preserving the confidentiality of conversations happening inside the pod.

The effectiveness of acoustic shadow formation relies on several factors, including the degree of curvature, wall's shape and material, and acoustic interaction with the surroundings. Additionally, fabrication limitations can significantly impact the feasibility of achieving optimal curvature and maximising sound performance. Therefore, striking a balance in the architectural design was crucial to overcome these challenges and create spaces that offered optimal acoustic conditions.

In conclusion

Acoustic shadow, the inverse of sound, plays a key role in the complex auditory processing of both space and movement, notably permitting the identification of the size and nature of objects and obstacles, and their speed, if moving, from sound alone. It is a key component in the evolutionary arsenal of auditory tools for perceiving and interpreting the environment. This has been turned to technological application in the use of characteristic shadows and their analysis, now increasingly with AI (artificial intelligence), to identify the unseen, such as submarines, underwater fish stocks, and internal lesions in medical applications.

For architecture, the concept also plays a pivotal role in both designing specialised listening environments, such as concert halls and in designing for noisy shared environments. In the first case, the game is to find ways to eliminate the shadows. In the second case, it is to create and exploit the shadows strategically. In particular, the part shadows play in the creation of speech privacy, renders the concept of the utmost importance in acoustic design. By selectively utilising the phenomenon of acoustic shadow, areas can be effectively shielded from the transmission of sound, ensuring confidential conversations and private interactions remain unheard by others. Conversely, shadowing can be used to shield areas for quiet focus from outside distraction and disturbance. The deliberate manipulation of sound reflections and the careful placement of barriers can help in establishing zones where speech is muffled or rendered unintelligible to those outside the designated area. Acoustic shadows serve as a powerful tool in maintaining confidentiality, fostering a sense of security, and enabling individuals to communicate freely without the fear of being overheard. Whether in open-plan offices, meeting rooms, or public spaces, understanding and harnessing the concept of acoustic shadow contributes significantly to creating environments that prioritise speech privacy and promote a sense of trust and confidentiality.

Notes

1 Dufour (2011).
2 Gibson (1979).
3 Clare and Holderied (2015).
4 Gordon and Rosenblum (2004, p. 88).
5 Dufour (2011).
6 'Exhibitions: Acoustic Shadows' (1999).
7 National Research Council (US) Committee to Review Results of ATOC's Marine Mammal Research Program. (2000).
8 Dufour (2011).
9 Langkau et al. (2012).
10 Connolly et al. (2022).
11 Matsuyama et al. (2022).
12 Hellier et al. (2009).
13 Gibson (1979).
14 Armit (1990).
15 Alberdi (2021).
16 Hua et al. (2010).
17 Peters (2010).
18 Peters et al. (2011, pp. 819–828).
19 Vyzoviti and Remy (2014, pp. 487–494).
20 Underwood and Zilka (2022, pp. 1–19).
21 http://www.rolandsnooks.com/#/meeting-pavilion/ last accessed 11 August 2023.
22 Williams et al. (2013, pp. 251–260).
23 Alambeigi et al. (2018, pp. 627–638).
24 Burry (1993).
25 Burry et al. (2013, pp. 176–186).
26 Alambeigi et al. (2016).

Bibliography

Alambeigi, Pantea, et al. (2016) Complex human auditory perception and simulated sound performance prediction: A case study for investigating methods of sound performance evaluations and corresponding relationship, in *CAADRIA 2016, 21st International Conference on Computer-Aided Architectural Design Research in Asia-Living Systems and Micro-Utopias: Towards Continuous Designing*, pp. 631–640.

Alambeigi, Pantea, Canhui Chen, and Jane Burry (2018) Negotiating sound performance and advanced manufacturing in complex architectural design modelling, in Klaas De Rycke, Christoph Gengnagel, Olivier Baverel, Jane Burry, Caitlin Mueller, Minh Man Nguyen, Philippe Rahm, Mette Ramsgaard Thomsen (Ed.), *Humanizing Digital Reality: Design Modelling Symposium Paris 2017*. Singapore: Springer Nature Pte Ltd., pp. 627–638.

Alberdi, Enedina, Miguel Galindo, and Ángel L. León-Rodríguez (2021) Evolutionary analysis of the acoustics of the Baroque Church of San Luis de los Franceses (Seville), *MDPI Applied Sciences*, 11: 1402. https://doi.org/10.3390/app11041402

Armit, Ian (1990) Broch building in northern Scotland: The context of innovation, *World Archaeology*, 21(3): 435–445. https://doi.org/10.1080/00438243.1990.9980118

Burry, Mark (1993) *The Expiatory Church of the Sagrada Família*. London: Phaidon Press Ltd.

Burry, Jane, Nicholas Williams, John Cherrey, and BradyPeters (2013). Fabpod: Universal digital workflow, local prototype materialization, in *Global Design and Local Materialization: 15th International Conference, CAAD Futures 2013*, Shanghai, China, July 3–5, 2013. Proceedings 15. Berlin Heidelberg: Springer, pp. 176–186.

Clare, Elizabeth L., and Marc W. Holderied (2015) Acoustic shadows help gleaning bats find prey, but may be defeated by prey acoustic camouflage on rough surfaces, *eLife*, 4: e07404. https://doi.org/10.7554/eLife.07404

Connolly, R. M., K. I. Jinks, A. Shand, M. D. Taylor, T. F. Gaston, A. Becker, and E. L. Jinks (2022) Out of the shadows: Automatic fish detection from acoustic cameras, *Aquatic Ecology*. https://doi.org/10.1007/s10452-022-09967-5

Dufour, Frank (2011) Acoustic shadows, *SoundEffects*, 1(1): 88–97.

'Exhibitions: Acoustic Shadows' (1999) (226) Art Monthly 34 (ISSN: 0142-6702, 2059-5255).

Gibson, James J. (1979) *The Ecological Approach to Visual Perception*. Hillsdale, NJ: Lawrence Erlbaum Associates.

Gordon, Michael S., and Lawrence D. Rosenblum (2004) Perception of sound-obstructing surfaces using body-scaled judgments, *Ecological Psychology*, 16(2): 87–113.

Hellier, Pierre, Pierrick Coupé, Xavier Morandi, and D. Louis Collins (2009) An automatic geometrical and statistical method to detect acoustic shadows in intraoperative ultrasound brain images, *Medical Image Analysis*, 2010 Apr; 14(2): 195–204. https://doi.org/10.1016/j.media.2009.10.007. Epub 2009 Nov 17. PMID: 20015675.

Hua, Ying, Vivian, Loftness, Robert, Kraut, and Kevin M. Powell (2010) Workplace collaborative space layout typology and occupant perception of collaboration environment, *Environment and Planning B: Planning and Design*, 37(3): 429–448.

Langkau, M. C., H. Balk, M. B. Schmidt, and J. Borcherding (2012) Can acoustic shadows identify fish species? A novel application of imaging sonar data, *Fisheries Management and Ecology*, 19: 313–322. https://doi.org/10.1111/j.1365-2400.2011.00843.x

Matsuyama, Momoko, Norihiro Koizumi, Yu Nishiyama, Ryosuke Tsumura, Hiroyuki Tsukihara, and Kazushi Numata (2022) An avoiding overlap method between acoustic shadow and organ for automated ultrasound diagnosis and treatment, in *2022 IEEE 11th Global Conference on Consumer Electronics (GCCE)*, Osaka, Japan, 746–747. https://doi.org/10.1109/GCCE56475.2022.10014062

National Research Council (US) Committee to Review Results of ATOC's Marine Mammal Research Program. (2000) *Marine Mammals and Low-Frequency Sound: Progress Since 1994*. Washington, DC: National Academies Press (US). https://www.ncbi.nlm.nih.gov/books/NBK225328/. https://doi.org/10.17226/9756

Peters, Brady (2010) Acoustic performance as a design driver: Sound simulation and parametric modeling using Smartgeometry, *International Journal of Architectural Computing*, 8(3): 337–358.

Peters, Brady, Martin Tamke, Stig Anton Nielsen, Søren Vestbjerg Andersen, and Mathias Haase (2011) Responsive acoustic surfaces: Computing sonic effects, in *eCAADe Conference Proceedings 2011: Respecting Fragile Places*, 819–828. http://www.rolandsnooks.com/#/meeting-pavilion/

Underwood, Jenny, and Leanne Zilka (2022) Fibre architecture: A soft simultaneity design practice, *Journal of Textile Design Research and Practice*, 10(1): 1–19.

Vyzoviti, Sophia, and Nicolas Remy (2014) Acoustically efficient origami-based partitions for open plan spaces, *Proc. eCAADe*, eCAADe32_, pp. 487–494.

Williams, Nicholas, Daniel Davis, Brady Peters, Alexander Pena De Leon, Jane Burry and Mark Burry (2013) FABPOD: An open design-to-fabrication system, *Open Systems* (CAADRIA 2013), 251–260.

3
TRANSFERRING THE IMPRESSION OF REAL AND IMAGINARY SPACES

Philip J B Jackson and Philip Coleman

Introduction

Natural sound propagates from the source via air as the medium to the receiver. The source of sound bears its character, timbre and directivity, the receiver too. The medium combines the direct path, including transmission and refraction, with reflections (as from shiny surfaces for light) and reverberation, which is diffuse (like daylight on a cloudy day). Reflections arise at interfaces with sudden changes in acoustic impedance that is determined by the material's compressibility and density. In air, the hard surfaces of stone floors and walls are especially reflective, whereas soft surfaces tend to absorb impinging sound. Reverberation relates to physical storage of sound energy within a space, which relies on having a closed volume and an absence of absorbing surfaces.

As intelligent perceivers of our environment, we humans can use details of the sound we receive to infer some of these properties of our environment, such as the position of a wall from the 3D location of a reflected sound image, the size and likely materials of a room, and the distance of the sound source by its loudness relative to that of the reverberation around us. These constitute perceptually important characteristics of any scenario, therefore, configuring the arrangement of an active protagonist relative to the listener within the scene.

It is for this reason that the use of reverb effects, which artificially provide this sense of the sound paths from the source to the receiver, is ubiquitous in audio production. When microphones are deliberately placed close to the sound sources of interest to capture the pressure waves in full with minimal interference, as they are in most media production, it falls to the reverb to recreate that impression of the space for the listener, to conjure an auditory image of that acoustic scene.

An adjacent technology in the development of audio media is so-called next-generation audio (NGA). Championed by standards bodies (European Broadcasting Union (EBU), Moving Picture Experts Group (MPEG), Audio Definition Model (ADM)) and driven into the mainstream by media technology giants (Dolby, Apple, Sony, Google, Samsung, etc.), NGA builds on the concepts of object-based audio. Rather than mixing sounds down together for a fixed set of loudspeakers, like the two (left and right) channels for stereo reproduction, each audio component is stored as an object with its level and position defined, so that the final mix can adapt to the user's audio system when the content is played. A popular analogy for this involves flat-pack furniture coming as a kit of parts with instructions and assembled in the user's home.

This flexibility of object-based audio confers several advantages on NGA:

- reproducible over any loudspeaker arrangement
- compatible to exchange between microphone and reproduction setups
- viable for headphone reproduction via binaural technology
- dynamic, so is applicable to wearable and portable devices with motion tracking, e.g., VR headsets, AR on mobiles and XR.

Another inherent opportunity with object-based media is the ability to adjust the content so that it better serves the needs of that user. The possibilities of media personalisation extend far beyond a playlist of recommendations with tailored advertising, including dialogue enhancement, immersion controls, style transfer, language translation, and sustainability features.

A shortcoming of the initial formulations of object-based audio however has concerned the question of reverb. In 2015, our research group proposed one approach to representing the room impression in the form of object-based reverb, namely the reverberant spatial audio object (RSAO).[1] This proposition combines physical and perceptual perspectives of sound in the audio signal and opens up virtual, remote, and on-demand listening to experience the sound of spaces in myriad ways, for spaces both real and imaginary.

The RSAO describes a representation of the spatial room impulse response (SRIR), encoding the key elements of the acoustic paths from a source to a receiver. The representation concentrates on characteristics of significant reflections and the overall reverberation of the space. Investigations with this representation have enabled us to test aspects of the entire media production pipeline, from the recording of the SRIR with certain arrangements of microphones in various rooms, through useful manipulations during production, methods for practical reproduction over loudspeakers and headphones, and evaluation of the listeners' experiences. We will now describe the methods of the RSAO and key outcomes of the tests we have performed, before discussing potential future ways in which it may be used.

Proposed methods

The RSAO, our object-based spatial reverb, has three key ingredients: the direct path from source to receiver, a set of early reflections, and the late reverberation. Rather than modelling the physics of the acoustical environment in precise detail, we describe the observed SRIR via a signal processing approach: the reverberant sound is the result of passing the source signal through a special filter. This filter, constructed in an object-based way, determines the loudspeaker or headphone signals required to create the spatial impression of the space.

Our pipeline begins by making acoustical measurements of the SRIR, as you may sample with a loud clap to get a sense of a room's dimensions. As anyone visiting a building site or an unfurnished property can attest, the acoustic reverberation of the spaces we inhabit through most of our daily lives is very different. Much of this difference can be attributed to the additional absorption from the surfaces and upholstery of the artefacts that share those spaces with us. Yet the fine details of the environment's boundaries can also significantly affect the spatial distribution and character of the reflections they produce. So, without compromising these significant effects, we go straight to acoustical measurement as a faithful account of a listener's experience of sound in the space. For us then, the challenge in developing this novel pipeline has been to ensure that these ingredients can be safely conveyed so as to replicate the intended room impression in the listeners' perception.

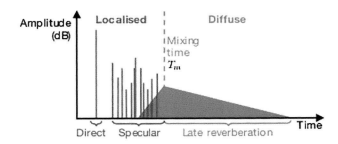

Figure 3.1 Idealised schematic of a room impulse response with direct sound, early reflections and late reverberation developing up to the mixing time and then decaying.

The object-based signal representation

The proposed RSAO representation appends encodings of the early reflections and late reverberation to that of the direct path, which corresponds to the line of sight from the sound source to the listener or receiver. Similar to the direct sound, the first reflection is often specular and has an associated direction of arrival at the listener. It also carries information in its timing and colouration along the way, providing cues to the material, relative position and orientation of the reflecting surface. The initial time delay gap of the reflection off the floor typically indicates the source distance or cues its elevation.

As the SRIR unfolds millisecond by millisecond (see Figure 3.1), further pulses arrive over time reflecting from one or more surfaces, as first-order or higher-order early reflections, imparting perception of the room's size. The reflections become more fuzzy and less localisable as the order increases, developing into diffuse reverberation around the third order. This transition into the late reverberation is demarcated by the so-called mixing time, which has a basis in both physical acoustics and auditory perception.

The late reverberation depends on the size of the room and characterises the absorption of the materials within the space, as the reflections repeatedly interact with its surfaces. Open doors and windows that bleed sound out cause reverberation to die away, while a chamber made of hard surfaces can reverberate for several seconds. Since surfaces' ability to absorb is influenced by both the geometry and the material itself, different frequencies may be affected at different scales, which we perceive through the hearing system's critical bands. Hence, the late reverberation is analysed in logarithmically spaced frequency bands, to yield parameters for the level and the rate of decay in each band.

Thus, in our object-based representation, the direct sound and early reflections are encoded as point source-like objects: copies of the direct sound with their own level, direction of arrival and colouration, whereas the late reverberation is encoded as a diffuse object with decaying envelope that is determined for each frequency band. The number of early reflections is a choice that depends not only on the acoustic environment in question, but also on the desired fidelity of the mid-order reflections and the computational resource available for a given application. The parameterisation process will be described in more detail below, after we have looked at some impulse response libraries of acoustic environments.

Libraries of SRIRs

In order to assess whether this RSAO representation is effective at conveying the sense of a space that we conceive as a room impression, we need to test a variety of different spaces. Beginning with rooms close to hand, we sought to create diversity by varying the size and properties of the spaces that we measured, but soon we looked to public datasets of SRIRs to determine the limits of our encoding, including large concert halls, specialist audio facilities and outdoor spaces.

In Coleman et al. JAES (2017), we tested a sound-treated audio lab, a concert chamber and two churches. For these experiments, we used a 48-channel bicircular microphone array with two nested circles each with 24 microphones to provide fine angular resolution in the horizontal plane. SRIRs were measured from multiple source locations at a central listening position.

In subsequent work,[2] we employed SRIRs from the University of York's OpenAIR dataset, which collates various measurements. Our tests included a sports hall, a nuclear reactor hall, a tunnel, a courtyard, a cliff landscape, the York Minster Cathedral and other stone buildings. These used a simpler type of receiver that is popular for spatial recordings, a sound-field microphone that combines four sensors to obtain first-order ambisonic signals, which correspond to the first-order spherical harmonics of the spatial sound field at the microphone. The signals are typically shared in 'B format' where the four W, X, Y, and Z channels relate to the omni-directional component followed by each of the three Cartesian axes. Adopting such a representation allows creators the opportunity to straightforwardly measure and encode additional rooms and spaces.

Further datasets were brought in to test for distance and make more controlled comparisons between rooms. Pori Hall (Aalto), Octagon Hall and Classroom (Imperial) and SurrRoom 1.0.[3] The last of these recorded SRIRs at a set of distances consistent across rooms, as well as binaural RIRs (BRIRs) measured with microphones installed in the ears of an acoustical head-and-torso simulator, a.k.a. a dummy head (see Figure 3.2), enabled formal listening tests to compare across rooms. These included comparison over distance and between binaural rendering of the reverb versus binaural recording of the room impulse response. We will say more about the results of those listening tests in the following section.

Extracting parameters of the room impression

An overview of the process for extracting the RSAO parameters from the measured SRIR is shown in Figure 3.3. The upper part (solid blue) extracts the parameters for each of the early reflections, up to a pre-determined number (typically 2–20, depending on the application); the lower part (dashed red) extracts parameters for each frequency band, whose number and frequency ranges are also pre-determined (typically 9 octave bands from 64 Hz to 16 kHz).

The parameterisation of the early reflections relies on detecting distinct pulses in the SRIR, identifying the strongest or most energetic ones, and then performing further analysis on the 2-ms segment of the SRIR, which is a multi-channel audio signal. Next, we find the reflection's direction of arrival: for the bicircular array, a high-resolution beamformer scans the horizon; for the B-format SRIR, we 'look' with a virtual cardioid. The direction that gives the maximum energy is selected and used to set the level. The steered output is then analysed to obtain linear prediction coding (LPC) coefficients of an autoregressive filter that fits the main peaks in its frequency response, a technique borrowed from the speech coding you might find on your phone, which here describes the colouration of that reflection. This process is repeated to extract the same set of parameters for each early reflection. Taken together, the parameters of these early reflections are intended to capture the nearest and most dominant features of the room acoustic, providing a sense of intimacy and presence within the scene.

For the late reverberation, the SRIR is divided into frequency bands by a bank of filters, each tuned to a specified frequency range. For example, with a bank of nine bandpass filters, each one covering a range of one octave, we obtain nine versions of the SRIR corresponding to each of the nine octave bands. The calculated mixing time is used to segment the late reverberation tail and set

Figure 3.2 Consistent distance SRIR and BRIR measurements in Lecture Theatre D.

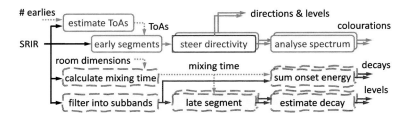

Figure 3.3 RSAO parameter encoding of SRIR into early reflection (upper, solid) and late reverberation (lower, dashed) parameters.

a datum to align the levels and decays across the bands. The levels are based on the energy in each band, while the rate of decay comes from the gradient of the line fitted to the energy profile of the reverberation tail over time, on a logarithmic amplitude scale. The late reverberation parameters speak most powerfully to the size of the surrounding space and the distance of the source, with variations across the frequency bands suggesting the character of the main materials that can make the space sound cold or warm, harsh or mild in nature.

Manipulating room impression in production

Our approach of encoding an SRIR that denotes specific source and receiver positions within the acoustic environment enables that same configuration to be encoded and, as we shall see, reproduced for the listener at another time, in another place. Yet, there can be many situations where it is desirable to adjust this real-world facsimile, such as when the source is to be placed at an alternative location, when some aspect of the room wants changing, or the listener moves, as for an interactive experience. Initial demonstrations of each of these types of situations have been trialled. In media production, the capability to easily make these adjustments is essential, for example, to bring elements of a scene together, to achieve dramatic effects and to maintain an illusion of reality with the experience. There is an opportunity with the RSAO parameters to develop a user interface for producers that connects a physical interpretation of the SRIR to the perceptual experience of the rendered reverb. Next, we will look at two aspects of rendering targeting reproduction over loudspeakers and over headphones, respectively.

Rendering the room impression over loudspeakers

Object-based audio was conceived with the intention of being reproducible over arbitrary loudspeaker arrangements. In practise, the most common loudspeakers lie on a circular, cylindrical, spherical or cuboid arrangement around the listener, providing various directions of arrival and a sense of envelopment. For a point source object, there are many available planning laws to relate the source position to the weights or gains for the loudspeaker signals. A popular choice is vector base amplitude panning to accommodate all directions. Another useful object type is a diffuse object, for which all the loudspeaker signals are decorrelated.

The direct sound and early reflections are treated as point sources in the RSAO representation. Hence, for each reflection, a copy of the source object is delayed and filtered and then panned onto the loudspeakers. The late reverberation is treated as a diffuse object. The decay envelope is constructed for the reverberation tail in each frequency band, combined across the bands, convolved with the source, and passed through a set of all-pass decorrelation filters. This is so that the loudspeaker signals are no longer correlated, which yields the impression of a diffuse sound field to the listener for the late reverberation.

The delay-filter-pan-and-sum operations of the early reflections tend to apply over a relatively short duration (<100 ms) from the first part of the SRIR and are not too demanding computationally. However, for highly reverberant spaces, the late tail can be several seconds long, which then causes a significant overhead. As a consequence, a suitable compromise must be sought according to the available computational resources and the number of RSAOs in the entire scene.

Rendering the room impression binaurally

For binaural reproduction, which aims to generate signals for headphone or earphone listening that nevertheless give a spatial impression of a scene within a space, a straightforward approach is to do this via so-called virtual loudspeakers, in two steps. The sound is first rendered to these virtual loudspeaker locations, as above, and then these loudspeaker signals are projected into binaural signals, using binaural or head-related impulse responses (HRIRs). A good result can be achieved with sufficient virtual loudspeakers and a high-quality dataset of HRIRs containing measurements recorded from the positions of those virtual loudspeakers with a binaural microphone.

One way to improve the quality further is to increase the number of virtual loudspeakers, but this improves performance incrementally with diminishing returns, at the cost of increased computation. Alternative methods to mitigate some of the most significant artefacts include the use of more sophisticated interpolation techniques by exploiting timing and other patterns observed in densely sampled HRIR datasets.

Another opportunity arises here owing to the projection of the virtual loudspeakers ultimately down to the binaural pair of signals, which for the diffuse sound may be achieved by directly decorrelating the binaural signal components. Note that the sound at low frequencies remains correlated between the two ears as the wavelengths are very large in relation to the listener's head, but as frequency increases, and the wavelength gets smaller, the level of the correlation resulting from the diffuse sound field increases.

We have now described the entire pipeline for the RSAO up as far as the listener's ears. The next stage and the proof of this representation of the room impression comes from experiencing the sound. To form some evaluations of the performance of this approach, we have conducted a number of formal listening tests in an attempt to understand the effectiveness of this description, its limitations and its potential for further uses and improvements.

Current outcomes

Having outlined the methods that support the RSAO's object-based encoding of the room impression, in this section, we review the current state of findings from investigations to evaluate the representation's utility. Formal listening tests were devised to assess various aspects of the represented room impression, such as comparisons across different rooms, judgements of source distance within a room and the so-called photocopier test that seeks to assure the repeatability of the entire pipeline. Additionally, as a proxy for these laborious and time-consuming participatory tests, perceptually relevant objective measures were examined that indicate the conformance in various standardised ways, such as signal-based measures of the reverberation time and listener envelopment. Inevitably, such assessments are limited and incomplete, yet they offer some evidence to indicate the potential of this approach to convey essential characteristics of the room impression within the NGA framework.

Objective measures of the room impression

Early work concentrated on detecting early reflections and using them to locate and visualise prime reflecting surfaces in relation to the sound source[4] using uniform and perturbed rectangular and circular arrays. We subsequently investigated first-order ambisonic microphones and a distributed ad-hoc array such as that typically used to record a live orchestral performance.[5]

The first few early reflections were seen to be detected from peaks in the group delay function on the SRIRs that were measured by the bicircular 48-channel microphone array with the swept-sine technique in the University of Surrey's Studio 1. Graphs showed an early reflection's direction localised using delay-and-sum beamforming, colouration being characterised with LPC filters, and spectrograms of reverberation decay in three frequency bands.[6]

After the initial proposition for an object-based representation of the room impression,[7] a wider survey of related literature contrasted our perceptually motivated signal-based representation with prior works that lacked either a compact description or a physically meaningful interpretation,[8] envisaging applications of the RSAO for sonic art, live recording and rendering on mobile devices.[9]

Following these opportunities for informal assessment of the RSAO with a few carefully selected acoustic environments, a wider set of objective tests was performed using the first-order ambisonic SRIRs from the OpenAIR dataset,[10] by adapting the early reflection direction finding to steer a cardioid, rather than the previously used beamformer.[11] Metrics used for the tests included the clarity index at 50 ms (C50), the early decay time (EDT), the 30-dB-decay reverberation time (RT30), and the inter-aural cross-correlation (IACC) applied to early and late portions of the rendered RSAO. Several refinements were identified. While most metrics compared the rendered RSAO favourably against values with the measured SRIRs, its sparser description of the early, first-order reflections and the cluster of second- and third-order reflections meant that the RSAO tended to underestimate the total sound energy in the room response for certain rooms. Overall, however, the distinct characteristics of the diverse environments were transferred from the places they were measured to the test environment via the RSAO.

In Chitreddy and Jackson Forum Acusticum (2020), the repeatability of the parameter estimation was evaluated using an approach referred to as the photocopier test, which is a reference to the results of feeding the output of the process back into the input to assess whether the process is repeatable. Parameters obtained directly from the SRIR measurements were used to render an impulse response over the 42-loudspeaker setup, which was again measured and encoded to give a synthesised parameter set. This process was repeated to yield parameters from measured, synthesised and re-synthesised SRIRs to demonstrate the consistency of the process, which was mostly kept within tolerances in the re-synthesised case, based on perceptually distinguishable distances.

From pictures to the sound of the space

Using an array of microphones to make acoustical measurements of a space is only one potential way to capture its character; a digital camera's array of pixels can capture spatial details optically. In 2017, an alternative line of investigation began to explore the potential of visual information to provide clues to the acoustics of a space,[12] which employed stereo photography, i.e., pairs of images, to extract 3D geometry of the environment. Later work harnessed computer vision and machine learning techniques to classify indoor materials from their visual appearance, which enabled the construction of acoustical simulations to compute an acoustical impulse response. Hence, simulated SRIRs could then supply the RSAO parameters for rendering. One benefit of this approach is that publicly available plugins (e.g., Steam Audio, Google Resonance) for a games engine (Unity) can render the experience for VR to provide interactivity, both to rotation and translation of the user, yielding six degrees of freedom (6DoF) interactivity with an immersive experience, illustrated in Figure 3.4. This extrapolation of the acoustics from the observed vantage point does not allow the user to explore the acoustics of a space actively beyond the scope of what it sees, yet it can complement the immersion with a stronger sense of presence through the user's interaction with the virtual space.

Formal listening trials

While objective measures provide a useful guide during the design and engineering of a complete and effective media pipeline, the ultimate test comes through assessing the experiences that users have with the technology. In respect of the RSAO, we are interested to answer questions such as:

- Does it convey the impression of the room whose parameters it encoded?
- Is the sound image correctly placed at the sound source?
- Does the RSAO function consistently across different playback equipment?
- Can the representation be useful creatively for modifying the room impression?

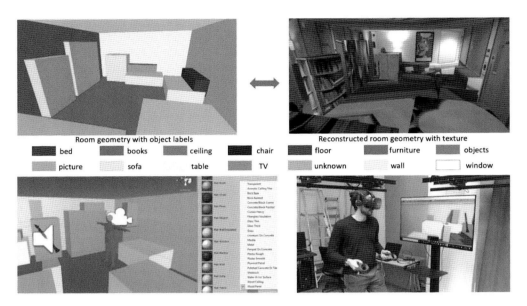

Figure 3.4 Illustration of visual characterisation of a meeting room for generating immersive sound with six degree-of-freedom interaction within a VR experience: (top, left) a perspective of the estimated geometry of cuboid objects with their colour-coded classification described by the key in the centre; (top, right) the same perspective overlaid with the visual appearance of the objects mapped from the camera images used for the classification; (bottom, left) the 3D environment and material properties translated into a games engine with a sound source (clarinet with loudspeaker symbol) and observer (grey manikin in a T-pose with a video camera symbol); (bottom, right) user wearing VR headset with headphones and hand-held controllers alongside a video display of the user's perspective.

Although all of our listening tests are finite and inevitably limited in some aspect, we describe a range of formal trials that have been conducted.

Flexibility of the representation

In Coleman et al. JAES (2017), the initial tenets of the RSAO supporting NGA were tested. One key promise of NGA is to render the content effectively no matter the loudspeaker positions or reproduction setup. Our listening tests showed that the RSAO conveyed the sense of room size and listener envelopment effectively across three different loudspeaker layouts. Furthermore, in NGA, it should be possible to edit objects; we showed that the RSAO could be used to impart a sense of distance. Using reverberation as a cue for source distance is typically a powerful cue, enabling greater depth in NGA audio content than by object location and level alone.

Spatial audio's third dimension

In Chitreddy and Jackson Forum Acusticum (2020), having encoded SRIRs from multiple distance measurements in a set of three rooms, listening test participants were invited to rate the apparent distance of the sound sources that they were presented in the audio booth, a sound-isolated laboratory with controlled critical listening conditions across the 300 Hz–8 kHz frequency range. The reproduction setup employed 42 loudspeakers arranged around the listener plus a subwoofer. This 42.1 setup sought to provide a high spatial fidelity of the rendered room impressions. Tests used a speech excerpt and a drum riff as the sound source. The main trends in the results confirmed that the perceived distance of the source increased with physical distance, as it had been hoped, which is attributed primarily to the energy ratio between the direct and reverberant parts of the sound signal. The distance ranges across the three rooms overlapped at only a few fixed distances, prompting the development of the SurrRoom dataset to give a larger consistent set of measurements in order to extend these tests.[13]

In Cieciura, Farheen and Jackson's Basic Auditory Science (2023), we used the SurrRoom dataset to explore the RSAO's ability to convey spatial audio's third dimension over headphones. This preliminary study invited 20 untrained listeners to rate the apparent source distance of a short clip of speech in a virtual room. Like the reference binaural RIR (BRIR) measurements, results for the RSAO showed a monotonic trend with physical distance and perceived distance in agreement, with significant differences between 1.5 m, 2.0 m, and 2.5 m. However, statistical analysis also showed significant differences between the RSAO and BRIR versions with the RSAO version appearing to sound around 20% nearer than with the reference BRIRs, pointing to an area for refinement of the RSAO encoding process.

Users' assessment of the room impression

In Pike, Cieciura and Jackson AES (2023), we exercised the SurrRoom dataset to test the RSAO's ability to transfer the impression of each specific room in turn. This test, which was developed during the COVID-19 pandemic, ran online allowing participants to use their own headphones at home. The test signals were therefore binaural, reproducing the spatial impression of each room through the combination of each early reflection from its direction of arrival together with the decorrelated diffuse late reverberation tail. This experiment put different rooms from the dataset head-to-head and pitted the RSAO directly against the BRIR measurements once again. For each test room, we selected the most similar and least similar rooms from the dataset, and then asked the participants to rate RSAO and BRIR versions of these rooms. Results from the 32 participants of this study indicated two main effects: first, confirming the ranking of similarity for the test, the next nearest and the farthest rooms; second, a modest yet significant difference between the BRIR and RSAO versions. On average, the size of the latter difference was less than the difference judged for the nearest alternative room. This suggests that the RSAO representation has some room for improvement, yet the impression it conveys gives a plausible rendition of the correct room more often than any other, within this dataset's range of rooms.

While participatory listening tests offer a limited set of results, and the findings reveal details of the encoding and the implementations that may be improved, the overall body of evidence supports the view that the RSAO has potential to translate the room impression. The pipeline can transfer this impression by encoding parameters from one of the various measurements of an acoustical environment to render an experience over one of a variety of reproduction setups, whether loudspeakers or headphones. Further developments could refine the consistency across a broad set of conditions and automate the kinds of adjustment most commonly applied in media production to arrange elements within a sound scene. Wider availability of such tools can facilitate listening comparisons to aid in the design and selection of spaces.

Future perspectives

We have described a pipeline that can transport the room impression, acquired from an acoustical measurement or photographs of a space, to listening at a different time and place, via the RSAO encoding. Various camera or microphone arrangements may be used to create a snapshot, and the sound rendered for loudspeakers or headphones, including head-tracking provides a responsive immersive environment. The essence of our object-based approach to encoding the acoustic environment comprises a description of prominent early reflections together with the time-varying colouration of the late reverberation tail. This compact representation forms a bridge between the environment's physical acoustics and the user's perceptual experience of it, and thereby opens opportunities for meaningful and creative alterations to the transferred room impression.

On the physical acoustics side, one can insert, delete, or adjust a strong early reflection to strengthen acoustic support, mitigate an unpleasant artefact or change the position of a sound source in relation to the listener. Such room compensation can aid with the clarity of speech, even out comb filtering effects or help with externalisation of a source with headphone listening, as well as for many other aesthetic reasons.[14] If listening to NGA in a large or reverberant space, the characteristics of that space can be accounted for in the audio rendering, avoiding the sense of one room inside another room when listening to the content. On the other hand, with careful design of a microphone system to feed the reverberation engine, existing spaces can be transformed to sound like new ones. For example, the late reverberation can be extended to transport an audience from a theatre to a church, on a new acoustical journey, reacting in real-time to any sounds emitted within the space.

The transmissibility of object-based audio continues to evolve as media technology giants and international standards bodies embrace the uptake of NGA (Dolby, DTS, Sony, MPEG, EBU, BBC, Google, Samsung). Meanwhile, tools for the production of NGA content for distribution over networks and media consumption are proliferating, with the capabilities becoming integrated also with mainstream and open-source production suites (Schweiger et al. IBC 2022). While several methods are available for dealing with reverb in these tools, further work is needed to provide fully functional and interoperable solutions.

Object-based audio is also prevalent, at a larger scale, at live events. Systems such as L-Acoustics L-ISA and d&B audiotechnik Soundscape bring object-based spatial audio to audiences of many hundreds or thousands of people. The parametric encoding can benefit these scenarios, by allowing the rendering of localised and diffuse reverberant content to adapt to a specific venue and the loudspeaker system installed there. Such scenarios can have significant time delays depending on the audience's position. Comparing the parametric encoding with knowledge of the venue's layout can enhance the experience by ensuring, for example, that early reflections never arrive before the direct sound.

We discussed above an approach for extrapolation that affords the user with re-orientation and translation freedoms to interact in the space. The spatial immersion combined with 6DoF interaction are compelling features within the whole family of alternative reality technologies (VR, AR, XR, etc.). Yet for moving sources, there is a need for tracking the source position and orientation. For VR environments or for re-animation of virtual characters from a user's movements, wearable tracking devices can help, whereas media production in other settings may benefit from markerless tracking and combine the advantages of both audio and video for this task.[15]

Here, we have concentrated on the implementation of the RSAO encoding of the room impression. In addition to refinements of the technical encoding, the future will bring flexibility and the multitude of ways users may harness it for personalisation. The use of flexible media over two-way networks enables the media to become smart and responsive. Beyond recommendations, personalisation can compensate for sensory or cognitive impairments; adapt for diverse styles, moods, and settings; and provide alternative commentary or language versions, while adjusting content for different devices or conditions. Reverb's ubiquity in media content assures its part to play in these modifications. In conclusion, we encourage you, dear reader, to keep your ears open to the spaces around you. It is an informative and fascinating aspect of our daily lived experience in the visually dominated modern world.

Acknowledgements

The authors thank the contributors to this research and funders.

Notes

1 Remaggi et al. AES Warsaw (2015).
2 Coleman et al. AES Berlin (2017).
3 Cieciura, Volino, Jackson AES (2023).
4 Remaggi et al. SSPD (2014); Remaggi et al. AES New York (2015); Remaggi et al. ICSV (2015); Remaggi et al. TASLP (2016).
5 Coleman et al. 142nd AES, Berlin (2017); Remaggi et al. AES (2018).
6 Remaggi AES Warsaw (2015).
7 Remaggi et al. AES Warsaw (2015).
8 Coleman et al. AES-DREAMS Leuven (2016).
9 Demonstrations were given in Leuven (2016), York IASS (2016) and Edinburgh DAFX (2017).
10 Murphy and Shelley, AES-USA San Francisco (2010).
11 Coleman et al. 142nd AES, Berlin (2017).
12 Kim et al. 3DV (2017).
13 SurrRoom 1.0 Cieciura, Volino, Jackson AES (2023).
14 Coleman et al. T-MM (2018); Menzies, Coleman and Fazi T-ASLP (2020).
15 Berghi and Jackson TASLP (2024).

Bibliography

[Berghi & Jackson TASLP 2024] Berghi, Davide, and Philip J.B. Jackson. "Leveraging visual supervision for array-based active speaker detection and localization." IEEE/ACM transactions on audio, speech, and language processing, Volume 32, 2023. Pages 984–995, ACM Digital Library https://doi.org/10.1109/TASLP.2023.3346643

[Chitreddy & Jackson Forum Acusticum 2020] Chitreddy, Sandeep, and Philip Jackson. "Source distance perception with reverberant spatial audio object reproduction of real rooms." In *Forum Acusticum*, Lyon, France. https://doi.org/10.48465/fa.2020.0883 pp. 2079–2086, 2020.

[Cieciura, Farheen and Jackson Basic Auditory Science 2023] Cieciura, Craig, Afshan Farheen, and Philip J.B. Jackson. "Three dimensions of space in object-based audio: perceived source distance." In *Basic Auditory Science*. UK Acoustics Network, p. 21, 2023.

[Cieciura, Volino, Jackson AES 2023] Cieciura, Craig, Marco Volino, and Philip J.B. Jackson. "SurrRoom 1.0 Dataset: Spatial Room Capture with Controlled Acoustic and Optical Measurements." In 154th *Audio Engineering Society Convention*. Helsinki, Audio Engineering Society, Paper 101, 2023. https://aes2.org/publications/elibrary-page/?id=22068

[Coleman et al. T-MM 2018] Coleman, Philip, Andreas Franck, Jon Francombe, Qingju Liu, Teofilo De Campos, Richard J. Hughes, Dylan Menzies, et al. "An audio-visual system for object-based audio: from recording to listening." *IEEE Transactions on Multimedia* 20, no. 8 (2018): 1919–1931. https://doi.org/10.1109/TMM.2018.2794780

[Coleman et al. AES Leuven 2016] Coleman, Philip, Andreas Franck, Philip J.B. Jackson, Richard J. Hughes, Luca Remaggi, and Frank Melchior. "On object-based audio with reverberation." In *Audio Engineering*

Society Conference: 60th International Conference: DREAMS (Dereverberation and Reverberation of Audio, Music, and Speech). Audio Engineering Society, Paper 2-3, 2016. https://aes2.org/publications/elibrary-page/?id=18071

[Coleman et al. JAES 2017] Coleman, Philip, Andreas Franck, Philip J.B. Jackson, Richard J. Hughes, Luca Remaggi, and Frank Melchior. "Object-based reverberation for spatial audio." *Journal of the Audio Engineering Society* 65, no. 1/2 (2017): 66–77. https://aes2.org/publications/elibrary-page/?id=18544

[Coleman et al. 142nd AES Berlin 2017] Coleman, Philip, Andreas Franck, Dylan Menzies, and Philip J.B. Jackson. "Object-based reverberation encoding from first-order Ambisonic RIRs." In 142nd *Audio Engineering Society Convention*. Berlin, Paper 9731, Audio Engineering Society, 2017. https://aes2.org/publications/elibrary-page/?id=18608

[Franck et al. JAES 2019] Franck, A., J. Francombe, J. Woodcock, R. Hughes, P. Coleman, D. Menzies, T. J. Cox, P. J. B. Jackson, and F. M. Fazi. "A system architecture for semantically informed rendering of object-based audio." *Journal of the Audio Engineering Society* 67, no. 7/8 (2019): 498–509. https://aes2.org/publications/elibrary-page/?id=20488

[Galindo et al. ICSV 2017] Galindo, Miguel Blanco, Philip Jackson, Philip Coleman, and Luca Remaggi. "Microphone array design for spatial audio object early reflection parametrisation from room impulse responses." ICSV 24 Proceedings, London, 2017.

[Kim et al. AES Berlin 2017] Kim, Hansung, Richard J. Hughes, Luca Remaggi, Philip J.B. Jackson, Adrian Hilton, Trevor J. Cox, and Ben Shirley. "Acoustic room modelling using a spherical camera for reverberant spatial audio objects." In 142nd *Audio Engineering Society Convention*. Berlin, Paper 9705, Audio Engineering Society, 2017. https://aes2.org/publications/elibrary-page/?id=18583

[Kim et al. Virtual Reality 2022] Kim, Hansung, Luca Remaggi, Aloisio Dourado, Teofilo de Campos, Philip J.B. Jackson, and Adrian Hilton. "Immersive audio-visual scene reproduction using semantic scene reconstruction from 360 cameras." *Virtual Reality* 26, no. 3 (2022): 823–838. https://doi.org/10.1007/s10055-021-00594-3

[Kim et al. T-MM 2020] Kim, Hansung, Luca Remaggi, Sam Fowler, Philip J.B. Jackson, and Adrian Hilton. "Acoustic room modelling using 360 stereo cameras." *IEEE Transactions on Multimedia* 23 (2020): 4117–4130. https://doi.org/10.1109/TMM.2020.3037537

[Kim et al. IEEE-VR 2019] Kim, Hansung, Luca Remaggi, Philip J.B. Jackson, and Adrian Hilton. "Immersive spatial audio reproduction for VR/AR using room acoustic modelling from 360 images." In *2019 IEEE Conference on Virtual Reality and 3D User Interfaces (VR)*, Osaka, Japan, pp. 120–126. IEEE, 2019. https://doi.org/10.1109/VR.2019.8798247

[Kim et al. APMAR 2019] Kim, Hansung, Luca Remaggi, Philip J.B. Jackson, and Adrian Hilton. "Spatial audio reproduction system for VR using 360 degree Cameras." In *The 12th Asia Pacific Workshop on Mixed and Augmented Reality (APMAR 2019) Proceedings*, Osaka, Japan, 2019.

[Kim et al. 3DV 2017] Kim, Hansung, Luca Remaggi, Philip J.B. Jackson, Filippo Maria Fazi, and Adrian Hilton. "3D room geometry reconstruction using audio-visual sensors." In *2017 International Conference on 3D Vision (3DV)*, pp. 621–629. IEEE, Qingdao, China, 2017. https://doi.org/10.1109/3DV.2017.00076

[Kim et al. Real VR-Immersive Digital Reality 2020] Kim, Hansung, Luca Remaggi, Philip J.B. Jackson, and Adrian Hilton. "Immersive virtual reality audio rendering adapted to the listener and the room." In Magnor, M., Sorkine-Hornung, A. (eds) *Real VR–Immersive Digital Reality: How to Import the Real World into Head-Mounted Immersive Displays*, pp. 293–318. Cham: Springer International Publishing, 2020. https://doi.org/10.1007/978-3-030-41816-8_13

[Menzies, Coleman & Fazi T-ASLP 2020] Menzies, Dylan, Philip Coleman, and Filippo Maria Fazi. "A room compensation method by modification of reverberant audio objects." *IEEE/ACM Transactions on Audio, Speech, and Language Processing* 29 (2020): 239–252. https://doi.org/10.1109/TASLP.2020.3036781

[Murphy & Shelley, AES-USA San Francisco 2010] Murphy, Damian T., and Simon Shelley. "Openair: an interactive auralization web resource and database." In 129th *Audio Engineering Society Convention*. San Fransisco, Paper 8226, Audio Engineering Society, 2010. https://aes2.org/publications/elibrary-page/?id=15648

[Pike, Cieciura and Jackson AES 2023] Pike, Chris, Cieciura, Craig, and Philip J.B. Jackson. "Plausibility of parametric spatial audio reproduction of B-format room impulse responses over headphones." In 154th *Audio Engineering Society Convention*. Helsink, Paper 10662, Audio Engineering Society, 2023. https://aes2.org/publications/elibrary-page/?id=22074

[Remaggi et al. AES Warsaw 2015] Remaggi, Luca, Philip J.B. Jackson, and Philip Coleman. "Estimation of room reflection parameters for a reverberant spatial audio object." In 138th *Audio Engineering*

Society Convention. Warsaw, Paper 9258, Audio Engineering Society, 2015. https://aes2.org/publications/elibrary-page/?id=17682

[Remaggi et al. ICSV 2015] Remaggi, Luca, Philip J.B. Jackson, and Philip Coleman. "Source, sensor and reflector position estimation from acoustical room impulse responses." *Proceedings of the International Congress* Sound Vibration (2015): 1–8.

[Remaggi et al. AES New York 2015] Remaggi, Luca, Philip J.B. Jackson, Philip Coleman, and Jon Francombe. "Visualization of compact microphone array room impulse responses." In *139th Audio Engineering Society Convention*. New York, Paper 218, Audio Engineering Society, 2015. https://aes2.org/publications/elibrary-page/?id=17894

[Remaggi et al. AES 2018] Remaggi, Luca, Philip J.B. Jackson, Philip Coleman, and Tom Parnell. "Estimation of object-based reverberation using an ad-hoc microphone arrangement for live performance." In *144th Audio Engineering Society Convention*. Milan, Italy, Paper 10028, Audio Engineering Society, 2018. https://aes2.org/publications/elibrary-page/?id=19424

[Remaggi et al. SSPD 2014] Remaggi, Luca, Philip J.B. Jackson, Philip Coleman, and Wenwu Wang. "Room boundary estimation from acoustic room impulse responses." In *2014 Sensor Signal Processing for Defence (SSPD)*, Edinburgh, UK, pp. 1–5. IEEE, 2014. https://doi.org/10.1109/SSPD.2014.6943328

[Remaggi et al. TASLP 2016] Remaggi, Luca, Philip J.B. Jackson, Philip Coleman, and Wenwu Wang. "Acoustic reflector localization: novel image source reversion and direct localization methods." *IEEE/ACM Transactions on Audio, Speech, and Language Processing* 25, no. 2 (2016): 296–309. https://doi.org/10.1109/TASLP.2016.2633802

[Remaggi et al. ICASSP 2018] Remaggi, Luca, Hansung Kim, Philip J.B. Jackson, Filippo Maria Fazi, and Adrian Hilton. "Acoustic reflector localization and classification." In *2018 IEEE International Conference on Acoustics, Speech and Signal Processing (ICASSP)*, Clagary, Canada, pp. 201–205. IEEE, 2018. https://doi.org/10.1109/ICASSP.2018.8462146

[Remaggi et al. ICASSP 2015] Remaggi, Luca, Philip J.B. Jackson, Wenwu Wang, and Jonathon A. Chambers. "A 3D model for room boundary estimation." In *2015 IEEE International Conference on Acoustics, Speech and Signal Processing (ICASSP)*, South Brisbane, Australia, pp. 514–518. IEEE, 2015. https://doi.org/10.1109/ICASSP.2015.7178022

[Remaggi et al. AVSUIM 2018] Remaggi, Luca, Hansung Kim, Philip J.B. Jackson, and Adrian Hilton. "An audio-visual method for room boundary estimation and material recognition." In *Proceedings of the 2018 Workshop on Audio-Visual Scene Understanding for Immersive Multimedia*, Association for Computing Machinery, New York, pp. 3–9, 2018. https://doi.org/10.1145/3264869.3264876

[Remaggi et al. AES 2019] Remaggi, Luca, Hansung Kim, Philip J.B. Jackson, and Adrian Hilton. "Reproducing real world acoustics in virtual reality using spherical cameras." In *Audio Engineering Society Conference: 2019 AES International Conference on Immersive and Interactive Audio*. York, UK, Paper 65, Audio Engineering Society, 2019. https://aes2.org/publications/elibrary-page/?id=20445

[Remaggi et al. ICA Aachen 2019] Remaggi, Luca, Hansung Kim, Annika Neidhardt, Adrian Hilton, and P.J. Jackson. "Perceived quality and spatial impression of room reverberation in VR reproduction from measured images and acoustics." In *23rd Proceedings of ICA*, Aachen, Germany, 2019. https://pub.dega-akustik.de/ICA2019/data/articles/001233.pdf

[Schweiger et al. IBC 2022] Schweiger, Florian, Chris Pike, Tom Nixon, Matt Firth, Bruce Weir, Paul Golds, Marco Volino, et al. "Tools for 6-DoF immersive audio-visual content capture and production." BBC White Paper WHP400.https://www.bbc.co.uk/rd/publications/tools-six-depth-of-field-immersive-audiovisual-content-capture-production, 2022.

[Woodcock et al. AES-Tokyo 2018] Woodcock, James, Jon Francombe, Andreas Franck, Philip Coleman, Richard Hughes, Hansung Kim, Qingju Liu, et al. "A framework for intelligent metadata adaptation in object-based audio." In *Audio Engineering Society Conference: 2018 AES International Conference on Spatial Reproduction-Aesthetics and Science*. Tokyo, Paper 11-3, Audio Engineering Society, 2018. https://aes2.org/publications/elibrary-page/?id=19637

4
BEYOND THE SWEET SPOT

Sound, space and emotion

*Raj Patel, Iain Forsyth & Jane Pollard
and Gerrie van Noord*

Raj Patel in conversation with Iain Forsyth & Jane Pollard

In the following exchange, leading creative sound specialist Raj Patel[1] and renowned interdisciplinary visual artists Iain Forsyth & Jane Pollard[2] elaborate on how their different backgrounds and shared curiosities underpin their collaboration, and how working together has influenced their respective practices. In the spirit of the iterative back-and-forth nature of their relationship, this text is presented as an ongoing conversation compiled from a series of online and face-to-face discussions as well as written reflections, interspersed with factual details on a selection of projects.

Raj Patel

My approach to sound is grounded in two experiences: playing instruments and the post-punk landscape of the late 1970s and 1980s, when I grew up. People were working out what to do with emerging technologies and making something new while not necessarily seeing barriers between one form of art or music and another. There was a lot of experimentation. My deep interest in music practice and theory, and experiences playing in spaces with great as well as poor acoustics led me to study Engineering Acoustics and Vibration. The result is that I am always thinking about how things sound, how they feel and how they impact experiences, individual as well as collective ones. I do think about physical aesthetics, especially through all my architectural design work, but that often makes me reference record covers or art. When I mention something to Iain & Jane, they will say: oh yeah, I bought that the week it came out. Or: I know the artist who did that. Sometimes I offer my perspective and they respond: we never thought of it like that!

Jane Pollard

We definitely have an imaginary shared history, of seeing the same bands. Particularly Iain and Raj, with Iain travelling to go to gigs that Raj was also at.

Raj

I trained under a generation of design architects, consultants and engineers that was very collaborative and for which everything was subject to question.[3] It was about bringing the best people together to try and make something good happen and create change. Then architecture became less about collaboration and more about the individual leading the project. Now many clients control the narrative. The same is true of making art. Certain artists, like Iain & Jane, are very open about their collaborators and what they bring to the process, but many aren't.

Iain Forsyth

In our post-Young British Artists (YBA) visual art education at Goldsmiths, the artist was the hero, the brand, the brains, the commodity. Collaboration happened of course but wasn't talked about. There's something very inauthentic about that. For us, it is about bringing forth relationships and being transparent about them. We encountered that approach in music, street and DIY culture. With people self-publishing fanzines, putting on club nights, and being respectful of others' contributions when they released a record, collaborators would be acknowledged and credited.

Our entire practice as artists has had elements of collaboration: beyond our inherent collaboration with each other there's always someone else.[4] It's usually a question that starts a dialogue: how can we do this? We've always valued working with people who have an extensive, often quite niche knowledge and are at the top of their game. That's also why slipping sideways from art into the music and film worlds has been quite comfortable for us because that way of operating was already part of our practice.

Art school makes you able to conceptually create value and you end up with a lot of people talking about how interesting something is. But talking about something doesn't make it inherently interesting if the initial substance is not conveyed to people looking at it and feeling it. What we experienced at art school was that there was little interest in any kind of emotional engagement with the work produced. And I don't think you can talk seriously about music and sound without thinking about emotion. We hear it and we feel it. It's visceral, and that's why it's so central to the human experience and our approach.

Raj

That's where we connected. The challenge I initially faced in my work with architects and designers was that I was talking about sound through drawings and sketches, diagrams, and numbers. A client had to have absolute faith that you had the right answer, or they needed persuading. In the mid-1990s the tools to persuade weren't readily available. They were very complex and expensive, with big physical scale models and acoustic testing. Computer models were still relatively primitive. The outputs were something you looked at and reviewed and from which you had to summarise the meaning and implications for the reader.

My colleague Neill Woodger and I started talking about how it would be ideal to create an environment in which you could bring people to hear and see spaces, and show how changes in design decisions, physically and aurally, would impact perception and emotion. This ultimately led to the development of the Arup SoundLab, an environment in which we could reproduce sound that we had measured or that we had created from a computer model into fully spatial (3D) audio. To be able to do that took five years because it relied on a series of aspects converging: not only being

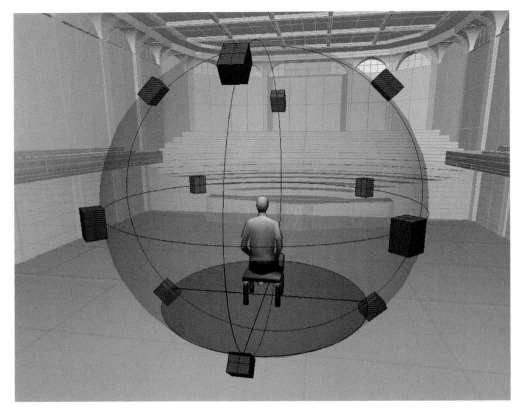

Figure 4.1 Arup SoundLab, 1998, conceptual rendering, listeners at the centre of an ambisonic sphere of loudspeakers, able to experience measured sound from a space, or sound created from a virtual model.

able to make it happen, but also being fast enough so that you could look at various options quickly in an active early design process and use it as a proactive design tool (Figure 4.1).

The first major step was to be able to listen to physical scale models using an ultrasonic source.[5] In early 1996, working on the new concert hall and cultural centre in Istanbul – a particularly complex project – a new piece of easy-to-use software called CoolEdit allowed simple scaling up and down of digital audio sounds. In addition, the price of high-quality analogue-to-digital sound cards started to drop below GBP 20,000. Using a sound card, the software and daisy-chaining all the computers in the office we could make various changes in the physical model and listen to them in real time, hear multiple positions in a space for one design iteration within a few hours.

At the time, there was a software encoding of ambisonic format in a piece of hardware by Lake DSP, but for most consultants the cost was prohibitive.[6] By the late 1990s, computers were able to process much faster and you could do everything needed on a desktop, and the first software encoders for ambisonics started to appear. When we launched the SoundLab soon after, we could make a computer model and test and retest multiple iterations in a process that took only a few days. What used to take months was greatly condensed and you could put clients, architects and consultants in the same room, let them listen to the same thing at the same time and agree on what everyone was supposed to do. It became a usable tool.

The big breakthrough was being able to let people have an emotional connection to what would happen in the building they'd designed. If you get acoustics right, no one comments, it's taken for granted that something sounds as it should. But if you get it wrong, most people react strongly. When we could sit people down in the SoundLab, many architects would say: I'm good with my eyes, not with my ears. Then they'd hear it and exclaim: well, that's obvious! They'd make an adjustment in their design because what they heard connected with them. The SoundLab enabled others to understand what they couldn't see or articulate. That completely shifted the dialogue and by the early 2000s, we communicated with clients entirely differently.

Jane

How do you convey aspects that are usually felt or experienced? You may agree on a sensibility, but you can't necessarily articulate it linguistically. We've certainly honed a way of talking about sound through our collaboration with Raj. When we first met, in 2005, we didn't understand what the SoundLab was capable of. We knew that you could take drawings and materials and then essentially project an architectural space and allow someone to experience a piece of music in it. But when Raj played us The Smiths in the Sistine Chapel we were like: wait a minute, can you record that as if it really happened? Recreate that in a totally different environment but hear the scale of that other space? At the time we had done a series of re-enactment projects that focused on the dissociation of experience and time.[7] We were curious about (re)creating the experience of sound in a specific location at a specific time. That was what essentially happened in our first collaboration, Silent Sound (2006) (Figure 4.2).

Iain

We had been doing research into Victorian demonstrations of new technology, science mixed with spiritualism. Two American brothers kept coming up: William and Ira Davenport. They used a cabinet on stage as part of their travelling show, which was presented as a spiritualist gathering but was essentially an escapology act. They toured the UK and performed a show at St. George's Hall in Liverpool in the 1860s.

Jane

The A Foundation commissioned us to do a site-specific performance piece that could translate into a longer-running installation. Looking at sites, we came across St. George's Hall and became obsessed with its beautiful circular room and the incredible history of people performing there, including the Davenports. We wanted to somehow merge their Victorian spectacle with contemporary technology and have an audience contemplate summoning the past to haunt the present.

We didn't want to turn the performance into a video installation as video cameras in the room would have changed the live experience. We commissioned Dr. Ciáran O'Keeffe, another regular collaborator, to write a text that would introduce the performance.[8] He told us about infrasound, frequencies the ear can't hear but that still physically affect you. The aesthetics and history of the hall informed a composition by J. Spaceman for a small ensemble, including 18 French horns situated around the balcony above and around the audience.[9]

We adopted the idea of the Davenport's cabinet and turned it into a soundproof vocal booth, within which Iain and I would repeat a phrase throughout the performance. That phrase was purported to be embedded as a subliminal message in the music the audience would hear. We used

Figure 4.2 Silent Sound, 2006, live 40-minute performance at St. George's Hall, Liverpool, with limited-edition CD and ambisonic installation at the A Foundation, Liverpool; Score: J. Spaceman (Spiritualized); Team: Raj Patel and Alban Bassuet at Arup Acoustics, Dr. Ciáran O'Keeffe, Charles Poulet, Andrew Bolton, Noah Rose, Jhon Bell at Metametric, and Dan Howard-Birt. A second live iteration with recording and production of a CD took place at Middlesbrough Town Hall in 2010 as part of the AV Festival.

ionisers to create a charged atmosphere. In the spirit of Victorian spectacles, we showcased contemporary technology, giving people a CD recording of the performance as they left. The subsequent installation translated that one-off event into a work that was accessible for several months.

Raj

Silent Sound was possibly one of the most complex recording projects in a live setting. Nobody had ever attempted to record full ambisonic spatial sound with an audience and then transfer that into binaural sound – the format for full spatial audio delivered over headphones – burn it onto CDs and give everyone a signed and numbered copy as they walked out. During the performance, my Mac G4 was installed in a mobile recording truck outside, with an external sound card to process the live ambisonic microphone in real time, and to four channels on the main desk, where it was recorded. The Central Processing Unit (CPU) was running at 97%–98% and we always had a concern that it might fail, so we were also recording with live close-microphones and an overhead stereo set as a back-up. But it worked! People were given a memento of that evening that sounded exactly as it had in the room and replicated the experience of being there. Creating that emotional connection was really important. The mobile recording unit was supplied and operated by Will Shapland, one of the most experienced location recording engineers in the music business. Will was particularly impressed by the ambience we were able to capture in the spatial recording and went on to use these techniques in numerous live concert recordings. Following the performance, we transferred the piece into a gallery installation with full ambisonic sound again, so people could experience the performance there. It was at the very edge of what the technology then allowed. A lot of our work together has been like that.

Iain & Jane

The installation of Silent Sound changed the way we work with space. We discovered that when some of your senses are deprived of stimulation and the active senses overcompensate, a new cognitive space opens up that consists of a negotiation between the ears, the brain and the imagination. This first project with Raj not only set the tone for our ongoing collaboration, but also set a high bar for how ambitious and precise we could be together.

Raj

There is a difference between creating a space that has a specific function – like a room in which to teach children with high-quality speech intelligibility or a concert hall to listen to music – and wanting to create a very singular emotion. You can change emotions very easily through shape, and as soon as you scale that up you change perception. That offers a lot of scope for playing around. You can do that in reality or in an imaginary way: you can build something with neutral acoustics and overlay it electronically with any space, or any number of spaces on top to create a series of emotions in sequence.

We have continued to push that. How do you want people to feel? What do you want them to hear? How do you want them to hear it? Iain and Jane would come up with new ideas and concepts through which the question shifted to: where could you take this from a sound perspective and what could we do that hasn't been done before? What will heighten the emotion or whatever else we want people to experience? Taking an audience on a journey became more and more important. Our working together often comes with certain conditions – of place, of space – but the evolution of what the work eventually becomes happens through our dialogue.

Iain

There is a kind of natural impulse to want to break things, to creatively tear things down to make something new. That is what makes making art exciting for me. If I smash it to pieces, what can I turn it into? Not that I'm suggesting smashing up the SoundLab, but metaphorically. When we were asked to present a work in a shipping container on the beach during Art Basel Miami 2009, a major international art fair that only lasts for a few days, we could have chosen a project that would have been easy to install. But we wanted to see if we could deliver the experience of Silent Sound in a dark confined space in such a bright location and transport people to a Victorian room in Liverpool through sound. To do that in literally no time was a challenge, but a lot of fun. A cornerstone of our collaboration with Raj has always been finding ways to challenge ourselves. To do things that we've never done before, and in some cases things that nobody else has ever done before.

Raj

Silent Sound in Miami was technically complicated because space for technology and time for getting the acoustics right were limited. As an acoustic consultant, you must know about low-noise ventilation. Designing a concert hall or a recording studio you should know how

Figure 4.3 Radio Mania, 2009, 29-minute two-channel stereoscopic video installation with ambisonic sound, presented at the BFI Gallery, London. Script: Kirk Lake; Music: Nick Plytas.

to isolate it from the outside world. When Iain and Jane mentioned they'd been invited to do something in Miami on the beach, my immediate response was: why don't we do Silent Sound? We needed to create a spatial sonic field that felt big enough to accommodate a high turnover of people. We thought we could do it, found the right people and made it happen. The collaborators you choose to stick with are those with whom you achieve something together that you could not have done alone.

Iain & Jane

Based on what we learned from Silent Sound, we began to think about how to combine audio with visuals to skew your experience of a space. Using stereoscopic 3D video and ambisonic spatial audio we wanted to construct a real-time, single-take, unedited experience for a gallery space, which was a simple rectangular black box at the short-lived British Film Institute (BFI) Southbank Gallery in London, to place visitors at the centre of a kind of virtual reality (Figure 4.3).

We found a standard 2D version of an early, now lost 3D film, Radio Mania (1922), in the BFI archives. We set about scripting a rehearsal for a re-making of the film, with a cast of actors and crew on a huge soundstage, which was to be projected life-size. To take that further, we wanted to shoot in two directions simultaneously. The rear screen showed a small band performing live, accompanying the action taking place on the other half of the sound stage. To record the sound, we used the same ambisonic microphone we had used with Silent Sound, which meant Raj and his team could create a soundtrack for the gallery installation. This effectively meant that sounds would be experienced in the same physical location that they had originated from. There was something about this experiment that felt akin to the rendering of a naturalistic perspective, where we were drawing on advanced technology.

Raj

The beauty of this installation was combining two full 3D films with full spatial audio, for the first time to our knowledge, with the result that as listener/viewer, you occupied a zone around the recording equipment and could turn around to look at and hear the room as if you were in it seamlessly.

Iain & Jane

After Nick Cave had written The Death of Bunny Munro, he asked us to produce the audiobook.[10] Most audiobooks are listened to on headphones and have a spoken narrative with maybe a bit of incidental music, which Nick wanted us to avoid. After the experience of producing a binaural version of Silent Sound on CD, working with the space in your head, between your ears, was an exciting creative proposition (Figure 4.4).

We started with questions. Could we sonically transport the listener to spaces in the narrative of the book? Could we spatially position Nick's voice? Could we borrow techniques used to enhance film sound, such as foley effects and a score, to embellish the narrative? This was the first time we became involved in arranging music for a piece, but we also created what would be called an FX track in film, using real-world sounds inserted at dramatically significant moments.

Figure 4.4   The Death of Bunny Munro, 2009, spatialised 3D audiobook, 8 hours. 15 minutes. Music by Nick Cave and Warren Ellis, published by Canongate in multiple formats, including a 7-CD version with a bonus DVD, and downloadable as an enhanced iPhone app. Recordings mastered by Will Quinell at Sterling Sound, NYC. Sound spatialised by Terence Caulkins, Anne Guthrie, Raj Patel and Ryan Biziorek.

Jane

The Death of Bunny Munro (2009) was the first project for which we used the exquisite corpse process: we'd write a line about something, then we'd write another line and imagined what the chapter should sound like. We would send it to Raj and his team in the US to work on the spatialisation. He'd respond with: you've said this about where it is, and then add a twist to it. They'd place Nick's voice in specific places with particular acoustic conditions. We did three samples composed in 3D in the SoundLab, listened to them on headphones and then sent them to Nick and Warren [Ellis].[11]

Raj

Whereas Silent Sound was about being in the so-called sweet spot, we wanted to expand the idea of a unique kind of virtual reality through sound, in which people could move around. Bunny Munro is probably the only project where you can deduce that you're in a concrete bunker that is this wide. Many of the locations in the book are real locations in Brighton, as is the protagonist's car, however, a lot of the narrative is based in an internalised emotional, psychological space. At one moment you are inside a Fiat Punto and when Bunny opens the window the sound alters in a certain way because his seat was on the right-hand side. But he is also having a mental breakdown and there is a pulsating sense that something is about to happen….

Jane

Bunny Munro is where we did most experimenting, where we flexed and stretched working between real spaces and that sort of imaginary space. It is effectively an audio-film that allows you to close your eyes and picture yourself in the movie of the book. There is precision in the replication of real spaces, but this movement between the real and imaginary is key. You hear sounds that pass you or move from behind to in front of you. We used the pan and zoom of sound in the way a film camera would.

Raj

It was like making a movie in a very short time span. Having to do 33 chapters, adding to up to over 8 hours, forced us to keep it interesting. We traded sections back and forth and had to make decisions quickly to be able to deliver in 26 days. Being in the same thinking space we also knew when to stop. Not every moment is taken up with sound, apart from Nick narrating. We haven't tried to fill every single gap. We could have carried on layering but there's something nice about the sparseness that creates a mood as well as space.

Jane

I don't think that we would have ventured into public art had it not been for our collaboration with Raj, because our work doesn't usually lend itself to that kind of scale. Soon (2011) responded to a square in the financial district in downtown Toronto and was commissioned for the city's Nuit Blanche, an annual festival. We worked in the shadow of the G7, which had taken place the year before, during which the city centre had been the focus of violent protests. There was therefore an

Beyond the sweet spot

Figure 4.5 Soon, 2011, 12-hour durational sound and light installation, commissioned for Nuit Blanche in Toronto. Curator: Nicholas Brown; Score: Scanner; Team: Raj Patel, Ryan Biziorek, Brian Stacy and Star Davis at Arup.

atmosphere, a tone to the area. You could have taken the constituent parts, the technology and the programming, put it somewhere else and it would have been a completely different piece. When looking at the site for Soon we wondered: how to create the feeling of something that could possibly happen? There were real constraints: it was an outdoor square surrounded by tall buildings and we had a limited budget. We were told that up to a million visitors might attend the event, during which they would be on the move, so we also had to think about how to catch people's attention immediately, without the scope to tell a story (Figure 4.5).

Raj

The original idea was to create a floating object without visible tension lines to the ground that would feel like some kind of alien visitation and have a sense of impending doom. We took all the measurements and looked at the different angles of the site. The problem was that if we couldn't properly secure the object and the weather conditions on that night turned bad, we wouldn't be able to go live. That was going to cost so much that it would take the entire budget.

It took some reorganising in our heads. In the end, we asked ourselves: what would happen if we took the object away but still do the rest? We came up with the idea for hand-operated searchlights off the top of a tall building, basically saying: this is a set, and if it's a set we've got to have lights. With lights you can make things appear as if they're quite solid, it's all about intensity and direction. Then we added sound. We used pan-tilt-zoom camera mounts, onto which we put big loudspeakers. The sound reflected off the surrounding buildings, making it move around the square.

Iain

Messages had gone out to all security agencies explaining that it was commissioned by the city, to make sure that everyone knew what the project was trying to do and that they shouldn't be worried. But at about 3 am, lots of police turned up and they doubled security. Despite all our best efforts to warn the authorities, the unease Soon generated was successful enough to cause genuine alarm.

Jane

We manipulated some of the lights so they seemed broken. When you see something like that your brain searches for a rational explanation and if you can't find one, unease kicks in. It was so simple but very effective. A lot of it was down to having to get maximum effect out of minimal options.

Iain

But people were loving it. We had groups doing communal rituals that had nothing to do with anything we'd planned. There was no performance. For quite a while the spotlights made people run away, which became a game. It now feels very indicative of our best work: it's good when you don't know where you are, when you are either looking at nothing, or you're having to look at everything to work out what it is you're experiencing.

Jane

We'd been saying for years that we wanted to find a musical artist whose work we could translate into a spatial sonic piece. More than any contemporary composer we know, Scott Walker pushed the boundaries of what's possible in recorded music. Then Bish Bosch (2012) came out. It was obvious this was a prime candidate to do something like that with. Scott's lyrics were difficult to pin to a specific visual reference and we knew that through an ambisonic installation, we could take his work further (Figure 4.6).

Iain

The spaces we've imagined in the SoundLab are reimagined in the audience's mind, which works differently when you put them on a screen. If we'd have made a film for Scott's record, we would be pinning those ideas to very specific visual information. By playing with the space of the sound, you never nail it quite so firmly to the ground. But trying to communicate about sound is like passing a bowl of water and hoping you don't spill too much. Scott's recording methods were dense and intricate, so spatially exploding these had to be done with great care and sensitivity. This was helped by being able to bring Scott's producer, Pete Walsh, onto the team. To start we produced 30-second samples of the five tracks that we eventually ended up working with and took them to Scott to listen to on headphones.

Raj

After hearing the samples, Scott said: this sounds really good. As long as Pete is involved, I'm on board. He only had one request: he wanted his voice in a particular place, so in all the pieces his voice is in one location. Other than that, he gave us free rein. Pete came with his samples, Terence Caulkins, who has worked on a range of Iain and Jane's projects, sat in the SoundLab for a week,

Figure 4.6 Bish Bosch: Ambisymphonic (with Scott Walker), 2013, ambisonic sound installation, 25 minutes, commissioned and realised at the Sydney Opera House for the Vivid Festival. Team: Charles and Cathy Negus-Fancey, Pete Walsh, Raj Patel, Terence Caulkins, Fergus Linehan and Ben Marshall.

and they did a track a day for four days. Iain and Jane came and listened on day 5. On the last day, I couldn't be here in person, so Terence was doing the mixing. On the final listen Scott came in. We all knew it was the moment. He had to do the final sign-off.

Jane

In terms of creative growth, it was exciting to step out of the real world of Bunny [Munro] and Brighton into what happens if you're on the back of a truck that is part of a travelling circus led by elephants? At one point you were on the back of an elephant in a kind of parade, in outer space, which then slid into very imaginative creative, more psychological spaces. With Bish Bosch, we really went for it.

To produce the 'script' we would imagine places, spaces and environments. We compiled photographs and sent Raj and Terence these pictures with some notes. We would arrange them in such a way that we felt they communicated something interesting. To receive that and translate it into maths is I guess what Raj and his team were doing.

Raj

When you receive something like that it tells you something about the mood, the place and the time. Some of that can be translated into architecture and design. When you say you're in an enormous old warehouse surrounded by oil drums, water dripping down the walls, puddles on the floor, you know what that sounds like and you can create that. That is about converting words into imaginative audio. Because we understand each other's languages now so well, we can create from that quite effectively. We've learned about each other's disciplines in the process of doing, but you can only push that when you've been doing it for a while. That was particularly important for Scott.

That's also where the music references come in. Do you remember the sound of this record, or what this band was like when they played those gigs? We don't have completely the same frame of reference, but we have good overlaps. That allows you to create a sound mood board and establish a framework. Terence and I would receive those images and sometimes add our own twist and then send the sound back without telling Iain and Jane what we'd done. They would respond: oh, you've now assumed that those elephants are on a sinking ship in the middle of the Atlantic Ocean. And we'd go: yes, that's exactly what we did.

Iain

We didn't get the commission until we were part-way through the process. At that point, we had only been thinking about the sound, not how to present it in a space. When we were asked to do something for the annual Vivid festival in Sydney, we wondered whether Bish Bosch could be it. And if it could, what would we do with it? The conversations around the presentation very much related to the fact that Scott didn't want to play live and travel. In our earlier work, we had worked with re-enactments of historical music performances, where the question was: how to stay as close as possible to the original intention? With an artist like Scott, a sound installation felt like a more authentic way of staying close to an authentic experience, even in his absence.

Jane

Normally we design the architecture of our installations, but this time we said to Raj: you decide, draw the sketches. Then we'll take those, interpret them and do something else with it. We embraced that intuitiveness of the process and pushed it by flipping it.

Iain & Jane

The methodology was very subtle here. We had to create an environment that wouldn't distract from the auditory experience, but we felt that a full blackout wouldn't be right. Audiences tend to react to darkness in quite specific ways, which can become intrusive. Instead, we designed a series of almost imperceptible lighting cues to accompany the sound. We also wanted to experiment with inverting the audience/performer dynamic, so the piece was experienced on the stage.

Raj

It's interesting thinking about intention and authenticity of experience. They're not necessarily the same, there's a gap. That's the grey area that I like to occupy when working with artists, where it's about what has come out of the intention as a creation that can be experienced in some shape or form.... Maybe the best word that captures what we're trying to work with is 'spirit', because that makes some allowance for things that happen that you don't intend, for fluke and happenstance, and speaks to an attitude.

Iain & Jane

Unlike previous ambisonic set-ups we'd worked with, which had speakers mounted in various locations within a structure, working at the Sydney Opera House opened up other possibilities. The audience would enter the dimly lit auditorium, be led to the stage, their eyes adjusting to the low light levels. The stage itself revealed nothing about the upcoming experience. One visitor described it as 'a theatre performance without the "theatre"', which resonated. Although the spatialising and programming all took place in New York, as usual, this time we also worked with specialists from Arup's office in Sydney, which was really useful because of their local knowledge and longstanding relationship with the venue. Raj and his team worked with loudspeakers in their inventory, suspending them from the theatre fly-tower rigging, in a geodesic dome arrangement. The whole installation had no recognisable architecture, in stark contrast to the Sydney Opera House itself, which is arguably one of the most recognisable pieces of music architecture in the world (Figure 4.7).

Iain

A lot of our work is about suggesting or manipulating an audience into a certain set of experiential conditions. It's like going back to old records that say: play at maximum volume! With Bish Bosch, there is sensory deprivation alongside incredible audio. Leading people into a dark space where there's very little to look at and to then get them to close their eyes … the pay-off for doing that is huge.

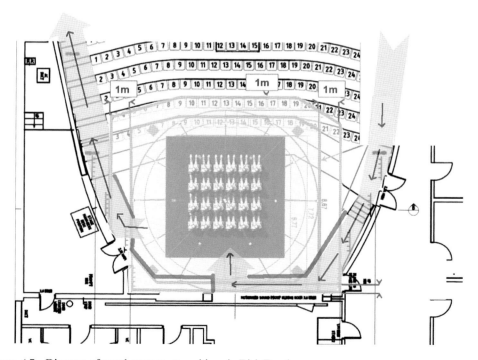

Figure 4.7 Diagram of vocal movement positions in Bish Bosch.

Raj

It's all about how you want your audience to experience something, what you are going to deliver to them and how are you going to get there. In a space like the Sydney Opera House, you have an expectation of what's going to happen. Bish Bosch took them on a journey into the unknown, turning that on its head. People realised they were not in the audience anymore but in the thing. That was a genius move.

Jane

At some stage, we wondered: what if we made our benches just tall enough that people's feet come off the floor? That sounds like such a simple thing, but if your feet aren't touching the floor, you hear depth underneath you in a way that can bring about real vertigo. It's this sort of thing that makes you realise that as an artist you can manipulate very simple factors that enable an incredibly powerful psychological effect.

Raj

It was a simple move, pulling yourself up onto the bench and letting your legs dangle. Because you couldn't feel the ground you got a sense of being on a rollercoaster, but in a fixed environment, without X or Y or Z axes movement. There is a moment when you're on the water and the audio is rolling backwards and forwards. I hate being on boats and I remember listening to it in the SoundLab, having to stand up and saying: we've done our job well here because I'm feeling quite sick now.

Emotion is an important factor in how we've thought about the work in our separate practices and in everything that we've done together: it's a crucial element. But there's also the actual physicality of the sound itself. We all experience it when we go to a loud music gig. You feel the vibrations, the intensity of it and the emotions that they've created. There's an evocation of that that we want to bring into our work, because you know what it feels like when you do it well.

What is technologically possible today wasn't when we first collaborated. In the SoundLab we brought a level of technical rigour and thinking to how those tools could be used to do some very specific things. Artists like Iain and Jane allowed us to twist all of that to do something that we might otherwise not have done. That goes back to the old adage that technology pushes art and art pushes technology. It's the constant pull and push and stretch between them that's critical. And a lot of the work that we've created has influenced others to push further too.

Compiled and edited by Gerrie van Noord.

Notes

1 For an overview of Raj Patel's work and Arup SoundLab, see https://www.arup.com/perspectives/soundlab (last accessed 25 September 2023).
2 For an overview of Iain Forsyth and Jane Pollard's practice, see https://iainandjane.com/ (last accessed 25 September 2023). Further information on all projects discussed can be found on the site.
3 See also Raj Patel, *Architectural Acoustics: A Guide to Integrated Thinking* (London: RIBA Publishing, 2020).
4 On their own collaboration, see "Iain Forsyth & Jane Pollard", in Ellen Mara de Wachter, *Co-Art: Artists on Creative Collaboration* (London: Phaidon, 2017), 64–69.

5. Ultrasonic sound is required because the models are scaled down from 1:1 to 1:50, so the sounds need to be scaled up by the same ratio in the audio range to allow accurate testing.
6. For an overview of Ambisonics, see https://en.wikipedia.org/wiki/Ambisonics (last accessed 25 September 2023).
7. On specifically this aspect, see Andrew Renton, "Do It Again, Do It Again (Turn Around, Go Back)", in *Perform, Repeat, Record*, ed. Amelia Jones and Adrian Heathfield (Bristol and Chicago, IL: Intellect, 2012), 493–510.
8. Dr. Ciáran O'Keeffe specialises in parapsychology and forensic psychology.
9. J. Spaceman is Jason Pierce, front man of the band Spiritualized.
10. Iain Forsyth & Jane Pollard have a longstanding working relationship with Nick Cave, resulting most notably in their 'musical documentary drama' 20,000 Days on Earth (2014).
11. Warren Ellis is a longstanding collaborator of Nick Cave, with whom Iain and Jane also work independently.

Bibliography

Ambisonics, https://en.wikipedia.org/wiki/Ambisonics (last accessed 25 September 2023).

Arup SoundLab, https://www.arup.com/ search SoundLab (last accessed 25 September 2023).

De Wachter, Ellen Mara, "Iain Forsyth & Jane Pollard", in *Co-Art: Artists on Creative Collaboration* (London: Phaidon, 2017), 64–69.

Forsyth, Iain & Jane Pollard, https://iainandjane.com/ (last accessed 25 September 2023).

Forsyth, Iain & Jane Pollard and Nick Cave, *20,000 Days on Earth,* 97 minutes (2014) Directed by Iain Forsyth & Jane Pollard. Written by Iain Forsyth & Jane Pollard, Nick Cave. Produced by Dan Bowen, Alex Dunnett, James Wilson. Cinematography Erik Wilson. Edited by Jonathan Amos. Music by Nick Cave, Warren Ellis. Production companies Corniche Pictures, British Film Institute, Film4 Productions, Pulse Films, Distributed byPicturehouse Entertainment. Release date 20 January 2014 (Sundance Film Festival), 17 September 2014 (United States).

Patel, Raj, *Architectural Acoustics: A Guide to Integrated Thinking* (London: RIBA Publishing, 2020).

Renton, Andrew, "Do It Again, Do It Again (Turn Around, Go Back)", in *Perform, Repeat, Record*, ed. Amelia Jones and Adrian Heathfield (Bristol and Chicago, IL: Intellect, 2012), 493–510.

5
ARCHITECTURAL ACOUSTICS OF THE SAGRADA FAMÍLIA BASÍLICA

Sipei Zhao and Mark Burry

Introduction to architectural acoustics

Architectural acoustics is the study of the behaviour of sound in built environments, including the design, construction, and evaluation of the acoustics of buildings and other structures.[1] It involves the use of principles and techniques from the fields of physics, engineering, architecture and psychology to understand and control the way sound travels and is perceived within a space. The primary goal of architectural acoustics is to create spaces that have good acoustical properties for the intended use of the space. In order to achieve this goal, architects and acousticians must consider a variety of factors, including the shape and size of the space, the materials used in the construction, and the placement of objects within the space. They may control sound absorption, sound reflection, and sound scattering to moderate the way sound travels within the space. Architectural acoustics plays an important role in a wide range of applications, including the design of homes, offices, schools, theatres, churches, and other public spaces. This section introduces the fundamental concepts in architectural acoustics, including sound generation and propagation, as well as room acoustics basics.

Sound generation

Sound in the context of architectural acoustics is a type of mechanical wave that travels through the air as fluctuations of pressure, which is usually created when a vibrating object causes the air molecules to vibrate. These vibrations travel through air in the form of a propagating sound wave, which can be detected by the human ear or by instruments such as microphones. There are many ways to generate sound in the air, among which human speech and musical instruments are the most common in architectural spaces.

Human speech

The human voice is a common source of sound, which is produced by the vocal cords located in the larynx (also known as the voice box) and amplified by the mouth and nose.[2] The vocal cords are made up of two folds of elastic muscle tissue that can be brought together or separated by the muscles in the larynx. When the vocal cords are relaxed and open, air from the lungs can pass through them, producing a sound known as voiced speech. When the vocal cords are tense and closed, the flow of air is blocked, producing a sound known as unvoiced speech. To produce different sounds, the vocal cords can be manipulated in various ways. For example, the pitch of the sound can be changed by adjusting the tension in the vocal cords. A higher pitch is produced when the cords are more tense, while a lower pitch is produced when they are more relaxed. The volume of the sound can be changed by adjusting the amount of air that is exhaled through the vocal cords. A louder sound is produced with a greater flow of air, while a softer sound is produced with a smaller flow of air. The sounds produced by the vocal cords are then modified by the shape of the mouth, nose, and throat to create the different vowel and consonant sounds of human speech. The brain controls the muscles in the larynx and the rest of the speech production system to produce the desired words and sounds.[3]

Musical instruments

Musical instruments generate sound through the vibration of an object, such as a string, a drumhead, or a reed. These vibrations create sound waves in the air, and the frequency, intensity and duration of the vibrations determine the characteristics of the sound wave, such as its pitch, loudness, and timbre. The characteristics of the sound produced by an instrument depend on the material and shape of the vibrating object. Some common ways that musical instruments generate sound include[4]:

- Strings instruments, such as guitars and violins, produce sound through the vibration of strings. The strings are typically made of a material such as metal, gut, or synthetic fibre, and are stretched between two points on the instrument. When the strings are plucked, strummed, or struck with a hammer or a bow, they vibrate and produce a sound wave.
- Woodwind instruments, such as clarinets and saxophones, generate sound through the vibration of a reed, which is a thin piece of flexible material (usually made of cane or synthetic fibre) that is attached to the mouthpiece of the instrument. When the player blows air into the mouthpiece, the reed vibrates and produces a sound wave.
- Percussion instruments, such as drums and timpani, produce sound through the vibration of a drumhead, which is a stretched membrane made of animal skin or synthetic material. When the drumhead is struck with a drumstick or a mallet, it vibrates and produces a sound wave.
- Brass instruments, such as trumpets and trombones, generate sound through the vibration of the player's lips as they blow air into the instrument. The player's lips act as a reed, vibrating as the air passes through them and creating a sound wave.

Sound propagation

The sound generated by a sound source, such as the human speech and musical instruments mentioned in the previous section, propagates through air, in the context of architectural acoustics, to the listener over time and spatial locations. The propagation of acoustic wave is described by the wave equation,[5]

$$\left(\frac{\partial^2}{\partial x^2}+\frac{\partial^2}{\partial y^2}+\frac{\partial^2}{\partial z^2}\right)p(x,y,z,t)-\frac{1}{c^2}\frac{\partial^2 p(x,y,z,t)}{\partial t^2}=0 \qquad (5.1)$$

where $p(x,y,z,t)$ denotes the sound pressure at the spatial location (x,y,z) at time instant t, and c is the speed of sound. In general, the speed of sound in air increases as the temperature increases and as the air density decreases. At a temperature of 20°C, the speed of sound in air is approximately 343 m/s.

Free-field propagation

In a free field with a uniform medium without boundaries or obstacles, the sound waves spread out in all directions at a constant speed from the sound source, similar to the way ripples spread out in a pond when a stone is dropped into it. This is referred to as a spherical wave because the wavefront is a sphere. Because the energy of the sound wave is spread out over an increasing area as it travels away from the source, the intensity of the sound wave decreases with the distance from the source, and the rate at which the intensity decreases is inversely proportional to the propagation distance. This is the distance inverse law, which is an important concept in the field of acoustics, as it is used to predict the loudness of a direct sound at different distances from the source. It is noted that the sound intensity not only decays with the propagation distance, but there is also an energy loss when sound propagates in the medium, although the attenuation of sound caused by air is usually negligible in the low-frequency range. For example, the sound pressure level is attenuated by approximately 1 dB per 1 km the sound wave travels at the frequency of 250 Hz; by contrast, for a sound wave at 4,000 Hz, the attenuation can be 24–67 dB per 1 km, depending on the temperature and humidity of the air.[6]

Effects of boundaries

The free-field model assumes there are no boundaries and obstacles for simplicity, but in practice, physical spaces are usually bounded and more acoustic phenomena occur due to the interaction between sound waves and structures. When the acoustic wave encounters a partition, such as a wall or a screen in buildings, some of the acoustic energy will be reflected back to the side of the sound source, some will be absorbed by the materials of the partition, and the rest will transmit to the other side through the partition. These wave phenomena are the underlying physics that dominate the acoustics in built environments and will be briefly introduced below.

Sound reflection is the process by which sound waves are reflected off a surface and the amount of sound reflection is characterised by the surface's acoustic reflection coefficient, which is defined as the ratio of the reflected sound energy to the incident sound energy, i.e., $R = E_r / E_i$, where E_r and E_i denote the reflected acoustic power and the incident acoustic power arriving at the boundary, respectively. The reflection coefficient is typically expressed as a decimal value between 0 and 1, with higher values indicating greater reflection. For example, a surface with a reflection coefficient of 0.9 reflects 90% of the incident sound energy, while a surface with a reflection coefficient of 0.1 reflects only 10% of the incident sound energy. The specific value of the reflection coefficient depends on the physical properties of the surface, including its roughness, smoothness, and composition, and the frequency of the sound wave. The reflected sound wave is not necessarily focused on the specular direction of the incident sound, and most reflectors in practical applications will also disperse acoustic waves to other directions.[7] To quantitate this effect, the scattering coefficient is defined as the ratio of the sound energy scattered to directions other than the specular direction to the total energy of all the reflected sound, i.e., $s_\theta = 1 - (E_s / E_r)$, where E_s is the specularly reflected acoustic energy, E_r is total reflected acoustic power, and the subscript θ denotes that the scattering coefficient is dependent on the angle of the incident sound. If measured in a diffuse sound field, the scattering coefficient is independent of the incident angle, and the subscript can be removed. Theoretically, s_θ takes a value between 0 and 1, where 0 means a totally specularly reflecting surface and 1 means a totally scattering surface.

Sound reflection and scattering are important concepts in the field of acoustics, as they play a role when sound propagates in different environments. The reflection of sound waves off surfaces can affect the perceived loudness of a sound, as well as its directionality and spatial characteristics. In some cases, it is desirable to control the reflection and scattering of sound in a space. In a recording studio, for example, it is important to reduce the reflection of sound to achieve a more accurate representation of the sound being recorded. In other cases, such as in a concert hall or theatre, it may be desirable to enhance the reflection of sound to create a more lively and immersive listening experience. In the Sagrada Família Basílica, for example, Gaudí adopted a combination of hyperboloids of revolution and hyperbolic paraboloids for many practical and aesthetic motivations, including diffusing sound to non-specular directions. While hyperbolic surfaces are expected to scatter sound more uniformly than planar geometries, the overall acoustic properties of an enclosure depend on many other factors, such as sound absorption and reverberation that will be introduced later in this chapter.

In the process of sound reflection, the sound waves cause the molecules in the material to vibrate, which results in two phenomena. The first is sound absorption due to heat dissipation and the second is sound transmission to the other side of the material. The degree to which a surface absorbs sound is characterised by the sound absorption coefficient. This coefficient is a measure of the fraction of the incident sound energy that is absorbed by the surface and is defined as $\alpha = 1 - (E_r / E_i)$. The sound absorption coefficient is widely used to design and optimise the performance of a variety of acoustic devices, such as acoustic panels. Materials and surfaces with high sound absorption coefficients (close to 1) will tend to absorb more sound, while those with low sound absorption coefficients (close to 0) will absorb less sound energy. The sound absorption coefficient of a material is influenced by a variety of factors, including the density, stiffness and thickness of the material, as well as the frequency of the sound waves. Some common materials with high sound absorption coefficients include foam, fibreglass, and other porous materials. These sound-absorbing materials can be placed on walls, ceilings, or other surfaces to reduce the intensity of reflections and improve the overall sound quality in a space. The interior wall surfaces of the Sagrada Família Basílica are mostly made of a variety of stones and brick tiles which are barely absorbing but highly reflective, making it a rather reverberant space.[8]

If the incident acoustic wave strikes on a thin and light structure, the sound wave may pass through it without being significantly affected. The sound transmission coefficient is a measure of the degree to which a material or surface transmits sound. It is defined as the ratio of the energy of the transmitted sound to the energy of the incident sound. A more widely used measure for sound transmission is the transmission loss (TL) that is defined as the logarithm of the reciprocal of the transmission coefficient, i.e., $TL = 10 \log_{10}(E_i / E_t)$, where E_t denotes the transmitted sound energy. The greater the TL, the more effectively the material or structure blocks the transmission of sound. Sound TL is an important consideration when designing buildings and other structures, selecting materials for soundproofing, and evaluating the acoustical performance of products such as windows and doors. Similar to the acoustic absorption and reflection coefficients, the TL of a material is affected by the mechanical properties of the material or structure and is usually a function of the frequency of the sound wave. If the structure is of finite size compared to the acoustic wavelength, another wave phenomenon known as diffraction will occur and the sound energy will bypass the structure. Acoustic diffraction must be considered in addition to sound transmission when calculating sound pressure levels behind obstacles such as sound screens.[9]

Room acoustics

Room acoustics involves the study of how sound waves interact with the surfaces, materials, and objects within a space, and how these interactions affect the quality of the sound within that space. While the preceding section discusses the sound wave interaction with a single surface or structure, in rooms or other enclosed spaces there are many boundaries and objects of various dimensions that are made from a large variety of materials. Therefore, it is impractical to track every reflection because the sound wave reflected by one wall will also be reflected by other walls before arriving at the receiver, thereby leading to multiple reflections.[10] Theoretically, the acoustics in any enclosure can be calculated by solving the wave equation with the appropriate boundary conditions for the specific geometry of the room, either analytically or numerically.[11] This leads to a description of the sound waves in terms of the modes of the enclosure. This modal decomposition approach is useful when the dimensions of the room are comparable to or shorter than the wavelength of the sound waves being considered. However, the analytical approach is applicable only to enclosures with a simple geometry, such as rectangular, cylindrical or spherical shapes. Moreover, with the increase in room dimension and sound wave frequency, the computational burden of the numerical approach will be overwhelming. As an alternative, the behaviours of medium to high-frequency sound in enclosures can be analysed with the statistical approach, where the detailed geometry of the room is less important. However, this approach is not as accurate at lower frequencies, where the wavelength of the sound waves is longer and the details of the room geometry become more important. The following section succinctly introduces both the analytical and statistical approaches.

Modal theory

The starting point of the modal theory is the wave equation in Eq. (5.1). In contrast to the free-field scenario, however, the boundary conditions in rooms must be taken into account. The multiple reflections from room walls, ceiling and floor interfere with each other and form many standing waves. These standing waves make the air in a room resonate at certain frequencies known as the 'natural frequencies', and result in different spatial distribution of the sound pressure within the room, which are called the 'mode shape'. The natural frequencies and mode shapes of a room are determined by the dimension and shape of the room and the material properties (e.g. reflective or absorbent) of the walls.

As a simple example, the natural frequencies of a rectangular room with rigid boundaries can be calculated as

$$f_N = \frac{c}{2}\sqrt{\left(\frac{n_x}{L_x}\right)^2 + \left(\frac{n_y}{L_y}\right)^2 + \left(\frac{n_z}{L_z}\right)^2} \quad (5.2)$$

where n_x, n_y and n_z are integers, L_x, L_y and L_z denote the dimension of the room, and c is the speed of sound. The corresponding mode shape at the natural frequency f_N is expressed as

$$\varphi_N = C\cos\left(\frac{n_x \pi x}{L_x}\right)\cos\left(\frac{n_y \pi y}{L_y}\right)\cos\left(\frac{n_z \pi z}{L_z}\right) \quad (5.3)$$

where C is a constant.

The acoustic mode in a room can be interpreted as superpositions of interfering waves travelling in various directions. The sound pressure at any spatial location (x, y, z) can be obtained by summing up all the modes in Eq. (5.3). The detailed derivation of the natural frequencies and mode shapes in Eqs. (5.2) and (5.3) are omitted here for brevity and can be readily accessed elsewhere.[12] It is noted that the above equations are applicable to rectangular rooms with rigid walls only, but the sound field of any room can be represented as the superposition of various acoustic modes. Although simple analytical solutions for the natural frequencies and mode shapes usually do not exist for the sound field in typical rooms with more complex geometries in life, numerical methods, such as the finite element method, can be employed to solve the wave equation for modes.

Modal density

It can be seen from Eq. (5.2) that there are an infinite number of natural frequencies in a rectangular room of any dimension but the distribution of natural frequencies depends on the room's dimensions. The modal density of a rectangular room, i.e., the number of acoustic modes per unit bandwidth, can be obtained as

$$n(f) = \frac{4\pi V}{c^3} f^2 + \frac{\pi S}{2c^2} f + \frac{L}{8c} \tag{5.4}$$

where $V = L_x L_y L_z$ is the room volume, $S = 2(L_x L_y + L_y L_z + L_z L_x)$ the surface area of the room, and $L = 4(L_x + L_y + L_z)$. At high frequencies, the first term on the right-hand side of Eq. (5.4) is dominant and the modal density is proportional to the room volume and the squared frequency. In the limiting case $f \to \infty$, this law of modal density is not only valid for rectangular rooms but also for arbitrary-shape rooms. This is understandable because any room can be conceived as being composed of many rectangular enclosures; the total modal density of the room is simply the superposition of the modal density of each individual enclosure. The modal theory is valid for any frequency. Furthermore, as shown in Eq. (5.4), the modal density increases with the square of the frequency, which will make the modal superposition method difficult to apply at medium to high frequencies. This is because when more than hundreds of complex terms are summed, the result becomes very sensitive to small errors in each term. Such errors are inevitable in practice. For example, the rooms might not be exactly rectangular and the actual room dimensions might differ slightly from that used in the model. These small errors accumulate to shift the natural frequencies and change the amplitude and phase of each mode shape to some extent, leading to incorrect sound pressure in rooms. Therefore, at high frequencies, statistical models are introduced to complement the modal theory for room acoustic analysis. One of the simple and widely known statistical models is the diffuse sound field model. In a perfect diffuse sound field, the sound pressure levels at any spatial location in the room are the same. The diffuse sound field is an ideal model, which is rarely seen in the real world.

Reverberation time

Both the abovementioned modal theory and statistical approach deal with the steady-state sound field excited by a stationary sound signal. If the sound signal ceases suddenly or the sound field is excited by a short-duration pulse signal, the sound does not stop propagating immediately. In addition to the direct sound that propagates straightforwardly from the sound source to the receiver, the sound waves reflected by the floor, ceiling, walls, furniture, etc. sustain for a certain time period before the acoustic energy is dissipated into heat. The specific time for the sound energy to fade away depends on the room dimension as well as the sound absorption properties of the interior materials.

To quantify the transient acoustic properties in rooms, the reverberation time (RT) is defined as the time required for the sound pressure level to fall 60 dB. The RT can be estimated by the Eyring equation[13]

$$T_{60} = \frac{0.161V}{-S\ln(1-\alpha)} \tag{5.5}$$

where ln denotes the natural logarithm, α is the average sound absorption coefficients, and S and V denote the total surface area and room volume, respectively. If all the room surfaces are perfectly absorbing, i.e., $\alpha = 1.0$, the RT T_{60} is 0. When the average absorption coefficient is small, i.e., $\alpha \ll 1.0$, $\ln(1-\alpha) \approx \alpha$ so the Eyring equation can be approximated by $T_{60} = 0.161V/S\alpha$. This is the Sabine equation that was originally derived by Wallace Sabine from measurements he made in a number of rooms at Harvard University.[12] The Sabine equation indicates that the RT is proportional to the room volume and inversely proportional to the overall surface area and the average acoustic absorption coefficient.

Reverberation is a part of the sound that one hears in rooms. A longer RT indicates a livelier, more resonant space, while a shorter RT indicates a more controlled, deadened space. The RT should be designed to cater for different purposes of the room. The RT in listening rooms should not exceed 0.5 s while the optimum RT for the current symphonic repertoire and chamber music is 1.8–2.1 s and 1.5–1.7 s, respectively. The optimum RT also depends on the room volume. Curve fitting from subjective listening results indicates the relationship between auditorium volume V and the RT as [12]

$$\log_{10} V = 5.72 + \log_{10} T_{60} - \frac{2.43}{T_{60}} \tag{5.6}$$

This relationship is for auditoriums only and may be different for other buildings subject to subjective listening tests.

Architectural acoustics measurements

Setting the scene architecturally

Before launching into an account of our measurements and interpretation of the architectural acoustics to enable remote audiences to have a virtual auditory experience of the Sagrada Família Basilica – one of the largest volumes for a religious building, or indeed, any building with a single main space, a little bit of history (Figure 5.1).

The Sagrada Família Basilica was originally a small parish church to be built on the outskirts of a growing Barcelona that commenced on site in 1882. Antoni Gaudí, the architect to whom the building is attributed as his magnum opus, inherited the plans for the building from the original architect a year later, when the former resigned over a professional matter. Gaudí devoted 43 years to the project until his tragic death in 1926. During this time, ambitions for the project grew in line with the levels of donation the building fund received, for it was and remains an entirely crowd-sourced project. By the time Gaudí died just one transept – the east arm of the building

Figure 5.1 The Sagrada Família Basilica interior in 2019.

had been completed, more or less, just a few percent of the building as a whole. By focussing on this one element, the Nativity Façade, Gaudí sought to give the donating public a sense of what was in store for them, for the building is truly original to the point of being extraordinary in look, scale, and urban presence. At 109 m, the transept towers gave an insight as to what the completed basilica would be like once completed with the central tower reaching 172.5 m, the tallest religious structure worldwide.

The interior was completed, and the vast scaffolding array was removed in 2010, and for the first time, an appreciation of the sound qualities in the immense volume of the basilica could be experienced. Over 90 m in length, 60 m wide, and 60 m high in the centre, the major RT is over 12 seconds, which is quite a challenge for an orator keen to project their voice with clarity to every corner of the building. Clearly, Gaudí had some insight into both the pitfalls of such a resonant space as well as the opportunities. He died before he could explain his theory, but we presume that his reasoning for placing the choir 15 m above the floor level and around the entire perimeter of the building interior partly recognised how the building would sound to the 7,000 people attending mass.

Room impulse responses

The RT is a simple indicator of the acoustic property of a room. It should be mentioned, however, that the sound pressure in a room varies with spatial locations of both the sound source and the listener; the corresponding RT is also different. For a detailed investigation of the sound field in a room, the room impulse response (RIR) is usually used by modelling the room as a linear system. According to the dynamics system theory, all properties of a linear system are included in its impulse response. Therefore, the RIR yields a complete description of the acoustic properties of a room, including the absorption, scattering, reflections, and reverberation. Due to the spatial variations of the sound field in a room, the RIRs measured at different locations are different. In the meanwhile, because the direct and reflected sound from the source arrive at different time, the RIR consists of many pulses at different time delays relative to the direct sound. The energies of the pulses on the temporal axis decrease gradually due to spherical spread of sound waves and absorption at the room surfaces, and the density of pulses increases due to the higher probability of reflections at larger times. An exemplary RIR measured in a small meeting room[14] is illustrated in Figure 5.2. According to the time of arrival, the reflected sound can be divided into early reflections, which arrive in the first 80 ms following the direct sound, and late reverberation, encompassing all the reflections that arrive after 80 ms. The early reflections are critical to acoustical intimacy and speech intelligibility while the late reverberation is important for some music perception such as envelopment.

The RIR only measures the acoustic properties of a room at a single point. However, a listener takes sound into the brain through two ears that are separated from each other by the head.

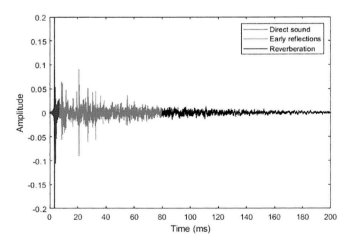

Figure 5.2 An exemplary room impulse response in a small meeting room.

Sound arrives at the two ears at different times with different amplitudes, leading to the perception of space itself. To account for this effect, the binaural RIR (BRIR) is utilised to characterise the acoustic transmission from a sound source to the two ears of a listener in a room. The BRIR captures not only the acoustic properties of the room but also the sound scattering by the human head, torso and ear geometry. By measuring the similarity in sound signals received by the left and right ears of a listener, the quantity interaural cross-correlation coefficient (IACC) is used to assess the quality of binaural recordings and to compare the spatial perception of different listening conditions. An IACC value of 1 means the signals are identical at both ears; a value of −1 means the signals are perfectly correlated but out of phase at the two ears; a value of 0 indicates that the signals are uncorrelated. The IACC can be calculated for the early reflections and late reverberations separately, with the former indicating the spaciousness and the latter relating to envelopment.

RIR measurement

The RIRs are fundamental to investigation of architectural acoustics as it contains all the information of the acoustic properties from the sound source to the receiving point. Therefore, accurate RIR measurement is critical to both design and evaluation of acoustic properties in any enclosed spaces. This section will introduce the RIR measurement techniques.

Measurement system

To measure the RIRs, a sound signal is required to excite the room and a sound recording device is needed to capture the signal. Traditionally, a balloon or a blank pistol is usually employed to generate an impulsive sound signal and a recorder is used to receive the sound signal. In recent years, with the development of digital technologies, more accurate, robust and flexible measurement techniques have been developed. A reproducible sound signal (see next section for details) is played through a sound source that is driven by a power amplifier. A microphone captures the signal, and a proper deconvolution method is used to calculate the impulse response from the output signal and the received signal. An exemplary measurement system is depicted in Figure 5.3, where a personal computer (PC) running an acoustic measurement software generates and sends the excitation signal (e.g., log sweep sine signal) to the loudspeaker through a power amplifier, and also receives the sound signal captured with a microphone, based on which the RIR is calculated. The audio interface connects the PC and the audio devices for data communication.

The omnidirectional sound source is placed at the usual position of the speakers or musician, e.g., on the stage of an auditorium. The sound source should be as omnidirectional as possible, otherwise the measurement results would depend not only on the position of the sound source but also on its orientation. To reduce acoustic radiation directionality of the sound source, multiple loudspeakers can be mounted on a spherical surface to vibrate in phase; typical examples in practice include the dodecahedron loudspeaker with 12 units, such as the omnidirectional loudspeaker (B&K Type 4292) in Figure 5.3, and the icosahedron loudspeaker with 20 units. The frequency response of the sound source should be as flat as possible with the phase distortion as small as possible in the frequency range of interest. The sound power of the source should be high enough to minimise the effect of any potential background noise during measurements and any nonlinear distortions should be avoided. Similarly, the frequency response of the microphone should also be flat and the phase and nonlinear distortions must be small to ensure accurate and precise measurements. Due to the physical propagation of the direct sound along the straight line connecting the sound source and the microphone, there is a delay between sound generation by the loudspeaker

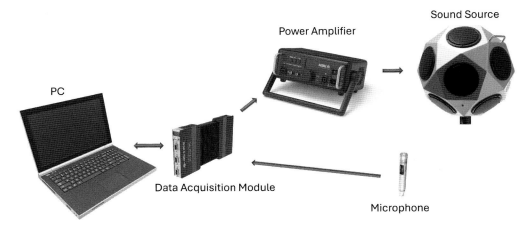

Figure 5.3 Diagram of a RIR measurement system.

and detection by the microphone. The inherent latency in the measurement system will introduce extra time delay, which should be compensated before measurements. The system latency can be measured via a prior loop-back test by directly connecting the system output to the system input.

In many room acoustics measurements, the number of receiving points is larger than the number of sound source points, as the audience is more than the speaker/singer/player. Therefore, the microphones need to be moved around after each test to measure the RIRs at multiple positions, which is time-consuming in practice. To improve measurement efficiency, an array of multiple microphones can be placed at multiple locations to measure the RIRs from the same sound source location simultaneously. In this case, different channels for the microphones should be synchronised to avoid errors in the measured time delays of the direct sounds at different locations from the same sound source. On the other hand, a single microphone at a point cannot measure the binaural characteristics of human hearing. To measure the BRIRs, two microphones need to be mounted at the two ears of either a human (known as listening subject recording) or an artificial head (known as dummy head recording). Mounting microphones in human ears can be intrusive and time-consuming. Therefore, a dummy head and torso simulator (HATS) with built-in microphones is widely adopted. Various commercial dummy head products have been developed, such as the 3Dio Free Space Binaural Microphone with two prosthetic, human-shaped ears, the Neumann KU-100 Dummy Head with two artificial ears on a dummy head, and the Brüel & Kjær 4128D/C HATS.

Excitation signals

The excitation signals in room acoustics measurements should be reproducible and have flat spectra to excite all the room acoustic modes. The traditional impulse sound signals from balloons or pistols may not have sufficiently flat spectra. Therefore, pseudorandom noise and sweep sine signals have been utilised as excitation signals.[15] The most widely used pseudorandom noise signal is the maximum length sequences (MLS), which are periodic sequences of integers. In the case of binary MLS, the integers are restricted to either 1 or −1. They are generated by n-stage shift registers and the period length is $N = 2n - 1$. The most important property of the MLS pseudo noise signals is that their Fourier transform has the same magnitude for all frequency components, thus their power spectra are flat and independent of frequency.[16] Another advantage of the pseudorandom noise signals over the traditional impulses from balloons and pistols is their high tolerance to background noise. Such periodic pseudorandom signals can even be used to measure the RIRs in auditoriums during actual performance by playing the excitation signal at an inaudible level according to the auditory masking effect. The signal-to-noise ratio can be significantly improved by averaging the measured RIRs over a long period, based on the assumption (usually valid in practice) that the RIRs of the room are not changing and the music/speech signals are uncorrelated with the periodic pseudorandom excitation signal.

Although the MLS pseudorandom noise signal is dramatically superior to the traditional impulse signals, it has several limitations. On the one hand, in any measurement using MLS as the excitation signal, nonlinear distortion, mainly induced by the loudspeaker, spreads out over the whole period of the recovered RIR. On the other hand, even under optimal conditions, the dynamic range of the measured RIR using MLS as the excitation signal is limited. The relatively low dynamic range may not affect the measurement of RT but will restrict the usage of the measured RIRs for other applications. For example, in Auralisation or virtual/augmented reality applications, the measured RIR will be convolved with some dry anechoic signals to generate virtual sound environment. In this case, a high dynamic range is desired because any abnormalities in the reverberant tail of an RIR are recognisable due to the high dynamic range of our auditory system and the logarithmic relationship between sound pressure level and perceived loudness.[17]

To overcome these shortcomings, sweep sine signals have been used as excitation signals in RIR measurements.[18] Sweep sine signals can be created either directly in the time domain or indirectly in the frequency domain. In the time-domain synthesis, the phase step that is added to the argument of sine expression is increased after each calculation of an output sample. The phase step can be increased by either adding or multiplying a constant value to generate the linear or logarithmic sweep sine signals, respectively. In either case, the time-domain synthesised sweep sine signals suffer from unwanted ripples in the frequency spectra due to the sudden switch-on at the beginning and switch-off at the end.[19] This problem can be avoided by synthesising in the frequency domain, where the sweep sine signal is obtained via inverse Fourier transform of an artificial spectrum with designed magnitude and phase. The magnitude spectrum must be white to generate a linear sweep sine signal and pink (decays 3 dB/octave) for a logarithmic sweep sine signal. The phase is calculated by integrating the group delay designed for either linear or logarithmic sweep sine signals. The main advantage of the sweep sine method is that it can separate the linear and nonlinear parts of the system response so that the nonlinear distortions can be suppressed in the RIR and can even be measured simultaneously.[20]

Room acoustics analysis

Once the RIRs are measured, the RT and many other room acoustics parameters can be calculated. To calculate the RT, the acoustic energy decay curve needs to be estimated from the measured RIR first. This can be achieved by the Schroeder backward integration method, which was first proposed by Schroeder.[21] It was proven that the average squared sound pressure (i.e., $\langle p^2(t) \rangle$) is related to the RIR (i.e., $h(t)$) by

$$\langle p^2(t) \rangle = \int_t^\infty h^2(\tau)d\tau = \int_0^\infty h^2(\tau)d\tau - \int_0^t h^2(\tau)d\tau \tag{5.7}$$

The RT is defined as the time that the acoustic energy decays 60 dB, hence it can be calculated from

$$\Delta L_p = 10 \log_{10}\left(\frac{\langle p^2(0)\rangle}{\langle p^2(T_{60})\rangle}\right) = 60 \text{ dB}$$

To measure the RT T_{60}, the excitation signal must be at least 60 dB higher than the background noise, which is difficult in practice. Therefore, considering the energy decay curve is linear in logarithmic scale, the RT T_{20} is usually measured instead, which is defined as three multiplies the time that the acoustic energy decays 20 dB. Similarly, the early decay time (EDT), which is defined as six multiplies the time for the acoustic energy to decay from −5 dB to −15 dB, is often used in practice. The RT may vary with spatial locations in an enclosed space because the sound energy decay curve in Eq. (5.7) is calculated based on the RIR measured at a specific position. The average RT over multiple spatial locations can be calculated to characterise the overall reverberance of a room.

Apart from the RT, other acoustic properties, such as clarity, loudness, warmth, etc., can also be analysed from the measured RIRs. The clarity of an acoustic space is a critical aspect that impacts the quality of sound experienced in that space. It is determined by the ratio of early sound energy to reverberant sound energy, expressed in decibels as C80.[22] The average C80 value is usually calculated using the energy in the 500 Hz, 1,000 Hz, and 2,000 Hz octave bands and at multiple positions within an enclosed space. The required level of clarity varies depending on the type of event being held in the space. During rehearsals, a C80 value ranging from 1 dB to 5 dB is often considered ideal, allowing for the clear perception of musical details. On the other hand, for concerts, a high degree of reverberance with a C80 value between −4 dB and −1 dB is usually preferred.

The concept of 'warmth' in room acoustics refers to the perceived quality of sound that results from the presence of low-frequency resonances in the room. It is a subjective quality that is difficult to quantify, but it can be estimated using various objective measures. One way to quantify the warmth is to calculate the ratio of low-frequency energy to the total energy in the RIR. Alternatively, the specific low-frequency resonances in the room can be analysed, and the strength of these resonances can be used to estimate the warmth of the room. The warmth of a room is a subjective quality that can be influenced by various factors, including the type of music being played, the size and shape of the room, and the presence of absorbing or reflective surfaces. As such, it may be necessary to conduct multiple measurements and use various analysis methods to fully understand the warmth of a room.

If the BRIRs are measured, the aforementioned IACCs can be calculated to evaluate the perception of spaciousness of a room. Therefore, the RIRs contain almost all the information about the room acoustics and various acoustic parameters can be calculated. It should be emphasised that all these acoustic parameters can each represent only one aspect of the acoustic properties of a space, and the acoustic quality evaluation is subjective and depends on listeners' unique auditory system and personal preference. The acoustic parameters should by no means completely replace subjective listening evaluation.

Case study: Sagrada Família Basilica

The problem from an architectural standpoint

The interior of the Sagrada Família Basilica is made from brick, natural and artificial stone, and mosaic. There are almost no flat surfaces with the walls and ceiling vaults being made from hyperboloids of revolution and hyperbolic paraboloids at a grand scale. We have hinted at the challenges for such a vast volume earlier in the chapter: an extraordinarily long RT rendering oratory impossible without electroacoustic assistance. This was apparent as soon as the interior was revealed in 2010, as were the remarkable and positive acoustic properties these same challenges offered to music and song. If we could understand these properties well enough a virtual acoustic environment could be created that would provide a robust and accurate simulation of how a sound from any designated position in the basilica would be received at any designated position elsewhere in the basilica.

With an earlier project, the Colònia Güell Chapel (1898–1912) Gaudí first introduced the architectural world to the hyperbolic paraboloid, well before using these surfaces for the basilica. He capitalised on the inherent structural and constructional benefits of these shapes and expanded his exploration to their acoustic properties. Delving into the essence of sound within a space, he examined how the concertina-like, curved walls might positively influence the chapel's interior resonance and disperse sound waves throughout the environment:

> We do not know if Gaudí had completed any special study of the acoustics of this church, but he was well aware that its structure, full of columns, pierced arches, broken volumes, and corners, was an effective way to avoid resonance. Furthermore, the hyperbolic paraboloid vaults disperse sound instead of concentrating it.
>
> We also believe, on the base of our own experience, that such an internal structure avoids, to a great extent, the echoing that could be produced in the absence of extensive enough absorbent surfaces.[23]

What are the effects of the profusion of hyperbolic surfaces on almost all the walls and ceiling vaults for the basilica: were Gaudí's assumptions correct? Had he found a way to offer acoustic warmth by deploying an unconventional geometry despite the hardness of stone and brick?

With appropriate measuring devices we aimed to be able to gauge the acoustic warmth of the Sagrada Família Basílica space sufficient for composers to be able to anticipate how their compositions would sound with accuracy, and for the organ builder to establish the best size and location for the instrument. Music could therefore be composed especially for the basilica with the basilica itself acting as part instrument through the acoustic properties of all its surfaces. As a final note, Gaudí had conceived of the entire building as a carillon with specially designed and scale-tested tubular bells for the 12 campaniles that comprise the three façades.

Experimental setup

The measurement system is the same as that in Figure 5.2, except that the microphone was replaced with a Kemar Manikin with two built-in microphones to record binaural signals. A PC with a B&K DIRAC Room Acoustic Software generated an exponential sweep sine (ESS) signal that was played back through a B&K OmniPower™ Sound Source (Type 4292-L). Before the measurements, we calibrated the microphone in the left ear of the Kemar Manikin using a B&K Sound Calibrator Type 4231.

We placed the sound source and the Kemar Manikin at various locations to measure the RIRs over the basilica, which are denoted as red and blue points, respectively, in Figure 5.4. We chose the sound source positions S1–S3 and the corresponding measurement locations M1–M19 at the ground level, S4 and M20–M27 on the 'cantoria' (the choir gallery that flows around the perimeter of the basilica) situated approximately 15 m above the nave floor level, and S5 approximately 45 m high from the ground. For the measurements at the ground level (M1–M19), the Kemar

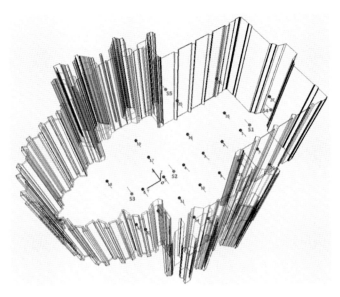

Figure 5.4 The sound source positions (points indexed with a starting letter S followed by a number) and the measurement locations (points indexed with a number).[24]

Manikin was oriented to face the apse, at the far end of the basilica axis, whereas for the measurements on the choir level, the Kemar Manikin was placed on a seat and oriented to face the front direction of the audience area. Because the Sagrada Família Basilica is symmetrical, we only measured the RIRs on the left side of the basilica (M1–M14) for the sound source positions S1–S3. For each pair of sound source and measurement location, we measured the RIRs three times with a 23.8 s ESS signal and averaged the results to suppress the effect of background noise. After each measurement, we moved the Kemar Manikin to the next measurement location and repeated the procedure at each location.

Reverberation time

Two typical RIRs, from the sound source position S1 to the measurement location M1 and M13, are shown in Figure 5.5. Because the measurement location M1 is close to the sound source position S1, the RIR in Figure 5.5(a) is composed of direct sound, strong early reflections, and late reverberation. By contrast, the measurement location M13 was far away from the sound source position, the direct sound and early reflections are hard to distinguish from the late reverberation sound in the RIR in Figure 5.5(b), noting the vertical axis in Figure 5.5(b) has been zoomed in for clarity.

The average RTs, the T_{20} and the EDT, in the Sagrada Família Basilica are presented in Figure 5.6, where the vertical bars indicate the standard deviation of the measurement results at different measurement locations. The EDT is similar to T_{20}, indicating a linear energy decay curve in the basilica. The T_{20} is the largest at 500 Hz and drops significantly at frequencies higher than 500 Hz, and the T_{20} at lower frequencies below 1,000 Hz fluctuates more significantly than that at higher frequencies above 2,000 Hz. The RT averaged from 500 Hz to 1,000 Hz is 12.0 s, consistent with previous measurements in churches built from brick and stone of a similar size to the Sagrada Família Basilica.[25] There is no optimal RT for such large-volume churches like the Sagrada Família Basilica (approximately 200,000 m^3) in the available literature.[26]

Architectural acoustics of the Sagrada Família Basílica

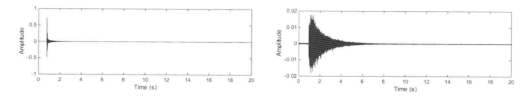

Figure 5.5 The RIRs measured in the left ear at (a) M1 and (b) M13 corresponding to the sound source position S1.

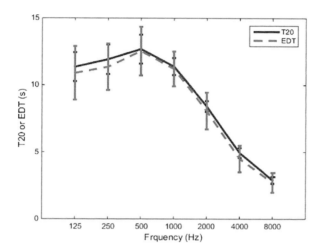

Figure 5.6 The average reverberation time (T_{20}) and early decay time (EDT) in the Sagrada Família Basílica.

Other objective metrics

Although RT is an important metric to characterise acoustics in enclosed spaces, it cannot reflect many other aspects of auditory perception, such as clarity and spaciousness mentioned in the previous section in this chapter. To explore these aspects, other acoustic parameters are calculated from the measured RIRs in the Sagrada Família Basílica. It is noted that only the results corresponding to the sound source positions S1, S2 and S3 that are on the ground level are shown here.

The subjective perception of clarity is related to the objective parameter C80 as mentioned earlier. The measured middle frequency of C80 (averaged from 500 Hz to 2,000 Hz) is shown in Figure 5.7(a), where the dashed lines indicate the optimal range for concert halls as a reference. In combination with Figure 5.4, Figure 5.7(a) shows that when the measurement location is close to the sound source, the value of C80 is higher because the early sound is dominant; when the measurement location is moved away from the sound source position, the value of C80 decreases significantly due to the dominance of late reverberant sound. On the other hand, the 80 ms timeframe corresponds to a 27.2 m distance with a sound speed of 340 m/s, which is smaller than the size of the Sagrada Família Basílica. Therefore, the sound energy of the early reflections is quite low, which also leads to relatively lower values of C80.

The auditory perception of spaciousness is represented by the parameter binaural quality index (1 − $IACC_E$). Figure 5.7(b) depicts the measured binaural quality index, averaging over the

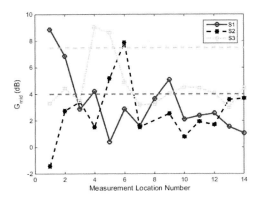 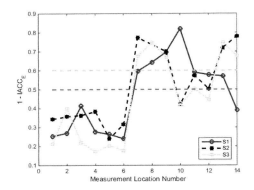

Figure 5.7 The measured middle frequency (500–2,000 Hz) (a) clarity (C80) and (b) binaural quality index (1 − IACC$_E$) at each measurement location corresponding to the sound source positions S1, S2, and S3. The red and green dashed lines indicate the optimal range for concert halls as a reference only.

frequency range of 500–2,000 Hz, in the Sagrada Família Basilica. It can be seen that all the values at the measurement locations M1 to M6 that correspond to the sound source positions S1 to S3 are below the typical satisfactory threshold. This is because these measurement locations and the sound source positions are all on the centre line of the Sagrada Família Basilica, leading to the high similarity of the sounds at the two ears. The values at other measurement locations are above the typical satisfactory threshold of 0.5, except those at M10 corresponding to S2 and at M10 and M12 corresponding to S3, which are 0.42, 0.42 and 0.45, respectively. The average value over all the measurement locations corresponding to all the sound source positions is 0.51, just above the typical satisfactory value, indicating a satisfactory spaciousness quality. The reference optimal range depicted by the dashed lines in Figure 5.7 is for concert halls, whose volumes are usually much smaller than the Sagrada Família Basilica. It is unclear whether these values can be used to assess the sound field properties in churches, especially in such a large church as the Sagrada Família Basilica, and further research is needed for a more comprehensive understanding.

Architectural implications

In many ways, the challenging acoustic performance of the Sagrada Família Basilica interior is a spectacular but singular matter. The vastness of the interior will seldom be matched and, in any event, interiors the scale of the basilica including stations, airports, and sporting stadia take advantage of electroacoustic systems. But this is not the point; for as long as cafés and restaurants emerge with an inadvertent acoustic clatter that thwarts civilised dialogue, architects and interior designers have a lot to learn, and Antoni Gaudí has provided some valuable insights to draw upon.

Summary

This chapter introduces the fundamental concepts in architectural acoustics. The sound generation mechanisms of human speech and common musical instruments are briefed first, followed by an introduction of sound propagation properties in both open and enclosed spaces. The room acoustics measurement techniques, for the RIRs in particular, are then reviewed. Finally, a case study for the acoustic measurements in the Sagrada Família Basilica is presented, with discussions on the measured RT, clarity and auditory spaciousness. It is expected that the chapter will provide an overview of fundamental concepts and measurement techniques for architects and designers.

Notes

1. Xiang, *Architectural Acoustics Handbook*.
2. Mahendru, "Quick Review of Human Speech Production Mechanism."
3. Mahendru.
4. Hartmann, *Principles of Musical Acoustics*.
5. Bies and Hansen, *Engineering Noise Control-Theory and Practice*.
6. Qiu, "Principles of Sound Absorbers"; Bies and Hansen, *Engineering Noise Control-Theory and Practice*.
7. Cox and D'Antonio, *Acoustic Absorbers and Diffusers: Theory, Design and Application*.
8. Zhao et al., "A Preliminary Investigation on the Sound Field Properties in the Sagrada Familia Basilica."
9. Zhao, Qiu, and Cheng, "An Integral Equation Method for Calculating Sound Field Diffracted by a Rigid Barrier on an Impedance Ground."
10. Kutterruff, *Room Acoustics*.
11. Jacobson and Juhl, *Fundamentals of General Linear Acoustics*.
12. Kutterruff, *Room Acoustics*; Jacobson and Juhl, *Fundamentals of General Linear Acoustics*.
13. Beranek and Mellow, *Acoustics: Sound Fields, Transducers and Vibration*.
14. Zhao et al., "A Room Impulse Response Database for Multizone Sound Field Reproduction (L)."
15. Stan, Embrechts, and Archambeau, "Comparison of Different Impulse Response Measurement Techniques."
16. Schroeder, "Integrated-Impulse Method Measuring Sound Decay without Using Impulses."
17. Muller and Massarani, "Transfer-Function Measurement with Sweeps."
18. Farina, "Simultaneous Measurement of Impulse Response and Distortion with a Swept-Sine Technique."
19. Muller and Massarani, "Transfer-Function Measurement with Sweeps."
20. Guidorzi et al., "Impulse Responses Measured with MLS or Swept-Sine Signals Applied to Architectural Acoustics: An in-Depth Analysis of the Two Methods and Some Case Studies of Measurements inside Theaters."
21. Schroeder, "New Method of Measuring Reverberation Time."
22. Beranek, *Concert Halls and Opera Houses, Music, Acoustics, and Architecture*.
23. Puig I Boada, I., L'Església de la Colònia Güell.
24. Zhao et al., "A Preliminary Investigation on the Sound Field Properties in the Sagrada Familia Basilica."
25. Cirillo and Martellotta, *Worship, Acoustics, and Architecture*.
26. Everest and Pohlmann, *Master Handbook of Acoustics*; Zhao et al., "A Preliminary Investigation on the Sound Field Properties in the Sagrada Familia Basilica."

Bibliography

Beranek, Leo. *Concert Halls and Opera Houses, Music, Acoustics, and Architecture*. New York: Springer-Verlag New York Inc., 2004.

Beranek, Leo, and Tim Mellow. *Acoustics: Sound Fields, Transducers and Vibration*. 2nd Edition, San Diego: Academic Press, 2019.

Bies, David Alan, and Colin H. Hansen. *Engineering Noise Control-Theory and Practice*. London: CRC Press, 2009.

Cirillo, Ettore, and Francesco Martellotta. *Worship, Acoustics, and Architecture*. Essex, UK: Multi-Science Publishing Co Ltd., 2006.

Cox, Trevor J., and Peter D'Antonio. *Acoustic Absorbers and Diffusers: Theory, Design and Application*. Boca Raton: Taylor & Francis, 2009.

Everest, E. Alton, and Ken C. Pohlmann. *Master Handbook of Acoustics*. New York: McGraw-Hill Inc, 2009.

Farina, Angelo. "Simultaneous Measurement of Impulse Response and Distortion with a Swept-Sine Technique." In *Proceedings of the 108th AES Convention*, 2000.

Guidorzi, Paolo, Luca Barbaresi, Dario D'Orazio, and Massimo Garai. "Impulse Responses Measured with MLS or Swept-Sine Signals Applied to Architectural Acoustics: An In-Depth Analysis of the Two Methods and Some Case Studies of Measurements Inside Theaters." *Energy Procedia* 78 (2015): 1611–1616. https://doi.org/10.1016/j.egypro.2015.11.236.

Hartmann, William M. *Principles of Musical Acoustics*. New York: Springer, 2013. http://link.springer.com/10.1007/978-1-4614-6786-1.

Jacobson, Finn, and Peter Juhl. *Fundamentals of General Linear Acoustics*. West Sussex: Wiley, 2013.

Kutterruff, Heinrich. *Room Acoustics*. Boca Raton: CRC Press, 2016.

Mahendru, Harish Chander. "Quick Review of Human Speech Production Mechanism." *International Journal of Engineering Research* 9, no. 10 (2014): 48–54.

Muller, Swen, and Paulo Massarani. "Transfer-Function Measurement with Sweeps." *Journal of the Audio Engineering Society* 49, no. 6 (2001): 443–471.

Puig I Boada, I. "L'Església de la Colònia Güell", Editorial Lumen, Barcelona, 1976, p. LXXIX.

Qiu, Xiaojun. "Principles of Sound Absorbers." In *Acoustic Textiles*, edited by Rajiv Padhye and Rajkishore Yayak, 245. Singapore: Springer Singapore, 2016.

Schroeder, Manfred Robert. "New Method of Measuring Reverberation Time." *Journal of the Acoustic Society of America* (1965): 409–414. https://doi.org/10.1121/1.1909343.

Schroeder, Manfred Robert. "Integrated-Impulse Method Measuring Sound Decay without Using Impulses." *Journal of the Acoustical Society of America* 66, no. 2 (1979): 497–500. https://doi.org/10.1121/1.383103.

Stan, Guy Bart, Jean Jacques Embrechts, and Dominique Archambeau. "Comparison of Different Impulse Response Measurement Techniques." *AES: Journal of the Audio Engineering Society* 50, no. 4 (2002): 249–262.

Xiang, Ning. *Architectural Acoustics Handbook*. Edited by Ning Xiang. Florida: J. Ross Publishing, 2017.

Zhao, Sipei, Eva Cheng, Xiaojun Qiu, Pantea Alambeigi, Jane Burry, and Mark Burry. "A Preliminary Investigation on the Sound Field Properties in the Sagrada Familia Basilica." In *Proceedings of ACOUSTICS 2016*, 1–8. Brisbane, Australia, 2016.

Zhao, Sipei, Xiaojun Qiu, and Jianchun Cheng. "An Integral Equation Method for Calculating Sound Field Diffracted by a Rigid Barrier on an Impedance Ground." *The Journal of the Acoustical Society of America* 138, no. 3 (2015): 1608–1613. https://doi.org/10.1121/1.4929933.

Zhao, Sipei, Qiaoxi Zhu, Eva Cheng, and Ian S. Burnett. "A Room Impulse Response Database for Multizone Sound Field Reproduction (L)." *The Journal of the Acoustical Society of America* 152, no. 4 (October 31, 2022): 2505–2512. https://doi.org/10.1121/10.0014958.

6
INTIMATE ACOUSTIC ENVIRONMENTS ON RECORD

Emil Kraugerud

Introduction

How can the presence of small-room acoustic ambience in music recordings contribute to feelings of intimacy? In sociology and psychology, intimacy is regularly referred to as something that pertains to certain aspects of communication and relations between people, and something that is expressed from one person to another, or to some inner personal realm.[1] This is also reflected in common conceptions of 'intimate sounds' in music recordings – that is, sounds that afford a perception of physical or emotional closeness to a recorded performer or persona.[2] Some scholars have also noted the sense of intimacy that may be afforded by the acoustics of small spaces. For example, Tor Halmrast suggests that early reflections imply that the listener shares the intimate space of a small 'box' with the performer.[3] For Halmrast, then, acoustic ambience in recordings can afford a sense of intimacy. In this chapter, I explore this acoustical sense of small-room intimacy in connection with its associations with domestic space, and in turn how that informs interpretation of music recordings. These explorations are based on concepts of the virtual recorded space,[4] referring to the perceived location of recorded instruments (whether recorded acoustically, electronically, digitally, etc.) in the depth, width, and height dimensions of a mix, and their recorded or emulated spatial surroundings. I end the chapter with a brief speculation on how these ideas may coincide with current and future tendencies in home production and online collaboration.

Acousmatic intimacy in recorded music

The backdrop for this chapter is my interest in the experience of intimacy from recorded musical sound, what I have conceptualised as acousmatic intimacy.[5] Whereas a lot of scholarly writing on intimacy experienced from recorded music has – to a greater or lesser extent – equated it with perceptions of physical proximity,[6] acousmatic intimacy is a particular emotional engagement that is afforded by physical proximity. Accordingly, I define acousmatic intimacy as

> a sensation of intimacy that may be experienced when listening to music recordings, when the origins of the sounds (musicians and instruments) are absent, [other than through their sonic presence,] and that is triggered by choices made in the production of the recordings.[7]

The concept thus comprises elements of physical sensation, emotional engagement, as well as the technological mediation of those through record production and reproduction.

The physical sensations that may lead to acousmatic intimacy can be categorised according to Joshua Meyrowitz's categories of experienced distance in TV images.[8] In his categorisation, mediated distance is the perceived distance between the viewer and the content (including people) of the image, whereas portrayed distance concerns the spatial relationships that are displayed within the image. He further divides portrayed distance into objective and subjective subcategories. Portrayed objective distance then refers to distance relations within a television image where the viewer is given the role of a detached observer, and portrayed subjective distance refers to images in which the camera assumes the point of view of a character and thus brings the viewer 'into' the image. Accordingly, acousmatic intimacy can be categorised in terms of a mediated intimacy between the listener and a recorded persona or as a portrayed objective intimacy between two or more recorded personas as experienced from either outside (objective) or inside (subjective) the recorded space.[9] These categories offer an inroad to understand how various forms of perceived physical closeness can afford feelings of intimacy.

In this chapter, I aim to broaden the understanding of acousmatic senses of intimacy by investigating how recording technology and technique can facilitate portrayed objective intimacy and how that in turn relates to ideas of intimate domestic space. As with music in general, acousmatic intimacy is technologically conditioned, and the phenomenon has developed in close interaction with technological development since the early days of recording. This chapter is particularly concerned with the ways in which recording technology has facilitated replication and simulation of realistic-sounding spatial ambience. Firstly, the development of condenser microphones that had increased dynamic range compared to their predecessors allowed for a recording room's acoustic response to be part of recordings (although that certainly led to an increased need to control acoustics). Secondly, various forms of dynamic range compression and expanders/gates have impacted recordists' control of that recorded acoustic response. Finally, various forms of reverb simulation – including acoustic echo chambers, electromechanical devices like plate and spring reverbs, and digital algorithm-based and convolution reverbs – have allowed recordists to construct both realistic and surreal spatialities for their recordings. One possible consequence of these developments is that the notion of the sound of domestic space has also entered the realm of music production, whether it is through recording in residential spaces or using plugin presets for music production software that promise the sound of living rooms or bathrooms.[10]

Domestic space as intimate

My notion of intimate domestic space is loosely based on Gaston Bachelard's phenomenological study The Poetics of Space,[11] which looks at the basic functions of the intimate spaces of our homes. Like Martin Heidegger in 'Building, Dwelling, Thinking',[12] Bachelard relies on the notion of the dwelling – homes are built not only in terms of physically setting up a building but also in terms of imbuing that building with personal experiences, secrets, memories, and dreams of past and future. The result of all of this is a safe place that provides shelter from the external world and affords an intimate relationship with it. This is, of course, problematic, given the fact that the home is not always felt as safe. Whereas Bachelard's idea of the dwelling appears to derive from his nostalgia concerning a happy childhood in his family's house, others may well have bad associations with their home. The dwelling, then, and its associated intimacy and refuge, should be considered an ideal home, not a real home, for the purposes of the following discussion. Michel de Certeau, Luce Giard, and Pierre Mayol, for example, describe the (ideal) home as follows:

> The territory where the basic gestures of "ways of operating" are deployed and repeated from day to day is first of all domestic space, this abode to which one longs to "withdraw," because once there, "one can have peace." One "returns to one's home," to one's own place, which, by definition, cannot be the place of others. Here every visitor is an intruder unless he or she has been explicitly and freely invited to enter.[13]

De Certeau, Giard, and Mayol thus associate the notion of the domestic space with a shelter from the demands of the outside world. Furthermore, they draw attention to the conflict between intimacy and intrusion, which emerges from the definition of intimacy as exclusivity – intimacy is what we share with only a selected few, and when someone else 'takes part' in it, they are seen as intruding.[14] Bachelard reminds us as well that the home is filled with memories. When we receive 'outsiders' there, we expose ourselves in a sense.

Bachelard also writes exclusively about the rooms in a house as intimate, whereas not everyone has experience with a house as such. Still, while the idea of the dwelling is complicated, as are the rooms (and acoustics) we associate with it, domestic spaces do differ from regular music performance spaces. Rooms in most Western working- and middle-class homes have relatively similar sizes, especially in terms of height, and the building materials are often wood, stone, plaster, or a combination of these. What we have in the rooms are also often similar: beds, sofas, tables, TVs, books, framed pictures, cupboards, windows, curtains, appliances, and so on. All these things affect the acoustics of the rooms in certain ways.

One of several ways to understand intimacy is thus to relate it to how the boundaries between public and private (and intimate) life are performed in relation to architecture. In that way, intimacy can also be related to certain sounds or types of acoustics. I am thinking specifically of domestic spaces such as living rooms, bedrooms, and kitchens, which tend to have no dedicated acoustic treatment. The sound of such a room would typically be made up of a preponderance of early reflections and a short reverb tail, and often acoustic artefacts such as comb filtering and frequency build-up – features that are not considered to be very suitable for musical activities (including performance, recording, and even record listening). These kinds of spaces share a similar relation between a strong spatial impression and a low sense of spaciousness. According to David Griesinger, spaciousness refers to the perception of being in a large and enveloping space, whereas spatial impression describes the degree to which we perceive that we are in an enclosed space:

> [O]ne might perceive very little spatial impression while listening in a small, highly absorbing room, but a highly reflecting room of the same size might give a strong spatial impression. However, by definition the sound of such a room would not be spacious.[15]

Domestic spaces, then, may signify intimacy not only through the acoustic impression of a small room but also through the interpretation of those acoustics as reminiscent of actual intimate domestic spaces.[16]

Intimate domestic space on record

A possible way to relate acousmatic intimacy to domestic spaces in recordings is to consider the extent to which the virtual space of a recording realistically resembles an actual intimate space. Some scholars taking a spatial approach to studying recorded musical sound have investigated the impact of what kind of spatial environment we hear as we listen to certain recordings, thus going beyond simply noting aspects such as a long or short reverb or echo.[17] In doing so, they ask what kind of space the reverb reminds us of (or does not remind us of). When we listen to recordings, we compare their virtual spatial environments to our experiences with actual spatial environments, consciously or otherwise.[18] Albin Zak, for example, points out that the room in which a performance is recorded may evoke certain associations.[19] When listening to a song recorded in a particular studio, as well, we may recognise the spatiality from another recording that was recorded in the same studio. Some recordists even record in 'non-studio' spaces in order to lend a signature ambience to the recording that may be easier to recognise than the ambience of a studio's dedicated recording room. For example, Zak cites an interview with producer Daniel Lanois where he says: 'I like the strange rooms that you find in old houses and buildings... You get a sound that's unique to that record, a sonic signature, if you like'.[20] This inclusion of specific kinds of acoustic spaces in recordings can also serve narrative purposes, and thus allow them to act as sonic markers.[21]

With the invention of electromechanical recording, recordings became capable of replicating room ambience, and, as Peter Doyle comments, 'of carrying [...] significant sonic information about the spaces in which they were made'.[22] The limited dynamic capabilities of previous recording methods, known as mechanical or acoustical recording, required a certain volume for sounds to be transferred to the record, which rendered most ambient sound inaudible. The importance of room ambience was new to recording, but it was not new to music in the 1920s. As Michael Chanan explains, different styles in the Western classical canon were made for and associated with different kinds of spaces, like cathedrals, salons, and concert halls. As such, these spaces imply certain social relations between performers on the one hand and between performers and listeners on the other, resulting in different performance practices.[23] The acoustical ambience that 'surrounds' a musical performance is therefore likely to influence our interpretation of that performance through the associations and expectations we have regarding particular types of acoustics.

A recording's ability to realistically replicate acoustics we associate with the home is central to the notion of intimate domestic space in recorded sound. This does not mean that performances must be recorded in such a space, whose features can be simulated in other ways, such as with digital reverbs. Halmrast argues for the importance of the effect of early room reflections on a sound's attack in determining what type of room the sound is in,[24] which in turn suggests that we can make assumptions about the size of the room, as well as the materials of which it is constructed, based on such auditory information. Furthermore, Denis Smalley's theory of source bonding[25] explains how a record listener might recognise aspects of sounds and reverbs and associate them to certain kinds of spaces, such as a room in the home, even when not aware of exactly what creates that ambience.

My previous points about how domestic space can be understood as intimate and how the acoustics of such spaces can contribute to affordances of acousmatic intimacy suggest that certain reverb characteristics implemented in a recording can contribute to a sense of intimacy. Moreover, while associations with the home are not always positive, the acoustics of those spaces can still be experienced as pleasant. For example, a study by Daniel Västfjäll, Pontus Larsson, and Mendel Kleiner found that sound samples with short and medium reverb times supplied higher levels of experienced arousal and induced more pleasant reactions than the same sounds with long reverb times.[26] Furthermore, Ana Tajadura-Jiménez and colleagues found that, when study participants were exposed to virtual acoustic rooms with varying sound source positions and acoustic properties, they considered the small rooms to be 'more pleasant, calmer, and safer than big rooms'.[27] Nicola Dibben comments that these findings imply 'that small virtual spaces will be preferred where the listener feels affinity for the singer/persona'.[28] Consequently, the acoustics of small spaces may – in addition to implying proximate and potentially intimate physical relations within the virtual recorded space – be experienced as pleasant and safe and will likely enhance or amplify the acousmatic intimacy already afforded by other sonic characteristics of a sound.

One way to approach the idea of intimate spatial environments in recordings is to consider the extent to which a recorded space might realistically resemble an actual intimate space. Any perception of an intimate space necessarily depends on coherence between one's conception of the sound of an intimate space and what one hears in the music. An understanding of intimacy as related to a small physical environment further implies proximity in that environment's necessarily short distances (between listener and performer, or between two or more performers). A striking example of this production effect is found in the acoustic version of Sigrid's 'Strangers'[29] (Strangers [Remixes], 2018), in which the voice sounds very proximate (the result of low-intensity singing close to the microphone in a breathy voice) while also featuring subtle and short reverb – we hear more of its early reflections than we hear of the reverb tail. Consequently, the initial feeling that the singer is very close to us is reinforced by the strong sense that we are in the same room as her. Three types of intimate distance relations are thus at play here: (1) the perceived proximity between the musician(s) (singer and piano player) and the listener suggests a mediated intimacy; (2) the small spatial environment suggests a portrayed intimacy between singer and piano player (or singer and instrument); and (3) the strong spatial impression of the spatial environment makes the portrayed intimacy subjective by virtually placing the listener in the same room as the performers. 'Strangers' thus exemplifies how a recording can be staged in ways that afford acousmatic intimacy in many senses.

For a more elaborate demonstration of the role of domestic space in affording intimacy, I now turn to David Byrne's recording of 'My Love Is You', from his eponymous solo album from 1994.[30]

Portrayed subjective intimacy in David Byrne's 'My Love Is You'

In the recording, variations in reverb characteristics provide sonic contrasts while also contributing to a sense of portrayed subjective intimacy in the recorded sound. This analysis will focus on how the location of sonic elements within the virtual recorded space along with the acoustic shaping of that space underlines the quotidian intimacy thematised in the lyrics. The lyrics portray an intimate relationship between the protagonist (as represented by Byrne's vocals) and a 'you'-character, addressing both the positive sides and the imperfections and vulnerabilities of both parts. This exposure of imperfections and vulnerabilities can be seen as part of a sharing of intimacy, in which the two characters are presented as equals. This aligns well with definitions of intimacy within sociology and psychology, where the concept involves awareness of an other's inner realm – including feelings and vulnerabilities that are normally not shared – as well as an exclusivity regarding with whom one shares those things.[31] As such, the term intimacy can describe a feeling of being attached to another person through exclusive recognition or the feeling of being recognised. Furthermore, the love in 'My Love Is You' is one that arises and abides on the partners' own premises, much like what Anthony Giddens describes as 'confluent love' – that is, love that is under continuous mutual renegotiation in terms of the partners' expectations and conditions.[32] Confluent love thus involves exposure and openness to imperfection, and as such, it matches the mentioned criteria for intimacy. Through the lyrics, then, the listener is presented with familiar themes of quotidian love, issues with which they likely have experience and which they may experience as related to their own quotidian reality. In turn, a feeling of relational closeness is afforded, an affordance that is strengthened by the spatial formation of the recording.

The arrangement consists of vocals and nylon-string guitar, as well as tuba, marimba, double bass, shaker, brushed snare drum, and cuíca. It expands gradually over the course of the recording, starting with the guitar playing a full verse (five bars) by itself, after which the vocals enter with the shaker for a verse. The second verse introduces the double bass and snare drum, and in the middle eight, the tuba appears. During the following marimba solo, the tuba is replaced by a cuíca before the previous grouping returns after the solo and continues through to the end of the song. Except for the marimba, all sounds seem to be placed in a similar, relatively dry space. At the same time, different elements appear as having been recorded at different distance from the microphones, resulting in different perceived proximities. The vocal and the guitar are quite close and performed quietly, as evidenced by the presence of fret noises and breath intakes, which otherwise tend to be quiet or inaudible in loud performances at a distance. The quiet performance also suggests that it was intended to be close and exclusive to the immediate surroundings of the performer (which happens to include the listener position). The other instruments are placed further back in the mix, but because of the apparent dimensions of the virtual recorded space, they are still relatively close to the listener position.

While this relative closeness alone is not necessarily enough to suggest a sense of mediated intimacy between listener and performers, the upfront placement underlines the quite individual role of the different instruments, especially voice, guitar, tuba, marimba, cuíca, and double bass. In turn, this suggests a case of multiple agency, a concept Edward Klorman has coined to describe musical passages or compositions that embody 'multiple, independent characters – often represented by the individual instruments – who engage in a seemingly spontaneous interaction involving the exchange of roles and/or musical ideas'.[33] Reading the recording of 'My Love is You' as an example of multiple agency underlines the sense of portrayed intimacy between recorded personas (represented by the instruments and voice). The shared recorded space emphasises this impression that the recorded personas are playing together – not just supporting the vocal performance – in the same room. Narrow panning of the different elements further suggests that the performers are placed relatively close together in the space. In that way, the recording evokes a portrayed intimacy between the instrumental personas that are placed closely together in the recorded space.

Moreover, the portrayed intimacy can be characterised as a subjective one if the listener is given the auditory impression of being in the room with the musicians at a live, though intimate, performance. This sense is strengthened by the entrance of the marimba in a separate acoustic space. While the other instruments are rather dry sounding, the marimba has a very pronounced, yet short, reverb on it, which makes it sound like it is playing from a neighbouring room. The strong spatial impression afforded by the marimba's reverb is atypical of commercial popular music and more readily associated with the acoustics of the home and, therefore, a domestic intimacy. In addition to signifying domestic intimacy, the reverb adds to the feeling of being in the room with the musicians as they are playing, which in turn adds a further dimension to the overall portrayed subjective intimacy of the song: Imagine, for instance, being in the living room of a house, surrounded by the rest of the band, and then the marimba chimes in from the kitchen with its solo. This combination of spatialities can be addressed via Denis Smalley's concept of spatial simultaneity,[34] which captures the impact of zoned spaces within the holistic space of the recording. Although spatial simultaneity sometimes points at the surreal,[35] it can also point at the very realistic, and even mundane, as is the case of 'My Love Is You'. Importantly, the spatial contrast introduced by the 'neighbouring' marimba space reinforces the immediacy of the other instruments and the space in which they are placed, precisely by offering a 'horizon' against which the closer space is measured. The spatial simultaneity thus affords a feeling of presence in the recorded space, and in turn, a perceived physical and relational closeness to the performers. In turn, the recording can be understood both as an expression of intimacy (thus emphasising the lyrics), as well as an intimate interaction between music and listener.

Acousmatic intimacy in the age of home studios

As scholars have previously shown, accounting for the type of acoustic space that is emulated in recordings is crucial to understanding how those recordings are interpreted by listeners.[36] In this chapter, I have demonstrated how certain spatial characteristics can add to a sense of intimacy in a recording. I have thereby moved beyond a common focus in popular music analysis on the perceived proximity of a sound when dealing with the concept of intimacy, while still also maintaining that perspective. The ways in which the shaping of the virtual recorded space contributes to acousmatic intimacy, as discussed here, have been facilitated by inventions and innovation in music production technology and recording practice, including the electrification of recording, new microphone and audio processing techniques, and not least new aesthetic approaches emerging with (either responding to or inspiring) technological development. Such innovation happens in relation to an audience that either accepts or reacts against new aesthetic approaches, and acousmatic intimacy is thereby under continuous negotiation. I will end this chapter with a brief speculation about what the role of domestic space in music production might look like going onwards, both in terms of the physical layout of studios as well as the virtual space of recordings.

The history of record production in the 20th and 21st centuries can be seen as a history of innovation in acoustics, audio electronics and aesthetic ideals, but also in communication. With the electrification of recording in the 1920s, which included the introduction of microphones to studios, recordists gained more control of the recorded sound, while their recordings also had less surface noise. The lowered noise floor also meant that the acoustics of recording rooms began to make a bigger difference in recordings, which in turn led to a greater need for controlling acoustics. Arguably, this development culminated in the 'dead' studio rooms of the 1970s that particularly affected the era's rock drum sounds (albeit with several notable exceptions). These non-reverberant rooms did not only lead to non-reverberant recordings, though. They also gave recordists an acoustic carte blanche that allowed increasingly meticulous space shaping with artificial reverberation techniques, and the ideal of the non-reverberant studio persists to a significant extent today. As for communication, roles and ways of working in the studio are under continuous negotiation. From mainly being owned by (or being part of) big record companies, through a time when they largely became independent businesses, studios have for a while now increasingly taken place in people's homes or as small project studios. Development toward more affordable computer technology has contributed to this trend, along with high-capacity file transfer over the Internet. Lastly, while the idea of Internet-based collaboration between studios is not new,[37] several digital audio workstations (DAWs) now incorporate online collaboration features, and some are even fully browser-based. Third-party suppliers like Audiomovers[38] and Splice[39] also offer services to simplify the flow of audio between computers and studios.

While there are still many limitations regarding how efficiently such collaboration tools can be used, it is worth thinking about what influence they may have on the sound of recorded music. For example, what happens when music is produced as collaborations between producers working from different home studios with less-than-ideal acoustic treatment?[40] I think we can assume two possible overlapping paths in terms of acoustic significance. One is a reduced utilisation of acoustic space in recordings, with extensive use of software instruments and direct input recording. This approach has – probably for other reasons than acoustic 'limitations' – taken place in electronic dance music (EDM)-inspired pop music for a while (and has another interesting result in that professional studios are often valued for their 'vibe' rather than acoustics). The second path would be the incorporation (intended or not) of acoustics from the collaborating home studios in the finished recording, potentially resulting in interesting cases of spatial simultaneity. A listener would then be presented with a collage of spaces that could – at least in some cases – be identified as intimate. Moving beyond this hypothesis would require data on how home producers deal with their acoustic situation as well as how they use such collaboration tools, if at all. There is, however, potential in the concept of acousmatic intimacy for explaining some of the musical outcomes of that hypothetical situation, where the producers' home environments are launched into the public sphere as a patchwork of intimate domestic spaces with their own associations, connotations, and relations.

Acknowledgements

This chapter is based on portions of my PhD dissertation, Come Closer: Acousmatic Intimacy in Popular Music Sound,[41] and was written as part of the project 'The Platformization of Music Production: Developer and User Perspectives on Transformations of Production Technology in the Online Environment (PLATFORM)', funded by the Research Council of Norway (FRIPRO) [grant number 324344].

Notes

1 See, e.g., Anthony Giddens, *The transformation of intimacy: Sexuality, love and eroticism* (Oxford: Polity Press, 1992); Kornelia Hahn, "Intimacy," in *Encyclopedia of social theory*, ed. George Ritzer (Thousand Oaks, CA: SAGE, 2005), 417; Christopher Lauer, *Intimacy: A dialectical study* (London: Bloomsbury, 2016); Cristina Miguel, "The transformation of intimacy and privacy through social networking sites" (1st Society of Socio-Informatics International Workshop for Young Researchers: Adoption of Social Networking, Maebashi, 2012); Jeffrey H. Reiman, "Privacy, intimacy, and personhood," *Philosophy & Public Affairs* 6, no. 1 (1976): 26–44; Richard E. Sexton and Virginia Staudt Sexton, "Intimacy," in *Intimacy*, ed. Martin Fischer and George Stricker (Boston, MA: Springer US, 1982), 1–20.
2 See, e.g., Nicola Dibben, "The intimate singing voice: Auditory spatial perception and emotion in pop recordings," in *Electrified voices: Medial, socio-historical and cultural aspects of voice transfer*, ed. Dmitri Zakharine and Nils Meise (Göttingen: V & R Unipress, 2013), 107–122; Peter Doyle, *Echo and reverb: Fabricating space in popular music recording 1900–1960* (Middletown, CT: Wesleyan University Press, 2005); Serge Lacasse, "The phonographic voice: Paralinguistic features and phonographic staging in popular music singing," in *Recorded music: Performance, culture and technology*, ed. Amanda Bayley (New York: Cambridge University Press, 2010), 225–251; Allan F. Moore, *Song means: Analysing and interpreting recorded popular song* (Farnham: Ashgate, 2012).
3 Tor Halmrast, "Attack the attack: Reverberation influences attack and timbre," *Institute of Acoustics* 40, no. 3 (2018): 345.
4 See, e.g., Ragnhild Brøvig-Hanssen and Anne Danielsen, "The naturalised and the surreal: Changes in the perception of popular music sound." *Organised Sound* 18, no. 01 (2013): 71–80; Moore, *Song means*, 21–44; William Moylan, "Considering space in recorded music," in *The art of record production: An introductory reader for a new academic field*, ed. Simon Frith and Simon Zagorski-Thomas (Farnham: Ashgate, 2012), 163–188.
5 Emil Kraugerud, "Come closer: Acousmatic intimacy in popular music sound." PhD dissertation (University of Oslo, 2020).

6 See, e.g., Doyle, *Echo and reverb*, 149; Simon Frith, "Art versus technology: The strange case of popular music," *Media, Culture & Society* 8, no. 3 (1986): 263–264, https://doi.org/10.1177/016344386008003002; Paula Lockheart, "A history of early microphone singing, 1925–1939: American mainstream popular singing at the advent of electronic microphone amplification," *Popular Music and Society* 26, no. 3 (2003): 376, https://doi.org/10.1080/0300776032000117003.

7 Kraugerud, "Come closer," 2.

8 Joshua Meyrowitz, "Television and Interpersonal behavior: Codes of perception and response," in *Inter/media: Interpersonal communication in a media world*, ed. Gary Gumpert and Robert Cathcart (New York: Oxford University Press, 1986), 258–259.

9 Kraugerud, "Come Closer," 38.

10 Waves's Renaissance Reverb plugin (https://www.waves.com/) includes presets labeled "Bedroom," "Bathroom," and even "Bathroom (guest)." Universal Audio (http://www.uaudio.com/) UAD RealVerb-Pro similarly includes a "Warm living room" preset.

11 Gaston Bachelard, *The poetics of space*, trans. Maria Jolas (Boston, MA: Beacon Press, 2014).

12 Martin Heidegger, "Building, dwelling, thinking," in *The domestic space reader*, ed Chiara Briganti and Kathy Mezei (Toronto, ON: University of Toronto Press, 2012), 21–26.

13 Michel de Certeau, Luce Giard, and Pierre Mayol, *The practice of everyday life: Volume 2: Living and cooking*, trans. Timothy J. Tomasik (Minneapolis: University of Minnesota Press, 1998), 145.

14 This sense of intrusion can be related to both the intimate sphere of the home and the intimate proxemic zone, see, e.g., Edward T. Hall, *The hidden dimension: Man's use of space in public and private* (New York: Anchor Books, 1990), 118.

15 David Griesinger, "The psychoacoustics of apparent source width, spaciousness and envelopment in performance spaces," *Acta Acustica united with Acustica* 83, no. 4 (1997): 721.

16 I should mention that precisely what kind of ambience along with its significance is very likely to vary culturally, with different traditions for constructing homes and with the different roles that the home occupies in those cultures. My arguments are mainly based on scholars from Europe and North America as well as my personal experiences from living in Northern Europe, and thus have cultural limitations. I would however think that some of my arguments here can be adapted to other contexts in subsequent studies.

17 See, e.g., Brøvig-Hanssen and Danielsen, "The naturalised and the surreal"; Ragnhild Brøvig-Hanssen and Anne Danielsen, *Digital signatures: The impact of digitization on popular music sound* (Cambridge: MIT Press, 2016), 21–59; Doyle, *Echo and reverb*; Emil Kraugerud, "Meanings of spatial formation in recorded sound," *Journal on the Art of Record Production* no. 11 (2017), https://www.arpjournal.com/asarpwp/meanings-of-spatial-formation-in-recorded-sound/; Serge Lacasse, "'Listen to my voice': The evocative power of voice in recorded rock music and other forms of vocal expression," PhD dissertation (University of Liverpool, 2000).

18 Brøvig-Hanssen and Danielsen, "The naturalised and the surreal"; Denis Smalley, "Space-form and the acousmatic image," *Organised Sound* 12, no. 1 (2007), https://doi.org/10.1017/s1355771807001665.

19 Albin J. Zak, III, *The poetics of rock: Cutting tracks, making records* (Berkeley: University of California Press, 2001), 100–101.

20 Zak, III, *The poetics of rock*, 101–102.

21 Eirik Askerøi, "Reading pop production: Sonic markers and musical identity," PhD dissertation (University of Agder, 2013). For other research that considers the importance of the specific types of spatial environments we hear in recordings, see Brøvig-Hanssen and Danielsen, "The naturalised and the surreal"; Brøvig-Hanssen and Danielsen, *Digital signatures*; Doyle, *Echo and reverb*; Simon Zagorski-Thomas, "The stadium in your bedroom: Functional staging, authenticity and the audience-led aesthetic in record production," *Popular Music* 29, no. 2 (2010): 251–266, https://doi.org/10.1017/s0261143010000061.

22 Doyle, *Echo and reverb*, 56.

23 Michael Chanan, *Musica practica: The social practice of western music from Gregorian chant to postmodernism* (London: Verso, 1994), 49.

24 Halmrast, "Attack the attack."

25 Denis Smalley, "Spectromorphology: Explaining sound-shapes," *Organised Sound* 2, no. 2 (1997): 110; see also Brøvig-Hanssen and Danielsen, "The naturalised and the surreal."

26 Daniel Västfjäll, Pontus Larsson, and Mendel Kleiner, "Emotion and auditory virtual environments: Affect-based judgments of music reproduced with virtual reverberation times," *Cyberpsychology & Behavior* 5, no. 1 (2002): 29.

27 Ana Tajadura-Jiménez et al., "When room size matters: Acoustic influences on emotional responses to sounds." *Emotion* 10, no. 3 (2010): 416, https://doi.org/10.1037/a0018423.
28 Dibben, "The intimate singing voice," 114.
29 Sigrid, "Strangers (acoustic)," in Strangers (Remixes) (Petroleum Records, 2018).
30 David Byrne, "My love is you," in David Byrne (Luaka Bop, 1994).
31 See, e.g., Lauer, *Intimacy*, 84–85; Miguel, "Short the transformation of intimacy," 10; Reiman, "Privacy, intimacy, and personhood," 32.
32 Giddens, *The transformation of intimacy*, 61.
33 Edward Klorman, *Mozart's music of friends: Social interplay in the chamber works* (Cambridge: Cambridge University Press, 2016), 122, emphasis in original.
34 Smalley, "Spectromorphology," 124.
35 Brøvig-Hanssen and Danielsen, "The naturalised and the surreal," 78–79.
36 See, e.g., Brøvig-Hanssen and Danielsen, "The naturalised and the surreal"; Brøvig-Hanssen and Danielsen, *Digital signatures*; Doyle, *Echo and reverb*; Lacasse, "'Listen to my voice'"; Moylan, "Considering space"; Zagorski-Thomas, "The stadium in your bedroom."
37 Paul Théberge, "The network studio: Historical and technological paths to a new ideal in music making," *Social Studies of Science* 34, no. 5 (2004): 759–781, https://doi.org/10.1177/0306312704047173.
38 See https://audiomovers.com/wp/listento/.
39 See https://splice.com/features/studio.
40 Such studios are often set up in an available room in the producer's home like a spare bedroom, or even as part of a main bedroom (as in bedroom studios). Such setups may come with limitations in space and budget to build acoustic treatment of professional studio quality, resulting in undesirable room reflections and other acoustic artefacts as described above.
41 Kraugerud, "Come closer."

References

Askerøi, Eirik. "Reading Pop Production: Sonic Markers and Musical Identity." PhD dissertation, University of Agder, 2013.

Bachelard, Gaston. *The Poetics of Space*. Translated by Maria Jolas. Boston, MA: Beacon Press, 2014. 1958.

Brøvig-Hanssen, Ragnhild, and Anne Danielsen. *Digital Signatures: The Impact of Digitization on Popular Music Sound*. Cambridge: MIT Press, 2016.

Brøvig-Hanssen, Ragnhild, and Anne Danielsen. "The Naturalised and the Surreal: Changes in the Perception of Popular Music Sound." *Organised Sound* 18, no. 01 (2013): 71–80. https://doi.org/10.1017/s1355771812000258.

Chanan, Michael. *Musica Practica: The Social Practice of Western Music from Gregorian Chant to Postmodernism*. London: Verso, 1994.

de Certeau, Michel, Luce Giard, and Pierre Mayol. *The Practice of Everyday Life: Volume 2: Living and Cooking*. Translated by Timothy J. Tomasik. Minneapolis: University of Minnesota Press, 1998. 1994.

Dibben, Nicola. "The Intimate Singing Voice: Auditory Spatial Perception and Emotion in Pop Recordings." In *Electrified Voices: Medial, Socio-Historical and Cultural Aspects of Voice Transfer*, edited by Dmitri Zakharine and Nils Meise, 107–122. Göttingen: V & R Unipress, 2013.

Doyle, Peter. *Echo and Reverb: Fabricating Space in Popular Music Recording 1900–1960*. Middletown, CT: Wesleyan University Press, 2005.

Frith, Simon. "Art Versus Technology: The Strange Case of Popular Music." *Media, Culture & Society* 8, no. 3 (1986): 263–279. https://doi.org/10.1177/016344386008003002.

Giddens, Anthony. *The Transformation of Intimacy: Sexuality, Love and Eroticism*. Oxford: Polity Press, 1992.

Griesinger, David. "The Psychoacoustics of Apparent Source Width, Spaciousness and Envelopment in Performance Spaces." *Acta Acustica United with Acustica* 83, no. 4 (1997): 721–731.

Hahn, Kornelia. "Intimacy." In *Encyclopedia of Social Theory*, edited by George Ritzer, 417. Thousand Oaks, CA: SAGE, 2005.

Hall, Edward T. *The Hidden Dimension: Man's Use of Space in Public and Private*. New York: Anchor Books, 1990. 1969.

Halmrast, Tor. "Attack the Attack: Reverberation Influences Attack and Timbre." *Institute of Acoustics* 40, no. 3 (2018): 335–346.

Heidegger, Martin. "Building, Dwelling, Thinking." Translated by Albert Hofstadter. In *The Domestic Space Reader*, edited by Chiara Briganti and Kathy Mezei, 21–26. Toronto, ON: University of Toronto Press, 2012.

Klorman, Edward. *Mozart's Music of Friends: Social Interplay in the Chamber Works.* Cambridge: Cambridge University Press, 2016.

Kraugerud, Emil. "Come Closer: Acousmatic Intimacy in Popular Music Sound." PhD dissertation, University of Oslo, 2020.

Kraugerud, Emil. "Meanings of Spatial Formation in Recorded Sound." *Journal on the Art of Record Production* no. 11 (2017). https://www.arpjournal.com/asarpwp/meanings-of-spatial-formation-in-recorded-sound/.

Lacasse, Serge. "'Listen to My Voice': The Evocative Power of Voice in Recorded Rock Music and Other Forms of Vocal Expression." PhD dissertation, University of Liverpool, 2000.

Lacasse, Serge. "The Phonographic Voice: Paralinguistic Features and Phonographic Staging in Popular Music Singing." In *Recorded Music: Performance, Culture and Technology*, edited by Amanda Bayley, 225–251. New York: Cambridge University Press, 2010.

Lauer, Christopher. *Intimacy: A Dialectical Study*. London: Bloomsbury, 2016.

Lockheart, Paula. "A History of Early Microphone Singing, 1925–1939: American Mainstream Popular Singing at the Advent of Electronic Microphone Amplification." *Popular Music and Society* 26, no. 3 (2003): 367–385. https://doi.org/10.1080/0300776032000117003.

Meyrowitz, Joshua. "Television and Interpersonal Behavior: Codes of Perception and Response." In *Inter/Media: Interpersonal Communication in a Media World*, edited by Gary Gumpert and Robert Cathcart, 253–272. New York: Oxford University Press, 1986.

Miguel, Cristina. "The Transformation of Intimacy and Privacy through Social Networking Sites." 1st Society of Socio-Informatics International Workshop for Young Researchers: Adoption of Social Networking, Maebashi, 2012.

Moore, Allan F. *Song Means: Analysing and Interpreting Recorded Popular Song*. Farnham: Ashgate, 2012.

Moylan, William. "Considering Space in Recorded Music." In *The Art of Record Production: An Introductory Reader for a New Academic Field*, edited by Simon Frith and Simon Zagorski-Thomas, 163–188. Farnham: Ashgate, 2012.

Reiman, Jeffrey H. "Privacy, Intimacy, and Personhood." *Philosophy & Public Affairs* 6, no. 1 (1976): 26–44.

Sexton, Richard E., and Virginia Staudt Sexton. "Intimacy." In *Intimacy*, edited by Martin Fischer and George Stricker, 1–20. Boston, MA: Springer US, 1982.

Smalley, Denis. "Space-Form and the Acousmatic Image." *Organised Sound* 12, no. 1 (2007): 35–58. https://doi.org/10.1017/s1355771807001665.

Smalley, Denis. "Spectromorphology: Explaining Sound-Shapes." *Organised Sound* 2, no. 2 (1997): 107–126.

Tajadura-Jiménez, Ana, Pontus Larsson, Aleksander Väljamäe, Daniel Västfjäll, and Mendel Kleiner. "When Room Size Matters: Acoustic Influences on Emotional Responses to Sounds." *Emotion* 10, no. 3 (2010): 416–422. https://doi.org/10.1037/a0018423.

Théberge, Paul. "The Network Studio: Historical and Technological Paths to a New Ideal in Music Making." *Social Studies of Science* 34, no. 5 (2004): 759–781. https://doi.org/10.1177/0306312704047173.

Västfjäll, Daniel, Pontus Larsson, and Mendel Kleiner. "Emotion and Auditory Virtual Environments: Affect-Based Judgments of Music Reproduced with Virtual Reverberation Times." *Cyberpsychology & Behavior* 5, no. 1 (2002): 19–32.

Zagorski-Thomas, Simon. "The Stadium in Your Bedroom: Functional Staging, Authenticity and the Audience-Led Aesthetic in Record Production." *Popular Music* 29, no. 2 (2010): 251–266. https://doi.org/10.1017/s0261143010000061.

Zak, Albin J., III. *The Poetics of Rock: Cutting Tracks, Making Records*. Berkeley: University of California Press, 2001.

Discography

Byrne, David. "My Love Is You," by David Byrne, track 8 on *David Byrne*, Luaka Bop, 1994, Tidal.

Sigrid. "Strangers (Acoustic)," by Martin Sjølie and Sigrid Solbakk Raabe, track 3 on *Strangers (Remixes)*, Petroleum Records, 2018, Tidal.

7
EMBRACING SUBTLETY
Reflections on an acoustic surface of glass

Zackery Belanger, Catie Newell and Wes McGee

Part 1: Excitement at the laboratory

The noisy acoustic test signal in Room Zero at historic Riverbank Acoustical Laboratory in Geneva, Illinois, USA comes in bursts. Over and over, it fills the reverberant chamber and decays smoothly into stillness, each pass reducing the uncertainty of the measurement. What is being measured is the absorption of architectural samples—felt, foam, chairs, or even light fixtures designed to absorb sound. In our case, the samples were smooth, shaped-glass surfaces, without the usual pores or fibres that have come to be identified with acoustic products. One by one the glass samples were placed in the centre of the laboratory floor while bursts of noise irradiated them in a diffuse field of sound (Figure 7.1).

We had been at the lab for hours testing a series of these glass surfaces, each a component pulled from one larger assemblage called Long Range that was the result of a few years of conceptualisation, design, and fabrication. It is a long expanse of dark green glass hexagons of varying depths and curvatures. Composed of double and pocketed layers, its shaping was designed to exhibit gradients of acoustic reflection, diffusion, absorption, and transmission. We chose absorption as an important behaviour to test for a few reasons. The architectural acoustics industry places a premium on absorption performance, and not coincidentally the most established and standardised testing methods are for absorption.[1,2] Also, given that these smooth panels look nothing like typical absorbers, that is the one property that would be least expected upon visual inspection. Absorption is the least trivial of the four gradients of behaviour, as reflection is expected from flat glass, diffusion is safely assumed from the prevalence of convex forms, and transmission is apparent and easily experienced by simply walking around the varied openings scattered along the surface of Long Range. Absorption, which is the removal of sound energy by conversion to heat, is more elusive and takes more rigour to detect. Of the four, it is the only one that requires extended exposure times, as reflection, diffusion, and transmission all deal primarily with what happens to sound when it first encounters a surface. Absorption occurs upon a sound wave's first encounter with a surface, but the testing standards do not measure only this brief interaction, opting instead to quantify the continuous removal of energy as the sound field interacts with the sample repeatedly and decays over time. When the cycles of noise ceased and the chamber fell silent for the day, our small group studied a stack of papers full of preliminary coefficients. Our glass surfaces exhibited

Embracing subtlety: reflections on an acoustic surface of glass

Figure 7.1 A portion of Long Range at Riverbank Acoustical Laboratories.

acoustic absorption that was higher than flat glass but relatively low by comparison to the acoustic products that dominate the industry. This was remarkable.

Concerning ourselves deeply with the behaviour of a surface in a laboratory can feel disconnected from the architectural spaces that we design, especially if those surfaces are meant as optional additions to any room, or no room in particular. The idea of application itself is strange, as if the behaviour of surfaces can be simply introduced or extracted away from a space. A property of a surface determined in the laboratory—in this case, acoustic absorption—is generally understood as unchanging and travelling with that surface. We claim that these surfaces have acoustic absorption, with the intention of adding them to spaces to lower reverberation into a predetermined target range. In this manner, acoustic products are understood to be engineered outsiders, often hindsight additions and agnostic to the description of a geometric space. In some ways, this strangeness is representative of a historic and important split that occurred between acoustics and architecture.

In the United States, this split can be traced historically; notably to the lecture hall at the Fogg Art Museum built by Harvard University in the 1890s. The bare, concave surfaces of its ceiling and walls forewarned of modernism, doing little to dissipate sound and leaving the space so reverberant that it was considered unusable. A young Harvard researcher named Wallace Clement Sabine was charged with the task of diagnosing and fixing the room. He succeeded by borrowing absorptive seat cushions from a nearby space and quantifying how much absorption each had to offer, along with the influence it would have on the space.[3] This humble study led to a deeper understanding of material absorption and spaces in general, and ultimately to an equation that now bears his name. Many practitioners still use Sabine's Equation today. It gives more mathematical weight to absorptive surfaces than reflective ones,[4] an approach which may have contributed to our tendency to think of applied, high-absorption surfaces as better and more genuinely acoustic than the less absorptive surfaces that emerge in the design process and typify our architectural spaces. Glass is widely considered a poor acoustic material that needs to be balanced with highly absorptive surfaces of pores or fibres. A goal of the Long Range project was to explore the acoustic potential of this 'poor' material and question how we can design with acoustics in mind using common architectural materials.

In step with Sabin's early developments in acoustics, modern design was stripping away three-dimensional ornamentation, which had always increased the complexity of room boundaries and helped to diffuse sound. Ultimately, the style reduced rooms to rectangular boxes of precise planes and reverberation became an increasingly common problem. Additive absorptive panels rose in desirability, in some ways inheriting, evolving, and magnifying the overlooked acoustically dissipative function of ornament.[5] It may be fitting that the reverberant chambers of the testing laboratories are themselves stark rectangular boxes, usually with just a few added diffusive surfaces. The stripped-down rectangular box has become a kind of troubled baseline.

For each sample tested in the laboratory, the result is a series of coefficients that indicate how absorptive that surface is across the range of frequencies audible to the human ear. The absorption coefficients are usually distilled into ratings of varying regional popularity: Noise Reduction Coefficient (NRC), Sound Absorption Average (SAA), and sound absorption rating (α_w), to name a few. The higher the rating, the more effective the absorber per unit area. This is how virtually all acoustic products are marketed as an entire industry competes for high-coefficient values. Manufacturers are the primary clients of the testing industry, as they have the best reason to endure the associated time and costs of the testing process. Each product needs to have a sample fabricated and shipped to the laboratory, where the tests proceed on different possible mounting conditions.[6] If the data indicate that changes need to be made to the product, then the process restarts. If the data are acceptable, then those numbers are distilled and work their way through marketing.

Most experts in acoustics have access to databases that list the absorption coefficients of common building materials—sealed concrete, metal decking, painted gypsum board, flat glass, etc.—but the ratings seen by most architects are limited to those that are marketed by acoustic product manufacturers. This contributes to the separation of acoustics from architecture, and the separation of acoustic products from the usual materials of our built environment. Manufacturers have made great strides in developing a range of product looks and deployment strategies, but the final installations frequently have a sense of addition and are poorly integrated. Exaggerating this problem is a lack of awareness in the architectural profession of the history and inner workings of these ratings and a dominant higher-is-better philosophy. If Manufacturer A has a product with a rating of 0.70 and Manufacturer B has a product with a 0.80, then Manufacturer B often wins with the 'better' product, even if the product with the lower coefficient would work well and is a better fit with the design of the space. Failed applications that are visually undesirable and fall short in performance are common.

In light of all this, it may be surprising that a small team would be excited to learn that their hard-earned assembly of custom-made kiln-slumped glass panels yielded relatively small absorption coefficients. But we were not aiming for high coefficients. The panels of Long Range were not designed to match the current product market or to be a new option used to correct the acoustics of a space. We were there to test for the possibility of shaping the rooms themselves, to study glass as an enclosure material with its intrinsic acoustic potential intentionally drawn out. The laboratory technicians, who have witnessed thousands of tests on as many different acoustic surfaces, were sure that these smooth glass panels would exhibit no discernable acoustic absorption at all. One also confessed (unofficially, of course) that these were the coolest-looking samples they had ever seen come through the laboratory. To place shaped glass—decidedly non-porous and non-fibrous—into a testing environment dominated by soft materials was a primary goal and motivation for Long Range. What excited us was the presence of absorption, regardless of magnitude. The low absorption coefficients were undeniably there, boldly suggesting a future where the materials that define spaces are intentionally shaped for intrinsic acoustic performance. A century-old facility that has shepherded countless acoustic products had just tested glass surfaces meant to question the very necessity of those products.

Part 2: Breaking the plane

In acoustics, it would not be an exaggeration to call the most subtle convexity profound. Planar surfaces have zero curvature, which means they reflect sound without altering its rate of dissipation.[7] Sound expands outwardly, and whatever sound energy is originally present dissipates as the wavefront grows and covers more and more surface area. When an expanding sound wave encounters a solid planar surface, it changes direction but its expansion and dissipation proceed in an uninterrupted manner. If it encounters a convex surface, however, its rate of dissipation will increase. This is where the shape of architecture begins to exert its influence (Figure 7.2).

Long Range comprises flat panes at one end, which yield to panes of gentle convexity, which themselves yield to panes of more and increasingly aggressive shaping. In the fabrication process, we discovered that our first attempt at gentle convexity appeared as too large a jump from the flat condition. That first attempt was already a minimum deviation given the fixed kiln stages of rising temperature, hold, and lowering temperature. Going deeper into that moment required the introduction of anticipation, a blurring of stages with a precisely timed early cutoff of the heating elements that allowed the temperature to rise and fall off momentum alone. The new panel of even

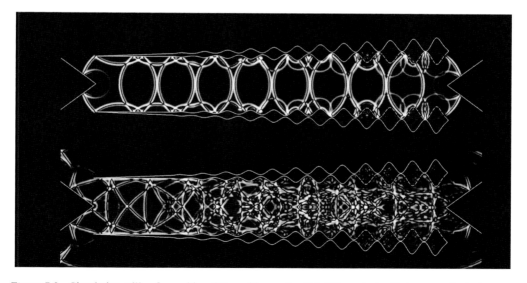

Figure 7.2 Simulation stills of sound in a 2D corridor made of rigid boundaries that progressively increase in curvature and perforated openings. The top still references an earlier time step. From left to right the energy is coherent, then dispersed, then reduced, indicating continuity between acoustic reflection, diffusion, and absorption.

Figure 7.3 The two-layered wall system of Long Range with shifting variations in curvature, openings, and direction of curvature.

gentler convexity lies between the flat and the first attempt. While the jump was softened considerably, it is clear that this halving could be repeated again and again and again. There is an extreme and important difference between flat and the first instance of curvature (Figure 7.3).

Optically, this difference is readily apparent because it takes glass from essentially invisible to visible. The transparency of glass is ideal for this kind of study because even the smallest adjustment in curvature creates a strikingly different reflective quality than that of a flat plane. All sides can be viewed simultaneously, and a single vantage gives a glimpse at the complexity of the form of the whole. Acoustically, though, the transition point between inert and influential warrants further study, as it is complicated by the relatively large wavelengths of sound and by testing methods that measure repeated rather than initial interactions. In any case, convexity is a doorway to acoustic design possibility.

Every component of Long Range begins from the same flat hexagonal glass plate. Slumping pulls the form continuously from primarily reflective to diffusive to absorptive to transmissive. One mechanism for absorption is Helmholtz resonance, which means that sound energy is absorbed when an air cavity with carefully determined openings responds intensely in a narrow band of frequencies.[8] For Long Range, the second layer of glass allowed for these cavities to be produced through the pairing of different slumped components. Auxetic cut patterns gave precise control over the number and size of openings that expose these cavities. There is also variation in the thickness of glass, which reduces as the slump depth increases. These tunable factors—volume of the air cavity, area of openings, and glass thickness—combine to make Helmholtz resonators that were intentionally accessed via the surface shaping of Long Range.

Because all components start from the same glass blank, and the slumping and auxetic unwindings preserve the amount of material in each panel, the weight does not change regardless of the resulting form.[9,10] At the laboratory, the obligatory weighing of each component required by the testing standard became an exercise in redundancy. Conceptually, Long Range explores form as a dominant factor contributing to differences in acoustic behaviour. Every panel is of identical material, perimeter condition, and mass to all other panels, and yet the resulting acoustic behaviour changes along its length as the arrangement exhibits different proportions of the traditional categories of reflection, diffusion, absorption, and transmission.

Deviations from the plane in the other direction[11] break to concavity rather than convexity, and this territory is just as rich. Concave surfaces have a generally poor reputation in acoustics because they form the domes and curved walls that can reverse the natural outward dissipation of sound, causing it to focus and yield strange amplification conditions. But whether and how an undesirable condition occurs depends on the curvatures involved and the location of the source and listener. Even severely concave surfaces manifest as diffusive if the sound has the space to pass through its point of focus and dissipate beyond that. Long Range incorporates some of these concave surface geometries, where the focusing effects can be heard only if the sound source and listener are in a very close and precise location. Beyond that particular spot, the surfaces contribute to the overall diffusivity and average surface depth of Long Range. Additionally, concave components increase the range of cavity sizes for Helmholtz resonance, so extended frequency ranges for absorption are possible.

Together, convex and concave curvatures encompass surface shaping that historically resulted from imprecisions of fabrication and intentional ornamentation.[12] Even today's advanced high-coefficient absorbers can be thought of as an aggressive mix of concavity and convexity if the shaping of pores and fibres are considered down to small scales. Placing typical acoustic reflectors, diffusers, and absorbers next to each other makes an increase in surface shaping complexity across these categories clear.[13] In some ways, the modern acoustic panel can be considered a descendant of architectural ornamentation with maximised complexity and influence per unit area. The increasingly aggressive shaping of Long Range from one end to the other was inspired by this kind of progression, though the range is intentionally limited to a smaller region of relatively mild shaping. Since so much of our built environment is flat planes and acoustic panels, the region near the breaking of the plane is largely unexplored in the context of contemporary design and fabrication possibilities.

Part 3: The sum of the parts

Designing for acoustics while considering visual implications is nothing new, and our team is far from the first to shape hard surfaces for sound, as the upper walls of most concert halls and the rear walls of most recording control rooms will readily attest. We are not aware, however, of any work that so intentionally uses surface shaping to access gradients and blur the boundaries of traditional acoustic behaviours.

Light's precision can be used for perceiving those geometric attributes of surfaces that matter for acoustics. Optical behaviours can be deployed as a reference to the anticipated acoustic performance. The design and fabrication process that produced Long Range seemed to bring the acoustic and visual into an inextricable single state. The relatively mild acoustic laboratory performance and its visual appearance emerged from the surface's geometry. Compared to clear flat glass, which is usually meant to be looked through, Long Range is meant to be looked at. Compared to flat planes which do not dissipate sound, Long Range is meant to dissipate in a controlled fashion. If its form were changed then both look and sound would shift accordingly; there is no altering one without the other. The entanglement of sonic and visual became a tool, where the correlation between form and acoustic behaviour was so clear that looking became an effective way to guide the acoustic design. Experienced acousticians sometimes encounter something like this when they visually inspect a new surface that has not been tested and still determine something about how it behaves acoustically. While these points may seem obvious, form-making with an acute awareness of simultaneous visual and sonic consequences is rarely invoked in architectural design

When we hear reverberation, we're listening to a slow removal of energy as sound waves repeatedly interact with room boundaries until they finally dissipate into silence. The speed of sound in typical interior conditions is about 343 metres per second, which means that sound fills spaces quickly and encounters every surface and object thousands of times in a fraction of a second. There are rare cases where first reflections from a particular surface are highly influential to our sonic experience, such as a spherical dome overhead which focuses a loud reflection back to our ears. The vast majority of rooms in the built environment, however, yield a much messier, diffuse experience. Lingering sound obscures the spoken word and makes communication difficult, so excessive reverberation is undesirable for most spaces.

Reverberation can be thought of as an accumulation effect, and in this sense, low absorption coefficients begin to have greater potential. A room with a polished concrete floor and flat glass walls needs a ceiling of high-absorption coefficient to keep the reverberation of that space within reason. Another way invites a range of contributions from all of a room's surfaces and opens up materials such as glass to the consideration of how their shape might do more. If every square metre does a little, then no square metre needs to do a lot, and a new generation of shaped spaces can emerge. It is in this sense that rooms created with a subtle influence per area could be considered extreme; every square centimetre is considered for its individual shape and how that shape contributes to the shape of the whole. The reverberation time of a space doesn't need to be the result of a mix of planar surfaces and acoustic panels. In Sabine's Equation, reverberation time is determined by the average absorption coefficient of the room as a whole,[14] and not specifically by the particular coefficient of any given surface. It is the result of the cumulative sonic influence of all of the boundaries of the room, including the objects within, and that result is only driven by high-coefficient surfaces when we choose to design that way.

Acoustic products are generally marketed in four categories—reflection, diffusion, absorption, and transmission. If the materials and shaping of rooms that emerge from the design process do not yield the acoustic environment the space needs, then dedicated acoustic panels of particular categories and quantities can be added to make up the difference. Long Range, however, is not a collection of acoustic panels that were each designed for acoustic specificity. It's not an arrangement of reflectors, diffusers, and absorbers, and it's not intended to fix rooms or to be applied in any fashion. It's a single surface that could itself be a room boundary, and from its shaping emerges a blurry, overlapping range of intrinsic acoustic properties. The surface amounts to more than the sum of its parts. By using a fabrication process that systematically draws these behaviours out of a traditional enclosure material, Long Range asks us to imagine not an empty rectangular glass box that warrants acoustic treatment, but a deformation of that box until it no longer needs such appliqué.

Absorption was an important goal because it is the most difficult behaviour to access when you start by breaking the plane. By testing a series of samples extracted from different regions of Long

Range, we also confirmed that the absorption coefficients were rising from one end of the surface to the other.[15] This rising corresponds to an increase in the depth and severity of glass shaping. Still subtle, this indicates that a certain amount of control is possible and that the proportions of energy being reflected, diffused, absorbed, and transmitted could be shifted relative to each other.

An intriguing fact lies hidden in the math that underpinned our laboratory testing experience, one which bolsters our excitement over the subtle values of the absorption coefficients of Long Range. When you dig into the math, you find that Sabine's Equation itself is used to determine those coefficients, and you find that they are not really true coefficients at all. One indication of this is that the testing standard allows coefficient values to exceed 1. True coefficients should never exceed 1, of course, because it would imply that a surface can absorb more than the available energy incident upon it. That would be enough to send most realms of physics into a crisis. In acoustics, though, this situation isn't quite crisis-inducing because these ratings are better understood not as coefficients, but as absorption per area. This means that higher coefficients are not better unless you want to minimise the coverage of your acoustic panels, and lower coefficients are not worse, but only if you embrace shaping every surface of the room.

Acknowledgements

The authors would like to recognise the support of Guardian Glass, the University of Michigan, and Arcgeometer LC in the creation of Long Range. We would also like to thank Elizabeth Teret and Emma-Kate Matthews for reviewing this essay.

Notes

1. *Acoustics—Measurement of Sound Absorption in a Reverberation Room*, BS EN ISO 354:2003 (International Organization for Standardization, May 2003), https://www.iso.org/standard/34545.html.
2. *Standard Test Method for Sound Absorption and Sound Absorption Coefficients by the Reverberation Room Method*, ASTM C423-23 (American Society for Testing and Materials International, May 2023), https://www.astm.org/c0423-23.html.
3. Wallace Clement Sabine, *Collected Papers on Acoustics* (Cambridge: Harvard University Press, 1922), 3–13.
4. Sabine, *Collected Papers on Acoustics*, 65–67
5. Zackery Belanger, *Acoustic Ornament* (Detroit: Arcgeometer, 2021), 26.
6. Mounting conditions can make a substantial difference in performance. For example, mounting a thin layer of felt on an airspace can double or even triple its rating.
7. This is a thought experiment that assumes ample dimensions and perfect rigidity. Dimensions and material cause significant variations in acoustic behavior, but the isolation of form as a driving acoustic factor was important in the design philosophy of *Long Range*.
8. Hermann Helmholtz, *On the Sensations of Tone* (New York: Dover Publications, Inc, 1954), 372–374.
9. Catie Newell, Zackery Belanger, and Wes McGee, "Shaping Glass for Acoustic Performance", *Glass Structures & Engineering* 7 (2022): 253–265, https://doi.org/10.1007/s40940-022-00187-9.
10. Catie Newell, Zackery Belanger, and Wes McGee, "Long Range: Intrinsic Acoustic Performance", in *Fabricate 2024: Creating Resourceful Futures*, ed. Phil Ayres, Mette Ramsgaard Thomsen, Bob Sheil and Marilena Skavara (London: UCL Press, 2024), 72–79.
11. Or simply turning a convex surface over. Whether concave or convex depends entirely on which side of a surface the observer is on.
12. For example, a hand-formed plaster wall and coffered ceiling versus machine-made and laser-installed painted gypsum board.
13. Belanger, *Acoustic Ornament*, 37.
14. Sabine, *Collected Papers*, 65–67.
15. Newell, Belanger, and McGee, "Shaping Glass," 263–264.

Bibliography

Acoustics—Measurement of Sound Absorption in a Reverberation Room, BS EN ISO 354:2003 (International Organization for Standardization, May 2003), https://www.iso.org/standard/34545.html.

Zackery Belanger, *Acoustic Ornament* (Detroit: Arcgeometer, 2021).

Hermann Helmholtz, *On the Sensations of Tone* (New York: Dover Publications, Inc, 1954), 372–374.

Catie Newell, Zackery Belanger, and Wes McGee, "Shaping Glass for Acoustic Performance", *Glass Structures & Engineering* 7 (2022): 253–265, https://doi.org/10.1007/s40940-022-00187-9.

Catie Newell, Zackery Belanger, and Wes McGee, "Long Range: Intrinsic Acoustic Performance", in *Fabricate 2024: Creating Resourceful Futures*, ed. Phil Ayres, Mette Ramsgaard Thomsen, Bob Sheil and Marilena Skavara (London: UCL Press, 2024), 72–79.

Wallace Clement Sabine, *Collected Papers on Acoustics* (Cambridge: Harvard University Press, 1922).

Standard Test Method for Sound Absorption and Sound Absorption Coefficients by the Reverberation Room Method, ASTM C423-23 (American Society for Testing and Materials International, May 2023), https://www.astm.org/c0423-23.html.

PART II

The psychology and physiology

Jane Burry

We are undeniably visual creatures and tend to privilege visual information over all other senses, However, hearing plays a powerful part in constructing or comprehending spatial experience, the localisation of sources of sound, their movement, our movement, our level of enclosure, and the size, geometry, and materiality of spaces. This is based not only on the direction but the comparative measures of direct sound and reflections reaching the ears. Binaural hearing with ears on either side of the head, allows not only phase differences between sound waves from the same source, arriving at the two ears, to be detected, but also the perception of the acoustic shadow of the head in distinguishing what reaches each individual ear. Such information allows a very clear differentiation between sounds coming from the left and right. Direct front and back sources are more subtle to distinguish, but here even the shape of the pinnae of the ear provides local changes to the sound information arriving to distinguish between equidistant, equal sound sources located frontally or behind the head. Instinctually, tiny movements of the head are employed to further articulate the spatial nature of what is heard. Everyone has a slightly individual hearing experience, while such precise localisation and spatialisation of sound emerge as a key survival tool for not only humans but all mammals.[1]

We hear sound in the frequency range from 20 to 20,000 Hz. Sounds in the lowest end of this spectrum for which the wavelength is 17 m, or in other words, much larger than the scale of our bodies, let alone heads, are very difficult to localise. Fortuitously this means when setting up spatial sound reproduction or amplification, the sub-woofer can be located just about anywhere, while the speakers for the higher frequency ranges interact with each other and our hearing in ways that make source direction highly critical to the outcome. 'Feeling' low-frequency sounds which set off vibration in parts of the body other than the ears, has become a familiar sensation in the era of mighty bass speakers installed in passing cars, where previously it brought terror from ballistics and earthquakes.

Just as we use horizontal and vertical visual references to ensure our balance and thus may struggle at sea or in other situations where there are no orthogonal references to the direction of gravity's pull, so with hearing, if all the reflective cues and clues with which we normally construct the space around us are removed, for instance in a fully absorbent anechoic chamber, or an unreasonably noisy echoey space, it is not uncommon to lose balance or experience vestibular challenges or nausea.

Psychologically, our interpretation of sound is influenced deeply by evolutionary and experiential factors. While, as humans, we are not as adept as owls at hunting in total darkness, we can nevertheless pick out an individual voice and follow its line of argument in a room full of loud competing voices in which the complexity of the longitudinal waveform (a pattern of continuously changing pressure in air molecules) and competing neural stimuli belie such a possibility. This reflects the lightning speed of reaction in hearing, the very short-term retention and singular ability to hang onto the information about the precise shape of the wave.[2] Our emotions are almost as intimately engaged with sound, both composed and primordial, as with more primitive senses like smell and taste.

The chapters in this section of the book all approach different aspects of the psychology and physiology of the sound of space and consider it at different spatial scales.

Commencing at the extremity, in Immersive Ambisonic Spatial Audio Design for Extreme Environments, Stuart Favilla considers designing for spatial hearing through technology augmentation, combining ambisonics and binaural synthesis. Here, extremity refers not to the scale of the spaces but to their adversity, for instance, their remoteness, hostility, alienness, danger, and inhospitality to life. Stuart considers how combining two existing technologies in novel ways can recreate spatial hearing and comprehension even where the medium of sound itself, air, is absent. He takes us on a fascinating journey under water, into space, into acoustically isolated frontline military action, and into the urban navigational and wayfinding experiences of people without sight or suffering cognitive impairment. It is a thought-provoking window on spatial hearing for both safety and richness of experience, and the technological opportunity to create it in extremis.

At the scale of Exterior Spaces, David Buck considers Landscape Design and Sound. This essay takes the reader on a panoramic and cinematic scan from the prehistory of music which 'could not be written down' through the progressive development of spatial and temporal music notation between the 9th and 13th centuries. The instrumentality in renaissance architecture of constructed perspective, humanism and harmonies gives ground to the poetic intertwining of rationalism and poetics in Enlightenment work such as that of Piranesi, Boulée, and Ledoux and thus to William Turner's expression of time and transition in painting, a segue to the temporal and indeterminate nature of landscape. Time is omnipresent in landscape, notably through continuous natural change in light and growth, and by experiencing it only by moving through it. A case is developed for a creative notation for landscape that draws upon the history of music.

Moving into the urban realm, Michael Fowler explores and unpacks the world of the fictional character Peter Lucian who alleviates his New York clients' psychological and physiological ailments by 'tuning their houses' according to a highly structured regime based on a musical map of the sounds of the entire city. Drawn from Michael Tyburski's 2019 film 'the Sound of Silence', there are musical and structural depths to be plumbed in this elegant exposé of sound, space, music, and psyche. Fowler draws out and follows the threads of musical Ursatz, or fundamental structure, composition, philosophy of sound, and acoustic ecology in a spatial web all his own.

Nina Garthwaite takes us indoors into the interior and some quite unusual, even serendipitous spaces and unique experiences in Imagining Together: A Cinema for Sound. Realising the potent intimacy of radio, a purely auditory medium, Nina writes about bringing together an audience in shared physical but nevertheless intimate settings, analogous to the experience of cinema for visuals. She charts the diverse and serendipitous adoption of spaces for In The Dark (ItD) a series of such events occurring over 13 years. Even at ItD's inception, radio was still real time, 'Mercurial' and not as universally accessible as now, in this time of burgeoning international podcasts. In this chapter, which explores the links between sound and inner imaginative experience, the focus is on both the imagination and the diverse, acoustically imperfect but distinctive contributions of the performance spaces themselves.

In Aural Diverse Spatial Perception: From Paracusis to Panacusis Loci, John Levack Drever considers aural diversity in what, in this book, we are referring to as Performative and Public Spaces. The specifics of our own physiology, our personal head-related transfer function, which we quickly learn and assimilate, together with our brain architecture, ensure we all hear differently from the perfectly symmetrical averaged head for which concert halls and other acoustics are designed. This is complemented by the highly distinct social and cultural aspects of hearing and sound expectation, or 'sensory anthropology'. While aural-diversity changes with health and age, there are boundaries to divergence where, for instance, careless acoustics impair function, such as aural spatial cues drowned out by the cacophony of a roaring hand dryer in an echoey space for a blind person. While exposing the auraltypical bias built into the standards and tests of hearing, and the relative lack of research in comparison to that into visual spatial perception, this chapter is, overall, a celebration of the extraordinarily diverse, nuanced, and liberating nature of spatial hearing.

Under a similar Performative and Public banner but from a very different point of view, Paul Bavister writes about designing performance spaces with emotions in On Sonic Growth and Form; Biometric evolution of sound and space. He considers both music driving the evolution of spatial design, and music evolved in response to space. He explores this through his own computational evolutionary experiments evolving both music and spaces, driven by the measured emotional responses of individual human participants. Through this work, he has been able to demonstrate that there are opportunities to innovate in both. It is possible to operate outside proscriptive acoustic standards based on our 20th-century acoustic understanding of apparently 'optimal' late 19th-century spaces and musical forms in ways that do not detract from the emotive experience of listeners. Cultural ossification can be overcome without risking the heroic design and compositional failures that have dogged past innovators.

Finally, Jonathan Tyrell considers Induction Loops, taking sound to the listening audience in a very different and enhanced technologically mediated way in Infrastructures of Inaudibility: The Spatial Politics of Assistive Listening. He considers some of the history and politics of initially accommodating, and latterly observing the rights of disabled people, as members of the travelling public, on public transport, specifically in relation to deafness and sonic technology. The induction loop is a singular intervention, signified visually, reliant on highly identifying proximal and oriented listening, and divorced from the spatial acoustics of the place in which it is sited. It highlights in both physiological and political terms the importance of authenticity and the inseparability of sound and space.

Notes

1 Grothe, B., Pecka, M., and McAlpine, D., 2009, Mechanisms of Sound Localization in Mammals Physiological. Reviews 01 JUL 2010. https://doi.org/10.1152/physrev.00026.2009.
2 Geisler, C.D., 1998, *From Sound to Synapse, Physiology of the Mammalian Ear*, Oxford University Press.

Bibliography

Geisler, C.D., 1998, *From Sound to Synapse, Physiology of the Mammalian Ear*, Oxford: Oxford University Press.
Grothe, B., Pecka, M., and McAlpine, D., 2009, Mechanisms of Sound Localization in Mammals Physiological. Reviews 01 JUL 2010. https://doi.org/10.1152/physrev.00026.2009

8
IMMERSIVE AMBISONIC SPATIAL AUDIO DESIGN FOR EXTREME ENVIRONMENTS

Stuart Favilla

Introduction

Growing evidence suggests our human ancestors communed at cave and canyon sites were chosen specifically for their unique spatial acoustics.[1] Archaeoacoustics researchers have demonstrated that many rock-art sites display unusual acoustic properties including resonances, reflections, and echoes.[2] Many of these unusual acoustic rock-art galleries require an uncomfortable climb to access, whereas others are adjacent to available flat rock faces which remain devoid of art. Other rock-art sites are more extreme, located outside, away from shelter and exposed to the elements, where the rock-art practitioners were no doubt vulnerable to predators. Yet our human ancestors were drawn to these sites, where they made their first repetitive markings on the rock face, intoning and chanting with their voices, listening to the otherworldly spatial resonances, echoes, and reflections.

Rock petroglyphs of Valcamonica depict unusual astronaut-like figures wearing what appear to be baggy coveralls, space helmets with radiating antennae and also each holding strange apparatus in their hands.[3] Hovering above is what looks to be a four-lobed landing ship. Local legends concerning the figures have likened them to haloed angels or saints holding crucifixes. Yet archaeologists have dated the petroglyphs to be pre-Christian, in fact, some date back to 8,000 years BCE (Figure 8.1).

The astronauts have been attributed to the Iron Age tribe known as the Camuni. Whether these figures depict ancient extra-terrestrial travellers to our planet earth is unlikely. Ceremonial dress, battle helmets, shields, tridents, and swords remain the current academic interpretations of the mysterious figures. The four-lobed spacecraft is considered to be a religious symbol. Known as the Camunian Rose it also adorns the regional flag of present-day Lombardy, Italy where the Valcamonica region is found.

The connection between rock-art sites and unusual acoustics is fascinating suggesting a vast evolutional role for spatial hearing, one incipient to rock art, language, music, and the formation of tools. Yet what if these figures did in fact depict astronauts from another world, stepping into an extremely unbreathable environment? In a fantastic hypothetical case, perhaps our extra-terrestrial astronaut visitors required pressurised space suits and helmets to survive the extreme atmosphere of the alien earth. An advanced spacefaring alien would perhaps also bring technologies to help

Figure 8.1 'The astronauts', c. 1,000 BC.

them hear in this new extreme atmosphere. If so, the technologies may have included a handheld Ambisonic microphone, a real-time audio processor capable of filtering an alien body or head-related transfer function (HRTF), and an in-suit playback system comprising speakers or other exotic transducers (e.g. headphones).

Ancient aliens and science fiction fantasy aside, extreme environments pose exciting challenges for audio designers working with Ambisonics and binaural synthesis technologies. As a species, we have adapted our hearing to our atmosphere. How then can we hear in an unbreathable atmosphere such as the surface of Mars or underwater where our spatial hearing does not work? Ambisonic and binaural synthesis technologies can also explore visceral and microscopic worlds, or the remote worlds of animals through the sensors of a drone. Extreme environments may present danger. Raging battlefields, military vehicles and aircraft, also expose personnel to extreme noise, impeding situational awareness critical for survival. Our cities, buildings, homes, and familiar environments we take for granted can be considered extreme by those of us living with disabilities or dementia. Here the hustle and bustle of a central railway station or the challenges of living indoors, highlights the precarious nature of life and the benefit spatial hearing technologies can bring.

Hearing space

Paramount to survival in the extreme environments of the past, our ancestors' spatial hearing evolved in and adapted to a complex acoustic world, determining the what and where information of sound in space.[4] Sentinel and vigilant, spatial hearing remains active even while we are asleep.[5] Evading predators, hunting prey, tuning in to our environment, our spatial hearing has supported and directed our evolution as a species. Blauert states;

sound events occur at particular times, in particular places, with particular attributes. The concept of spatial hearing acquires its meaning in this context. Therefore, nearly all auditory events excluding perceptual events such as ringing in the ears and tinnitus are spatial in nature.[6]

Our hearing and comprehension of space develops from birth in correlation with our other senses including vision and touch. Gradually as we grow, we also learn to hear and understand the complex spatial information provided by the environment around us. At distances of further than a few metres, known as extrapersonal space, our spatial hearing supports our vision or becomes a dominant sense when our vision fails.[7] Sound waves beyond a few metres reach our ears as planewaves yet closer in we may perceive spherical wave-fronts. Close up and within arms-reach, lies peripersonal space, where nearfield spatial hearing supports our tactile sense, communication and perception of speech. In this range, we may perceive spherical wave-fronts and infer the distance to the source based on our bodily interactions.

Our hearing can discriminate amplitude, frequency, and spectra of sound waves. It is also very sensitive to timing information discerning asynchronous clicks for example to just a couple of microseconds (10–6 seconds).[8] Our spatial hearing can localise the direction and distance to a sound source laterally and vertically. It can comprehend the context of sound in space providing us with a sense of our surroundings, the size of a room for example or the contours of a landscape based on reflections, scattering, and diffusion. Spatial hearing can segregate streams of multiple sound sources allowing us to focus on one conversation amongst a large crowd of people talking. Spatial hearing is also a dynamic experience where we learn to listen, adjust the position of our head and body and tune with our surroundings.

The what and where of spatial hearing has been studied for more than a century.[9] The 'what' studies have focused on the perception and understanding of speech and other information from the complex auditory scene around us. The where-research has focused on spatial localisation seeking to understand both binaural and monaural localisation. Localisation research has identified a range of theories, auditory cues and perceptual thresholds for spatial hearing. An in-depth discussion describing this research in detail is beyond the scope of this chapter so a summary table of relevance to Ambisonic reproduction via dynamic binaural synthesis is presented in Table 8.1 below.

One way to record what we hear spatially is to place tiny microphones within our own ears, also known as binaural recording. Dummy-head microphones can also be employed, designed to approximate a human head geometrically or anthropometrically (based on averaging sets of human head and ear shapes). Both of these techniques impart HRTFs complex filter functions generated by bodily reflections from the torso, shoulders, head, and complex comb-filtering of the pinna and ear canal. Like fingerprints, HRTFs are individual, relative to a large range of variables including ear shape.

However, recording binaurally and with a dummy head both have their limitations. As each of us has developed our spatial hearing through our own ears, listening to a binaural recording made through another person's ears or through an expensive head and manikin such as the KEMAR (Knowles Electronics Manikin for Acoustic Research) which anthropometrically averages 4,000 U.S. Air Force trainees,[10] the resultant spatial sound recordings can have mixed results. Dummy-head microphones typically remain stationary during a recording and despite reproducing many of the cues listed in Table 8.1, the fixed-head nature of the recordings limits localisation, introducing front and back errors. In-ear microphones too are fixed to the position of the listener's head during the recording.

Table 8.1 Summary of spatial auditory cues[11]

Spatial auditory cue	Description of spatial auditory cue	Range	Signal processing
Interaural Delay Times (ITD)	Human difficulty hearing phase differences in sine waves above 1.5 kHz.	Sound frequencies below 1.5 kHz	Ambisonic encoding and binaural synthesis via *head-related transfer function* (HRTF) convolution filtering
Interaural Level Difference (ILD)	The head effectively blocks sound waves with wavelengths smaller than roughly 0.2 m (1.7 kHz). Soundwaves lower than 1.7 kHz will diffract around the head and enter both ears. Soundwaves higher than 1.7 kHz, localised to either the left or right of the listener, will be attenuated correspondingly by the head.	Sound frequencies above 1.7 kHz	
Echo threshold	The minimum time delay between a source and its reflections where we perceive repetitions or echoes.	1 ms for impulse sounds Above 50–80 ms for speech and music	BRIR Binaural impulse response filtering Using HRTFs captured in reverberant rooms
Localisation dominance and fusion	A direct sound gains localisation dominance over its reflections and appears to the listener as one auditory event or fusion.		
Summing localisation	If the sound and reflections reach the ear in less than 1-ms delay between them fusion occurs. However, the reflections codetermine the localisation of the sound.	<1 ms	
Peripheral auditory processing also known as head-related transfer function (HRTF)	Encompasses reflections from the torso, shoulders, head and spectral filtering of the pinnae to spatially encode audio.	40 Hz–10 kHz	Binaural synthesis
Head movements	Small movements of the head improve both absolute localisation and spatial resolution.	Head movements of 5–10 degrees	Immersive or *Dynamic* binaural synthesis
Binaural decoloration	Short reflections of less than 3ms combine with the direct sound creating comb filter effects that colour sound. The position of spectral peaks will differ between two ears also while the head moves.	<3 ms, above 3.5 kHz	*Dynamic* binaural synthesis
Other environmental cues	Atmospheric and ground effects	Variable	Preserved in Ambisonic soundfield recordings

Binaural recordings can also be achieved through digital signal processing. Utilising binaural recordings of full-spectrum noise, HRTFs can be approximated, and binaural synthesis can utilise these to process a mono audio signal, with no spatial information whatsoever, into a binaural sound with a precise spatial location relative to the listener via stereo headphones.

There are many approaches to binaural synthesis each having subtle differences. Interpolation and mixing of HRTFs, windowing techniques, blurring, edge diffusion, and specialised binaural equalisation are just some of the rendering techniques used by current commercial products. Some methods utilise detailed libraries of thousands of KEMAR manikin HRTFs whereas some manufacturers offer personalised measurements. HRTFs can also be recorded in reverberant spaces reproducing the spatial reverberations of a room. This technique is known as binaural room impulse response (BRIR) filtering.

Ambisonics

Ambisonics enters the stage here as a method able to record a 360-degree soundfield, preserving the direction of planewaves to sound sources, their reflections and much of the environmental effects in the recording. Binaural synthesis then provides the overlay filtering of the peripheral auditory system, interaural level differences (ILDs), interaural time difference (ITDs), HRTFs, etc. which results in a convincing binaural spatial sound image in stereo headphones. Ambisonic soundfield recordings can be processed with relative ease prior to synthesis for example rotations of the entire soundfield relative to the listener's perceived position.

Dynamic binaural synthesis[12] further extends this, mapping sensor systems to track the movements of a listener's head. Head movements can be mapped to rotate the Ambisonic soundfield effectively freezing the spatial sound image for immersive listening. Listeners are then free to move their head and direct their hearing, exploring the immersive audio scene. Dynamic binaural synthesis creates rich immersive interactions where localisation is also improved, allowing small head movements to improve absolute localisation, localisation resolution and perception of moving sound sources.[13]

Michael Gerzon's Ambisonics patent first described a tetrahedral array of four microphones and a process of playback and reproduction via as few as three loudspeakers.[14] One can see the extension of stereo intensity-based, coincident microphone techniques into three dimensions. Since its inception, Ambisonics has developed into a versatile 3D audio branch technology, capable of recording and reproducing a 360-degree soundfield.[15]

Ambisonic technologies include (i) specialised microphone arrays, (ii) software or hardware technologies that encode microphone signal array geometries into spherical harmonics, (iii) spherical harmonic software domain processing techniques, (iv) software and/or hardware playback systems for decoding from spherical harmonics back into multichannel or binaural audio, and finally (v) playback systems including speaker arrays (physical and virtual) including auditoria.

The resurgence of Ambisonic technologies has been largely driven by immersive interactions, virtual and augmented reality, and immersive media underpinned by dynamic binaural synthesis.[16] Despite the emergence of higher-order Ambisonic (HOA) microphones into the recording industry, the application of intensity-based, coincident, tetrahedral, first-order Ambisonic microphones remains popular with many commercial products now available on the market. Ambisonics is available in a wide range of media formats including streaming formats such as Youtube's 360 virtual reality (VR) Audio. Other contenders for 360 immersive audio including the 3DIO quad binaural microphone arrays have fallen from favour as audio designers struggled to counter the changes in colouration and localisation blur that occur when switching or crossfading between sets of ears facing different directions.

Extreme spaces
Unbreathable

We watch a video[17] while listening over headphones, as a diver swims towards an underwater camera, bubbling a jet of air from their regulator valve. We hear bubbles clearly in 3D spatial audio as the diver swims towards us and passes behind. The spatial nature of the sound is strongly evident in the example and we can also hear the reverberations of bubbles in the pool. We hear differences in the spread and colour of the bubbles at the end of the pool compared to close proximity.

One of the most surprising developments of Ambisonic microphones has been their underwater application. The recorded example in the above-linked video was developed in Finland by Delikaris-Manias et al. and consists of a tetrahedral array comprising hydrophones mounted upon a short steel pole.[18]

Despite the uncanny resemblance of the researchers' underwater array to the so-called tridents of the Valcamonica astronauts, the application demonstrates how an underwater intensity-based, coincident Ambisonic microphone signal can be encoded and processed via binaural synthesis utilising HRTF samples recorded in air. The result is a binaural 3D spatial audio recording allowing us to hear directionally underwater. The authors have also included in their solution a detailed method of their spatial encoding taking account of underwater pressure and three-dimensional particle velocity to ascertain the direction of arrival of wave-fronts accurately. These calculations have been used to implement a directional audio coding utilising DirAC.[19] The size of the microphone array was increased to compensate for the speed of sound in water.

One can easily envisage a practical, wearable, head-mounted microphone complete with single board computer for processing and underwater earbud headphones. Divers could wear such a device to determine the direction of incoming sounds or provide situational awareness relative to surf, reefs, shorelines, and water craft which human underwater hearing does not otherwise provide.

The same technologies could also be deployed in space suits to improve hearing in the extremely thin carbon dioxide (CO_2) rich Martian atmosphere? NASA's Perseverance rover provided the first ever sound recordings on Mars[20] and despite the disappointment of the Supercam microphone being mono, recordings of wind, the rover driving, the rover's helicopter and even a laser zapping sound, have provided valuable data relating to how sound may travel in the Martian atmosphere. The speed of sound on Mars is slower (240 m/s) but the composition (96% CO_2) and density of the atmosphere severely limit how far sounds will travel. Higher frequencies are drastically affected and although filters may be utilised to recolour or enhance sound arriving at a microphone, more specialised frequency boosting would be required to pick up sound at distance.

One advantage Ambisonics has is its ability to decode into virtual speaker arrays and then resample again prior to binaural synthesis taking place. Once the spherical harmonics are decoded to virtual speakers arrayed in a hemisphere, cubic or other 3D array, the virtual speakers themselves can be attenuated up or down, thus distorting and warping the soundfield. Prior to decoding to virtual speakers, the Ambisonic signal may be resampled into higher orders. Higher orders can then

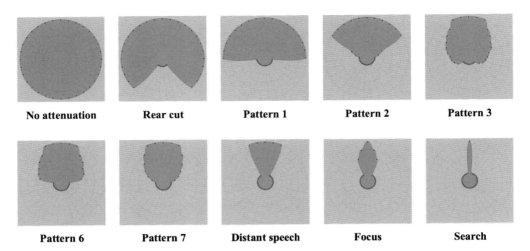

Figure 8.2 2D virtual speaker mappings for first-order microphone soundfield warping and beamforming as viewed from above. The outer circle represents a boost of +96 dBV while the inner circle is zero volume, courtesy of the author.

decode to more virtual speakers which may each be gain adjusted or even turned off allowing for tighter directivity patterns to be achieved (see Figure 8.2 Virtual speaker mappings for soundfield warping and beamforming). This is a very simple description of soundfield processing via virtual speaker arrays.

Soundfield processing is limited by computing power and the demands of a real-time application. Offline processing may render soundfields created through thousands of speakers over hours or even days whereas real-time applications mapping 3D soundfields to 2D planar loudspeaker mappings for horizontal beamforming may be executed within 20–40 ms. As such one might imagine a Martian hearing solution mapping a single Ambisonic microphone signal to dozens of virtual speaker arrays, each dedicated to a specific frequency band. This way higher frequency signals could be boosted facing forward for the astronaut, whereas the sound behind their ears could be utilised for situational awareness and include low middle and bass transmission. Attenuating virtual speaker levels (as much as +−60 dBV) can also be achieved readily in real-time, so a Martian space helmet may include a dial or knob for the astronaut to engage acoustic search beams to discern a sound at a distance. Wide and fan-shaped virtual speaker mappings could be used for close-up work tasks.

Remote and unfamiliar

Sound travels much faster and further in water. Although concert speaker arrays and specialised auditoria can provide audiences with stunning spatialised music, the speed of sound limits their effective size. A totally impractical solution to staging an Ambisonic stadium rock festival would be to stage it underwater. Entertaining this idea will require all the festival-goers to have wearable head-mounted binaural synthesis technology and underwater earphones.

For over a century, we have been listening spatially underwater over large distances. Passive sonar has been employed to hear ships and submarines since the Great War (1918).[21] Instead of dummy-head microphones, hydrophones attached to the bow or hull of the ship provided a similar head-shadowing effect, with early headphone operators learning to detect and localise U-boats.

Oceanic sound can travel extremely far. SOFAR, stands for the 'sound fixing and ranging channel'.[22] Descending deep into the ocean the speed of sound increases with pressure until temperature decreases to a point where the speed of sound slows. The point where the speed of sound is at a maximum is known as the surface sonic layer depth (SLD) and resides at approximately 0.8–1 km depth.[23] The surface of the SLD forms the bottom reflective layer of the acoustic duct, or waveguide for sound to travel, reflecting between the SLD and the surface. Other channels also exist for higher frequencies, but the SLD waveguide will radiate sounds including ships, submarines and even whales for thousands of kilometres.

Although the defence technologies are kept secret, we may assume that Ambisonics and HOA[24] hydrophone arrays are deployed in the depths of our oceans and not just by the Americans. In 2017, the People's Republic of China unveiled two undersea deployment platforms powered by what appears to be hydrogen fuel cells.[25] Facing upwards appears to be some form of Ambisonic hydrophone array. According to Trevithick, the Challenger Deep programme was developed by the state-run Chinese Academy of Sciences and deployed the two deep oceanic acoustic sensors at the southern end of the Marianas' trench and another near the island of Yap. Both sensors boasted a range of roughly 1,000 km allowing them to monitor ship and submarine movements surrounding the major strategic U.S. naval base at Guam.

In wild contrast to the underwater Ambisonic hydrophone detection and localisation of ships is the detection of bats on the wing. Bats emit wideband (including ultrasonic) frequency-modulated signals for echo-location for flying, avoiding obstacles and hunting prey.[26] Lee et al. developed a tiny Ambisonic microphone to localise and track the flights of bats. To capture the high-frequency biosonar sounds (20–80 kHz range) the researchers discovered they had to make a tiny microphone, which performed very well. As the researchers were also interested in the Bat's lower frequency modulating tones the microphone was also tested down to 50 Hz where it also performed very well.

Theoretically, the perfect Ambisonic microphone array would be point-sized. However, molecules of air (or other medium) still have to physically move a tiny microphone's parts and as microphones decrease in size, so too do their performance and signal-to-noise ratios. These factors have so far limited the development of tiny arrays, however size does matter.

Spatial aliasing occurs when the distance between the capsules is larger than half the wavelength of a target frequency. Spatial aliasing introduces ripples, bulges, and distortions in plots of spherical harmonics[27] impacting the performance and introducing directional errors.[28] Spatial aliasing affects most commercial Ambisonic microphones in the frequency range above 8–12 kHz with smaller arrays such as the Sennheiser Ambeo performing better.[29]

Lee et al. also adopted the HARPEX (high angular resolution planewave expansion) signal processing method. HARPEX is an up-mixing process that aims to re-assign time-frequency-domain processing bins.[30] It is a method that has developed from earlier work by Pukkli[31] that aimed to up-mix stereo (L,R) sound images into 4 channel B-format (W, Z, Y, X) Ambisonics. Pukkli's method involved dividing the signals into diffuse and non-diffuse components, then creating a synthesis for the Ambisonic diffusion while non-diffuse components employed pair-wise vector-based amplitude panning.

HARPEX utilises a similar approach[32] whereby frequency components of the soundfield are decomposed into two planewaves which are then reconstructed and presented to virtual loudspeakers. Results demonstrated that first-order Ambisonics decoded through HARPEX utilising pair-wise panning to an octagon speaker array performed well. Due to the required size of the processing window, HARPEX has a processing latency of approximately 240 ms at 44.1 kHz. This can be reduced, however, through processing at higher sampling rates.

Hazardous

Military personnel operate in heavily armoured vehicles, in hostile environments and in extreme conditions for extended durations. Inside typical contemporary armoured vehicles, the crew operate in cramped noisy environments shutoff from the outside world. This poses significant challenges for armoured vehicle crews, especially in dangerous close combat environments, urban or crowded civilian settings.

During operation, vibration passes through to the interior spaces in armoured vehicles exposing crews and passengers to unsafe noise levels.[33] Noise exceeding a personal daily exposure of 85 dB(A) SPL or a sound peak level of 140 dB(A) SPL is known to be extremely harmful to hearing.[34] Consequently, enclosed protective headphone systems are worn and can significantly reduce noise exposure levels.

In addition to the harmful effects on vehicle crews, noise significantly limits the intelligibility of speech and other sounds.[35] Extra attention is required to comprehend speech in noisy conditions adding to the crew's cognitive load, which in military circles is termed combat load. Therefore, noise is both stressful and fatiguing to crews.

Crews of typical armoured fighting vehicles, including the Abrams tank and M113 armoured personnel carrier, are frequently exposed to extreme noise levels from 98 up to 119 dB(A) SPL[36] which may be impacted by the type of road surface and the engine cooling fan noise. Noise levels also peak in intensity when all the crew hatches are closed. Over the long operational history of the M113, little attempt has been made to improve these conditions. Other armoured vehicles within the North Atlantic Treaty Organisation (NATO) include the A3 family (M113), Leopard, Challenger, Warrior and M109 vehicles, all exhibiting unsafe noise levels.[37]

A range of specialised headphones have been designed for military use[38] including conventional enclosed headsets (single hearing protection) earplugs for speech communications only (communication earplugs) powered headsets with noise cancellation systems (active noise reduction or ANR systems) headsets with protective ear-cups enclosing communication earplugs (double hearing protection) and bone conduction headphones which leave the ear canal exposed for tactical surveillance and special operations. Despite the level of protection offered by helmets and ear protectors, extreme noise can still significantly mask hearing entering the mouth, nose cavities or via transcranial or transmitted through the body.

In a high-noise environment, one may assume that external active noise cancellation methods may be applied to reduce the noise of the vehicles themselves. In reality, this is rarely achievable and especially when there are environmental reflections. Either way systems such as this typically require many powerful speakers and therefore are not practical to install on most military vehicles. Software ANR methods such as least mean square filters, can be applied to an Ambisonic audio stream reducing unwanted noise. Currently, these algorithms work best for lower frequencies under 500 Hz but similar results may be achieved by simple filter attenuation.

Ambisonic soundfield warping and beamforming can also be used to remove unwanted noise. Figure 8.3 demonstrates an example of a first-order Ambisonic recording of a human speaker located 10 m away from an extreme engine noise source. The microphone itself is located above the vehicle engine idling at 102 dBA SPL. Utilising a beamforming approach much of the vehicle noise is eliminated, clearly revealing the previously masked speech formants.

Immersive Ambisonic spatial audio design for extreme environments

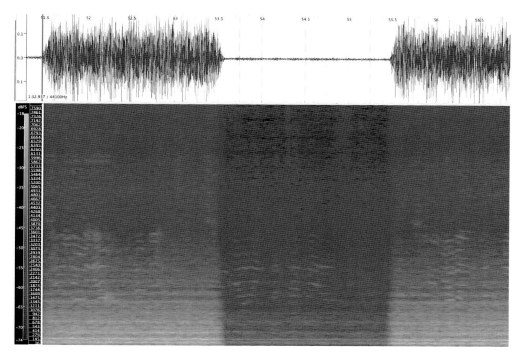

Figure 8.3 Normal speech at a distance of 10 m, Ambisonic beamformer off (full noise) and on, revealing speech and formant patterns.

Precarious

Our cities host many important services, businesses and activities for people living with low vision and blindness. Intrinsic to safe access to these is the support of independent orientation and mobility. Audio cues are already provided at elevators and at pedestrian crossings but many other large indoor areas including busy central train stations and shopping malls can place high cognitive demands and stress on people with vision disabilities.

Research undertaken by the City of Melbourne Council highlights the concentration required while navigating and the importance of avoiding peak hours, large crowds, and crowded trains.[39] These situations can also be very difficult for guide dogs and the masking effects of noise can be so great that the audible sounds from traffic lights can be drowned out. Because of this people with low vision and blindness may deviate hundreds of metres on their journey to cross the road at a safe location. Light and wayfinding guides such as high visibility pathway markers are also used where possible as many legally blind people still have some vision, blurred, greatly restricted in field or peripheral, that they can use during daylight. Many though become profoundly blind at night.

As a result, many people living with low vision and blindness meticulously plan their journeys in advance. To assist familiarisation and on-foot situational awareness, an immersive audio app was developed featuring rich Ambisonic audio recordings made while walking safe pre-planned routes of City locations including the major thoroughfares and platforms of Melbourne's Flinders Street Railway Station. The technologies for the project were developed prior to the arrival of

Apple AirPods[40] and therefore explored a range of dynamic headtracking software solutions before developing its own headtracking device.[41] Blind and low-vision participants acknowledged the immersive quality of the audio and enjoyed being able to direct their hearing to different sounds in the scene. The project experimented with overlaid speech labels in the scene.

Effective navigation, orientation and mobility technologies are still emerging for people living with low vision and blindness. Google's VR Audio SDK incorporates Ambisonic formats for street recording to be utilised through Google Maps but as yet there are still too few 3D audio recordings to accompany the VR map and street views. Augmented reality technologies such as the Microsoft Soundscape[42] app demonstrated great promise for this field. Soundscape's technologies included real-time binaural music guides and audicons for navigation. Despite promising pilots and several positive studies with blind and low-vision groups, Microsoft have ceased development and the app is no longer available. Although it is not clear what method of binaural synthesis is utilised, it is likely the music and audio files were spatialised in real-time via Ambisonics prior to binaural synthesis.

People living with dementia share similar mobility and wayfinding challenges but within the context of their own home. Many people are living with moderate to advanced dementia in care homes which offer access to outside areas, gardening activities and day trips. However, during the recent COVID epidemic, many of these activities have been curtailed and excursions cancelled. More and more centres are reliant on their activities staff to provide entertainment and relief to extreme boredom, the impacts of which can bring depression, agitation and problem behaviours. Noise and certain sounds can also bring on these behaviours too. Despite the popularity amongst residents of films such as the Sound of Music and the parade of André Rieu DVD concerts, the constant exposure to these programmes can also affect staff negatively.

Although immersive audio design for dementia is emerging, so too is the divergent use of headphones in residential care settings. Armchair travel as it is generally known, is a style of single person or group activity that is currently popular in dementia care. It can be directed by activity care staff and include maps, books, videos, music, and singing or scaffolded with technology such as a tablet app where care staff, a family member or volunteer staff member uses Google Earth to visit places of birth and life to stimulate the sharing of experiences, promote reminiscence and positive interactions.[43] Silent disco systems also leverage technologies that allow groups of dementia participants to dance together to their own individual playlists.

The effects of music in moderate to advanced dementia have also been well documented, decreasing the range of problem behaviours such as aggression, agitation, wandering, repetitive vocalisations and irritability.[44] The benefits a particular song or track may bring to one person may have no impact on another. Survey research identified a readily recallable sweet spot or 'reminiscence bump' to be the popular music heard during teenage years, between the ages of 13 and 19.[45] Spatial audio immersion music has yet to make an impact in dementia care but care centre's music collections usually include ambient nature CDs, for example, rainforests, ocean sounds, and birdsong and these have broad appeal.

Headphone training rituals are important for introducing the use of headphones to residents as headphone cords can pose safety hazards. While conducting research at a Melbourne-based care centre, the author observed how a care staff used an mp3 player and headphones to mitigate agitation and anxiety. Residents had their own playlists preloaded onto a mp3 player with corded headphones. The staff member took care to introduce the technology first tapping the resident gently on the shoulder, then resting the headphones there, adjusting, and then tapping the residents' head very gently prior to securing them. The staff member referred to this sequence of actions as 'headphone training'. Once the headphones were secure and the music playing, the effect was miraculous instantly dispelling the anxiety and agitation.

Bluetooth cordless headphones can mitigate risk but also introduce complexities of Bluetooth pairing for staff to manage. Headphones can isolate listeners heightening the experience and increasing the benefits of audio. Bone conduction headphones are lightweight and also allow for situational awareness. Headsets with transducers placed above the mandibular bones, for example, Shockz,[46] have been used by runners and cyclists to listen to music and also hear traffic. The headphones leave the ear completely uncovered and carry sound vibrations through the skull vibrating the bones of the middle ear. Bone conduction headphones pair via Bluetooth but are much more lightweight, although easy to become lost. Spatial audio works very well with Bone conduction headphones, however there is noticeable attenuation of the bass.

Where it is not possible to use headphones or speaker arrays to present immersive audio, ultrasound directional speakers can provide useful and novel solutions. For example, the author has used two Ultrasonic Acouspade[47] directional speakers to present immersive binaural Ambisonic recordings. The Acouspades use music to carry an ultrasound wave at 34 kHz. This produces a silent sonic beam that may extend right across a room or to a chair fixed in position. Binaural immersion is possible once a stereo binaural signal is beamed to each ear respectively. Upon reaching the pinna, reflections spill the carrier music into their original frequencies and the immersive 3D audio illusion is magically revealed to the listener. Ultrasound speakers have very little bass and begin to roll off at 500 Hz. Still, there is enough spectral content to achieve vivid spatial effects. Utilising HARPEX[48] and a Sennheiser Ambeo[49] microphone, immersive wildlife recordings clearly reveal sounds in front and behind, from above and below.

Extreme futures

The versatility of Gerzon's original vision of Ambisonics, the tetrahedral microphone array and the flexibility of encoding to spherical harmonics and decoding to loudspeakers, will continue to offer novelty and application amongst today's spatial sound technologies. Together with the development of dynamic binaural synthesis, up-sampling, soundfield processing and noise reduction techniques, first-order Ambisonics will continue to provide spatial sound designers with useful and affordable tools. Therefore, we may expect to see their continued application in remote, unfamiliar, alien, precarious, hostile, unbreathable, hazardous, dangerous, and life-threatening environments.

But where to next? As microelectromechanical (MEMs) microphones increase in performance and decrease in physical size, there is clear potential for Ambisonic microphones to become much smaller and cheaper. The potential for remote listening via robust streaming devices will play an important role in wildlife and habitat monitoring, including rivers, seas, and oceans. A grid of Ambisonic sensors may provide methods for measuring the accumulation of noise in urban environments, aiding our understanding of its impacts on well-being. The shared agency of human and machine listening needs to be developed together with novel visualisation or multisensory displays. No doubt we will see before long consumer hydrophone arrays for recreational diving affording divers with underwater spatial hearing. As binaural synthesis can translate spatial hearing from air to water, there is clear application for future space industries and hearing technologies for interplanetary atmospheres. The technique may be theoretically applied to other radical environments, liquids, molten metals, brain EEG signals, interstellar gravitational waves perhaps? Where there is an energy distribution that can map to spherical harmonics and events or data that can be sonified, ambisonic spatial hearing may be brought to bear.

In some distant future, the submersible landing ship sinks towards a vast underwater mountain range beneath the oceans of Callisto, one of three of Jupiter's moons discovered to hold an ancient and intelligent life. The aquatic astronaut pressurises in an armoured suit built for extreme pressure, propelled into the icy blackness by silent tiny sensory robotic cilia. Held forward a communication staff, tipped with a tiny tetrahedral array shines a beacon of light as Earth's astronaut emissary searches the silent blackness for signs of life. Firstly, it is heard, its direction clear, resonant chanting and intoning, music and welcome. The astronaut follows and enters a vast resonant cave.

Notes

1. Díaz-Andreu and García Benito, "Acoustics and Levantine Rock Art: Auditory Perceptions in La Valltorta Gorge (Spain)."
2. Scarre, Lawson, and Lubman, "Archaeoacoustics."
3. Giarelli, L. 'Antropomorfi detti astronauti' Zurla area Natural Reserve of Ceto, Cimbergo and Paspardo, Nadro https://en.wikipedia.org/wiki/Rock_Drawings_in_Valcamonica#/media/File:Antropomorfi_detti_astronauti_(a)_-_R_1_-_Area_di_Zurla_-_Nadro_(ph_Luca_Giarelli).jpg
4. Roginska and Geluso, *Immersive Sound*.
5. Blauert, *Spatial Hearing: The Psychophysics of Human Sound Localization*.
6. Blauert, *Spatial Hearing: The Psychophysics of Human Sound Localization* "all auditory events." 2
7. Spiousas, et al., "Sound Spectrum Influences Auditory Distance Perception of Sound Sources Located in a Room Environment."
8. Rasch, "Synchronization in Performed Ensemble Music."
9. Wade, "Early Studies of Binocular and Binaural Directions."
10. Burkhardt and Sachs, "Anthropometric Manikin for Acoustic Research."
11. Kohlrausch et al., "An Introduction to Binaural Processing."
12. Ludovico and Pizzamiglio, "Head in Space: a Head-Tracking Based Binaural Spatialization System."
13. Leung et al., "Head Tracking of Auditory, Visual, and Audio-Visual Targets."
14. Craven and Gerzon, "Coincident Microphone Simulation Covering Three Dimensional Space and Yielding Various Directional Outputs."
15. Frank and Zotter, *Ambisonics*.
16. ICSA Xchange, "Existing Formats: A Survey of Existing Formats for Storage and Streaming of Ambisonic Soundfields.."
17. https://www.youtube.com/watch?v=3WARepl3lEg.
18. Delikaris-Manias et al., "Real-Time Underwater Spatial Audio: a Feasibility Study."
19. Politis, "Parametric Spatial Audio Processing of Spaced Microphone Array Recordings for Multichannel Reproduction."
20. NASA Science, *Sounds of Mars*, https://mars.nasa.gov/mars2020/participate/sounds/
21. Bruton and Coleman, "Listening in the Dark: Audio Surveillance, Communication Technologies, and the Submarine Threat during the First World War."
22. Navy Supplement to the DOD Dictionary of Military and Associated Terms.
23. Helber et al., "Evaluating the Sonic Layer Depth Relative to the Mixed Layer Depth."
24. Higher-order Ambisonics, microphones have increased directivity.
25. Trevithick, "China Reveals It Has Two Underwater Listening Devices Within Range of Guam.".
26. Lee et al., "High-Frequency Soundfield Microphone for the Analysis of Bat Biosonar."
27. Kurz and Frank, "Comparison of First-Order Ambisonics Microphone Arrays."
28. Frank and Zotter, *Ambisonics*.
29. Bates et al., "Comparing Ambisonic Microphones – Part 1"; Bates et al., "Comparing Ambisonic Microphones – Part 2."
30. Berge and Barrett, "High Angular Resolution Planewave Expansion."
31. Pulkki, "Directional Audio Coding in Spatial Sound Reproduction and Stereo Upmixing."
32. Berge and Barrett, "High Angular Resolution Planewave Expansion."
33. Pääkkönen and Lehtomäki, "Protection Efficiency of Hearing Protectors against Military Noise from Handheld Weapons and Vehicles."
34. Department of Defence NOISE LIMITS criteria standard ACGIH 2003, MIL-STD-1474D.

35 Van Wijngarrden and Soo, "Protecting Crew Members against Military Vehicle Noise."
36 Jaklitsch, "A Summary Study of Noise and Vibrations in the M113." 1–20.
37 Van Wijngarrden and Soo, figure 2. Weighted interior noise levels 2004, 7.
38 Van Wijngarrden and Soo, figure 5. Categories of hearing protectors, 11.
39 Riordan and Potter, *Understanding Accessibility for Sensory Disabilities.*
40 Apple Airpods, https://www.apple.com/au/airpods/
41 McCarthy et al., "Towards an Immersive AuditoryBased Journey Planner for the Visually Impaired."
42 Microsoft Soundscape, https://www.microsoft.com/en-us/research/product/soundscape/
43 Upton et al., "Evaluation of the Impact of Touch Screen Technology on People with Dementia and Their Carers Within Care Home Settings."
44 Ihara et al., "Results From a Person-Centered Music Intervention for Individuals Living with Dementia."
45 Rao et al., "A Focus on the Reminiscence Bump to Personalize Music Playlists for Dementia."
46 https://shokz.com/ bone conduction headphones.
47 Acouspade Directional loudspeaker by Ultrasonic, https://www.ultrasonic-audio.com/.
48 Berge, "High Angular Planewave Expansion."
49 Sennheiser Ambeo VR Mic, https://en-au.sennheiser.com/microphone-3d-audio-ambeo-vr-mic.

Bibliography

Acouspade Directional Loudspeaker by Ultrasonic (2011), https://www.ultrasonic-audio.com/
Apple Airpods, https://www.apple.com/au/airpods/
Bates, Enda, Marcin Gorzel, Luke Ferguson, Hugh O'Dwyer, and Francis Boland. "Comparing Ambisonic Microphones – Part 1." In *AES International Conference on Sound Field Control Guildford*, UK (2016).
Bates, Enda, Seán Dooney, Marcin Gorzel, Hugh O'Dwyer, Luke Ferguson, and Francis Boland. "Comparing Ambisonic Microphones – Part 2." In *the Proceedings for the 142nd Audio Engineering Society Convention Berlin*, Germany (2017).
Berge, S, and N Barrett. "High Angular Resolution Planewave Expansion." *Proceedings of the 2nd International Symposium on Ambisonics and Spherical Acoustics*. May 6–7 (2010): 1–6.
Blauert, Jens. *Spatial Hearing: The Psychophysics of Human Sound Localization*. Cambridge: MIT Press (1997).
Bruton, Elizabeth, and Paul Coleman. "Listening in the Dark: Audio Surveillance, Communication Technologies, and the Submarine Threat during the First World War." *History and Technology* 32, no. 3 (July 2, 2016): 245–68.
Burkhardt, J M, and R M Sachs. "Anthropometric Manikin for Acoustic Research." *Journal of the Acoustical Society of America* 58 (1) (1975): 214–222.
Craven, Peter Graham, and Michael Anthony Gerzon. "Coincident Microphone Simulation Covering Three Dimensional Space and Yielding Various Directional Outputs." *United States Patent Application* No. 593, 244 (1977): 1–6.
Delikaris-Manias, Symeon, Leo McCormack, Ilkka Huhtakallio, and Ville Pulkki. "Real-Time Underwater Spatial Audio: A Feasibility Study." Aalto Acoustics Lab, Aalto University, Espoo, Finland, 23–26 May 2018. *Audio Engineering Society, 144th Convention* (2018).
Department of Defence Design Criteria Standard: *NOISE LIMITS* (1997) ACGIH 2003, MIL-STD-1474D.
Díaz-Andreu, Margarita, and Carlos García Benito. "Acoustics and Levantine Rock Art: Auditory Perceptions in La Valltorta Gorge (Spain)." *Journal of Archaeological Science* 39 (12) (2012): 3591–3599.
Frank, Matthias, and Franz Zotter. *Ambisonics: A Practical 3D Audio Theory for Recording, Studio Production, Sound Reinforcement, and Virtual Reality*, Cham, Switzerland, Springer Nature(April, 2020).
Giarelli, L. 'Antropomorfi detti astronauti' Zurla area Natural Reserve of Ceto, Cimbergo and Paspardo, Nadro, https://en.wikipedia.org/wiki/Rock_Drawings_in_Valcamonica#/media/File:Antropomorfi_detti_astronauti_(a)_-_R_1_-_Area_di_Zurla_-_Nadro_(ph_Luca_Giarelli).jpg
Helber, Robert, Charlie N Barron, Michael R Carnes, R A Zingarelli. "Evaluating the Sonic Layer Depth Relative to the Mixed Layer Depth." *Journal of Geophysical Research* 113 (2008): C07033, doi:10.1029/2007JC004595.
ICSA Xchange. "Existing Formats: A Survey of Existing Formats for Storage and Streaming of Ambisonic Soundfields." https://ambisonics.iem.at/xchange/fileformat/existing-formats (2015) retrieved May 2023.

Ihara, Emily S, Catherine J Tompkins, Megumi Inoue, and Sonya Sonneman. "Results from a Person-Centered Music Intervention for Individuals Living with Dementia." *Geriatrics & Gerontology International* 19 (1) (2018): 30–34.

Jaklitsch, Fanz. "A Summary Study of Noise and Vibrations in the M113 Armored Infantry Carrier." *Report for the Economic Engineering Branch Advanced Systems and Concept Research Division Research and Engineering Directorate, U.S. Army Tank-Automotive Center* (1964), 1–20.

Kohlrausch, Armin, Jonas Braasch, Dorothea Kolossa, and Jens Blauert. "An Introduction to Binaural Processing." In Blauert, J. (eds) *The Technology of Binaural Listening*. Modern Acoustics and Signal Processing. Springer, 1-32. Berlin, Heidelberg (2013).

Kurz, Eric, and Matthias Frank. "Comparison of First-Order Ambisonics Microphone Arrays." *In the Proceedings of the ICSA 2015. Third International Conference on Spatial Audio* (18th September 2015), 1–9.

Lee, Hyeon., Michael J Roan, Chen Ming, James A Simmons, Ruihao Wang, and Rolf Müller. "High-Frequency Soundfield Microphone for the Analysis of Bat Biosonar." *The Journal of the Acoustical Society of America* (2019), 1–10.

Leung, Johahn, Vincent Wei, Martin Burgess, and Simon Carlile. "Head Tracking of Auditory, Visual, and Audio-Visual Targets." *Frontiers in Neuroscience* 9 (January, 2016): 1627.

Ludovico, L A, and D Pizzamiglio. "Head in Space: A Head-Tracking Based Binaural Spatialization System." In proceedings of *Sound and Music Conference* 2010 (Barcelona, Spain), 369–376. https://mds.marshall.edu/wdcs_faculty/31/

McCarthy, Chris, Tuan Dung Lai, Stu Favilla and David Sly. "Towards an Immersive AuditoryBased Journey Planner for the Visually Impaired." *OZCHI 2019 ACM International Conference Proceeding Series* (2019) Fremantle WA, Australia, December 2–5, 2019, 387–391.

Microsoft Soundscape: A Map Delivered in 3D Sound, https://www.microsoft.com/en-us/research/product/soundscape/

NASA Science, *Sounds of Mars*, Mars mission Perseverance (2020), https://mars.nasa.gov/mars2020/participate/sounds/

United States Government Us Navy. Navy Tactical Reference Publication NTRP 1-02 Navy Supplement to the DOD Dictionary of Military and Associated Terms June 2012. Independently Published by the Department of the U.S. Navy (June 2012).

Pääkkönen, R, and K Lehtomäki. "Protection Efficiency of Hearing Protectors against Military Noise from Handheld Weapons and Vehicles." *Noise & Health* 7 (26) (2005): 11–20.

Politis, Archontis. "Parametric Spatial Audio Processing of Spaced Microphone Array Recordings for Multichannel Reproduction." *Journal of the Audio Engineering Society* Vol. 195 (2015) Doctoral Thesis Aalto University Publication Series.

Pulkki, Ville. "Directional Audio Coding in Spatial Sound Reproduction and Stereoupmixing." In *28th International Conference of the AES: The Future of Audio Technology–Surround and Beyond*, Pitea, Sweden (2006), 1–8.

Rao, C, J Peatfield, K McAdam, A Nunn and D Georgiva. "A Focus on the Reminiscence Bump to Personalize Music Playlists for Dementia." *Journal of Multidisciplinary Healthcare* 14 (2021): 2195–2204.

Rasch, R A. "Synchronization in Performed Ensemble Music." *Acustica* 43 (1979): 121–131.

Riordan, Ashlee and Katie Potter. *Understanding Accessibility for Sensory Disabilities. Technical Report*. City of Melbourne, THICK Publishing (2015), 3–35.

Roginska, Agnieszka, and Paul Geluso. *Immersive Sound*. New York, Taylor & Francis (2017).

Scarre, Chris, Graeme Lawson, and David Lubman. "Archaeoacoustics." *The Journal of the Acoustical Society of America* 121 (4) (2007): 1819.

Sennheiser Ambeo VR Mic (2016), https://en-au.sennheiser.com/microphone-3d-audio-ambeo-vr-mic

Shockz. "Bone Conduction Headphones Range." https://shokz.com/

Spiousas, Ignacio, Pablo E Etchemendy, Manuel C Eguia, Esteban R Calcagno, Ezequiel Abregú, and Ramiro O Vergara. "Sound Spectrum Influences Auditory Distance Perception of Sound Sources Located in a Room Environment." *Frontiers in Psychology* 8 (June, 2017): 1097.

Trevithick, Joseph. "China Reveals It Has Two Underwater Listening Devices Within Range of Guam." Media Piece Presented by *The War Zone* (June, 2019) 30, https://www.thedrive.com/the-war-zone/17903/china-reveals-it-has-two-underwater-listening-devices-within-range-of-guam

Upton, Dominic, Penney Upton, Tim Jones, Karan Jutlla and Dawn Brooker. *Evaluation of the Impact of Touch Screen Technology on People with Dementia and Their Carers Within Care Home Settings*. UK: Commissioned by Department of Health West Midlands, University of Worcester, Worcester (2011), 5–25.

Van Wijngarrden, Sandor, and James Soo. "Keynote 1 - Protecting Crew Members against Military Vehicle Noise" Meeting Proceedings RDP for the *Habitability of Combat and Transport Vehicles: Noise, Vibration and Motion* – 10th April 2004, Prague (2004), 1–18.

Wade, Nicholas. "Early Studies of Binocular and Binaural Directions." *Vision* 2 (1) (2018): 13.

Wood, Katherine C, and Jennifer K Bizley. "Relative Sound Localisation Abilities in Human Listeners." *Journal of the Acoustical Society of America* 138 (2) (2015): 674–686, https://www.youtube.com/watch?v=3WARepl3lEg

9

IN AN OPEN (MUSIC) FIELD. SPACE AND TIME NOTATION FOR REPRESENTING LANDSCAPE

David Buck and Carla Molinari

The origin of music notation

The use of directionality, of higher on the page to mean higher in the voice, was already built into these accent signs long before they were used for music. And since the other axis, left to right, was already built into the science of writing to indicate passage of time, all that was left was to combine these two axes, vertical and horizontal, to represent sound and time at once.[1]

In art and architecture, as in music, the systems of notation developed throughout history are varied and complex. The attempt to rationalise bodily experiences and find unambiguous, safe and reliable rules capable of regularising those more subjective and ephemeral sensations of our existence is inherent in human nature. Investigating the under-researched territory between music and landscape notation, in this essay we focus on the early medieval period of Western Art music neume notation – in campo aperto. In examining the period of music notation before it became codified into providing temporal and pitch precision, it provides alternative readings of the role that music notation might play in developing new representational methods for landscape space. In doing so it builds upon existing critiques of the limitations of the perspective by Evans, Damisch, and others, creating alternative methods for the inclusion of the essential temporality of landscape material and experience.[2] By providing novel non-orthogonal alternatives to landscape representation, where materiality and time can be measured by space, it also draws attention to un-noticed and under-theorised correlations between musical and landscape space.

Our records of music date back 40,000 years to fragments of early wind instruments in Slovenia showing that 'being sound and shaped time, music begins'.[3] There had of course been music before notation but as the Spanish theologian Isidori of Seville had noted of this earlier history, 'unless sounds are remembered by man, they perish, for they cannot be written down'.[4] When in the early 9th century, the first practical way of recording music through writing was invented, 'it defined neither the pitches of specific notes in a melody, nor the intervallic relations between successive notes'.[5] This research focuses on the earliest period of Western Art music notation from which records survive and in particular on the archives of St. Gallen in Switzerland which holds the oldest extant music manuscript in the world of neume notation: manuscript 359 held in their

Stiftsbibliothek.[6] Dated from 922 to 926 and credited to Hartmann's time as abbot, this 178-page compendium is only 280 by 125 mm in size. This is a notation that guides voices to works already known, that 'tells how to sing, not what the song was',[7] and the central premise of these early music notations was 'the idea that written space represents time was automatically built in'.[8] These St. Gallen neumes are also unique in that amongst the earliest surviving historic records they 'preserve neumes with performance nuances'.[9] These early examples use neumes, thought to derive from the ancient Greek πνεῦμα pneuma ('breath') or νεῦμα neuma ('sign'), and were termed cheironomic or in campo aperto – in an open field. The idea that these early neumes were derived from choirmaster's hand gestures – hence cheironomic – was given currency by Andre Mocquereau[10] in the late 19th century, while Levy has more recently argued that other possible sources including Alexandrian prosodic signs, punctuation signs, or Byzantine ekphonetic notation.[11] Parrish states that it is generally held today that the 'direct origins off the nuemes lay in the accentuation sign of Greek and Roman literature, ascribed to Aristophanes of Byzantium'.[12] Whatever their precise origin and termed non-diastemmatic, these neumes predated the later development of horizontal staff lines that later allowed for the more accurate notation of pitch, and were located on the manuscript without visible staff lines (although Everist has noted that not all staffless neumes were unheightened). In the southern French Aquitanian notation dry point lines scored into the surface of the paper provide 'an axis around which pitches could be oriented',[13] and so could not be correlated to precise pitches. A single neume could however at that time represent a single pitch or a series of pitches all sung on the same syllable. These early neumes could show not just single sounds but those in combination, and whether the following notes rose or fell.[14] Although they did not attempt to specify the pitches of individual notes, the intervals between pitches within a neume, nor the relative starting pitches of different syllables' neumes, these neumes did indicate changes in pitch and duration within each syllable. Later in the 11th century, Beneventan neumes – from the southern Italian churches of Benevento north of Naples – were written at varying distances from the text to indicate the overall shape of the melody and were called 'heightened' or 'diastematic' neumes, which showed indicatively the relative pitches between neumes. The notion of diastemmatic came from the Greek word for 'interval', so neumes were carefully 'heighted', that is, 'placed at various distances from an imaginary line representing a given pitch, according to their relationship to that line'.[15] These diastematic neumes lasted until the end of the 10th century when 'the imaginary line about which diastematic neumes where placed became a real one'.[16] Shortly after this, between one and four staff lines – an innovation traditionally ascribed to the Italian Benedictine monk Guido d'Arezzo[17] – clarified the exact relationship between pitches by marking one line in a different colour as representing a particular pitch, usually C or F. But the paradigm of high and low – the primary element in the notation of pitch and represented in notation by movement up and down the page – has formed a basic principle of music writing in the West ever since.[18] There was, unsurprisingly given Europe's diverse liturgical traditions, a proliferation of different systems in the 10th century, which may have resulted from 'the practice of taking copies from dictation; and perhaps as well an inherent licence of the gestural method which encouraged scribes to personalise the mimetics of chant notation and shape their own gestural forms'.[19] The rise of one new notation system did not immediately remove the others and as Parrish has noted 'some manuscripts were still being written *in campo aperto* at St. Gall in the fourteenth century'.[21]

So neume notation spanned a range of approaches to delineating how sounds and words were related, 'some following a contour which seems to match the pitched notation, others ignoring pitch contour altogether, and still others going positively against the pitch-contour patterns in their use of vertical positioning in the space above the text'.[20] In addition to the 'normal' neumes themselves, there were also neumes specifically in some St. Gall manuscripts that allowed correlation to ways of rhythm c (celeriter or fast), t (tenere or hold) tb (tenere bene or hold well), and for melody l (inferius or go down) and st (statim or immediately) a (altius or rising pitch).[21] It might be presumed that the rising and falling of the neumes might allude to the pitches being sung but actually this is not the case, rather as Susan Rankin has argued,

> the Sankt Gallen scribe's habit of beginning the neumes for each syllable at a position vertically immediately above that syllable has a direct relation to this emphasis on correct delivery, on reading and singing which enable the listener to understand the text, and, in consequence, on a specific association between words and parts of a melodic line.[22]

This is likely to have also been influenced by the simple mechanics of writing. A line of rising symbols would have allowed a scribe to more naturally write a series of neumes without moving one's hands.

Later between the tenth and 13th centuries, the nuemes became modified to more squarish shapes that we would recognise in modern music notation but still without time values. This was primarily the result of being placed on staff lines and the desire for a system of symbols that would allow time values to be independent of the words, so that time could flow continuously. It was during the 13th century that musica mensurate (or cantus mensurabilis) as it was known, included the creation of symbols for rest (silences without composed sounds), so that time could now flow continuously through music and its notation. In his treatise Ars Cantus Mensurabilis from 1,260, Franco of Cologne in referring to time without sounds wrote

> time is the measure of a sound's duration as well of the opposite, the omission of sound, commonly called a rest. I say 'rest is measured by time', because if this were not the case two different melodies – one with rests, one without – could not be proportionally accommodated to one another.[23]

By the 13th century, 'the difference shaped neumes had been reduced to essentially two note values – long and breve, so their relationship to each other could only be understood in context'.[24] Now time could flow continuously through sounds and silence. Preceding the development of the perspective by over two centuries, early music had already developed a notation in which aspects of time and material could be represented.

Representing space

The St. Gallen monastery, founded in 612 as a hermitage, quickly became one of the main ones in Europe, also thanks to the School – Scriptorium – founded by Pepin the Short in the 9th century, where the study of the liberal arts flourished. In an extraordinary coincidence linking our earliest records of music and architectural history, the Italian humanist Poggio Bracciolini discovered Vitruvius's De architectura in 1,414, the only surviving manuscript of ancient architecture and the textual basis for Renaissance architecture (Figure 9.1).

The only document of its kind dating back to the Middle Ages, this plan represents an ideal monastery. The buildings, drawn in red ink, show a complex system of spaces composed in a practical way, with attention to the different uses, made explicit by written notes in brown ink. The drawing perfectly occupies the entire parchment in a delicate and improbable interlocking of spaces, further underlining the utopian nature of this project. Compared to the forma urbis of Rome, probably the most famous ancient plan of all, the plan of St. Gallen is the representation of

Space and time notation for representing landscape

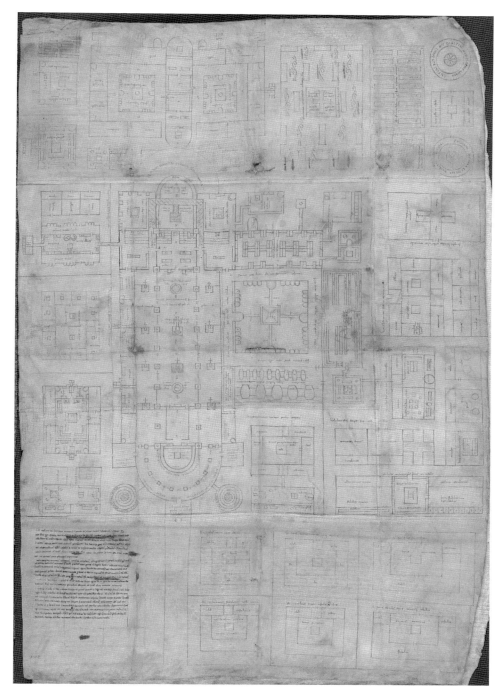

Figure 9.1 Two outstanding products have been created at the Scriptorium: St. Gallen neumatic notation, one of the most advanced forms of open-field notation for Gregorian chant discussed earlier, and the St. Gallen plan.

an ideal project, an exemplary monastery never built, and in this, it reveals a highly rational design, which does not have to reckon with the topographical and morphological reality of the context, but instead with the abstraction of spaces. In this sense, the St. Gallen's plan and the notation are clear and rare examples of a slow evolution in the search for fundamental laws to codify sensory experiences. Although still linked to an imprecise and primitive notation system of space and music, especially considering the final products, the intent instead, the methodological research behind both of these examples, clearly reveals a study aimed at transforming the representation of the empirical data according to rational rules and principles.

The lengthy process of research and transformation of the methods of notation of space culminated in the fifteenth Florentine century when, according to Hubert Damisch and Erwin Panofsky, among others, perspective was born.[25] Although there are several examples of perspective drawings before the 15th century, in this specific period, in Italy, scholars of art and a little later of architecture theorised and defined the rules for representing the third dimension of space. Leon Battista Alberti, in his 'De Pictura' (1435), illustrates for the first time in history the rules to follow for painting volumes in perspective; a few decades later, Sebastiano Serlio was the first to explore the practices for architectural design. The perspective will revolutionise the pictorial method and the architectural composition as a possibility of designing spaces and understanding and decoding reality. In this sense, it is interesting to note that both Panofsky in 'Perspective as Symbolic Form' [*Die Perspektive Als 'Symbolische Form'*] (1927) and Damisch in his 'The Origin of Perspective' (1994), two essential texts for the critical understanding of the perspective, have the same fundamental point of departure. Panofsky underlines the similarities of different systems of spatial notation and identifies their origins in an attempt at verisimilitude of art; Damisch, instead, with his structuralist approach, traces underlying and common patterns despite the different social groups. However, both theorists specify that perspective is an artificial construct; for both scholars, perspective is a human product, an abstract representation system that manipulates vision and reduces it to a series of geometric laws, and as such, it is correct to speak of this system by identifying the specific places and times of its invention. As pointed out by Margaret Iversen, Damisch goes further and, in his positivistic critique of an abstract and artificial system, identifies what appear to be more than admirable constraints.[26] The limits of perspective have been widely discussed recently, especially in its architectural and landscape transposition. Indeed, perspective was born as a manipulative system of reality to be able to represent the three-dimensionality of space in two dimensions and, therefore, inside the borders of a canvas. When the same principles are used in space's physical and tangible composition, any intention of involving the fourth dimension of time is necessarily lost. As underlined by Robin Evans in his illuminating parallelism between the process of translating texts and the transformation of architectural drawing into the project: 'To translate is to convey. It is to move something without altering it […] things can get bent, broken or lost on the way'. In this sense, any system of representation is, and can only be, an approximation.[27]

In regard to landscape representation, Joseph Mallord William Turner perfectly embodies the perspective's dilemma. A magnificent, revolutionary painter who paved the way for art made of light and matter and slowly detached himself from the need for figurative realism, Turner from 1807 and for almost 30 years was Professor of Perspective at the Royal Academy in London. Following Andrea Fredericksen, Turner was exceptionally passionate about the theme of perspective and studied its principles and origins at length. For his lessons, he made 170 tables, including various diagrams and drawings. Apparently, Turner 'also cautioned future architects in his audience against "designs of architecture [that] are but splendid drawings when destitute of practicability by an over-indulgence of fanciful combinations"'.[28] In these words, we find perspective's limits, the complexity, and, if desired, the inconsistency. An artist like Turner, who explores alternative dimensions within his canvases, forgetting any rational rule of construction of space, warns architects of the problems of projects related to drawings that exaggerate in imagining perspective combinations until they become impossible. In the United Kingdom, this is the glory period of the drawings of Giovanni Battista Piranesi, but also Joseph Michael Gandy. It is the moment when the ideal consciously refuses to become real and follows imaginative paths. In Piranesi's prisons, as well as in Gandy and Soane's Bank of England, perspective is instrumental in the sublimation of the object. It is no coincidence that some of the 107 drawings made by Turner for his lessons seem to be inspired precisely by the Roman master, even if they lack any aspiration to represent the spatial complexity and impossibility for which Piranesi is so famous. Therefore, like most people of the time, Turner seemed to bind architecture to a dimension of practicality and feasibility, basing himself on a boundary between art and architecture that was instead crumbling. On the other hand, Piranesi has no interest in drawing reality, his landscapes are mostly invented, and the rules of perspective in his works are often denied. An example for all is the view of the Colosseum from 1757 to 1761 (Figure 9.2).

Landscape has like architecture broadly borrowed from the perspective methods of representing time and space in spite of its acknowledged limitations. The English landscape architect James Corner noted in 1993 that the temporality of landscape was not just from its nature as a living biome, but also from our motion through it, and that 'there is a duration to experience'[29] more profoundly than in many architectural settings. The challenges for representation are also compounded by our allographic relationship with the landscape itself, separating us from the

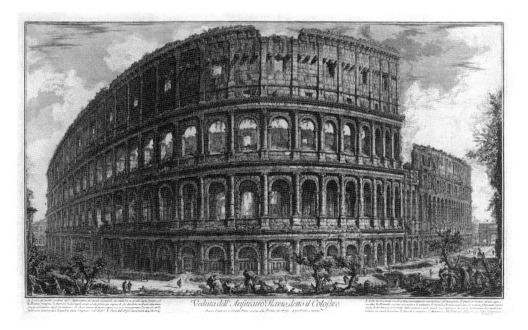

Figure 9.2 A distorted perspective, intentionally wrong, which exaggerates the urban presence of the subject, eliminating any form of symmetry of the oval and trying to escape the finiteness of the volume. Piranesi portrayed a complex landscape, which could not be defined by only one point of view and is subject to multiple times, in a similar way to the multiple scenes painted by the Picturesque artists of the period.

materiality of landscape and 'its abstractness with respect to the actual landscape experience'.[30] In this sense, the limitations of perspective and particularly its inability to render time add further weight to the need for alternative representational methods for landscape that early music notations seem so fruitful to support.

Transcribing an autumn afternoon

To return to further study of a neumatic notational approach to landscape we examined the sounds of a contemporary journey through Rousham Garden in Oxfordshire. Originally designed by William Kent in 1748, this regarded Picturesque landscape is noted for Kent's consideration of the auditory aspects of journeys through it, including a limestone pavilion called the Temple of Echo to amplify sounds of the River Cherwell passing by, and by the use of water in fountains, streams and rills to augment the visual sequences that the topography and sculptures provide.

Our earlier studies[31] identified nine auditory and visual sequences that contained both moments of approach in motion and moments of contemplation at rest. We now focused on two segments of this overall route through the landscape, the approach to the Octagon Pond and cascades, and the journey from it to the Temple of Echo. We were following in the footsteps of John Clary, the head gardener when the project was complete in 1748, who had written and described the spatial sequence to the clients in London, encouraging them to visit more frequently. In doing so our journey traces physically and symbolically the garden as originally designed. We drew upon neumes' method of distinguishing between simple and compound sounds, between rising and falling tones, and the basic rhythmic value of virga and punctum, between whole and half 'note' values. Rather than the secondary symbols from Medieval music that correlated to notions of rhythm, for our research we instead developed new symbols to denote the direction and acoustic horizon of the sounds we encountered, emphasising their spatial character, which was not needed for early music performances, with static singers and audience. These spatial symbols, like those of the early neumes, are not abstract as they may seem, but rather follow an iconic pictorialist approach, seeking to simply represent the parameter graphically (Figure 9.3).

The landscapes 'neumes' were drawn and redrawn in a process that bears witness to the actions of transcription. Containing both drawing and erasure these landscape notations embody their own history on the surface, this time of experimentation conflated with the sounds themselves. This exploratory method of drawing and erasure is similar to the 'historical use by composers of "tabula compositoria", or erasable slates'.[32] These timber boards had one end fashioned into a handle a response to the expense of paper production, and allowed for the writing down of new compositions which could then once complete be copied by scribes into the necessary sections for voices and instruments.

This is similar to composite notational systems within the collection of St. Gallen's Stiftsbibliothek, including Cod. Sang. 530 from 1517. In this notation, the descant is in mensural notation with note heads on a five-line staff, while the three other vocal parts are indicated with alphabet letters and rhythmical symbols below it. This and other composite notations with measured, alphabet, and rhythmic symbols reveal how music notation could provide models for drawing upon different notations from music (and beyond) to make a new notation. In Cod. Sang 359, page 7, we see the neumes rise and float further away from the text as they progress from left to right along each line. If the rising lines that these neumes were attached to were starting points in an aural journey segment, rather than denoting pitch, might we also consider our landscape notations to be more closely correlated to the starting points of segmented journeys? Perhaps what matters in landscape auditory experience is less the actual duration, than the sense of time that our movement through the landscape and the sounds we create and encounter as a spatio-temporal experience.

Space and time notation for representing landscape

Figure 9.3 These auditory symbols were located within the space and time of the visual sequences of Kent's design.

Beyond rhythm

Piranesi, and also the French Neoclassical architects Claude Nicolas Ledoux (1736–1806) and Etienne-Louis Boullée (1728–1799) 'were explicit in their rejection of mathematical reason as the structure of architectural theory. Betraying an authentic existential anguish, they struggled to transform theory into an explicit metaphysics that explained the meaning of architecture through a poetic discourse'.[33] In this dilemma between reason and the poetics of architecture, which nowadays seems to fuel the debate on design criticism, the rules of perspective, therefore, prove to be inadequate. The artificiality of this construct of representation of space is unable to involve the temporal dimension, which, as already established by the German philosopher Gotthold Ephraim Lessing (1729–1781) in his famous Laocoon; or, On the Limits of Painting and Poetry, is the quintessential realm of poetry, but also, as brilliantly reviewed by Rosalind Krauss, it is the extended field where art, landscape, sculpture and architecture meet. In this sense, neumatic notation, as well as some ancient methods of representation of space in relation to music, such as the Aboriginal songlines, can suggest new possibilities for understanding and interpreting the experience of architecture and landscape, finally including other sensory dimensions in addition to the purely visual ones. In the imprecision, and in the dynamic relationship inherent in these methods, we find the richness, complexity, and probably the only alternative to the reconciliation between poetics and reason.

In drawing analogies from theoretical projects from Piranesi to Duchamp and including Eisenstein's works in film montage, we can draw lines that connect our studies from music to landscape and architectural representation experiments that also resist the domination of perspectival rendering of space and time. As Perez-Gomez noted,

> some of the most outstanding works of architecture, such as examples by Gaudi and Le Corbusier, subverted the reductive instrumentality of architectural representation and

also aimed at transcending the enframing vision. These powerful works unveil the true potential of architecture in a postmodern world.[34]

Returning, to conclude, to Piranesi, we identify the strength of representation in his drawings hidden among the missing spaces, among the possible open folds that the viewer finds himself having to imagine. Therefore, it does not seem a coincidence that Russian director Sergei Eisenstein (1898–1948), author of the theory of cinematic montage and a great scholar of compositional processes, dedicates particular attention precisely to Piranesi, dwelling on the 'spatial distortions and interpenetrations, the smashed continuities, and the multiplication of vanishing points – in short, the explosions – [that] create disorientation as well as pathos and ectasis',[35] but also to ancient systems of notation and representation. Indeed, in his attempt to trace an archaic origin of cinema and of the method of composition by montage, Eisenstein analyses Auguste Choisy's perspectives in the same article in which he dwells on the cinematic potential of some Egyptian drawings, compared with Miro paintings and drawings of children. In these examples, the picture is reduced to an essential representation, almost primitive signs that take on value because they are placed in sequence. In reading and understanding the drawing, the human mind connects the traces and reconstructs the movement and experience of space.

This research might work to stimulate further discussion on the landscape notational and compositional potential of early medieval Western Art music, not just for landscape's varied sounds, but for alternative readings of landscape temporality that go beyond the conventional analogies and readings of landscape 'rhythms'. Perhaps our work is analogous with the 20th-century composer American Earle Brown who 'wished his music to be free of the "rigidity of the image" that he saw as a product of the nineteenth century'.[36] More open notation allows for the potential for alternative/multiple readings of landscape's auditory character, and so new compositional possibilities. Perhaps these new landscape notations provide a snapshot of a dynamic space in motion. Earle Brown in his experimentations with alternative readings of notating time, defines 'time notation' as 'durations extended in space relative to time, rather than expressed in metric symbols as in traditional notation'.[37] Our notation is not simply a static representation but is a means of analysing two connected spheres – sounds performed in a religious environment, with sounds experienced in a natural one. This process of transcription allows us to literally draw out potential relationships between the two, bringing a degree of empiricism to what might at first appear as disparate fields. As the English composer Hugo Cole (1917–1995) noted in regard to early music notation that 'it is not part of an inevitable order, but a tool that we ourselves can modify, replace and develop'.[38] The analogy between them is also strengthened by understanding that the early music notations contributed to the growth of a musical culture that had previously relied on memory and recall, the new written culture supporting the existing auditory one. The role of sound in landscape experience has been noted since the early contributors to the 18th-century Picturesque including William Gilpin (1724–1804) and Uvedale Price (1747–1829) included it in their treatise, and these new notations might allow us – like the in campo aperto notation – to function not a means of sound production, but to 'remind the reader of sounds that he (sic) has heard'.[39] Freed from the necessity to be involved in sound production, our notation might be like the precedents examined here, 'therefore free to assume any form'.[40]

In functioning as a system to support auditory recall and analysis, this and other new landscape notations can bring a valued and heightened understanding of the auditory component of landscape experience. In doing so we make two contributions, firstly to a better understanding of landscape temporality, and secondly through addressing the under-researched and under-theorised area of landscape sound. In this sense, it draws parallels with Eisenstein's notation who in seeking to understand how sound and vision might correlate in film, created new compositional methods for what was then in his own specific cinematic realm, still an open field. Notations can therefore be not just about representing time, but a temporal structure for analysis, investigation, proposition and thought.

It is also noteworthy that the early neumes were not writing notes per se, rather they were writing motion, the perception of space and direction between relative notes as they were to be performed. This also implies a method which can help us in the notation of landscape time – to develop a system of relative direction and relative auditory space, creating a means to better understand the relationships between auditory landscape events, a kind of plastic unfolding experience in which the sequence of sounds and spaces, can be first understood and then later designed. Rather than the tonal and temporal precision that later music notation created, which in regularising temporal movement into metronomic events brought a form of precision, the early neumes provide an alternative reading of time, less rigid but closer to our experience of an unfolding experience.

If sounds can be connected to words – as in neume notation – then surely a similar system can be used to connect sounds to visual sequences. The text and sound relationship of early music notation, with plastic correlations but linked by a sequential liturgical narrative and a less metronomic sense of time, might in fact be a cinematic prototype that could be the precedent for Eisenstein's vertical montage which has recently been transcribed into landscape auditory and visual-spatial sequences.[41] Eisenstein saw the means to design the sequential inter-relationships between auditory and visual components in film as a central compositional tool for cinema. Although his notation at first sight appears to show direct correlations between sound and visual sequences, the score for his 1944 film Ivan the Terrible for example included a score by Sergei Prokofiev in which the accents – both visual and auditory – did not match in a metronomic manner, but provided instead a sensual and organic counterpoint. These 'correspondences' as Eisenstein described them, not only guided the technical construction of the film but made it possible for viewers to experience new thoughts and feelings. If Eisenstein's notation allowed for visual and auditory narratives to be analysed and composed, their spatio-temporal relationship to music has also been recognised by others, including the Hong Kong architectural educator Tung-Yiu Stan Lai, who in exploring Eisenstein's 'dialectic montage' noted that 'the sequence of shots as a temporal organization with rhythm is in fact analogical to the rhythm in spatial organization of a path. This further reminds us of the musical nature in space'.[42]

Medieval music notation was developed to support the Roman emperor Charlemagne's musical culture across a rapidly expanding territory, allowing music to spread beyond the geographic and practical limits of its earlier entirely oral culture. By focusing our research on non-diastemmatic neumes when the pitch was relative to adjacent pitches rather than absolute better correlates to landscape's un-tempered sounds. It also resonates with landscape time which is also relative rather than rhythmic. So often discussions of time in landscape focus on the cycles of diurnal or seasonal change which although they provide a meta-structure for organising one aspect of landscape experience, also emphasise a metronomic notion of landscape time: regular and predictable, repetitious and fixed. And yet our experience of moving through landscape spaces could be considered much more closely correlated to the specific spatial, acoustic and temporal details in individual journeys through a landscape. The scale of landscape temporality is like the St. Gallen example discussed earlier, shorter moments that un-fold in interrelated sequences.

By studying music notation from the period when both pitch and time were plastic allows our notation to function as Cole noted as 'both a graphic sketch map and a detailed symbolic representation [...] simultaneously'.[43] Our notations can function as the early notations did, stimulating our auditory recall and imagination, rather than creating prescriptive rules to follow for a defined 'performance'. Free of the need for metric or tonal precision, landscape notation might instead become a tool to focus our attention to qualitative aspects of landscape's auditory experience, allowing us to design – to literally draw and draw out – aspects of sound that inform our experience. Linking music and landscape notation beyond shared temporal qualities is also their common allographic function. As Nelson Goodman notes in Languages of Art, allographic arts require notation as their representational method in order to transcend the limitations of time and where the works are ephemeral. Of all the temporal practices that use notation – including dance, film and architecture – music provides both longest and broadest history for landscape of representing time, space and material in a dynamic array. Notation's function as a visible means of representation between conception and realisation allows for previously un-noticed relationships between musical and landscape time to be explored. Landscape can benefit from the early development of music notation to understand how to develop new notational languages and from this new compositional means.

We might end with the reminder of the English composer Cornelius Cardew (1936–1981), that 'notation and composition determine each other'. These new landscape notations might – as Paul Griffiths reminds us – reveal in spite of their open ambiguity – a field of space and time that connects us to the expanded temporality of music.

Music, being immaterial, touches on the immaterial – on the drift of thought and feeling, on divinity and death. Music, as sound, can represent the auditory world: the moan of wind, the repeated whispers of calm waves, the calls of birds. Music as idealised voice […] can sing or sign, laugh or weep. Music, as rhythm, can keep place with our contemplative rest and our racing activity. Music, in preceding through time, can resemble our lives.[44]

Notes

1 Thomas Forrest Kelly (2015) *Capturing Music the Story of Notation*. New York: W.W. Norton and Company, p.43.
2 Margaret Iversen (2005) "The Discourse of Perspective in the Twentieth Century: Panofsky, Damisch, Lacan". *Oxford Art Journal*, vol. 28, No. 2, pp.191–202.
3 Paul Griffiths (2006) *Concise History of Western Music*. Cambridge: Cambridge University Press, p.1.
4 Isidori of Seville in W.M. Lindsay (ed.) (1911) *Isidori hispalensis episcopi, etymologiarum sive Originum libri*. Oxford: Clarendon Press, lib.III, xv, p.2.
5 Susan Rankin (2011) "On the Treatment of Pitch in Early Music Writing". In *Early Music History*, volume 30. Cambridge University Press, pp.105–175, 105.
6 St. Gallen, Stiftsbibliothek, Cod. Sang. 359: Cantatorium (https://www.e-codices.unifr.ch/en/list/one/csg/0359).
7 Thomas Forrest Kelly (2015) *Capturing Music the Story of Notation*. New York: W.W. Norton and Company, p.12.
8 Thomas Forrest Kelly (2015) *Capturing Music the Story of Notation*. New York: W.W. Norton and Company, p.15.
9 Mark Everist, *The Cambridge Companion to Medieval Music*. Cambridge: Cambridge University Press, p.27.
10 Andre Mocquereau (1889) Paleographie Musicale, series i, 1, p.96fT.
11 Kenneth Levy (1987) "On the Origin of Neumes." In *Early Music History*, volume 7. Cambridge University Press, pp.59–90, 62–64.
12 Carl Parrish (1957) *The Notation of Medieval Music*. London: Faber and Faber, p.4.
13 Mark Everist, *The Cambridge Companion to Medieval Music*. Cambridge: Cambridge University Press, p.21.
14 The main neumes were for single notes *virga* for a rising inflection, *punctum* for a falling, for two note neumes *podatus* for low to high movement, *clivis* for high to low, for three note neumes *scandicus* for three ascending notes, *climacus* for three descending notes, *torculus* for low-high-low, *porrectus* for

high-low-high, and for compound neumes *scandicus flexus* for three rising and one falling, *porrectus flexus* for high-low-high-low, *torculus resupinus* for low-high-low-high, and *podatus subpunctis* for low-high-low-low.

15 Carl Parrish (1957) *The Notation of Medieval Music*. London: Faber and Faber, p.9.
16 Carl Parrish (1957) *The Notation of Medieval Music*. London: Faber and Faber, p.9.
17 Carl Parrish (1957) *The Notation of Medieval Music*. London: Faber and Faber, p.9.
18 Susan Rankin, "On the Treatment of Pitch in Early Music Writing". In *Early Music History*, volume 30. Cambridge University Press, pp.105–175, 109–110.
19 Kenneth Levy (1987) "On the Origin of Neumes". In *Early Music History*, volume 7. Cambridge University Press, p.90.
20 Susan Rankin (2011) "On the Treatment of Pitch in Early Music Writing". In *Early Music History*, volume 30. Cambridge University Press, pp.105–175, 140–141.
21 Carl Parrish (1957) *The Notation of Medieval Music*. London: Faber and Faber, p.11.
22 Susan Rankin (2011) "On the Treatment of Pitch in Early Music Writing". In *Early Music History*, volume 30. Cambridge University Press, pp.105–175, 168.
23 Oliver Strunk (1965) *Source Readings in Music History: Antiquity and the Middle Ages*. New York: W.W. Norton and Company, p.140.
24 Jane Alden (2007) "From Neume to Folio: Mediaeval Influences on Earle Brown's Graphic Notation". *Contemporary Music Review*, Vol. 26, No. 3/4 June /August, pp.315–332, 320.
25 Hubert Damish (1994) *The Origin of Perspective*. Cambridge: MIT Press.
26 Margaret Iversen (2005) "The Discourse of Perspective in the Twentieth Century: Panofsky, Damisch, Lacan". *Oxford Art Journal*, Vol. 28, No. 2, pp.191–202.
27 Robin Evans (1997) *Translations from Drawing to Building and Other Essays*. London: Architectural Association, p.154.
28 Andrea Fredericksen (2004) *Vanishing Point the Perspective Drawings of J.M.W. Turner*. London: Tate.
29 James Corner (2002) "Representation and Landscape". In Simon R. Swaffield (ed.), *Theory in Landscape Architecture: A Reader*. Philadelphia: University of Pennsylvania Press, p.147.
30 James Corner (2002) "Representation and Landscape". In Simon R. Swaffield (ed.), *Theory in Landscape Architecture: A Reader*. Philadelphia: University of Pennsylvania Press, p.146.
31 David Buck and Carla Molinari (2022) "A Picturesque Vertical Montage: Auditory and Visual Sequences at Rousham Garden". *Landscape Research*, vol. 47, No. 5, pp.523–538.
32 D. Schmidt (2000) "Schrift – Graphik-Bild: Zur Notation in Earle Browns *December 1952* und dem Zyklos *Folio*". In W. Budday (ed.), *Musiktheorie: Festschrift Fur Heinrich Deppert zum 65, Geburtstag*. Tutzing: Hans Schneider, pp.183–207, 200.
33 Alberto Pérez-Gómez (1982) "Architecture as Drawing". *JAE*, Vol. 36, No. 2, p.7.
34 Alberto Pérez-Gómez and Louise Pelletier (1992) "Architectural Representation beyond Perspectivism". *Perspecta*, Vol. 27, p.39.
35 Steven Jacobs (2016) "Eisenstein's Piranesi and Cinematic Space". In *Aspects of Piranesi: Essays on History, Criticism and Invention*. University of Ghent: A&Sbooks, p.145.
36 Jane Alden (2007) "From Neume to Folio: Mediaeval Influences on Earle Brown's Graphic Notation". *Contemporary Music Review*, Vol. 26, No. 3/4 June/August, pp.315–332, 317.
37 Earle Brown (1975), "Notation". In *Twenty-Five Pages*. Toronto: Universal Editions (Canada) Ltd.
38 Hugo Cole (1974) *Sounds and Signs: Aspects of Musical Notation*. London: Oxford University Press, p.151.
39 Susan Rankin (2011) "On the Treatment of Pitch in Early Music Writing". In *Early Music History*, volume 30. Cambridge University Press, pp.105–175, 111.
40 Hugo Cole (1974) *Sounds and Signs: Aspects of Musical Notation*. London: Oxford University Press, p.147.
41 David Buck and Carla Molinari (2022) "A Picturesque Vertical Montage: Auditory and Visual Sequences at Rousham Garden". *Landscape Research*, vol. 47, No. 5, pp.523–538.
42 T.S. Lai (2011) "Eisenstein and Moving Street: From Filmic Montage to Architectural Space". *Design Principles and Practices: An International Journal*, vol. 4, no. 6, pp.383–400, 388.
43 Hugo Cole (1974) *Sounds and Signs: Aspects of Musical Notation*. London: Oxford University Press, p.26.
44 Paul Griffiths (2006) *Concise History of Western Music*. Cambridge: Cambridge University Press, pp.3–4.

Bibliography

Alberti, Leon Battista, 1404-1472. De La Peinture = De Pictura (1435). Paris: Macula, Dédale, 21AD.
Alberto Pérez-Gómez (1982) "Architecture as Drawing". *JAE*, Vol. 36, No. 2, 7.
Alberto Pérez-Gómez and Louise Pelletier (1992) "Architectural Representation beyond Perspectivism". *Perspecta*, Vol. 27, 39.
Andre Mocquereau (1889) "Paléographie Musicale". Berne: H. Lang, 1889, 1, p.96fT.
Andrea Fredericksen (2004) *Vanishing Point the Perspective Drawings of J.M.W. Turner*. London: Tate.
Carl Parrish (1957) *The Notation of Medieval Music*. London: Faber and Faber.
Dörte Schmidt (2000) "Schrift – Graphik-Bild: Zur Notation in Earle Browns December 1952 und dem Zyklos Folio". In W. Budday (ed.), *Musiktheorie: Festschrift Fur Heinrich Deppert zum 65, Geburtstag*. Tutzing: Hans Schneider, 183–207.
David Buck and Carla Molinari (2022) "A Picturesque Vertical Montage: Auditory and Visual Sequences at Rousham Garden". *Landscape Research*, Vol. 47, No. 5, 523–538.
Earle Brown (1953) "Twenty-Five Pages for 1 to 25 Pianos". Edition Peters, June 1953.
Erwin Panofsky (1927) *"Die Perspektive Als "Symbolische Form"*. Leipzig: Teubner.
Hubert Damisch. The Origin of Perspective. Translated by John Goodman. London, England: MIT Press, 1994.
Hugo Cole (1974) *Sounds and Signs: Aspects of Musical Notation*. London: Oxford University Press.
Isidori of Seville in William M. Lindsay (ed.) (1911) *Isidori hispalensis episcopi, etymologiarum sive Originum libri*. Oxford: Clarendon Press, lib.III, xv, 2.
James Corner (2002) "Representation and Landscape". In Simon R. Swaffield (ed.), *Theory in Landscape Architecture: A Reader*. Philadelphia: University of Pennsylvania Press, 146, 147.
Jane Alden (2007) "From Neume to Folio: Mediaeval Influences on Earle Brown's Graphic Notation". *Contemporary Music Review*, Vol. 26, No. 3/4 June /August, 315–332.
Kenneth Levy (1987) "On the Origin of Neumes". In *Early Music History*, volume 7. Edited by Iain Fenlon. Cambridge: Cambridge University Press, 59–90.
Margaret Iversen (2005) "The Discourse of Perspective in the Twentieth Century: Panofsky, Damisch, Lacan". *Oxford Art Journal*, Vol. 28, No. 2, 191–202.
Mark Everist, *The Cambridge Companion to Medieval Music*. Cambridge: Cambridge University Press. 2011.
Oliver Strunk (1965) *Source Readings in Music History: Antiquity and the Middle Ages*. New York: W.W. Norton and Company.
Paul Griffiths (2006) *Concise History of Western Music*. Cambridge: Cambridge University Press.
Robin Evans (1997) *Translations from Drawing to Building and Other Essays*. London: Architectural.
St. Gallen, Stiftsbibliothek, Cod. Sang. 359: Cantatorium (https://www.e-codices.unifr.ch/en/list/one/csg/0359).
Steven Jacobs (2016) "Eisenstein's Piranesi and Cinematic Space". In *Aspects of Piranesi: Essays on History, Criticism and Invention*. Edited by Dirk De Meyer, Bart Verschaffel, and Pieter-Jan Cierkens. Ghent: University of Ghent: A&Sbooks, 145.
Susan Rankin (2011) "On the Treatment of Pitch in Early Music Writing". In *Early Music History*, volume 30. Edited by Iain Fenlon. Cambridge: Cambridge University Press, 105–175.
Tung-Yiu.S. Lai (2011) "Eisenstein and Moving Street: From Filmic Montage to Architectural Space". *Design Principles and Practices: An International Journal*, Vol. 4, No. 6, 383–400.
Thomas Forrest Kelly (2015) *Capturing Music the Story of Notation*. New York: W.W. Norton and Company.

10
LEND ME YOUR EARS

Michael Fowler

For Antonella

Ursatz

In an early scene of Michael Tyburski's 2019 film The Sound of Silence we find protagonist and 'house tuner' Peter Lucian (played by Peter Sarsgaard) in Central Park at the Naumburg Bandshell with an array of tuning forks laid out in front of him. We learn that as a house tuner of considerable repute, and whose work has already found the pages of The New Yorker, Lucian alleviates his client's psychological and physiological ailments by tuning their apartment's soundscapes via their numerous appliances. At the Bandshell, Lucian is busy calibrating an acoustic theory of New York City that he has derived from harmonic analyses of field recordings of various neighbourhoods.

The theory is expressed through an auditory map that hangs on the wall of his darkly lit bunker apartment and is a meticulous and vast collection of small, pinned notes of musical staves replete with triads (chords) that associate with particular streets, parks, playgrounds, and buildings in the city. These musical notations are the indices of his auditory recordings and seem to be a function of the character's brilliant auditory imagination and capacity to engage an intense listening focus which enables infinite recourse into the aesthetic qualities and nuances of the smallest sonic detritus.

What Lucian appears to be constructing is a diatonic musicalisation of place and lived experience, the tools of which are various trios of tuning forks – representing first degree (I – tonic), third degree (III – mediant), and fifth degree (V – dominant) – that map the anchoring harmonies of each borough of New York City as major or minor triads in root position. This is telling given that we later learn that he is a trained music theorist, and a traditionalist who seems intent on perpetuating, or at least adapting, the influential methods of the Austrian music theorist Heinrich Schenker (1868–1935) and his concept of a fundamental musical structure or Ursatz. But in Lucian's case, we are taken well beyond the rarefied instances of the Western musical canon, and into the auditory nooks and crannies of domesticity.

Like Lucian's seeming championing of a structuralist approach to identifying the hidden tonal scaffolds of the urban soundscape, Schenker too was concerned in pinpointing how Meisterwerke emerged from a simple process of a fundamental line or Urlinie unfolding over a predictable lower voice Bassbrechung (bass arpeggiation) for which Grundtöne (fundamental notes) anchor a work in a particular Tonraum (tonal space). Guided by the motto Semper idem, sed non eodem modo (always the same, but not in the same way), Schenker's motivations regarding his analyses of a select subset of Western music emerges from a conservative politics[1] in which an objective science is proposed as the identifying mechanism in which to emulate and conserve the secrets of German Meisterwerke.

Schenkerian analysis is a graphic approach to the study of musical scores in which reductive processes are employed that privilege established functional hierarchies. David Beach identifies three basic premises on which Schenker's theory is built: these are that 'melodic motion at deeper levels progresses by step', that 'some tones and intervals such as the dissonant seventh require resolution', and that there is a distinction between 'chord and harmonic step (Stufe)'.[2] A Schenkerian graph, or Urlinietafel, reveals the Urstaz of the score in question in the form of an a-rhythmic cantus firmus that is considered a means in which to illuminate how the forked trio of foreground (Vordergrund), middleground (Mittelgrund), and background (Hintergrund) seemingly construct or perhaps tune a Tonraum.

As a reductionist approach, Schenkerian graphs are used to reveal what Felix Salzer calls 'the organic coherence of a composition taken as a whole', and their ability to 'delineate and explain the function of each progression or each motion of the detail (foreground) in relation to the overall musical structure (middle-ground and background)'.[3] This organicism is posited as natural (that is, from Nature), and built on the notion of repetition as the underlying principle of the Motiv as an emergent property of the world beyond music:

> Man repeats himself in man; tree in tree. In other words, any creature repeats itself in its own kind, and only in its own kind; … Thus a series of tones becomes an individual in the world of music only by repeating itself in its own kind; and, as in nature in general, so music manifests a procreative urge, which initiates this process of repetition.[4]

Here, Schenker draws on the structure of the overtone series as the successive division of a root tone to produce the octave, perfect fifth, and so on (Figure 10.1). It thus provides a utility for which we can 'derive for our tonal system not only the major triad, C:g:e1 (1:3:5), but also the minor triad, e2:g2:b2 (10:12:15)'.[5] Schenker starts here from the fundamental tone of C2 (which he labels C), then divides the fundamental by three to generate G3 (labelled g), then by five to create E4 (e1 or the first E to emerge from the series when starting with C2) – hence the ratios 1:3:5. This method of division also produces a minor triad. A division of the fundamental C2 (C) by ten produces E5 (labelled e2), by twelve produces G5 (labelled g2) and by fifteen produces B5 (labelled as b2) – 10:12:15. Thus Lucian's collection of tuning forks in The Sound of Silence and his search for what he describes as the 'fundamental constants' of the New York City soundscape resonates with Schenker's argument of the importance of the overtones 3 and 5 and the characterisation of the major triad as 'a conceptual abbreviation of Nature'[6] given the ease in which it can be derived from the overtone series.

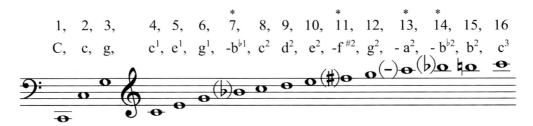

Figure 10.1   Notation of the overtone system after Schenker (Harmony, 6).

But the foundations of Schenkerian theory and its accompanying graphic language are not only an argument for identifying an aesthetic apogee in Western music, but equally a mode for auditory awareness or Fernhören (structural or deep listening). Pieter C. van den Toom calls Fernhören an engagement in 'the mind's ears and eyes more completely in the analytical process, [in order] to ensure that a sense of the materials was not lost to that process'.[7] Like Schenker's search for the Ursatz, Peter Lucian similarly asks if the urban soundscape too is at some level organised according to a hidden aesthetic order that is capable of greatly influencing our moods and behaviours. During fieldwork at the Naumburg Bandshell, Lucian carefully strikes each of his chosen tuning forks until we hear the resultant chiming triad – a resonant G-major. He describes this key to his client and love interest Ellen (Rashida Jones) as 'the sound of nostalgia', or what Christian Schubart in Ideen zu einer Aesthetik der Tonkunst similarly called rustic, idyllic, and lyrical.[8] The significance of Peter's use of tuning forks here lies in their connection not only to a musical context in which a fixed vibrating object provides a control frequency against which one can tune an instrument (or compare a secondary sound to), but similarly within the medical context of otology, and the use of tuning fork tests to assess unilateral conductive hearing loss (in the middle ear), and unilateral sensorineural hearing loss (in the inner ear).

Audition

What then does Peter Lucian listen to in the New York City soundscape, and do his trio of tuning forks act as conductors of some hidden information stream? We first draw the distinction that Jean-Luc Nancy makes between 'listening' and 'hearing'. Nancy speculates on a philosophy that has superimposed upon the act of listening 'something else that might be more on the order of understanding'.[9] This entails that to listen is to seek a subject and 'something (itself) that identifies itself by resonating from self to self',[10] for which we can discover the form, structure and movement of self in terms of an 'infinite referral [renvoi]'.[11] As Adrienne Janus suggests:

> Nancy's listening subject, if it is a subject at all, tends to dissolve, to fuse with and to absorb all those elements of self and world that might otherwise be termed "objects," in other words, all objects, insofar as they resonate, tend to become listening subjects.[12]

This further leads to Nancy's notion of the sonorous present and how listening must be considered not as a metaphor for access to the self, but as the reality of this access. This presents a conceptualisation of hearing that 'is not embedded in real spaces but in an otherworldly conception of space once hearing takes place'.[13] For Casey O'Callaghan, Nancy's renvoi and the search for a subject must be in the presence of sounds that are the immediate (that is non-mediated) objects of auditory perception. Here, O'Callaghan qualifies sonic objects as being requisitely 'public' (that is, intersubjectively perceived) for them to be qualified as heard. In contrast to both the approaches of Locke[14] and Robert Pasnau,[15] O'Callaghan argues against considering sounds as secondary qualities or sensible attributes of objects, but rather particular individuals that possess audible qualities such as pitch, timbre, and loudness.[16] This means that an auditory hallucination heard only by an individual is an auditory experience, though lacks a shared auditory perception, thus rendering it an unheard event. In The Sound of Silence, we encounter Tyburski subtly playing with this notion through the characterisation of protagonist Peter Lucian. While visiting the New York Public Library we find Lucian busy studying an open piano score. As Lucian reads the notation, Tyburski's sound design integrates the aria Terribile d'Aspetto from Mozart's 1771 opera La Betulia liberata, in doing so positioning what we assume to be Lucian's inner auditory hallucination as a deeply embedded auditory experience of infinite referral in which we seem to be immersed in Lucian listening to himself listening. Here we can turn again to Nancy:

> it is not a hearer [auditeur], then, who listens, and it matters little whether or not he is musical. Listening is musical when it is music that listens to itself. It returns to itself, it reminds itself of itself, and it feels itself as resonance itself.[17]

Certainly, O'Callaghan argues for 'a distinction between genuinely hearing or perceiving a sound and enjoying an auditory experience' because to be 'auditorily aware of anything requires being aware of sound'.[18] In Lucian's case though, and for many musicians, auditory recreations are what Carl Seashore recognises as distinct musical imagery and a special case of auditory awareness:

> [The musician] creates music by "hearing it out," not by picking it out on the piano or by mere seeing of the score of any abstract theories, but by hearing it out in his creative imagination through his "mind's ear." That is, his memory and imagination are rich and strong in power of concrete, faithful, and vivid tonal imagery.[19]

Timothy Hubbard and Keiko Stoeckig also argue for identifying the experiential quality of a musical image as its essence – that is, 'the creation of an image seems to be a re-creation of the sensory experience, the qualia'.[20] The work of composer David Dunn seemingly emerges from this discourse through his instrumentalisation of a performance practice that is a real-time searching out and cataloguing of sound objects and auditory images.

In his composition Purposeful Listening in Complex States of Time (1998), Dunn asks a solo interpreter to perform 20 acts of listening according to a detailed score in 20 different outdoor environments of low-level ambient sound. The modes of listening are diverse and include focusing on sounds emanating from particular spatial locations (sky level, body level, ground level) as well as sounds that are occurring in real time, those that are remembered, and those that are to be imagined as occurring in the future. We find parallels here to Peter Lucian's activities in The Sound of Silence. That is, the quest of the listening subject to reveal the embedded ratios of 5:6 (minor third), 4:5 (major third), and 3:2 (perfect fifth) which consolidate Schenker's observation that 'every tone is possessed of the same inherent urge to procreate infinite generations of overtones', which is 'in no way inferior to the procreative urge of a living being'.[21] The trios of tuning forks are then the conductors of what Nancy calls the 'permission, the elaboration, and the intensification of the keenest disposition of the "auditory sense"'[22] that musical listening affords. As Dunn notes, music relies not only on the perception of sound in time, but the perceiver who is engaged in both organising that perception and assigning it meaning. Beyond this is the realisation that this capacity takes place regardless of the intention of a composer or the specific nature of sounds occurring in an environment. It is the nature of perception that is the fundamental ground from which all music arises and not its materials, structures, or communicative intent.[23]

For Kenneth LaFave, music is thus something like 'a sonic experience that manages to be both a phenomenon and knowable 'thing-in-itself' at one and the same time'.[24] That is to say that although vibrations in the air are registered by our ears as objects of auditory perception, this does not mean that they immediately lend themselves to possessing existential status as such. This can equally be said of music in which 'just because certain sounds are registered by ears as music, over and above sound, does not lend special existential status to music as such'.[25] They require, as Dunn suggests, a perceiver who organises perception and assigns meaning.

Lend me your ears

Such auditory tactics are inevitably linked to John Cage's (1912–1992) problematisation of the traditional relationship between performer, audience member, and composer that was first explored in 4′33″ composed in 1952. The premiere is well documented,[26] and consisted of pianist David Tudor performing a tacit over three movements which were indicated through opening and closing the lid of the piano. Though the work ran for 4 minutes and 33 seconds and consisted entirely of ambient sounds from within and outside Maverick Concert Hall (Woodstock, NY), Cage later indicated in the 1967 Second Tacet Edition that the work could be for any number of players and any length of time. Rather than relying on active gestures of the performer that are directed toward musical instruments, machines, or other sound-producing objects, a passive gesture simply allows a time frame to be observed, and the resulting observed sounds to constitute the work. Cage's shift toward the demarcation of a rarefied musical space via an auditory awareness (or perhaps Fernhören) is certainly what composer Peter Ablinger (b. 1959) draws on in his Hörstücke (Listening Pieces). For Ablinger, the focus is similarly 'not so much [on] that which is heard (the object), but the hearing itself (the subject)'.[27]

In a reprise of Nancy's notion of listening as a search for the subject through renvoi, Ablinger's Listening Piece in Four Parts was staged in 2001 at four locations in California (Dockweiler State Beach, Baldwin Hills, Downtown LA, and Palm Springs Trail Station). The work consisted of Ablinger setting up 20 foldable chairs in a quasi-concert formation for around two hours at each location. Participants mostly included the artist and his wife, with Ablinger recalling that 'the chairs were removed after about 2 hours at each place. But the 4 places remain – now as a piece of music – for all who are aware of this fact' (Figure 10.2).[28]

Figure 10.2 Part three of Peter Ablinger's Listening Piece in four parts (2001). Parking lot at 4th Street/Merrick Street, Los Angeles. Copyright Peter Ablinger, 2001. Reproduced with permission.

Unlike Peter Lucian's urban soundscape Ursatz that is built on identifying relationships between Stufen (scale degrees) in a hierarchy as a natural function of the overtone series, and thus replete with a well-defined musical semiotics, Cage, Dunn, and Ablinger actively seek to reject the ideals of modernism. This allows them to embrace anarchic structures so that a transience of meaning becomes a feature rather than a flaw. As Cage argues, as a composer he must

> free his music of a single overwhelming climax. Seeking an interpenetration and non-obstruction of sounds, he renounces harmony and its effect of fusing sounds in a fixed relationship. Giving up the notion of Hauptstimme [primary voice], his "counterpoints" are superimpositions, events that are related to one another only because they take place at the same time.[29]

But directing us away from a Hauptstimme seems to deliver us back to the Naumburg Bandshell with New York City as Hörstuck (audio piece). We are audience then to the urban soundscape as a rich musical tapestry of sources emanating from people, transportation, nature, and technology – Cage's '"counterpoints" as superimpositions'. Here, the juxtaposition of various systems within the urban landscape has a resultant sonic component whose order emerges from the stratification of acoustic niches. These niches can be thought of as auditory streams: that is, perceptual units of an auditory event as a single, multi-facetted happening, for which a stream 'serves the purpose of clustering related qualities, … and acts as a center for our description of an acoustic event'.[30]

It is intriguing then that in The Sound of Silence, the intricate network of auditory streams in NYC is associated with absolute moods or feelings via an obfuscated yet enduring diatonic reality. For example, Lucian describes the Financial District's fundamental 'D-minor' harmony as 'fast-paced, frenetic and reckless'. It appears that the protagonist's tuning forks, as conduction devices of the New York City soundscape, seem to confirm the utility of Schenkerian ideology as a preserving frame of the aesthetic lineage of the Baroque-era 'doctrine of the affections' or Affektenlehre. In what Herbert Schueller calls a 'Cartesian theory based on the assumption that emotions are definite in character, concrete in form, and separable in the mind and in fact',[31] Affektenlehre became a means in which 18th Century composers could utilise particular musical figures that would represent the passions and affections. Lucian seems to be mirroring here Scheuller's characterisation of Affektenlehre as putting 'human nature under rationalistic observation',[32] as we can intuit from the following dialogue in the film:

Peter: 'An invisible system, but powerful nonetheless. Sounds, that in a sense guide people through the city. I mean they are not conscience of it, but it's there, and its different in all parts of the city…'
Ellen: 'So each part of the city has a different chord?'
Peter: Well, every part of the city has its own instructive atmosphere. Its impossible to disconnect the sound from the collective state of mind. It effects the people, you understand?… Why do people feel the way they do, why do people act the way they do, there is a reason … there is an order here.

Tuning

The notion of the tuning of a soundscape, that is, the number and spacing of a soundscape's tones or pitches, and how this may influence the urban condition has been an avenue of investigation of the acoustic ecology movement since the late 1960s. Like Peter Lucian, the founders of the movement, R. Murray Schafer, Barry Truax, and Hildegard Westerkamp also usurp the aesthetics of Western Classical Music as a scaffold in which to document, analyse, and represent acoustic phenomena of the everyday. As Schafer asserts:

> I am going to treat the world as a macrocosmic musical composition. This is an unusual idea but I am going to nudge it forward relentlessly. The definition of music has undergone radical change in recent years… Today all sounds belong to a continuous field of possibilities lying within the comprehensive dominion of music.[33]

Emerging from the post-war economic boom, acoustic ecologists forged an approach to the sustainable amelioration and tuning of noise that did not simply frame it as an engineering problem, but one that required a relentless redress through what Truax calls the 'communicational model'.[34] Indeed Schafer's hawkish evocations of the problem of noise pollution from the 1970s have been vindicated through studies documenting the effects of elevated noise levels.[35] The psychological and physiological dimensions of this shift are subtly portrayed by Tyburski in the Sound of Silence through Lucian's remarkable listening ability. This ability seems to be connected to the fact that for most of the film, he is wearing noise-cancelling headphones or industrial earplugs.

The high-energy urban soundscape of New York City thus embodies Peter Lucian's shadow in The Sound of Silence, and is presented as a character in the film, though one in which he does not come into conflict with what Schafer would identify as the 'indiscriminate and imperialistic spread' of its 'vulgarities'.[36] Instead, Lucian is keen to simply decode its order to reveal something of its 'universal constants'. Tyburski even provides a short back story to the urban soundscape through inter-cut historical footage of site visits of the city's Noise Abatement Commission from 1929. As Emily Thompson explains, the commission

> transformed public perceptions of the problem of noise by scientifically demonstrating the power and pervasiveness of that problem. It heightened New Yorker's awareness of noise and it educated them to listen in new ways.[37]

The commission's work involved utilising innovative scientific measuring devices for recording sound pressure levels throughout the city using the new metric of the decibel (dB). Certainly, elevated sound levels in the modern city have become more than a historical Leitmotiv, and have gained the attention of both urban planners and architects. In 2013, Lewis et.al reported that up to '15% of the US population aged 20–69 (26 million people) may have NIPTS [noise induced permanent threshold shift] caused by excessive exposure' to sounds in the workplace or leisure environment.[38] An increase in NIPTS of as little as >10 dB at 4,000 Hz over the course of even a few years can cause a 'permanent decrement in hearing acuity with substantial social and economic ramifications'.[39] Indeed, Gershon et.al have previously reported that in New York City, mass transit use is associated with temporary threshold shift symptoms which is a potential long-term risk to city commuters.[40]

But Tyburski's characterisation of Peter Lucian in The Sound of Silence is not one of an activist acoustic ecologist who confronts urban sound sources and their 'vulgarity' or physiological dangers. Rather, the protagonist is an artisan who tunes his client's appliances and devices to restore a natural harmony in which the Grundtöne heard within their homes finds diatonic agreement to the Ursatz of the exterior urban environment. Peter explains to his client Ellen upon discovering that her toaster produces an 'E-flat', that it must be rectified and tuned to a 'E-natural' given that the fundamental pitch in her surrounding neighbourhood is a 'C', and moreover that her fridge is a 'G'. Tuning the toaster to produce a major triad – Schenker's conceptual abbreviation of Nature – rather than a minor one becomes Lucian's compositional figure as site-specific intervention born of an Affektenlehre designed to relieve Ellen of her restlessness and depression. As Lucian explains, the key of C-minor is 'mundane', and a 'key of resignation'. Indeed, Schubart considered C-minor as evoking unhappy love, longing, pining, and sighing.[41]

Lucian's tuning here though also points towards what acoustic ecologists espouse as the desired shift from lo-fi soundscapes to hi-fi ones. As Truax asserts, the increasing deployment of technological systems in the built environment has promoted 'standardization and uniformity, right to the micro level of hums and broad-band [white] noise', which in turn produces low-fidelity soundscapes. Truax argues that from an 'ecological standpoint, the hi-fi [high-fidelity] soundscape is populated by many individual 'species' which are the result of local conditions'.[42] But as a true Schenkerian, Lucian's analyses of New York City's soundscape identify Grundtöne (in the form of the Stufen 1, 3, 5) as in service of a hidden Ursatz, and thus the scaffold against which all other sonic objects should be tuned.

Here, we find a connection between Schafer's concept of the urban 'keynote' and the importance of Schenker's Grundtöne in dictating the shape and qualities of the Tonraum:

> Keynote is a musical term; it is the note that identifies the key or tonality of a particular composition. It is the anchor or fundamental tone and although the material may modulate around it, often obscuring its importance, it is in reference to this point that everything else takes on its special meaning. Keynote sounds do not have to be listened to consciously; they are overheard but cannot be overlooked, for keynote sounds become listening habits in spite of themselves.[43]

Like Lucian's intimation of a fundamental Ursatz of New York City, keynotes become the currency for which Schafer identifies 'the soundscape as the "universal" composition of which we are all composers'.[44] This also connects Lucian's musical map designating Gundtöne in New York City to the way in which Schafer et.al sketched the 'character' of the soundscape in the town of Skruv in Sweden in Five Village Soundscapes (Figure 10.3):

> In Europe the current is 50 Hz giving a musical pitch of approximately G sharp. It is this fundamental that sounded from the cardboard factory. The other pitches resulted from the presence of strong harmonics. An additional feature of the Skruv soundscape was the piping F sharp of the train whistles. The resulting aggregate of pitched sounds was a dominate ninth chord, quite in tune.[45]

In a similar vein to Schenker's identification of Stufen that defines a Tonraum through an unfolding over time, keynotes bind together various acoustic communities (that is, auditors who perceived sounds through proximity) and their ensuing superimposed 'counterpoints' of everyday sonic objects.

Silence

It seems that at the diegetic level, the everyday sonic encounters of Tyburski's characters in The Sound of Silence do not draw hard distinctions between auditory events that can be qualitatively distinguished as either noise or sound. This non-binary also extends to the notion of silence as a container, presumably empty of sonic objects, versus a perceptual experience of quietude. When

Lend me your ears

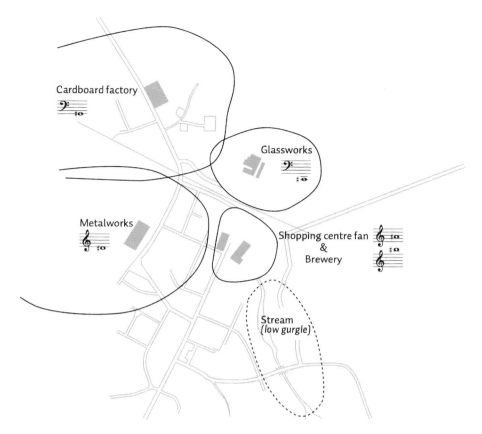

Figure 10.3 Auditory map of Skruv. Image courtesy of Barry Truax used with permission.

Peter introduces Ellen to the construction site of an impressive high-rise apartment of a wealthy electrophysiologist client, he describes his role as acoustic consultant in its design in terms of the quest for 'true electrical silence'. Here, the concrete pillars that run vertically through the space as the 'bones' of the building reach down to bedrock in order to avoid the conduction of sound from the street, thus isolating the apartment's interior soundscape. But in a seeming contradiction to the notion of 'true silence', we later surmise that Lucian's praxis is something that emerges from his philosophical position that silence can be tuned because it is actually 'full of sounds'.

In the communicational model of Truax, sound is something that distinguishes itself because it produces a relationship between an auditor and an environment.[46] That is, meaning in such a system is an emergent property and establishes what Nancy calls 'an order of understanding'.[47] By contrast, noise has a negative mediating effect, and possesses what Jacques Attali identifies as an alienation mechanism:

> A noise is a resonance that interferes with the audition of a message in the process of emission. A resonance is a set of simultaneous, pure sounds of determined frequency and differing intensity. Noise, then, does not exist in itself, but only in relation to the system within which it is inscribed: emitter, transmitter, receiver.[48]

Here, we can draw a subtle distinction between the work of bodies such as the Noise Abatement Commission to ameliorate excessive sound pressure levels of urban sound sources (what Truax identifies as engaging the 'energy transfer model'[49]), and the identification of noise as an irritant to communication as the first step towards the implementation of 'positive' soundscape design.[50] For Attali, noise can manifest not only as an intermediating sonic object or device that disseminates the object, but similarly as the auditor, and thus be a disruption of a social process. This draws Attali into a fascinating yet extreme argument that 'noise is violence' given its disruptive qualities (which indeed at higher sound pressures can be dangerous, though not fatal): 'To make noise is to interrupt a transmission, to disconnect, to kill. It is a simulacrum of murder'.[51]

In this framework, noise does not simply represent the negation of meaning, but rather imparts censorship, rarity, and moreover causes ambiguity given its interdiction on the source auditory stream. Attali further argues that 'music is the channelization of noise, and therefore a simulacrum of the sacrifice'.[52] Music thus becomes the syntactic structure that gives noise form, and aligns to what Peter Lucian similarly understands in The Sound of Silence: that the tuning of noise to create music according to the Schenkerian code, structures communities and is a mirror to the world, a repository, and a 'social score'.[53] In Attali's view, 'music sublimates the violent or carnivalesque noises of daily life',[54] while concurrently enabling subversive content: it thus engages both the institutional and material political economy. We might then align the craft of house tuning and the New York City Ursatz to what Attali argues is the explicit function of music as a 'reassuring', in which 'the whole of traditional musicology analyzes music as the organization of controlled panic, the transformation of anxiety into joy, and of dissonance into harmony'.[55]

In accepting Lucian's dictum that 'the silence is full of sounds', we can further ask what the auditory semiotics of such a container are? Indeed, one of Cage's most documented anecdotes about what led him to conceive of 4'33" is found in his seminal collection of essays and poetry A Year From Monday:

> It was after I got to Boston that I went into the anechoic chamber at Harvard University … Anyway, in that silent room, I heard two sounds, one high and the other low. Afterward I asked the engineer in charge why, if the room was so silent, I had heard two sounds. He said "Describe them." I did. He said, "The high one was your nervous system in operation. The low one was your blood in circulation."[56]

As a specialised room used in acoustic calibration and measurement, an anechoic chamber's materiality and geometry are designed in such a way as to negate the typical auditory experience found in any other room. As an anti-room or acoustically 'dead' space, there exists no secondary or tertiary sound reflections, and therefore no discernible reverberation signature. When sounds are produced within the space, the experience of them is only through fleeting direct sound.

Despite the anechoic chamber being inscribed with what Attali would call noise given its disruption and interdiction of sources that are introduced into it, Cage recounts how the immutable force of the body propagates an auditory experience in the form of two sounds. If there exist sounds in a room that ostensibly suppresses all sounds, then Cage concludes that sounds themselves must be universal, and moreover, seemingly independent of our will to subjugate them.

Whether we are auditors in an anechoic chamber or concert hall, for Stephen Davies, there exist three possible listening modes that we can engage regarding 4'33". One method is to simply hear the sounds of 4'33" solely for their aesthetic interest as sonic objects, without regard to any musical frameworks of value, and thus simply attend to their properties as they unfold in time. A second method is aligned to what Attali recognises as Cage's announcement of 'the disappearance of the commercial site of music: music is to be produced not in a temple, not in a hall not at home, but everywhere'.[57] This produces a conceptual revision in which the sounds of the every day become music 'only because the standard notion of music is undermined and rejected'.[58] But for the house tuner and defender of the Ursatz Peter Lucian, the sounds that constitute 4′33″ enact Davies' third listening mode in which they are approached

> as if they are musical or in relation to the musical (as traditionally conceived). This approach involves regarding the sounds that happen as if they are products of intentions of the kind composers usually have. It is to hear them as tonal (or atonal), as developing or answering earlier sounds, as (if appropriate) melodies, chords, and the like. Moreover, this mode of listening is to be historically grounded, as all musical listening is.[59]

Thus, when Lucian deploys Fernhören in order to identify Grundtöne in his client's appliances and devices, and subsequently tunes them to the established code of the Ursatz, he is composing in the sense that Attali identifies composing as a collective sensemaking which seeks to identify and give voice to signals emerging from noise.

Composing is thus something that attempts to 'create our own relation with the world and try to tie other people into the meaning we thus create'.[60] We can think of house tuning then in light of what Attali identifies as a potential of composition to be a solution to the alienation of labour and repression of desire, and through what Robin James calls a 'deregulatory neoliberalism' in which the recycling of noise into signals produces an ecological resilience.[61] This resilience further enables subjects to optimise creative capacities through an investment in self-intensification. The body is still present too, as Attali attests that composing is

> to exchange noises of bodies, to hear the noises of others in exchange for one's own, to create, in common, the code within which communication will take place. The aleatory then rejoins order. Any noise, when two people decide to invest their imaginary and their desire in it, becomes a potential relationship, future order.[62]

But in order to frame the broader auditory semiotics of The Sound of Silence, we return once again to the Naumburg Bandshell with tuning forks in hand, and draw on what Helmholtz describes in his 'Proof of Ohm's Law' from On the Sensations of Tone as a Physiological Basis for the Theory of Music. Here, Helmholtz speaks of a device in which a tuning fork is fastened to a thick wooden board with rubber feet which is then placed on a table. The rubber feet prevent sound being conveyed to the table, though the sound of the fork bought into vibration is still relatively weak. Helmholtz further suggests that

> if the prongs of the fork be bought near a resonance chamber of a bottle-form, of such a size and shape that, when we blow over its mouth, the air it contains gives a tone of the same pitch as the fork's, the air within this chamber vibrates sympathetically, and the tone of the fork is thus conducted with great strength to the outer air.[63]

In The Sound of Silence, Tybursk casts Lucian as not only an artisan seemingly engaged in Attalian composition, but like Helmholtz, an experimentalist and empiricist. We can think of the vibrating forks in Helmholtz's experiment then as the conductive apparatus Lucian exploits to tease out information about the Ursatz of New York City's boroughs. This process generates a 1:1 transmediation between the tuning forks and the soundscape in the form of an auditory map replete with ever-finer ratios of acoustic division that construct a Tonraum, and accordingly associate to an Affektenlehre as tuning utility. We find then that Helmholtz's resonance chamber morphs from the body of a bottle-form to the body of the built environment and urban landscape itself. If there are true sounds in the silence to be found, then they must be accessed through the calibration of this body that is the resonance chamber, through Fernhören, and through the transformation and tuning of its component noise into a music of, in, and for the everyday.

Notes

1 For an extensive treatment see Nicholas Cook, *The Schenker Project: Culture, Race, and Music Theory in Fin-de-siècle Vienna* (Oxford: Oxford University Press, 2007). 140–198.
2 David Beach, *Schenkerian Analysis: Perspectives on Phrase Rhythm, Motive and Form* (London: Routledge, 2019), 3.
3 Heinrich Schenker, *Five Graphic Analyses* (Mineola, NY: Dover Publications Inc., 1969), 14, 16.
4 Heinrich Schenker, *Harmony*, ed. Oswald Jonas, trans. Elizabeth Mann Borgese (Chicago, IL: University of Chicago Press, 1954), 6.
5 Schenker, *Harmony*, 22.
6 Schenker, *Harmony*, 28.
7 Pieter C. van den Toom, *Music, Politics, and the Academy* (Los Angeles: University of California Press, 1996), 95.
8 Christian Schubart, *Ideen zu einer Aesthetik der Tonkunst* (Vienna: Bey J. V. Degen, 1806), 380.
9 Jean-Luc Nancy, *Listening*, trans. Charlotte Mandell (New York: Fordham University Press, 2007), 1.
10 Nancy, *Listening*, 9.
11 Nancy, *Listening*, 9.
12 Adrienne Janus, "Listening: Jean-Luc Nancy and the 'Anti-Ocular' Turn in Continental Philosophy and Critical Theory." *Comparative Literature* 63 (2011): 194.
13 Ingrid Skykes, *Society, Culture and the Auditory Imagination in modern France: The Humanity of Hearing* (New York: Macmillan, 2015), 11.
14 John Locke, *An Essay Concerning Human Understanding* (Oxford: Clarendon Press, 1698/1975), 9–14
15 Robert Pasnau, "What Is Sound." *Philosophical Quarterly* 49 (1999): 304–324.
16 Casey O'Callaghan, *Sounds* (Oxford: Oxford University Press, 2007), 15–17.
17 Nancy, *Listening*, 67.
18 O'Callaghan, *Sounds*, 13.
19 Carl E. Seashore, *Psychology of Music* (New York: Dover, 1967), 5–6.
20 Timothy L. Hubbard and Keiko Stoeckig, "Representation of Pitch in Musical Images." In *Auditory Imagery*, ed. Daniel Reisberg (New York: Psychology Press, 1992), 199.
21 Schenker, *Harmony*, 28–29.
22 Nancy, *Listening*, 27.
23 David Dunn, *Purposeful Listening in Complex States of Time*, 1998, 3. http://static1.1.sqspcdn.com/static/f/288545/5477319/1264209903233/Plicsot.pdf\%3Ftoken\%3DSPX5LGQDjV7rGvTggs3sIqpLyE8\%253D
24 Kenneth LaFave, *The Sound of Ontology: Music as a Model for Metaphysics* (London: Lexington Books, 2018), 64.
25 LaFave, *The Sound of Ontology*, 29.
26 Kyle Gann, *No Such Thing as Silence: John Cage's 4'33"* (New Haven, CT: Yale University Press, 2010).
27 Peter Ablinger, "Hörstücke," https://ablinger.mur.at/hoerstuecke.html
28 Ablinger, "Stühle," https://ablinger.mur.at/docu01.html\#4parts
29 John Cage, *A Year from Monday* (Middletown, CT: Wesleyan University Press, 1963), 22.
30 Albert S. Bregman, *Auditory Scene Analysis: The Perceptual Organization of Sound* (Cambridge: MIT Press, 1990), 10.
31 Herbert M. Schueller, "'Imitation' and 'Expression' in British Music Criticism in the 18th Century." *The Musical Quarterly* 34, no. 4 (October 1948): 546.
32 Scheuller, "'Imitation' and 'Expression'," 547.
33 R. Murray Schafer, *The Soundscape: Our Sonic Environment and the Tuning of the World* (Rochester, VT: Destiny Books, 1977), 5.

34 Barry Truax, *Acoustic Communication* (Westport, CT: Ablex Publishing, 2001), 11.
35 These include greater incidences of arterial hypertension, myocardial infarction, heart failure, and stroke. See Thomas Münzel et al., "Environmental Noise and the Cardiovascular System." *Journal of the American College of Cardiology* 71, no. 6 (2018): 688–697, and Eunice Y. Lee et al., "Assessment of Traffic-Related Noise in Three Cities in the United States." *Environmental Research* 132 (2014): 182–189.
36 Schafer, *The Soundscape*, 3.
37 Emily Thompson, *The Soundscape of Modernity: Architectural Acoustics and the Culture of Listening in America, 1900–1933* (Cambridge: MIT Press, 2004), 164.
38 Ryan C. Lewis, Robin R. M. Gershon and Richard L. Neitzel, "Estimation of Permanent Noise-Induced Hearing Loss in an Urban Setting." *Environmental Science and Technology* 47 (2013): 6393–6399.
39 Lewis et al., "Estimation of Permanent," 6393.
40 R. R. M. Gershon, M. F. Sherman, L. A. Magda, H. E. Rilet, T. P. McAlexander and R. Neitzel, "Mass Transit Ridership and Self-Reported Hearing Health in an Urban Population." *Journal of Urban Health* 90, no. 2 (2012): 262–275.
41 Schubart, *Ideen zu einer*, 377–378.
42 Barry Truax, "Soundscape Composition as Global Music: Electroacoustic Music as Soundscape." *Organised Sound* 13, no. 2 (2008): 104.
43 Schafer, *The Soundscape*, 9.
44 Truax, "Soundscape Composition," 103.
45 R. Murray Schafer, Bruce Davis and Barry Truax, *Five Village Soundscapes* (Vancouver: A.R.C. Publications, 1977), 233.
46 Truax, *Acoustic Communication*, 94.
47 Nancy, *Listening*, 1.
48 Jacques Attali, *Noise: The Political Economy of Music*, trans. Brian Massumi (Minneapolis: University of Minnesota Press, 1985), 26–27.
49 Truax, *Acoustic Communication*, 5.
50 See Jian Kang and Brigitte Schulte-Fortkamp eds., *Soundscape and the Built Environment* (Boca Raton FL: CRC Press, 2016).
51 Attali, *Noise*, 26.
52 Attali, *Noise*, 26.
53 Attali, *Noise*, 9.
54 Josh Epstein, *Sublime Noise Musical Culture and the Modernist Writer* (Baltimore, MD: John Hopkins University Press, 2014), xxvi.
55 Attali, *Noise*, 27.
56 Cage, *A Year from Monday*, 134.
57 Attali, *Noise*, 136–137.
58 Stephen Davies, *Themes in the Philosophy of Music* (Oxford: Oxford University Press, 2003), 15.
59 Davies, *Themes in the Philosophy*, 14.
60 Attali, *Noise*, 134.
61 Robin James, "Neoliberal *Noise*: Attali, Foucault & the Biopolitics of Uncool." *Culture, Theory and Critique* 55, no. 2 (2014): 142.
62 James, "Neoliberal *Noise*," 143.
63 Hermann von Helmholtz, *On the Sensations of Tone as a Physiological Basis for the Theory of Music*, trans. Alexander J. Ellis (London: Longmans, Green and Co., 1895), 55.

Bibliography

Ablinger, Peter. "Hörstücke" (Listening pieces). Accessed August 9, 2023. https://ablinger.mur.at/hoer-stuecke.html

Ablinger, Peter. "STÜHLE/*CHAIRS*" *Weiss/Weisslich 14 and 29, Chairs, Listening Places, Chair Projects since 1995* Accessed August 3, 2024. https://ablinger.mur.at/docu01.html

Attali, Jacques. *Noise: The Political Economy of Music.* Translated by Brian Massumi. Minneapolis: University of Minnesota Press, 1985.

Beach, David. *Schenkerian Analysis: Perspectives on Phrase Rhythm, Motive and Form.* London: Routledge, 2019.

Bregman, Albert S. *Auditory Scene Analysis: The Perceptual Organization of Sound.* Cambridge: MIT Press, 1990.
Cage, John. *A Year from Monday.* Middletown CT: Wesleyan University Press, 1963.
Cook, Nicholas. *The Schenker Project: Culture, Race, and Music Theory in Fin-de-siècle Vienna.* Oxford: Oxford University Press, 2007.
Davies, Stephen. *Themes in the Philosophy of Music.* Oxford: Oxford University Press, 2003.
Dunn, David. *Purposeful Listening in Complex States of Time*, 1998, 3. https://www.davidddunn.com/~david/Index1.htm
Epstein, Josh. *Sublime Noise Musical Culture and the Modernist Writer.* Baltimore, MD: John Hopkins University Press, 2014.
Gann, Kyle. *No Such Thing as Silence: John Cage's 4'33".* New Haven, CT: Yale University Press, 2010.
Gershon, Robyn R. M., M. F. Sherman, L. A. Magda, H. E. Rilet, T. P. McAlexander, and R. Neitzel. "Mass Tr. "Mass Transit Ridership and Self-Reported Hearing Health in an Urban Population." *Journal of Urban Health* 90, no. 2 (2012): 262–275.
Hubbard, Timothy L. and Keiko Stoeckig. "The Representation of Pitch in Musical Images." In *Auditory Imagery*, edited by Daniel Reisberg, 199–235. New York: Psychology Press, 1992.
James, Robin. "Neoliberal *Noise*: Attali, Foucault & the Biopolitics of Uncool." *Culture, Theory and Critique* 55, no. 2 (2014): 138–158.
Janus, Adrienne. "Listening: Jean-Luc Nancy and the 'Anti-Ocular' Turn in Continental Philosophy and Critical Theory." *Comparative Literature* 63 (2011): 182–202.
Kang, Jian and Brigitte Schulte-Fortkamp eds. *Soundscape and the Built Environment.* Boca Raton, FL: CRC Press, 2016.
LaFave, Kenneth. *The Sound of Ontology: Music as a Model for Metaphysics.* London: Lexington Books, 2018.
Lee, Eunice Y. et.al. "Assessment of Traffic-Related Noise in Three Cities in the United States." *Environmental Research* 132 (2014): 182–189.
Lewis, Ryan C., Robin R. M. Gershon and Richard L. Neitzel. "Estimation of Permanent Noise-Induced Hearing Loss in an Urban Setting." *Environmental Science and Technology* 47 (2013): 6393–6399.
Locke, John. *An Essay Concerning Human Understanding.* Oxford: Clarendon Press, 1975.
Münzel, Thomas et.al. "Environmental Noise and the Cardiovascular System." *Journal of the American College of Cardiology* 71, no. 6 (2018): 688–697.
Nancy, Jean-Luc. *Listening.* Translated by Charlotte Mandell. New York: Fordham University Press, 2007.
O'Callaghan, Casey. *Sounds.* Oxford: Oxford University Press, 2007.
Pasnau, Robert. "What Is Sound." *Philosophical Quarterly* 49 (1999): 304–324.
Schafer, R. Murray. *The Soundscape: Our Sonic Environment and the Tuning of the World.* Rochester, VT: Destiny Books, 1977.
Schafer, R. Murray, Bruce Davis and Barry Truax. *Five Village Soundscapes.* Vancouver: A.R.C. Publications, 1977.
Schenker, Heinrich. *Five Graphic Analyses.* Mineola, NY: Dover Publications Inc., 1969.
Schenker, Heinrich. *Harmony.* Edited by Oswald Jonas and translated by Elizabeth Mann Borgese. Chicago, IL: University of Chicago Press, 1954.
Schubart, Christian. *Ideen zu einer Aesthetik der Tonkunst.* Vienna: Bey J. V. Degen, 1806.
Schueller, Herbert M. "'Imitation' and 'Expression' in British Music Criticism in the 18th Century." *The Musical Quarterly* 34, no. 4 (October 1948): 544–566.
Seashore, Carl E. *Psychology of Music.* New York: Dover, 1967.
Skykes, Ingrid. *Society, Culture and the Auditory Imagination in modern France: The Humanity of Hearing.* New York: Macmillan, 2015.
Thompson, Emily. *The Soundscape of Modernity: Architectural Acoustics and the Culture of Listening in America, 1900–1933.* Cambridge: MIT Press, 2004.
Truax, Barry. *Acoustic Communication.* Westport, CT: Ablex Publishing, 2001.
Truax, Barry. "Soundscape Composition as Global Music: Electroacoustic Music as Soundscape." *Organised Sound* 13, no. 2 (2008): 103–109.
van den Toom, Pieter C. *Music, Politics, and the Academy.* Los Angeles: University of California Press, 1996.
von Helmholtz, Hermann. *On the Sensations of Tone as a Physiological Basis for the Theory of Music.* Translated by Alexander J. Ellis. London: Longmans, Green and Co., 1895.

11
IMAGINING TOGETHER

Nina Garthwaite

Imagining together: a cinema for sound

In The Dark started in 2010. The aim was to create a space a bit like a pop-up cinema, but for sound. I wanted to experiment with playing audio that had originally been created for radio broadcast, podcast, and sound art, to a gathered listening audience. I was interested in a particular type of audio, which defies easy definition, where recorded sound, music, and voice are used to create poetic evocations, flights of fancy and to tell stories – the audio equivalent of film, poetry, or literature.

At that time, putting radio into a group listening setting was an unusual proposition. The concept of watching films together in a public communal space was a very familiar one. Even solitary activities like reading had a communal dimension through literary festivals, poetry readings, and book clubs. But radio was not a medium that had been translated into public space in that same way.

This was pre-podcast boom, and even online 'listen again' functions were in their infancy. Most people listened to the radio via a radio: a rolling programme that you would tune into, at home while doing the dishes, driving, maybe bubbling in the background at work. People thought less of individual programmes and more of stations which made the idea of listening events a confusing concept. As someone once responded to my idea: 'but If people want to listen to the radio, they can listen to the radio!'.

They had a point. The radio was already something special, a medium that was also a companion that manifested in our imaginations and reached into our most private moments. Perhaps the prospect of sitting in a room with other people and listening to a curated selection of programmes divorced from their wavelengths of origin seemed odd and unnecessary.

But while we watched films from all around the world, read books in translation and art and music were unfettered by borders, when it came to broadcast radio, we were curiously domestic. Now, American audio is part of our listening in the UK, but back then, even English language audio from other countries was unknown to most living here. Different sonic languages existed, and we never encountered them. And, the relative absence of online archives at the time meant that radio was ephemeral. There was something about rummaging through the archives and bringing gems from the past and around the world together in a space that felt like it could offer something.

Plus, there was something magical about the idea of bringing people together just to listen that was perhaps related to the nature of sound.

In 1959 Donald McWhinnie wrote: 'The spoken word is hard to catch, it is gone as soon as it is formed; you hear it, but can you capture it?'[1] In a similar mode, over 50 years later, Danish producer Rikke Houd describes sound as 'spacious, fluid, moving. Sound lets us travel far into our minds, in time and space, between the real and unreal, the seen and the unseen, the waking and the dream, the now and the then....'[2] The mercurial nature of sound seems to have driven its aesthetics, particularly within creative radio. This vision sees audio producers as Hermes-like, crossers of boundaries, not just conveying voices and music and sound but evoking spirits in the space between them, bringing them over to the other side.

This symbolic connection is also rooted in radio technology. Until recently you had to 'tune in' to the radio, search for the frequency where voices would rise out of the static. Poet and broadcaster Seán Street describes how on Christmas Eve 1906, a time when radio transmissions were used primarily for morse code, the Canadian inventor Reginald Fessenden conducted an experiment in the transmission of speech and music. The experimental broadcast was picked up by ships at sea. Street says, 'it was not too great an exaggeration to say that some believed the source was supernatural; the timing, the content and the apparent technical miracle of the whole thing was overwhelming'.[3]

And while creative radio tends to dance with factual content, poets were among the first to explore the creative potential of sound, further pushing it in the direction of the imagined and the abstract. Perhaps thanks to the early Dada experiments with sound poems, there has always been a tendency towards absurdism. Within audio drama, the imaginative possibilities of the empty space of the studio where sounds could appear out of nowhere and multiple realities could drift into one another meant those such as Louis MacNeice, Dylan Thomas and Samuel Beckett were drawn to the possibilities of creating work for sound.

Whether this association is essential, an artefact of its history, or just a story that we tell, the symbolic resonances between radio and the manifestation of imaginative – possibly spirit – worlds are entangled. Through this lens, a cinema for sound offered the potential for not just an informative but an other-worldly experience. I wanted to see the characters and landscapes that appeared inside me when I listened to the radio swirling around the room, hanging from the light fittings, crawling across the floor. An invisible picture palace that was real and imagined and where the boundaries between inner and outer worlds were blurred: I wanted to imagine together.

The first in the dark

Things get a little less dreamy when they get practical. There was not – and still isn't – such a thing as a cinema for sound. After trawling the streets of London, it didn't take long to abandon the idea of trying to find the perfect space.

The first event was held as part of the London International Documentary Festival at The Hub Kings Cross – a hybrid work space/venue. There was a bar, which seemed important, the lights were dimmed but not off and we arranged the seating cabaret style to break up the sense of a front-facing visual focal point. I think there were also some bean bags but mainly chairs. As this was quite a conceptual leap for the average punter, the audience was mainly made up of audio producers.

It was curated by Alan Hall of Falling Tree Productions, a maker whose work I admired from listening to BBC Radio 4. As well as an inventive and poetic radio producer, he was deeply engaged in the international audio community and knowledgeable about audio beyond the domestic

UK airwaves. He had also, it turned out, recently held an 'in the dark' event for the American radio duo the Kitchen Sisters. It was his use of this term that gave In The Dark its name, evoking something of the shadowy projections I imagined dancing across the walls as we listened.

After a few shorts from around the world, we listened to a full 40-minute documentary in silence, together.[4]

Aside from the piece itself, the main memory of future In The Dark collaborator Leo Hornak is the sound of the London buses outside. What I remember about the space was how, despite the higgledy piggledy cabaret seating, during the course of the listening everyone had slowly oriented themselves to face the front of the room. No ghostly figures swinging from the rafters. Our imaginations had chosen to stick to convention.

A review on the BBC Radio 3 blog picked this up. It quoted the experience of one listener who described it as looking 'as if we were all gazing up at an invisible screen watching the programmes in our heads'.[5] While it wasn't quite the un-boundaried, dissolving of self and space that I had dreamed of, I think I consoled myself with this more generous version of events. At least we were, still, imagining together.

When I gave my initial intro at the event, I spoke of In The Dark as 'we' not 'I'. I didn't know who 'we' were yet. But a group quickly assembled.[6] There was something about radio and the possibility of podcasts that seemed to excite the younger generation I was a part of – the idea that there might be something new to be discovered in it. And it is very much together we began to explore what else this space might be.

The spaces

In 13 years, In The Dark has never held a listening event in the 'ideal' space – if such a thing exists. Ranging from cinemas to cemeteries, we have held events in around 200 locations around the world. With no established 'normal' or shared culture around events with no visual focal point, we – and our audiences – have often found ourselves having to make unusual negotiations between space, sound, and the experience of being sat in a room with other people. And yet each imperfect venue has added something to the shared experience. While admittedly it hasn't always been ideal, sometimes the collaboration of sound and space has been uncanny.

In 2014 we curated a selection of programmes around the theme of water.[7] Crammed into a barge in Battersea we were all immersed in the listening, eyes closed or gazing at the light flickering off the chandeliers (this was a well-decorated barge). A piece started playing which involved a great storm at sea. Just as the action reached its climax our boat, which was safely moored and had been as still as if it were on dry land, suddenly began to rock quite violently. The chandeliers rattled as giggles turned into gasps. The storm – real and imagined – passed fairly quickly but it was the moment of the night everyone wanted to talk about. Many asked if we had planned it. We hadn't.

But while it's fun to play directly with sound and space, most of the time we're simply looking for an affordable (or free) space that will have us. Rooms above pubs feature heavily. And it is amidst the buzzing of drinks fridges, floors pulsing with music from downstairs bars and crumby pub speakers that some of the most transformative collective experiences have occurred.

One event that stands out was in 2012 when the very established International Features Conference was due to be held in London, hosted by the BBC. New up-starts in the audio world, we weren't invited to be a part of it, so we set about creating a fringe event. Wanting to attract an audience that was already signed up for four days of solidly listening to radio, we did the only thing we could: theme it on sex.

Sofia Saldanha was the lead curator and her bottom line was that there was no point in being embarrassed: the audio had to make you feel something. To say that she listened to every accessible piece in existence that vaguely touched on the theme is probably not stating it hard enough. At the time there were very few programmes that dealt with sex unflinchingly. A few years later the popularity of podcasts like The Heart would transform audio into a medium that speaks confidently about sex, but in 2012 most audio was broadcast radio and even today radio is much more demure than its visual counterpart. So, Sofia listened to everything she could get her ears onto and would bring shortlisted selections to my living room for the team to listen together.

In those early days, we'd meet once a month, drink a lot of wine, and listen. Some of my happiest early memories are in fact here, in the domestic space with newfound friends. In my far too small living room, one of us would be on the floor, another perched on a foot stool, the noisiness and arguing about what we were going to play falling into a transportive silence when someone said: 'have you heard the piece?' and we all stopped to listen.

Hearing it again today, the sex-themed set feels both less sophisticated and much bolder than anything that could exist now. Much of the final selection were early outputs of those venturing into the world of podcast, experimenting with form. They were often rough around the edges which I think meant the vulnerability of the subject is felt in the materiality of the sound. This was a shift, in a way, from the aesthetics of the first events that honoured 'quality' audio. Instead, it was showcasing a tentative and fearless venturing into the possibilities of audio in a world without gatekeepers.

On the night we piled into the upstairs room of The Clachan pub – the only central London pub near the conference that would accommodate us on a low minimum bar spend. The speakers weren't great, but they worked. By this time, we had a more general audience coming so with the conference goers too we ended up having to run the set twice. The queue running through the pub was a happy surprise for the pub landlord who still didn't understand what we were doing.

There were bodies squeezed next to each other, and this time there were pews that made orienting around to the front of the room more difficult. The piece started with a sound composition by one of our team Ed Prosser, made from audio ripped from pornography. The avoidance of eye contact, let alone physical contact, was a very hard thing to achieve.

The audio we played included 'Afternoon Delight' by Kaitlin Prest[8] for a (pre-The Heart) radio show in Canada called Audio Smut. It featured the actual recorded sound of female masturbation. It also included 'Prepared to Love' by Karl James,[9] a very explicit account of a first sexual experience. The shifts in feeling in the room as one piece gave away to the next were palpable. And while the content that we played was often emotive, funny, touching, it was unrelentingly visceral. Where I had wanted to invite our inner worlds to dance together, Sofia's vision asked them to go one step further – to enter into each other's bodies, to generate heat. This was not so much a channelling of spirits, but a physical possession.

Afterward, it felt as though the audience was pulsing with a new energy. The possibilities of this medium and how we could play with the experience of a gathered audience felt like it shifted that night.

And yet with a limitless budget, it is unlikely that such an event would have ever been held in an upstairs room of a pub. But I'm also not sure a more 'fitting' venue would have been as interesting. I can't remember why we didn't venture into a Soho sex shop, but I suspect it was partly about not wanting to over-egg the pudding, to make the event feel too staged – or even held – by the space. If the audio was strong, it would be enough. And it was.

Exploring the inner experience of sound

I've often wondered whether it's true that sound has a particular capacity to engage our inner worlds, to inhabit our bodies. All art forms, when they're doing what they do right, are a collaboration between tangible outer materials and our intangible inner worlds. Yet even filmmakers extol the particular quality of sound. Robert Bresson said:

> The ear is profound, whereas the eye is frivolous, too easily satisfied. The ear is active, imaginative, whereas the eye is passive. When you hear a noise at night, instantly you imagine its cause. The sound of a train whistle conjures up the whole station. The eye can perceive only what is presented to it.[10]

Over the years I have explored this capacity of sound through events but also through teaching. Many of the exercises I have developed are a way to help students observe the effect of sound on their inner worlds, inspired by some psychological research.

In 2014–2016, I was fortunate to be a collaborator for Hubbub, a two-year interdisciplinary residency at the Wellcome Collection in London. During the residency, we were introduced to Russell T. Hurlburt, a Professor of Psychology at the University of Nevada. His assertion was that the subjective nature of inner experience has meant that: 'the science of psychology has largely abandoned the careful study of inner experience'.[11] As a result, he created a method called Descriptive Experience Sampling (DES), designed to capture 'high fidelity' glimpses of 'pristine inner experience'. He says:

> By 'inner experience' I mean thoughts, feelings, sensations, tickles, seeings, hearings, and so on, anything that appears directly before the footlights of consciousness.... Inner experience can be of internal events (such as seeing in imagination the cloud billowing from the World Trade Center) or of external events (seeing the billowing cloud on a real TV).[12]

His method involves a participant wearing an earpiece which signals a 'beep' at random times. In those moments the participant must make a note of any inner experiences in the moment before the beep – as best they can. They then bring these notes to an interview where Hurlburt teases out the nature of those inner experiences.

Doing a mini-version of the experiment with Hurlburt, our group had a glimpse into the variety of different ways we describe and perceive our inner experience. As he interviewed us about our inner moments before the beep, we discovered that a simple thought such as 'what am I going to eat for lunch later?' would seem to manifest in much more elaborate and peculiar ways than the first reporting of it might suggest – an image of the supermarket aisles, for example, or the imagined taste of a possible choice of sandwich.

Hurlburt has categorised these different types of experience as Inner Speech; Partially Worded Speech; Unworded Speech; Worded Thinking; Image; Imageless Seeing; Unsymbolised Thinking; Inner Hearing; Feeling; Sensory Awareness; Just Doing; Just Talking; Just Listening; Just Reading; Just Watching TV; Multiple Awareness.[13]

As a teacher of audio, I experimented with Hurlburt's techniques as part of my classes. Initially, I simply replicated the experiment as best I could to help students consider their own, and potential listeners' imaginative experiences. I also began to explore how pre-existing audio pieces manifested in their imaginations.

After listening to a short audio piece which takes place in a car, I ask the students where they are, imaginatively. They are often surprised and delighted, not just at their own imaginings but at others' variations. Those that are more visual in the mind's eye can see things out of the window of the imagined car. When I ask where they are in the car, some will realise that they are in the back seat, and that is different from the another student who is in the front seat, but even more delightfully they may discover that they are floating in the air of the car. And often, when a surprising twist happens in the piece, they leap – first into the protagonist's body and then onto the pavement – from where they watch, confused, as the car drives away. There is nothing explicit in the spoken word of the piece to direct these perspective shifts, but something about the combination of sound and the impact of the events unfolding that precipitate them.

Another exercise I made deals directly with the experience of actual sound in the classroom (or at home for online classes). Beginning with listening to all the sounds in the space, and then picking one sound to focus on, I guide the group to write a stream of consciousness responses to a series of prompts, describing the shape of the sound, the texture, the smell, the taste, and on and on until we get to images, emotions, and then finally any memories evoked by the sound.

The environmental sounds are often fairly common – car tires on a wet road, street diggers, water running through pipes, faint sound of the neighbour upstairs, indeterminate humming, possibly air con. We read out and compared the texts. Often, they start in these familiar places but quickly they develop into strange imaginings and finally into particular – and often vivid – emotional memories. A recent student created this beautiful text from the exercise, and I think it is a good illustration of how a sound in our environment can expand internally:

> Buzz of computer, anonymous cracking in the walls. Crows in the distance, off to the right. Scratching of the pen. The low thrum of a summer rain. You, turning in the bed. Steady continuous, a-rhythmic or – no – pulsing. A steady thrum. A gentle, a wall, a sheet, a wall, between us and the world. A numbing. A delicate sweet smell. Wet on warm grass, warm skin. A bitter taste. Endless grey days spent in bed. Days of longing and melancholy and cigarettes and coffee cups and stains on the sheets, and birds in the distance and other beds and other lovers and headaches and bitter coffee in the morning and out your door and up the stairs and into my flat and longing and longing and watching the jaybirds outside the window in the morning. Why am I sad? And you, and the rain. It rained that whole April outside the window.[14]

McWhinnie says:

> The radio listener – if his mind works in this particular way – will clothe the sounds he hears with flesh and blood; and since he has to find his images in his own experience and imagination they will be images which belong to him in a special way; in fact he will create them on the basis of the aural stimulus offered.

While audio may not uniquely inspire the imagination, it does seem a particularly direct method of entering a vivid inner space. Listening collectively then, might indeed be an eccentric proposition. Let us imagine together. Let us be in our intimate inner world which is infinite, in this confined and clunky room, together.

The listening body

While to this day In The Dark is a happy negotiation with the imperfect reality of space, in 2017 and again in 2019 artist and audio producer Eleanor McDowall and I were given the chance to imagine what a cinema for sound *could* be during a residency at the Barbican.[15] Although it was five years later, both of us remained inspired by the visceral nature of Sofia's sex-themed set and decided to explore the physical relationship with sound by theming the residency around 'The Listening Body'.

As part of our research, we put out an informal survey on social media asking audio-makers and listeners to reflect on the way they listen and to imagine what form a 'cinema for listening' might take. The results were interesting. Of 133 respondents 40% liked to be still – lying in bed before sleep, on the sofa or in a bath. 37% associated focussed listening with movement (walking, driving, running).[16]

The focus on movement was particularly interesting to me in the context of a previous experience of a listening tent at the Latitude Festival where we decked out the space with swings.[17] While I'm not going to pretend this was a complete success – place swings in a forest in a festival and you can guarantee chaos will ensue – there were moments of quiet, where the drunken punters subdued. What emerged was a dreamy sonic retreat where the awkwardness of where to look was finally resolved as eyes moved into soft focus and the natural movement of the world around them lulled them into their inner worlds while letting them remain connected visually and physically to the space. I think there were moments during this where the experience of feeling the spirit of sound in *and* around us were achieved, the gentle rocking perhaps taking us back to a time where our boundaries were fluid.

In The Barbican, we revived this idea with swinging pod-chairs in an otherwise fairly blank scratch space within which we conducted a variety of experimental events, as well as having a running listening set throughout the day for people to drop into.

Within these experiments we played with movement in a more direct way, holding an event entitled 'Dancing to the Radio'. It was run in collaboration with audio producer Jess Shane who'd happened to have the same idea at around the same time all the way over in Canada. Together we curated a set of spoken word audio that we felt leant itself to movement (our research involved dancing around at home). We then invited an audience of around 30 people to a fairly drab conference room in the Barbican. And, for an hour, we collectively interpretive-danced while listening to spoken words.

It was a strange and wonderful experience with audience members letting go, seemingly, of any self-consciousness. At any given point there were people being still, people moving wildly, squirming along the floor or leaping in the air. It was probably the closest we've ever got to the embodying of the spirits of radio and I still can't quite believe anyone came and then entered into the exercise with such gusto.

As another part of the exhibit, we also asked visitors to the space to imagine their dream listening cinema. 'A planetary observatory', 'a womb-like space', 'a ferris wheel' were some of the suggestions.[18] Floating featured heavily. At the end of the exhibition, I'm not sure we were any closer to having a practical vision of what a cinema for sound might be like. But perhaps that isn't the point.

Sometimes I wonder if the cinema for sound is meant to be an imagined space, a place of possibility and therefore for doing things that wouldn't normally seem sensible. I think the reality of a practical, physical, lived-in world of conventions, awkward bodies and averted gazes has become a compromise that I have come to actively enjoy. And while some of our activities have explored what a cinema for sound might be like, the imperfect compromise frames a big part of what I think I enjoy most in collective listening events. Yes, a floating cinema in a pool of clouds would be lovely, but as a voluntary organisation what we usually have to settle for is a room with people in it on rickety chairs.

When we are forced to contend with the physical world while reaching for something imagined, we are forced to hold hands to get there. The noticeable misalignment between intention and physical world is not something most people experience often and so we are forced to be open and vulnerable in a way we aren't used to and so we rely on each other more to get there. The possibility of collaboratively imagining what this space *could be* allows us to confront and wrestle with our existing spaces. In their failure to meet requirements, it makes those environments visible in a way that we otherwise tend to ignore.

Two years later, at a second Soundhouse iteration, some of these real-world limitations were brought into focus as we took steps towards considering accessibility as part of the imagination of a listening cinema.

Eleanor McDowall invited poet Raymond Antrobus as a contributor, posing the question 'How might we translate sound to make it more accessible to the eye?' Inspired by the artist Christine Sun Kim who has created alternative texts to standard closed captions,[19] Raymond created a 'found poem' to explore his experience of sound in translation.[20] Alongside this, Eleanor produced transcripts for all the work played, offering a subjective description of the sound design and music, as well as the spoken word, using tactile or visual imagery, as opposed to sonic language to translate the audio into text.[21] Here is an excerpt of one of El's transcripts:

> Large industrial machinery creaks uneasily with itself. It moans like a metallic whale. A hard cut. Traffic roars into the scene. Voices out in the world. A switch is flicked. The outside world stops. Vibrations return. The outside world returns. Voices. The moaning of a metallic whale. We fall back into the metal belly – alone. Suddenly – traffic carves across – high heels walk through the foreground. The metal belly eats the outside world and we are left in the large reverberant space. It stills. A vibration starts to move. It blooms. A ceiling full of pulsating neon lights. Air is rushing into the space – as if pushed through a pipe – a gas filling every corner. Music – a blooming.[22]

What strikes me in these transcripts, Raymond's Poem and Christine Sun Kim's work that has inspired them, is the resonance with the DES-inspired student texts that attempt to capture the inner experience of sound in words. The embodiedness – the melding of senses in the way it takes form innerly. They bring that which is often attributed to sound back into a verbal, imagistic, poetic space – which is where many of the early radio experiments emerged from. For all the mythology around sound, we are, ultimately, engaging in a space that is universally shared – an inner world more vast and amorphous, un-boundaried and imaginative, than any physical space could offer. The symbolism around sound is simply a direct reminder of the chasm between inner and outer experience, and the constant attempts to bring that inner experience into tangible being.

And while the pandemic meant that our discussions around physical space remained hypothetical, our consultant Dr Jess Hayton complicated our understanding of the cinema of sound as a listening space through considerations of visually impaired visitors.

In UK legislation, the Equality Act (2010) requires 'reasonable adjustment' to physical space and Jess spends much of her focus on what form these 'reasonable adjustments' might take to existing designs. However, she argues that in an ideal world inclusive design would consider all users from the beginning.

In her essay, she posed the provocation that

> a space dedicated to the art of sound might seem a natural source of enjoyment and escape for a person with impaired vision. However, we need to accept the space which we are escaping to. How can we be sure that this sanctuary will be a positive, safe, secure environment to find distance from the world beyond?[23]

Her questions around the overlapping of recorded sound and 'real world' sound posed challenging questions about the implications of these imperfect spaces. In a space where convention is still undefined, Jess' call for inclusive design is a creative provocation. Might a sound cinema be a context where we really could imagine a new type of space together?

After the pandemic, I was on the brink of bringing In The Dark to a close. I felt that my time in this messy and exploratory place might have come to a natural conclusion. But a young producer I met through teaching, Talia Augustidis, has chosen to take it on and has already held three events in London. In Bristol too, where we have a sister branch, the events have revived.

Epilogue: listening with Sofia

The last time I saw Sofia,[24] it was the end of July 2022 in Lisbon. She was unwell, undergoing treatment for cancer, exhausted and not able to go out much. I stayed three days with her, and we spent almost the entire time lying on her bed, listening to podcasts while drifting in and out of sleep. There's a Gregory Whitehead quote about the listener that comes to mind when I remember this time. He says:

> The radio listener, in all her protean moods, is as slippery, unstable, ghosted as la radia herself: in the bath, or in a doctor's waiting room, in the car, on the move, somewhere else, probably thinking about something – else – some other place, some other voice. Not us. So on both sides of the call and response, then, the listening situation is viscous, fluid, indefinite, only the potential for a crossing – no promises.[25]

Listening to audio allowed us to be together, to be engaging with something together, be present in space with each other while also being able to disappear into our own inner worlds when we needed to. How much were we listening? How much were we imagining? How much were we dreaming, worrying, thinking, feeling? How in our bodies were we? How close? How far apart? When listening together, there is only the potential for a crossing – no promises.

While I was there, I also told Sofia about this chapter (I had yet to put pen to paper). She said, 'oh audio is all about sound and space for me'. She explained that now, even when she is making her own work for radio, it is a room full of people listening together that she has in mind.

She also told me a story about a series of listening sessions she held in Lisbon for a piece of her own work. After one of the events a woman came up to her and said, 'it was so beautiful, but you know I have to say, I preferred the film version you played last time'.

No matter how much Sofia tried to convince her that she definitely hadn't made a film, the woman was sure that she had – intricately describing one of the scenes that, sure enough, was from Sofia's sound piece. Finally, the woman pulled over a friend of hers as a witness but her friend settled it – no, there was no screening, it was audio last time too.

Sofia was laughing as she told me this story, unpretentious about sound and unworried by the conceptual merging of mediums, just delighted at the real and imagined thing that had taken place. I hold onto her openness of her spirit as I write.

Last month, back in London, sitting in another of these imperfect spaces, a guest, rather than an organiser, I was moved by the new possibilities and energies it created room for. The audio was – fittingly – themed around the idea of liminality – a state of transition from one stage to the next.[26] The set involved a hybrid recorded audio piece and live musical performance. And as rickety benches were packed up at the end of the night, I was grateful that the experimentation continued.

And just as poetry merged into early creative sound and a listening cinema might now re-manifest sound as writing, listening to the radio together might involve dancing, and a boat can be an accidental participant in the action, perhaps it is ok if listening cinemas manifest, for some people, as imaginary screens in the sky.

Acknowledgements

This chapter is dedicated to the love, laughter, imagination, heart, guts, poetry, and ears of Sofia Saldanha who passed away on Christmas Day, 2022. In The Dark will forever be indebted to her bold and mischievous spirit.

It is also dedicated to my fellow explorers in the dark:

Connor Walsh, Leo Hornak, Mair Bosworth, Ed Prosser, Katharina Smets, Mike Williams, Rosanna Arbon, Emily Knight, Delores Williams, Nija Dalal, Clare Salisbury, Talia Augustidis, Caitlin Hobbs, Alan Hall, Eliza Lomas, Olivia Humphries, Siobhan Maguire, David Waters, Sophie Anton, Andrea Rangecroft, Sam Grist, Eleanor McDowall, Rosa Eaton, Hugh Chignell, Francesca Panetta, Laura Irving, Mugabe Tureya, Pete Hazell, Thalia Gigerenzer, Jo Tyler, Phil Smith, Sarah Stolarz and the many others who have been a part of In The Dark over the years.

Notes

1 Donald Mc-Whinnie, *The Art of Radio* (London: Faber and Faber, 1959), 21.
2 Rikke Houd, *Things I've Learned* (Strange and Charmed, 2017), https://www.strangeandcharmed.co.uk/rikkehoud
3 Seán Street, *The Poetry of Radio* (London and New York: Routledge, 2012), 24.
4 Liam O'Brien and Michael Kane, *Might Mac* for RTÉ's Doc on One. The following three events, also curated by Alan, featured Berit Hedemann from Norway, Chris Brookes from Canada, and Nina Perry from South London.
5 Abigail Appleton, *Hurrah for Adventurous Radio Feature-Making*, BBC Radio 3 Blog, January 28, 2010, https://www.bbc.co.uk/blogs/radio3/2010/01/hurrah-for-adventurous-radio-f.shtml
6 Connor Walsh, Leo Hornak, Ed Prosser, Sofia Saldanha and Mair Bosworth.
7 *The Deep End*, curated by Clare Charles and David Waters for In The Dark, Tamesis Dock, London, February 24, 2014.
8 Kaitlin Prest "Afternoon Delight" Audio Smut, 2011. https://www.theheartradio.org/audio-smut/2015/1/2/afternoon-delight
9 Karl James "Prepared to Love" The Dialogue Project 2010. http://understandingdifference.blogspot.com/2010/07/prepared-to-love.html
10 Susan Sontag, *Against Interpretation* (New York: Farrar, Straus & Giroux, Inc., 1964) 62.
11 Russell T. Hurlburt, "Exploring Inner Experience", accessed January 8, 2016, http://hurlburt.faculty.unlv.edu//sampling.html
12 Hurlburt, "Inner Experience".
13 Hurlburt, "Inner Experience".
14 Georgia Walker, "Untitled" (unpublished text and audio work, September 10, 2022), typescript and audio. Georgia also worked with sound to craft this text into a very poetic and evocative audio piece.
15 Part of the "Level G" programme, curated by Siddharth Khajuria.
16 *Soundhouse: The Listening Body*, various contributors, Barbican 2017, https://sites.barbican.org.uk/thelisteningbody/
17 The idea of set designer Samara Tompsett.
18 *Soundhouse: The Listening Body*, various contributors, Barbican 2017, https://sites.barbican.org.uk/thelisteningbody/
19 Christine Sun Kim *[Closer Captions]* for Pop-up Magazine, https://www.youtube.com/watch?v=tfe479qL8hg

20 Raymond Antrobus, *Poem With Captions*, *Soundhouse: Intimacy and Distance*, Barbican 2020, https://sites.barbican.org.uk/soundhouse-poemwithcaptions/
21 Eleanor's interest in transcripts was Inspired by Constellations' workshop on collaborative transcript creation.
22 Eleanor McDowall, *Transcripts*, *Soundhouse*, Barbican, 2020, https://sites.barbican.org.uk/soundhouse-transcripts/
23 Jess Hayton, *Sonic Architecture and Inclusive Design* for *Soundhouse: Intimacy and Distance*, Barbican, 2020, https://sites.barbican.org.uk/soundhouse-sonicarchitecture/
24 The curator of our 2012 sex-themed event. Sofia subsequently moved back to Portugal and set up In The Dark Lisboa.
25 Gregory Whitehead, "Bewitched, Bothered and Bewildered" August 29, 2012, https://gregorywhitehead.net/2012/08/29/bewitched-bothered-bewildered/
26 *Liminality* curated by Bridey Addison-Child, organised by Talia Augustidis for In The Dark, January 16, 2023.

Bibliography

Appleton, Abigail. *Hurrah for Adventurous Radio Feature-Making*, BBC Radio 3 Blog, accessed January 28, 2010. https://www.bbc.co.uk/blogs/radio3/2010/01/hurrah-for-adventurous-radio-f.shtml
Barbican, *Soundhouse: Intimacy and Distance*. https://sites.barbican.org.uk/soundhouse/
Barbican, *Soundhouse: The Listening Body*. https://sites.barbican.org.uk/thelisteningbody/
Chignall, Hugh. *British Radio Drama, 1945–63*. London: Bloomsbury Publishing, 2021.
Houd, Rikke *Things I've Learned*. Strange and Charmed, 2017.
Hurlburt, Russell T. "Exploring Inner Experience". Accessed January 8, 2016. http://hurlburt.faculty.unlv.edu//sampling.html
In The Dark, "Past Events", accessed May 2, 2023. http://www.inthedarkradio.org/category/events/past-events/
Mc-Whinnie, Donald. *The Art of Radio*. London: Faber and Faber, 1959.
Sontag, Susan. *Against Interpretation*. New York: Farrar, Straus & Giroux, Inc., 1964.
Street, Seán. *The Poetry of Radio*. London and New York: Routledge, 2012.
Sun Kim, "Christine [Closer Captions] for Pop-Up Magazine". https://www.youtube.com/watch?v=tfe479qL8hg
Whitehead, Gregory "Bewitched, Bothered and Bewildered", accessed August 29, 2012. https://gregory-whitehead.net/2012/08/29/bewitched-bothered-bewildered/

12
AURAL DIVERSE SPATIAL PERCEPTION

From paracusis to panacusis loci

John Levack Drever

Introduction – (re)prioritisation of hearing

Spaciousness is closely associated with the sense of being free. Freedom implies space; it means having the power and enough room in which to act.

(Tuan, 2003)[1]

Human biology and allied fields have a tendency to presuppose a multimodal subject who intuits space predominantly via binocular vision. This hierarchy of the senses, with fractional input from hierarchy of the senses hearing, touch, balance, and proprioception, offers discreet spatial information to help reaffirm the visual scene. Hearing comes to the fore when sight is compromised: 'in darkness, fog, dense smoke, forests, blizzard – and of course for blind and visually impaired people'.[2] Flipping this paradigm, what follows is an unabashed (re)prioritisation of the sense of hearing vis-à-vis spatial acuity.

It goes without saying that sound is intrinsically spatial and site-specific. The spatiotemporal journey from a sound's cause to its reception by the outer and/or inner ear is generative, contingent on its frequency content and the quality of the surfaces and mediums it encounters including the air which also acts as a frequency-specific absorber. Spatial hearing can perceive some of that story, informing us of a source's dimensions and relative distance: a large object far off or a small object close up. A source may be in a fixed position, moving or diffuse and ubiquitous, such as the surf on the beach. Spatial hearing is not limited to localisation (i.e., perceived perception of the location of a sound source), it can tell us about the surrounding topography: height, width and depth, open air or enclosed, or a combination.

Significantly, deviating from the standard auraltypical textbook positing of a perfectly calibrated pair of symmetrical ears positioned at the sides of the cranium – a normative model that has been ineluctably inscribed into our audio technologies and the fabric of the environments we have fashioned and inhabit, informing who gets to hear what: signal, noise, ambience – this chapter will offer a fuller account of spatial hearing within the frame of the nascent discourse of auraldiversity.[3,4]

Origin of spatial hearing

Hearing: an evolutionary-derived sense, primed within an immersive acoustic horizon for a world in which humans were preyed on by the likes of sabretooth tiger (or where humans were the predator) within the dense undergrowth. Our lack of earlids bears testament to this existential primordial role. I'm reminded of the astonishing auditory spatial alertness of my children, with only a few months of extrauterine, non-aqueous acoustical experience. Placing the sleeping baby down in the cot, I would painstakingly attempt to extract my reassuring hand and step backwards out of the room towards the door, soundless, like a Ninja circumventing a Japanese nightingale flooring system. Through countless failures, I had familiarised myself with the location of the loose floorboards and how much weight each could bear. Yet, I was doomed to failure, with the tiniest of creaks, signalling to the infant their abandonment, resulting in a bout of crying. Notwithstanding the hubbub of pedestrians and traffic on the exterior flanking two sides of the room, the infant, at this early stage of development implicitly knew which sounds were generated from its immediate reality, demonstrating highly operational spatial auditory perceptual skills.

Field of hearing

Hypothesising on sound, space, and hearing goes back to the 1920s with the foundational work of Finnish geographer, Johannes Gabriel Granö. Revolutionary for a geographer of his day, he placed the human senses at the centre of his approach to comprehending the landscape, comparing the capacity for distance perception that the array of senses has to offer. In a thought experiment, he swapped the visual metaphor of the field of vision for the field of hearing, but emphasised sight as the primary sense of garnering distance information: 'auditory phenomena can be localized only to some extent, in that we can estimate the direction in which the source of a sound is situated binaurally, but our ears do not provide much information on its distance'.[5]

Acoustic space

In the heady days of media theory, Edmund Carpenter and Marshall McLuhan demarcated human sensory experience as either acoustic or visual, with touch being contiguous with hearing. In the realm of acoustic space, that preceded the visual medium of writing, they conjectured that human experience was 'boundless, directionless, horizonless, … the world of emotion, primordial intuition'.[6] They conceptualised a hypothetical, unimpeded notion of spatial hearing, which, unlike visual space which is

> flat, about 180 degrees, […] pure acoustic space is spherical'[7]: We hear equally well from right or left, front or back, above or below. If we lie down, it makes no difference, whereas in visual space the entire spectacle is altered.[8]

R. Murray Schafer, with the World Soundscape Project, furthered the notion of acoustic space, through the articulation of soundscape. Schafer was quick to point out the acoustical foibles of this precondition of unhindered spherical omnidirectional configuration, bringing into play the physics of 'refraction, diffraction, drift and other environmental conditions',[9] where sounds 'inhabits space rather erratically and enigmatically'.[10]

Symmetrical hearing

Aping Carpenter and McLuhan's geometry, the psychoacoustics expert Jens Blauert took a Euclidian schema to spatial perception, proposing a sphere with an 'assumed symmetrical' human head following an 'international standard for measurement of the skull' as the model.[11] He divided this sphere into a horizontal, frontal and median plane using the eye sockets, ear canals and noise bridge as guides. Blauert's schema provides a system of spatial coordinates of azimuth (left–right, front–back, and between), elevation (up–down), and distance.

This perfect geometrical schema brings to mind the proportions and symmetry, golden ratios abound, of Vitruvian Man as expressed by Leonardo da Vinci following the architectonics of Vitruvius. In the picture's caption de Vinci states his take on the ideal facial proportions: 'The distances from the chin to the nose and the hairline and the eyebrows are equal to the ears and one-third of the face'.[12] Frustratingly, in this most iconic of illustrations, the pinnae – the hollows, folds and grooves of the wonderful flappy, and sometimes hairy, cartilage at the side of our head – are obscured by Vitruvian Man's long curly locks.

Azimuth

When an external sound arrives at the entrance to the two auditory canals, at any specific moment, there may be subtle differences between the two signals. The brain is able to perceive some of these small changes. For example, our perceptual mechanism is sensitive to the speed of sound. The time difference of one mid-frequency ranged sound with a distinct envelope arriving at an angle of 90° to travel to the ear on the other side is about 0.65 ms. If the sound wave is smaller than the dimensions of the cranium, we have the capacity to read the different stages of the wave cycle as it is received by each ear. We struggle with localisation of low-frequency sound as the wavelengths are long in comparison to our heads, depriving us of those spatial cues. The brain also gauges the loss of acoustic energy as the sound passes from one ear to the next. The combination of these skills provides azimuth information.

Elevation

When sound passes from one side of the head to the other, it diffracts creating a kind of acoustic shadow. The head, shoulders and pinnae act as a 'complex direction-dependant filter',[13] spectrally filtering incoming sound. This is referred to as the head-related transfer function (HRTF). This plays a fundamental role in elevation. It is predominantly a monoaural signal working on the vertical axes.

To complicate the principles above, in reality our face and pinnae are asymmetrical, and unique, notwithstanding one's piercings and adornment of the helix, rook or tragus. Moreover, throughout our lives, the pinnae, which are larger in men, will grow and droop.[14] Experiments have shown that we have learnt our own HRTFs; if the morphology of those features suddenly change, say if we hypothetically swapped the pinnae for another's, our spatial perception would be temporarily impaired, yet can be learnt after a few weeks.[15]

The dummy head

With the demand for reliable and repeatable measurements, dummy heads (or Kunstkopf), are customarily used to record human spatial hearing that incorporate the human vagaries of the HRTF in place of an actual human head. Stefan Krebs, who charted the historical development of the Kunstkopf from 1957 to 1981, observed:

> All three groups took it for granted that artificial heads were replicas of male human heads – indeed, in a classic case of engineers modelling the world from their own reality, in two instances the artificial heads were even equipped with the researcher's own pinnae.[16]

There are a number of commercial dummy heads available today with an omnidirectional mic positioned at the entrance to the ear canal. The dimensions and characteristics are based on an average adult head and pinnae. Neumann's Dummy Head KU100 claims: '…a truly immersive binaural sound experience with headphones. Although it uses only two channels, its spatial depiction appears three dimensional and shockingly realistic'.[17] Algorithms that simulate binaural effects are more commonly used, calibrated with generic data on head and pinnae dimensions.

Ear wiggling

Of course, mobility allowing, we move our head to help us localise sound more precisely, but our ears pretty well stay in the same position related to the rest of our head, unlike some species. I'm infinitely fascinated with my dog's dexterous pinna movements, with change in position conveying mood; change in shape appearing bigger to adversaries; a tool for batting away flies; and the spatial hearing ability of pinpointing. Some lucky people have been blest with the vestigial ability of ear wiggling, exploited for its comedic value by the likes of Stan Laurel, and an appealing physiological alien attribute sported by Vulcans and Na'vi.

Lateralisation

Spatial auditory perception is not exclusively about the externalisation – perceiving sound sources outside of one's head. A source can be perceived as spatially located inside the head, called lateralisation. My most intense experience of lateralisation was the sonic experience of an MRI scan – a standard test for those with unilateral tinnitus and hearing loss to check for acoustic neuroma. I could pinpoint the shifting focus of high-pitched tones, with Ryoji Ikeda-like aesthetics, sequentially progressing through the inside of my skull, a highly spatial but fully internalised experience, bypassing the outer-ear mechanism.

Envelopment

The auditory spatial concept which momentarily places hearing on a pedestal above other senses is that of envelopment, the term preferred by acousticians, or immersion, synonymous with VR marketing. Char Davies, whose pioneering immersive interactive VR environment installation with 3D computer graphics and interactive 3D sound, Osmose (1995), which tracked the 'the immersant's' breath and balance, expressed the feeling as being, 'spatially encompassed and spatially surrounded'.[18]

Non-binaural spatial cues

There are a range of sonic cues that can inform us of the spatial environment that are not predicated on binaural hearing. The Doppler effect can be experienced mono-aurally and still indicates an object's speed and movement to or away from one's position. There are some categories of environmental sound that we can't help but be spatially placed where our mapping of the world expects them to be located: song birds in the mid to high-frequency range, such as Wren or Blackbird, sing from above; planes fly overhead.

Cocktail party effect

Unlike the controlled acoustic environment of the psychology experiment which tends to use simple uncluttered sinusoidal tones on headphones, our soundscapes are complex; multiple sounds compete for our attention. Spatial auditory competence is also about navigating that contrapuntal ochlophonic world where we depend on comprehending a specific voice within the crowd. The ability to focus one's auditory attention on a single sound source in an acoustic environment with multiple sources is known as the cocktail party effect.[19] As well as timbre and pitch, azimuth perception provides important spatial information to help separate them out. Whilst the urban soundscape is not micro-managed like a film soundtrack, it is revealing how a masterful sound designer such as Walter Murch expertly works to the supposed threshold of aural cognition:

> never to give the audience more than two-and-a-half things to think about aurally at any one moment. Now, those moments can shift very quickly, but if you take a five-second section of sound and feed the audience more than two-and-a-half conceptual lines at the same time, they can't really separate them out. There's just no way to do it, and everything becomes self-cancelling.[20]

Back in our complex real-world soundscape, where we could be exposed to more than 20 simultaneous voices in an enclosed echoic corridor, waiting room or carriage with high levels of residual noise, the cocktail party effect is pushed to its limits. A 'verbal chiaroscuro' is produced where 'dialogue is only partly intelligible'.[21]

Lombard reflex

A noisy atmosphere may also result in an exaggerated application of the Lombard reflex.[22] This is the habitual judgement and continual adjustment of diction in relation to the distance of the desired recipient and the quality of the prevailing sonic environment. An adaptive feedback loop of utterance-listening-utterance (known as vocal sidetone) is formed, resulting in changes not only in vocal intensity (which in an ochlophonic soundscape, can lead to runaway sound levels), but also in pacing, the realignment of spectral peaks and oral articulation to facilitate lip-reading. For mobile phone (a mono medium) use in noisy public transport, the sidetone (a level of the signal of the user's voice being feedback to their speaker) may be masked, blocking auditory feedback and creating a further need for speakers to raise their voice levels, once again contributing to the crowding of the soundscape.

Sonic proxemics

Spatial hearing also has a role to play in how we regulate personal space, known as proxemics.[23] Porteous distinguishes between categories of 'intimate', 'personal', and 'social' distance.[24] Intimate distance, which he suggests ranges from actual physical touch to 8 inches away from a person, maybe appropriately occupied during, for instance, 'wrestling, making love and comforting'.[25] Utterances within this space are generally kept to a whisper. Personal distance equates to a zone between 8 inches and 2 feet away, roughly an 'arm's length'.[26] Within such proximity, cultural norms tend to prescribe that vocalisation is maintained at a moderate speaking level. For social distance (from between 4 and 7 feet), voices tend to be louder. Shouting, though, may create problems of 'social-space violation'.[27] In certain contexts, raising the voice may be regarded as an attempt to enhance the status of the speaker, increasing the extent of his or her acoustic arena and penetrating the personal space of those nearby.[28]

Free field and diffuse field

In the acoustic lab, two contrasting, hypothetical, spatial environments are utilised: free field and diffuse field:

Free field refers to an environment where sound radiates out from its source, uninterrupted, nether to return, 100% absorption. This has the same acoustic effect as flying in the sky devoid of surfaces, a sonic continuum. An approximation of the free field is fashioned in the soundproof anechoic chamber, where only direct sound is supported. This idealised space allows for hyperlocalisation, 'a perceptive effect ... that irresistibly focalizes the listener's attention on the location of emission'.[29]

In contrast, in the diffuse field, there is no acoustical differentiation between direct and reflected sound, sound is omnidirectional, as sound waves travel in every direction, achieving a contiguous space of equal sound pressure throughout. This is the product of 100% reflection, as approximated in a reverberant chamber. A diffuse field provokes the ubiquity effect, 'an effect linked to spatio-temporal conditions that express the difficulty or impossibility of locating a sound source',[30] such as the voice of the celebrant in a Gothic cathedral.

The joy of spatial hearing

The increasing trend to compose and simulate beguiling surround sound environments with VR, gaming, cinema, and electronic music isn't completely new. This joy of hearing the sounding out of space was set into motion from the firing of brass canons on Lake Ullswater in the Lake District summoning the spirit of Wordsworth – 'Such a variety of awful sounds, mixing and comixing, and at the same moment heard from all sides, have a wonderful effect on the mind'[31] – to the spontaneous calling by children in caves and tunnels. It is also a reminder of how sonic spatial traits are learnt, socially determined, context-specific and practised habitually:

> Since listening with understanding depends on culture, rather than on the biology of hearing, auditory spatial awareness must be considered the province of sensory anthropology. To evaluate aural architecture in its cultural context, we must ascertain how acoustic attributes are perceived: by whom, under what conditions, for what purposes, and with what meanings. Understanding aural architecture requires an acceptance of the cultural relativism for all sensory experiences.[32]

Whose spatial hearing?

So far we have explored the important role in speech perception and intelligibility, our sense of place, navigation, agency, mobility and sociability, and ultimately survival that auditory spatial perception can play, presented in pretty much a textbook manner, but as you may have noticed the specificities and caveats have been building up – such as the presupposed symmetry and the standardising of the head and pinnae features – making one question, whose spatial hearing does this model actually represent?

This gap between the individual and the idealised is exemplified in Don Ihde's ground-breaking first-person, phenomenology, Listening and Voice: Phenomenologies of Sound first published in 1976. It cogently charts out some fundamental ground on spatial hearing. Thirty years on, Ihde published the second Edition, with the addition of a new chapter where he tells us of his attempts to embody and adapt to hearing aids due to age-related hearing loss. He recalls how, in Listening and Voice, he exulted in his faculty of perceiving sound that, 'is simultaneously experienced as both surrounding and directional. Hearing aids, however, cannot simply match this phenomenon'.[33] This marked shift in one person's hearing, which is typical of many, exemplifies the need for a more fluid definition of hearing: auraldiveristy.

Auraldiversity's seemingly unradical proposition can be simply articulated: we all hear differently. This is not just divergence with others per se, but with ourselves, momentarily, with bilateral temporal threshold shifts or the kind of unilateral transitory pops and buzzes we may experience when flying, and the more long-lasting, such as the developmental stages of auditory cognition in infancy, age-related hearing loss, autism, post-traumatic stress disorder (PTSD), dementia, misophonia, hyperacusis, and tinnitus. Increasingly these differences are being mediated by personal audio technologies such as noise cancelling headphones, bone conduction hearing aids, and cochlear implants.

Auraldiversity is an immediately relatable proposition and an inevitable universal predicament, that we tacitly know, embody, and expertly practice. When we start to address auraldiversity as sonic practitioners and theorists, however, the givens, the orthopraxy, the standards and regulations, summed up in one word, the norms, are fundamentally challenged. Jonathan Sterne penned a helpful user's guide to Impairment Theory.[34] He starts by defining impairment: 'A physical limit is experienced as an impairment when a person has a point of comparison beyond that limit'.[35] Auraldiversity embraces impaired spatial perception which goes by the medical term paracusis loci, and goes beyond, incorporating spatial hearing that exceeds the prescribed normative model, which we could call panacusis loci, such as echolocation.

Auraltypical perception

The terms auraldiversity and auraltypcial consciously echo the established concept of, neurodiversity, and for some, auraldiversity indivisibly intersects with the neurodiversity. To transpose the standard definition of neurotypical, I assert that: auraltypical people tend to impose their understanding of normal hearing on everyone else as correct and natural.[36]

As auraltypical traits are so deeply ingrained in our pedagogic and scientific underpinnings and hence, the culture at large, it is hard to listen beyond that way of thinking and practising, even when one's own hearing is either less or more than typical.

Textbooks reinforce a pre/proscribed model of hearing, which is characterised by a series of metrics which give measurable boundaries to the auraltypical paradigm. Stated as truths, without condition, qualification, or limitation, they cover audibility and pain thresholds, frequency sensitivity, and even specific dimensions of the ear canal. Of most significance for this discussion is the factoid that humans have binaural hearing and can localise sounds in space around them, with angular resolution down to 1°.

High-speed hand dryers in resonant and reverberant toilets

The concept of auraldiversity crystalised whilst evaluating the wide range of users negatively impacted on by the sound of high-speed hand dryer.[37] The tenor of my report was on aurally divergent loudness perception resulting in discomfort and anxiety for users in this undervalued but necessary space of the public realm or workplace. Important spatial issues had also been raised by blind and visually impaired people, whose hearing was not impaired but who were being robbed of their ability to harness sonic spatial cues inhibiting their orientation: 'Anything which masks ambient sounds could be a problem for a person with sight loss, to a greater or lesser degree'.[38] The high-acoustic energy broadband noise emitted from the strip of air travelling more than 100 mph, colliding with other fast-moving strips, and the casing of the hand dryer and the hands, consequently reverberating in often boxy, highly resonant and echoic space, was akin to an empty glass, being instantly filled of water – a wall of sound, devoid of perspective.

Otologically normal

The A-weighted decibel scale, which has been treated as the acoustic gold standard for much of the 20th century,[39] is used to characterise human hearing's frequency response, where 0 dB(A) is the threshold of hearing and 140 dB(A) equals pain.[40] On scrutinising what underpins this metric, we find an increasingly finessed version of Fletcher and Munson's equal-loudness contours of 1933, enshrined into standard ISO 226:2003, titled, Normal Equal-Loudness-Level Contours.[41] The standard is predicated on a sample endowed with 'otologically normal' hearing. That is a:

> person in a normal state of health who is free from all signs or symptoms of ear disease and from obstructing wax in the ear canals, and who has no history of undue exposure to noise, exposure to potentially ototoxic drugs or familial hearing loss […] within the age limits from 18 years to 25 years inclusive.[42]

In the standard, this circumscribed sample of young adults with acute '20/20' hearing, stands in for all hearing types. Similar exclusive groupings are commonly used in psychology and acoustic engineering experiments aping a WEIRD (White, Educated, Industrialised, Rich, Democratic) approach to sampling.[43,44]

Binaural listener

Detailing the specifications of the otologically normal person in ISO 226:2003 is the stipulation of binaural listening. Binaural means a person who has two matched audio inputs, at either side of their head, both inputs receiving and registering the same signal from a sound source directly in front of them identically. Binaural experiments have been central to the scientific understanding of auditory spatial perception so much so that the received opinion of auditory science is that: 'Spatial hearing is almost-entirely underpinned by "binaural" hearing'[45] – the 'almost' referring to the perception of sound on the extremes of the vertical axes (i.e. elevation) which tend to be monoaural.

Gascia Ouzounian expounds that the predominance of the binaural listener, emerging out of the 19th century: 'He was produced through the emerging science of spatial hearing, the development of binaural and stereo technologies, and the experience of hearing theatre and music transmitted in three dimensions […]'.[46]

The binaural listener is part and parcel of the auraltypical paradigm, and thus presents a substantial normative auditory monolith that requires questioning. Human spatial perception with regards to auditory perception has been taken as read, as a universal ideal which afford access to auditory space as a continuum that is experienced symmetrically.

'Suitably equipped receptor'

Crucially, touchstone tomes on concert hall acoustic design, Beranek (2010) and Barron (2010), placed the audience at the centre of the listening experience, articulating nuanced yet measurable criteria – reverberation time; clarity; intimacy; liveness; warmth; loudness of direct sound; reverberant level; balance; diffusion; ensemble; apparent source width.[47,48] Despite this granularity, however, there is no room for variation in the listener's hearing capacity, predicating the listening experience on, in words of uncompromising modernist composer Milton Babbitt: 'a suitably equipped receptor'.[49]

Beyond the concert hall, this prescribed listenerhood is indelibly inscribed into our audio formats; the stereo image is predicated on the notion of the sweet spot – a spatially rich phantom image, conjured precisely at the tip of a triangle. Along with the symmetry between loudspeaker and listener, it requires an otological normal, binaural listener – 'the relative stability of a singular auditory perspective continues to characterize stereo reception and the (still) modern stereo listener'[50] – as do the numerous iterations of 'immersive' '3D' 'surround sound'.

Paracusis loci

Despite the centrality to human experience – and increasingly, audio technologies – there is a dearth of extant research on spatial hearing impairment. I read this as an indication of the gross underappreciation of this ability compared to say studies on visuospatial skills, spatial memory loss in Alzheimer's or proprioception. It was a pressing concern, however, in Poitzer's foundational text on otology, Lehrbuch der Ohrenheilkunde from 1878. In the section titled Paracusis Loci, the text acknowledges that even with normal ears, humans are not great at judging the direction of a sound, but it provides some potentially existential archaic advice for military surgeons on the risks of unilateral hearing impairment:

> For in outpost service during the night, so important in war, when it is the duty of the advanced sentry to observe any movements in the direction of the enemy's camp which can only be perceived by the sense of hearing, ill consequences might easily arise in case of one-sided dullness of hearing of the sentry by his forming an erroneous judgment as to the direction of the sound.[51]

Asymmetrical hearing

We are often told of the astounding powers of neuroplasticity, where over time our brain can grow and reorganise itself, for example adjusting for asymmetrical hearing loss. But this should not be taken for granted. Responding to my first articulation of the ideas in this chapter at the Aural Diversity Network's fifth Workshop, the hearing therapist Lena Batra recollected the challenges of a sudden onset of asymmetrical hearing loss in addition to her existing hearing condition:

My hearing loss developed symmetrically in childhood, revealing itself more fully by the age of 7 or 8. So I became adjusted to that, in the sense of it being a relatively 'normal' existence with all the usual challenges that come with having a significant hearing loss.

But I will never forget the impact of having an accident in 2007, which left me with a whiplash injury and knocked out a part of my hearing on the left. This additional loss was not definable on pure tone audiometry, but nonetheless, it absolutely blew my perception of spatial hearing out of the water.

The change in my perception was phenomenal and the fact that it did not show up on an audiogram really did nothing to help me, in terms of gaining an understanding of what had happened.

But in effect, what was going on was the fall out, if you like, from a brain that is used to working with similar levels of input from both ears, and being able to shape its internal auditory landscape, based on the subtle and necessary differences in input between the two ears – to a brain that could no longer work well with what it was perceiving, because what it was now receiving from one ear was so vastly different from the other.

Even if the change in hearing on the left was not measurable, the perception of that sound was now completely different and that carried significant consequences for my spatial perception.

I think this is something that we often do not hear enough about – the effect of asymmetry of hearing. Particularly with a sudden onset, where the brain has not yet had the benefit of any natural neural plasticity, that ability to begin to realign itself.[52]

'See spatially through my ears'

At the same workshop, sound artist, Hugh Huddy, who has gradually acquired severe sight loss, talked of the necessity of contextually specific high-quality acoustics affording a kind of 'auditory seeing':

> I think that as my eyesight disappeared, […] I actually seek out physical spaces that optimise rather than reduce masking, so I've got the highest level of auditory seeing so that I can see spatially through my ears. Feels like a very visual thing and maybe the reason why my eyes move, and look, because that's is where I'm experiencing it. It could be my visual brain rewiring.[53]

The lecture hall was long and narrow, and the lectern was positioned in the middle of one of the long sides. As I was using the installed public address system, the direct sound signal of my voice was eclipsed by the stronger signals of my amplified voice coming from both sides of the reverberant room. For Hugh my voice was rendered 'de-spatialised', depriving him of valuable acoustic cues needed to indicate my actual physical presence. When it came to him to ask me a question, he didn't know where in the room to address me: 'Microphones do mess things up for me as do reflective spaces and anything that gets in the way of the position of a sound source'.[54]

Occlusion

Unilateral hearing impairment can be caused by occlusion, a blockage between the ear canal and the middle, such as a build-up of ear wax, or inflammation. This has the auditory effect of bone conduction hearing becoming dominant, with a more boomy low-frequency heavy quality, with the amplification of internal mouth sounds and voice directed to the inner ear.

We can selectively block our ear canals to create an occlusion effect by pressing the tragus inwards (a vestigial ear plug). Or we can create a more fluid mix of fluctuating between closed and open. The swimming stroke, front crawl, activates what Avital Ronell calls periscopic listening:

> one ear is submerged under water.... In the meantime, the other ear exposes itself to the 'outside', making itself capable of hearing the din of a different register of noises, which it receives before turning down. It exchanges places vaguely comparable to outside and inside with the other ear [...] it would appear that the ears are indeed operating stereophonically, attending to double sonic events, receiving and shutting out, responding to the varied calls of air and water pressures. Sometimes an inmixation of the two distinct states can take place, for example, when the ear retains water.[55]

Hyper-asymmetry

The participatory performance practice developed by sound artist, Iris Garrefls, Lauschen – a German term, meaning close listening or eavesdropping – invites the audience to use sound cones, to listen in on her itinerant vocal improvisations. The cones are fashioned rudimentary from rolled-up paper. Garrelfs regards them as, 'a device to re-orient an audience's listening experience and open new aural perspectives'.[56] It provokes a hyper-asymmetric spatial hearing perspective, contradicting the spatial cues received by each ear and like a dog's pinna movements, encourages the searching out of new spatial relations, in tandem with the performer's intimate peregrinations around the audience.

This kind of unilateral spatial experience is pertinent to the sculptural work of Michael Snow. Prompted by his father's increasing sight loss, he challenged the binocular prevalence of sculpture in Monocular Abyss (1982). It invites one eye to spy into the 'abyss' – the pitch-black insides of the geometrical object, provoking a reassessment of ocular perceptual habits.

The leading ear

Aside from marked asymmetrical hearing loss, just as some of the population is righthanded and the other lefthanded, the dominance of a particular side is normal – we tend to have an instinctual preference to use the same ear for mobile phone use – ambidextrous hearing is an exception. The pioneering audiologist, Alfred Tomatis called this the leading ear. From basic experiments in the early 1950s, he surmised: 'there is a preferential ear, designated to execute the more special and more precise control functions, and endowed with an acquired functional dominance [...]'.[57]

Panacusis loci

One of Superman's many super abilities is his exceptional capacity for the cocktail party effect: at ease, he can segment one stream of audio from in infinitesimal mix of troubled voices in true 360° spherical fashion at a distance. Equally impressive is the highly honed skill of echolocation (i.e., sonar), not only reserved for bats and whales. From cane tapping or mouth clicks the prevailing environment can be sounded out. The spectral transformation in the reflections of these consistent transient sounds imparts spatial information on surfaces and shapes. This technique has been exquisitely practised and expounded by Danial Kish, who now trains both blind and sighted people.

He charts the limits of his spatial perception through what he calls 'flash sonar': '360° view; my sonar works as well behind me as it does in front of me; it works around corners; it works through surfaces; generally a kind of fuzzy 3-dimensional surface geometry'.[58] This skill however is predicated on competent binaural hearing and high-frequency sound perception.

Aurally diverse listening positionality

In the sound studio, we are encouraged to hone our listening skills, to unreservedly trust them unreservedly, placing them at the centre of this engineering process – primacy of the ears – with our heads precisely positioned in the sweet spot.[59] Increasingly over the past 30 years, I have had to counterbalance the acoustic energy levels in my mixes. The visual feedback is telling me of a slight asymmetry, a weighing towards the left. This can be accounted for by my conductive hearing loss on my right sight, coupled with constant high-frequency tinnitus on the same side, notwithstanding the predictable presbycusis of a man in his early 50s.

Just as we learn about the developments in our own hearing we need to be evaluating and shaping the built environment for an aurally diverse population. In this regard, an explorative mode of listening positionality should be adopted. Dylan Robinson developed this approach within the discourse of decolonial music studies which expansively includes within its purview: 'self-reflexive questioning of how race, class, gender, sexuality, ability and cultural background intersect and influence the way we are able to hear sound, music, and the world around us'.[60] To put this into action we need to develop methods to co-compose, co-design and co-create with the aurally divergent in the centre, to fashion a spatially equitable environment.

Within the current array of auralization and surround sound tools, including elevation, in acoustically pristine soundproof studios, it is tempting to eschew the limitation of human spatial perception and to mix for smooth and boundless horizontal and vertical coordinates centred on a perfect sphere with the listener's unflinching head at the centre. I remember the astute spatial guidance from my first electro-acoustic music lecturer, Andrew Lewis: 'You are not composing for owls!'. I return to this remark in my own pedagogic work, extending the imperative to pose the question: for whom are you composing, designing or shaping the world?

Spatial hearing and freedom

In conclusion, I want to re-emphasise the centrality of spatial hearing, whatever the disposition of your sensorium. In the chapter's epigraph, I cited Yi-Fu Taun, who links space with freedom and agency. A similar sentiment is expressed by John M. Hull in his audio diary, imparting an aural Road to Damascus moment, through spatial hearing of the most common of everyday sound, rain. Not a wall of white noise, rather individual droplets of differing intensities, imparting discrete spatial information of depth and distance, sounding the surfaces. This revelation came at a difficult time when he was experiencing a sense of bereavement due to his increasing sight loss: 'the rain gives a sense of perspective and of actual relationships of one part of the world and another […]'.[61]

But spatial perception can also be conjured up and savoured in the imagination, devoid of external stimulus. Dictated through the movement of one eyelid spelling out each word, in The Diving-Bell and the Butterfly (1998), Jean-Dominique Bauby shared his experience of paralysis and sensory privation following a massive stroke. With regards to hearing, he lost the ability to receive sound on the right, leaving only his left ear which amplified and distorted the everyday sounds of the hospital: 'Heels clatter on the linoleum, carts crash into one another, hospital workers hail one another with the voices of stockbrokers trying to liquidate, radios nobody listens to are turned on'.[62] Yet his 'diving bell' response to external stimulus was countered by the perspicacity of his auditory imagination, with the exquisite, subtlest of spatial experiences:

> Far from such din, when blessed silence returns, I can listen to the butterflies that flutter inside my head. To hear them, one must be calm and pay close attention, for their wing-beats are barely audible. Loud breathing is enough to drown them out. This is astonishing: my hearing does not improve, yet I hear them better and better. I must have the ear of a butterfly.[63]

Acknowledgements

Special thanks to Lena Batra (www.lenabatra.com), Hugh Huddy and the Aural Diversity Network (https://auraldiversity.org/).

Notes

1 Tuan, *Space & Place*, 51.
2 Clarke, "Music, space and subjectivity", 92.
3 Drever, "The case for auraldiversity".
4 Drever & Hugill, *Aural Diversity*.
5 Granö, *Pure Geography*, 126.
6 Carpenter & McLuhan, *Explorations in Communication*, 207.
7 Carpenter & McLuhan, *Explorations in Communication*, 70.
8 Carpenter & McLuhan, *Explorations in Communication*, 67.
9 Schafer, "Acoustic space", 84.
10 Schafer, "Acoustic space", 84.
11 Blauert, *Spatial Hearing*, 14.
12 Magazú et al., "The Vitruvian Man…", 760.
13 Moore, *An Introduction*, 229.
14 Ferrario et al., "Morphometry of the normal human ear".
15 Van Wanrooij & Van Opstal, "Relearning sound localization".
16 Krebs, "Testing spatial hearing…", 232.
17 Neumann, "Dummy head: KU100".
18 Davies quoted in Dyson, *Sounding New Media*, 107.
19 Cherry, "Some experiments on the recognition of speech".
20 Kenny, "Walter Murch".
21 Chion, *Film: A Sound Art*, 496.
22 Lane & Tranel, "The lombard sign…".
23 Porteous, *Environment and Behaviour*, 58, after Edward T. Hall.
24 Porteous, *Environment and Behaviour*, 58.
25 Porteous, *Environment and Behaviour*, 37.
26 Porteous, *Environment and Behaviour*, 37.
27 Porteous, *Environment and Behaviour*, 37.
28 Porteous, *Environment and Behaviour*, 58.
29 Augoyard & Torgue, *Sonic Experience*, 59.
30 Augoyard & Torgue, *Sonic Experience*, 130.
31 Rée, *I See a Voice*, 54.
32 Blesser & Salter, *Spaces Speak, Are You Listening?*, 18.
33 Ihde, *Listening and Voice*, 246.
34 Sterne, *Diminished Faculties*, 193.
35 Sterne, *Diminished Faculties*, 194.
36 Drever, "The Case for Auraldiversity".

37 Drever, "Sanitary soundscapes".
38 Dan Pescod, RNIB, quoted in Telegraph 2013.
39 Dan Pescod, RNIB, quoted in Telegraph 2017.
40 See Chapter 4 of Sterne's *Diminished Faculties* for a historically informed critique of pain thresholds
41 ISO, *Acoustics – Normal equal-loudness-level contours*, 1.
42 ISO, *Acoustics – Normal equal-loudness-level contours*, 2.
43 Henrich et al., "The weirdest people in the world?".
44 Drever & Cobianchi, "Auraltypical acoustics?".
45 Culling & Akeroyd, "Spatial hearing".
46 Ouzounian, *Stereophonica*, 35.
47 Beranek, *Concert and Opera Houses*.
48 Barron, *Auditorium Acoustics*.
49 Babbitt, "Who cares if you listen?", critiqued in Drever, "'Primacy of the Ear'"
50 Théberge, *Living Stereo*, 21.
51 Poitzer, *Lehrbuch der Ohrenheilkunde*, 202.
52 Batra, *Aural Diversity Network*.
53 Huddy, *Aural Diversity Network*.
54 Huddy, *Aural Diversity Network*.
55 Ronell, *The Telephone Book*, 193.
56 Garrelfs, *Lauschen*.
57 Tomatis, *The Ear and Language*, 115.
58 Kish, *TED: How I use sonar to navigate the world*.
59 Drever, "'Primacy of the Ear'".
60 Robinson, *Hungry Listening*, 10.
61 Hull, *On Sight and Insight*, 27.
62 Bauby, *The Diving-Bell and the Butterfly*, 104.
63 Bauby, *The Diving-Bell and the Butterfly*, 104.

Bibliography

Augoyard, J.-F., & Torgue, H., *Sonic Experience: A Guide to Everyday Sounds*. Trans. McCartney, A. & Paquette, D. Montreal. Kingston: McGill-Queen's University Press, 2005.
Babbitt, M., "Who cares if you listen?" *High Fidelity*, Feb, 1958. https://artasillumination.wordpress.com/wp-content/uploads/2015/02/who-cares-if-you-listen.pdf, Accessed 3rd August 2024.
Barron, B., *Auditorium Acoustics and Architectural Design*, 2nd Edition. Abingdon: Spon Press, 2010.
Batra, L., *Aural Diversity Network, Workshop 5*. University of Leicester, Leicester, 2023.
Bauby, J.-D., *The Diving-Bell and the Butterfly*. London: Fourth Estate, 1998.
Beranek, L., *Concert and Opera Houses: Music, Acoustics and Architecture*, 2nd Edition. New York: Springer, 2010.
Blauert, J., *Spatial Hearing. The Psychophysics of Human Sound Localization*. Cambridge: The MIT Press, 1996.
Blesser, B., & Salter, L.R., *Spaces Speak, Are You Listening? Experiencing Aural Architecture*. Cambridge: The MIT Press, 2007.
Bregman, A.S., *Auditory Scene Analysis: The Perceptual Organization of Sound*. Cambridge: The MIT Press, 1990.
Carpenter, E., & McLuhan, M. (eds.), *Explorations in Communication*. Boston, MA: Beacon Press, 1966.
Cherry, E.C., "Some experiments on the recognition of speech, with one and with two ears". *Journal of the Acoustical Society of America*, 1 Sep; 25(5): 975–979, 1953.
Chion, M., *Film: A Sound Art*. Trans. Gorbman, Claudia. New York: Columbia University Press, 2009.
Clarke, E., "Music, space and subjectivity". In G. Born (Ed.), *Music, Sound and Space: Transformations of Public and Private Experience* (pp. 90–110). Cambridge: Cambridge University Press, 2013.
Culling, J.F., & Akeroyd, M.A., "Spatial hearing". In Christopher J. Plack (Ed.), *Oxford Handbook of Auditory Science: Hearing*. Oxford Library of Psychology, Oxford University Press, Oxford, 2010.
Drever, J.L., "'Primacy of the Ear' – But Whose Ear?: The case for auraldiversity in sonic arts practice and discourse". *Organised Sound*, 24(1), Cambridge University Press: 85–95, 2019.

Drever, J.L., "Sanitary soundscapes: The noise effects from ultra-rapid 'ecological' hand dryers on vulnerable subgroups in publicly accessible toilets". In *Proceedings, AIA-DAGA 2013, the Joint Conference on Acoustics*, Merano, 2013.

Drever, J.L., "Sound art: Hearing in particular". In Jane Grant, John Matthias and David Prior (Eds.), *The Oxford Handbook of Sound Art*. Oxford: Oxford University Press, 2021, pages xi–xxvi. https://doi.org/10.1093/oxfordhb/9780190274054.002.0006

Drever, J.L., "The case for auraldiversity in acoustic regulations and practice: The hand dryer noise story". In *Proceedings of the 24th International Congress on Sound and Vibration*. London, 2017.

Drever, J.L., Cobianchi, M., & Rosas Pérez, C., "Auraltypical acoustics? A critical review of key standards and practices". In *Conference Proceedings, InterNoise22*. Glasgow, Scotland, pp. 5993–6767, 2022.

Drever, J.L. & Hugill, A., *Aural Diversity*. Abingdon: Routledge, 2022.

Dyson, F., *Sounding New Media: Immersion and Embodiment in the Arts and Culture*. Berkeley: University of California Press, 2009.

Ferrario, V.F., Sforza C., Ciusa V., Serrao G. & Tartaglia G.M., "Morphometry of the normal human ear: A cross-sectional study from adolescence to mid-adulthood". *Journal of Craniofacial Genetics and Developmental Biology*, Oct–Dec; 19(4): 226–233, 1999.

Garrelfs, I., "Lauschen". Accessed 26th November 2023, http://irisgarrelfs.com/lauschen.

Granö, J.G., *Pure Geography*. Baltimore, MD: The John Hopkins University Press, 1997.

Henrich, J., Heine, S.J., & Norenzayan, A., "The weirdest people in the world?". *Behavioral Brain* Sciences, June; 33(2–3): 61–83; discussion pp. 83–135, 2010.

Huddy, H., *Aural Diversity Network, Workshop 5*. University of Leicester, Leicester, 2023.

Hull, J.M., *On Sight and Insight: A Journey into the World of Blindness*. Oxford: Oneworld Publication, 1997.

Ihde, D., *Listening and Voice: Phenomenologies of Sound*, 2nd Edition. Albany, NY: State University of New York Press, 2007.

ISO, *Acoustics – Normal Equal-Loudness-Level Contours ISO 226:2003*. Geneva: International Organization for Standardization, 2003.

Kenny, T., "Walter Murch: The search for order & picture". Accessed 26th November 2023, https://filmsound.org/murch/waltermurch.htm.

Kish, D., "How I use sonar to navigate the world". Accessed 26th November 2023, https://www.ted.com/talks/daniel_kish_how_i_use_sonar_to_navigate_the_world.

Krebs, S., "Testing spatial hearing and the development of Kunstkopf Technology, 1957–1981". In Viktoria Tkaczyk, Mara Mills, and Alexandra Hui (Eds.), *Testing Hearing: The Making of Modern Aurality* (pp. 213–242). New York: Oxford University Press, 2020.

Lane, H.L., & Tranel, B., "The Lombard sign and the role of hearing in speech". *Journal Speech Hearing Research*, 14: 677–709, 1971.

Magazù, S., Coletta, N., & Migliardo, F., "The Vitruvian Man of Leonardo da Vinci as a representation of an operational approach to knowledge". *Foundations of Science*, 24: 751–773, 1999.

Moore, B., *An Introduction to the Psychology of Hearing*, 4th Edition. San Diego, CA: Academic Press, 2000.

Neumann, "Binaural Dummy Head KU100". Accessed 26th November 2023, https://www.neumann.com/en-en/products/microphones/ku-100/.

Ouzounian, G., *Stereophonica: Sound and Space in Science, Technology, and the Arts*. Cambridge: MIT Press, 2020.

Poitzer, A., *Lehrbuch der Ohrenheilkunde*. Stuttgart: Enke, 1878.

Porteous, D.J., *Environment and Behaviour: Planning and Everyday Urban Life*. Reading, MA: Addison-Wesley Publishing Company, 1977.

Rasmussen, S.E., *Experiencing Architecture*. Cambridge: MIT Press, 1997.

Rée, J., *I See a Voice: A Philosophical History*. London: Harper Collins, 1999.

Robinson, D., *Hungry Listening: Resonant Theory for Indigenous Sound Studies*. University of Minnesota Press, Minnesota, 2020.

Ronell, A., *The Telephone Book: Technology, Schizophrenia, Electric Speech*. New Ed edition, University of Nebraska Press, Nebraska, 1991.

Schafer, R.M., "Acoustic space". In Seamon, D., and Mugerauer, R. (Eds.), *Dwelling, Place and Environment*. Dordrecht: Springer, 1985.

Sterne, J., *Diminished Faculties: A Political Phenomenology of Impairment*. Duke University Press, Durham, NC, 2021.

The Telegraph, UK newspaper online 18 July 2013 "Super fast hand dryers 'as loud as a road drill', research suggests". Accessed 4th August 2024: https://www.telegraph.co.uk/technology/10187164/Super-fast-hand-dryers-as-loud-as-a-road-drill-research-suggests.html

Théberge, P., Devine, K., & Everrett, T. (eds.), *Living Stereo: Histories and Cultures of Multichannel Sound.* New York: Bloomsbury, 2010.

Tomatis, A., *The Ear and Language*. Trans. B.M. Thompson. Norval: Moulin, 1996.

Tuan, Y.-F., *Space and Place*. Minneapolis: University of Minnesota Press, 2003.

Van Wanrooij, Marc M., & Van Opstal, A. John, "Relearning sound localization with a new ear". *Journal of Neuroscience*, 1 June, 25(22): 5413–5424, 2005.

13

ON SONIC GROWTH AND FORM; BIOMETRIC EVOLUTION OF SOUND AND SPACE

Paul Bavister

Introduction

Architects and designers are used to developing bespoke and unique solutions for clients that result from a design journey. The destination is unknown; the journey is defined by an iterative process of question and experiment that results in project that is of itself and a true and unique reflection of the requirements of the client brief and the knowledge and skill of the design team. Architect Frank Gehry is often quoted as saying, 'If I knew where I was going, I wouldn't do it. When I can predict or plan it, I don't do it'.[1] The ambition of a creative design team is to produce something innovative, challenging and ground-breaking that leads the design conversation going forwards.

Yet there is a contradiction in the aspirations of architects and designers and concert venue requirements, in that there are rigid conventions at play in the development of spaces for sound that can override the creative processes used by architects and creative engineers. The acoustic brief for the room is apparently sacrosanct and cannot be questioned, and that the role of the design team often appears to be reduced to framing the hall with public spaces and cladding. The shape, volume, and materiality of the interior of the hall are determined by inviolate performance factors and international standards that are seemingly unbreakable by both architect, acoustician and client. There is some merit in this, concert hall design is notoriously difficult to meet appropriate standards, and history is dotted with heroic failures by even the most august designers. Look at the saga of Barenek's New York Philharmonic Hall as an example of a highly talented design team providing a widely derided hall. Sticking to the standards offers the best opportunity to avoid failure and public disgrace, as well as possibly developing a hall that is successful and loved by the audience and orchestras alike. However, the result of such codification is too often a predictable room that delivers adequately and professionally and a little anodyne.

This chapter represents a series of experiments undertaken at University College London (UCL) in 2021 and 2022 that used biometric evolutionary computational processes to develop or reverse engineer spaces and music (Figure 13.1). The chapter will discuss the work in the context of acoustics and architecture and will introduce the concepts of biometric evolutionary computation. It will then describe the processes involved in generating emotional responses interfacing with evolutionary systems. It will also briefly describe the experiments undertaken and discuss the limitations and successes of the processes.

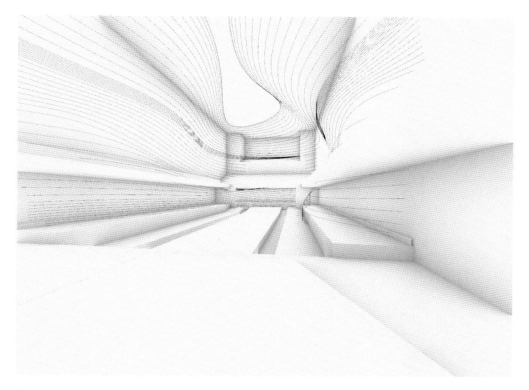

Figure 13.1 A biometrically evolved volume overlaid over a digital model of the Musikvereinssaal.

Background

The history of the development of acoustic spaces, from Greek amphitheatre to the modern concert hall, has been by and large one of chance and trial and error.[2] Acoustic science is a field in its infancy at little over 100 years old. This study regards historic acoustic design as an iterative evolutionary process with little or no acoustic input other than success of an event in subjective terms, leading to the development of the acoustic spaces we know today.

It can be argued that the consolidation of acoustic precedent occurred in the 1930s, when a series of acoustic typologies were defined and established.[3] Such established conventions were based on the 'golden age' of auditoria in the late 19th century, which saw the construction of Vienna's Musikvereinssaal, the Amsterdam Concertgebouw, Boston's Symphony Hall and the Konzerthaus Berlin. The study of such halls allowed acousticians to define and quantify the appropriate volumes and surface textures of the chamber hall and symphony hall to establish a set form, generating adequate support for typical musical genres. Such an approach is great for the concertgoers who demand to see late 19th-century repertoire at its finest, but not so good for the development of contemporary and new musical forms.

Innovation in room shape and technology has continued unabated throughout the 20th century, giving rise to new typologies of halls, such as the fan-shaped hall and the vineyard terrace hall (Forsyth 1985), all of which make for fantastic experiences for an audience, but in sub-optimum acoustic conditions. After 100 years of innovation since Sabine and the quantification of late energy and reverberation, we are still genuflecting towards shoebox halls and vineyard halls as a benchmark for success,[4] leading to a possible ossification of form over the 20th century. Acousticians have suggested that halls should be made for the ears but not the eyes.[5] And therein is a point: not to degrade the architecture, but to maintain a sense of acoustic optima rather than an uncomfortable mix of differing criteria, balancing sight lines, occupancy, and sound.

The clear development of precedents generates an 'optimum' space, a mother ship that defines all that is good or 'right' about a space. This 'optimum' is a clear block to further development, resulting in preconceptions that are 'unbreakable', the Wiener Musikvereinssaal and Stradivarius violins are case studies in point. Will there ever be a better Stradivarius violin? Any difference in tonality or playability will likely be viewed as the negative rather than the positive.

The ossification of hall design is only a part of the story, and it maybe argued that the history of music is also the history of the spaces in which it has been performed – and if one of the elements ceases to evolve, then so does the other. Haydn wrote for a considerable number of different spaces, generating a series of evolving soundscapes that matched site (Mayer 1978). In its purely acoustic form, music previously needed a fluid and wide-ranging set of spaces to evolve into. If this set of spaces is fixed, then it is possible that fixed spaces will not give music any agency to evolve and adapt. This is theoretically possible, music is evolving and growing, but thanks to technology and amplification, music no longer depends on space to give it form. Music can exist decoupled from space, existing in the minds of headphone listeners and the controlled environments of cars.[6]

Evolutionary design

Architecture is a living, evolving thing…. Our culture's striving towards civilisation is manifested in places, houses and cities it creates. As well as providing a protective carapace, these structures also carry symbolic value, and can be seen as being continuous with and emerging from the life of those who inhabit the built environment.[7]

Since the publication of Charles Darwin's 'On the Origin of Species' in 1859, there have been considerable leaps in understanding the mysteries of the planet, and how we have come into being. Many words have been written on the biological mechanics of evolutionary systems, on genetics and DNA strands, but such logic can also be directly applied to other fields of endeavour. The same rules apply: ideas of variation, randomness, fitness criteria, success and replication all have a place in any historical re-evaluation, should that history contain enough epochs or generations to start showing the patterns and rhythms of an evolutionary process.

For an evolutionary system to work, there needs to be an environment or context for the system to work or work against. In the case of biological systems, the oceans of our nascent planet were the original context or playing field for a primitive life form to gain a foothold or perish. A starting population of entities, whether they are single-cell life forms or occupied spaces, holds the key to success or failure. Should one of this initial population not have the requisite capacity to survive in its environment, it will remove itself from the gene pool and play no further part in the process. However, should the entity have the capacity to survive, it will seed the next population. Thus, the survival of the fittest goes on to dominate the field – until, of course, the field changes.

Should the context of the evolutionary system change, then the system must adapt, via mutation, or die. Such adaptations throw up novel answers to the contextual problem and are part of the colour and variance of evolutionary systems.

Over many generations or epochs, a system can evolve something as complex as the human brain without there having been any initial design or organising principle.

In the case of spaces for music, a similar process has been undertaken, in an apparent and a seemingly random series of performance spaces; meeting rooms and back rooms of public houses have seeded an initial population of spatial performance typologies or genotypes. The success or failure of this initial population has gone on to either seed the next population of performance spaces or fade into obscurity, withdrawing from the gene pool.

My work is based on significant developments in the field which have sought to address the development of auditoria and optimised acoustic and musical strategies, by using biometric sensing and virtual acoustics. These are outlined below.

There are examples of optimisation processes for acoustic spaces or musical performances using evolutionary processes such as genetic algorithms that aim to 'evolve' a space for acoustics or sound and music.[8] These algorithms use a predetermined set of criteria or 'fitness functions' to drive an iterative process to reach a goal, be it a better acoustic space or music that suits a different criterion. There are other examples of optimisation that rely on an objective criterion as inputted by a user, such as reverse-engineering of a hall from an impulse response.[9] These optimisation strategies all use a standard measure of success (such as how a room sounds to experienced ears), existing acoustic measures from classic halls, or ISO 3382 standards to develop an 'optimised' result.

However, if the answer to an optimisation strategy is already there – such as the established measures of a classic hall, or an accepted acoustic standard – the value of an optimisation process is called into question. Generating a multi-objective search that balances acoustic metrics with material concerns, for example, will save money but is unlikely to generate a sufficient result that questions or further develops the original established source material.

Evolutionary search processes: basic concepts

A significant amount of work in sound acoustics and architecture uses evolutionary algorithms, and before these projects are reviewed, it is important to understand what they are and how they work, as well as understanding the benefits and problems associated with them. Since the development of digital plugins such as Galapagos, Octopus and Biomorpher for McNeel's Rhino 3D modelling software, a series of projects have been undertaken using evolutionary processes, introducing architects and designers to powerful new optimisation tools.

An evolutionary algorithm (EA) can be described as a stochastic metaheuristic optimisation and search methodology, meaning that the algorithm starts off with random integers that are then selected and honed as to their fitness for a given solution. The potential solution to a problem is assumed to be represented by an individual, and each individual is represented by a set of parameters made up of a string of integers. Each integer is fundamental to the outcome of the search and is linked to a facet of a practical application. This is a system of representation as it will define the outcome in 'real world' terms. The most common form of integers used in EAs is binary (using 0 and 1), but as the early stages of this work are wholly in Max/MSP – an object-oriented coding platform – normal numbers can be used. I have used integers from 1 to 20 as an initial population. These are then scaled to represent aspects of the area to be searched or optimised.

Following the initial generation of integers, the output of the initial value of individual solutions is weighted according to how close it has reached a solution. This is often referred to as a fitness function. When an evolutionary process is taking place, the more successful solutions will yield better-quality offspring, and a better outcome will emerge from the process.

Processes

The language of evolutionary computation borrows heavily from the biological processes that inspired it, thus each integer used is called a gene, and a string of integers or genes that represents a solution is referred to as a chromosome.

Each chromosome will generate its own offspring; the genotype is part of the initial population that will seed subsequent generations, and the phenotypes are emergent phenomena of child chromosomes that will go on to populate the search space.

To establish an evolutionary system, the resulting phenotypes are judged against fitness to the search solution. The proximity to the solution is key, and the phenotypes that are closest to the anticipated answer seed the next generation. This is undertaken by a crossover function, where the two parent chromosomes are taken and split at a predetermined crossover point, with each side of the split being recombined or swapped with the other half, forming a new chromosome entirely based on the parent integers.

To add variation and diversity to the system, it is often useful to mutate a chromosome to a degree. This is undertaken by randomly changing the value of an integer or a series of integers to subtly change the system. Mutation can be of benefit if there is not enough diversity to improve on a problem, and the system is at risk of getting stuck and not progressing.

The termination of an evolutionary search can be defined by whether a solution has been found or cannot be bettered by a fixed number of iterations. This type of search relies on a clear idea of what the solution ought to be. If the answer to a given problem is unknown, it is often better to use different fitness functions (see interactive evolutionary systems below.)

The duration of the process is defined by the 'search space' which is a term made up of the possible combinations of the initial integer set. Thus, if a chromosome is made up of ten integers with each integer having a possible value of 1–20, the possible combination of solutions is 1,020. Systems that can self-evaluate can make significant progress in large search spaces, undertaking the generation and evaluation of each epoch in milliseconds; this leads to very efficient search operations and quick answers.

The key steps involved in a genetic algorithmic search are:

1. Initialisation: an initial population of candidate solutions generated randomly.
2. Evaluation: fitness value of each of the candidate solutions is evaluated against a set of criteria.
3. Selection: the candidate solutions that are closest to the fitness function are chosen to seed the next generation.
4. Recombination: the chosen solutions are split and combined creating better solutions or offspring (if the EA is a genetic algorithm).
5. Mutation: a solution is modified on a percentile scale to generate more genetic diversity in the system.
6. Replacement: the resulting offspring of the selection, recombination and mutation processes replace the parent solutions.
7. Repetition: steps 2–6 are then repeated until the search criteria are met or a limit on epochs or generations is reached.

Interactive evolutionary programming

There are a multitude of differing approaches to evolutionary systems, but typically, most use an autonomous evaluation of the fitness function to move on to the next epoch. In computational

Biometric evolution of sound and space

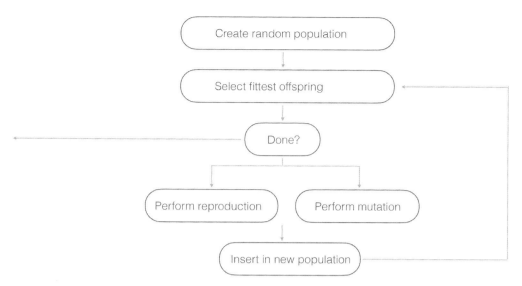

Figure 13.2 Evolutionary programming.[10]

terms, this can be a very quick analysis of the current set of genotypes, and then an evaluation. This system allows huge search spaces to be evaluated over short periods of time, which is useful when there is a clear goal in mind, and the fitness function for each generation is clear, i.e. when the programmer knows what the answer to the algorithm is supposed to be, but not sure of the approach to get there.

The approach used in this work is that of an interactive EA, where a human user evaluates fitness to seed the next generation. This is useful in establishing an approach where the answer is unknown and can vary according to different operators and users. Interactive evolutionary searches are often used in creative applications to generate new forms of media that may not have been foreseen by the system user.

There are drawbacks to this approach; in an automated EA, the search space can be vast and may require as many as 10,000 generations to find an acceptable solution. An interactive EA, on the other hand, requires a human to review the current generation of phenotypes and then make a judgement call on the degree of success. This takes time, and only a fraction of the search space can be covered at any one time. There is also a risk that the operator will change their mind and start feeling that another route is preferable to the current path.

Interactive evolutionary algorithms (IEA) are slow, subject to user fatigue and inconsistency and do not cover a wide search space, but they do allow for unique interpretations of a given search criteria in a way that a computer cannot. One method of addressing the inherent issues with interactive genetic algorithms or IEAs is to reduce the possible generations that may lead to an optimal solution. This can be achieved by not having the crossover function (Figure 13.2). Such a move will retain an element of evolution in the search but could reduce the number of generations from tens of thousands to just double figures. This will drastically impact the amount of search space covered, but with more users generating less phenotypes, a suitable search space can be covered. It is critical that the mutation function is relatively high, as without it there will be limited genetic diversity in the system, and a solution will never be found.

A second key aspect to optimising an interactive evolutionary search is to reduce the thinking time for a user. A typical genetic algorithm (GA) requires recombination of parent chromosomes, which requires a user to select two parents that share a relative level of fitness to the original search problem. This takes time in the analysis of each phenotype solution. A way of optimising this is to use a single-parent genotype from which to generate multiple offspring,[11] Use of single-parent EAs significantly reduces the thinking time for a user, and selection can be made consciously, by inputting a selection into a system, or unconsciously, via sensors on the body.

Despite the drawbacks, interactive EAs are useful in establishing a direction of travel, and with multiple users can show commonalities in preferences that may not show an answer but could show where a future answer may lie.

Biometric sensing

Research has been undertaken in the psychophysiological response of the body to stimuli and the impact of lateral reflections in auditoria.[12] The research showed that an increase in lateral fraction increased the psychophysiological response in a listener. This research is fundamental in looking at the impact of external stimuli on a listener, but more importantly, that such a reaction is not a conscious response to stimuli but an unconscious response; it is what the body is telling us, shorn of the dogma of external conventions and pressures such as external contexts and subjective influences.

It is possible that these two ideas of evolutionary computation and biometric sensing could work together to produce a definitive answer to a time-old question of what listeners really like, over what listeners think or say they like: defining an objective view over a typically subjective view. This work addresses the imbalance of creative reciprocity between sound and space, and uses technology to reinforce and reinstate the relationship between the two?

The two experiments described here use psychophysiological reactions in the body as a fitness function to an evolutionary process, evolving space into an acceptable envelope for sound, and sound into an appropriate fit into a given space, this is the definition of Biometric Evolutionary Computation (Figure 13.3).

The two experiments illustrated here were undertaken in the UCL sound lab in 2021 and 2022. Listeners were seated in an acoustically treated sound lab in a first-order ambisonic sound reproduction system. In the first experiment, they were played fragments of classical music in an evolving virtual acoustic space where the azimuth, elevation, and distance of an acoustic reflector were optimised to suit the listeners' psychophysiological response, detected via dermal response on the first and second fingers of the non-dominant hand (because the participant is less likely to move it during the tests). In the second experiment, the listener was immersed into an evolving sonic space whose rhythms and timbres were evolving to form a 'soundscape' that generated 'emotional change' in the participant.

The output of the evolved space experiment was a 'room' that evolved purely from the psychophysiological response of the participant to a selected piece of music. This room was developed parametrically in Rhino 3D modelling software using the Grasshopper plugin. Each participant developed a room specific to their own taste (Figures 13.5, 13.6 and 13.7).

The output of the evolved soundscape experiment was a three-minute audio file that was evolved from the psychophysiological reaction of the listener to evolved sounds in one of four preselected sites and convolved in real-time in the sound lab (Figures 13.4). This generated a sequence of timbres, rhythms, and pitches that whilst not 'music' in a narrative and structural sense, nevertheless allows an observer to analyse tempo structures and note onsets in the context of architectural space.

EFP 2 - Mozart
Emotional Response: 20

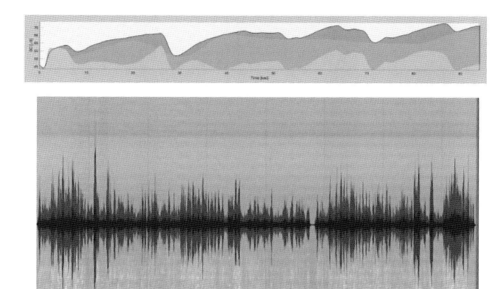

Figure 13.3 GSR data from a response to Mozart with waveform and spectrograph analysis.

The experiments were undertaken in a controlled environment with ethics approval from UCL. Each participant was questioned at the end of the test to triangulate the results and ensure that the emotional change detected in the participant was the result of positive emotional stimuli as opposed to a negative.

The galvanic skin response (GSR) output of all the tests was analysed in Matlab's Ledalab GRS convolution software that looked at the dermal response output. The output of the virtual acoustic simulation in the evolved space from music experiment was analysed in Marshall Day's IRIS software, and the acoustics of the Rhino model were analysed in Arthur van der Harten's Pachyderm software and compared. The output of the evolutionary soundscape experiment was analysed in IRCAM's Partiels music analysis software (Figure 13.8) and that output was further analysed in IBM's SPSS Statistical Analysis software.

The main findings or output of the experiments showed that it is possible, via an evolutionary process driven by psychophysiological response, to optimise an acoustic space that differs from the ISO 3382 standards without a drop in subjective appeal. The tests also showed that it is possible to replicate the intuitive response of a musician in a space, in that of a non-musician, due to the observation of significant correlations between late reverberation, clarity and tempo in the analysis of the results.

Figure 13.4 Spaces used in the evolutionary soundscape experiment, from top left; Entrance to Barbican Cold Stores, Charterhouse Cloisters, Fabric Nightclub and the Barbican Foyer (all images Solen Fluzin).

Musical limitations

It should be noted that this work concerns itself with spaces designed for acoustic music in Europe during the last millennia, the Western canon and the 'golden age' of auditoria, where attributes of a host space such as volume and material conditions can be considered integral to generating an emotional response in a listener. This may exclude music from Asia for example, where music was

Biometric evolution of sound and space

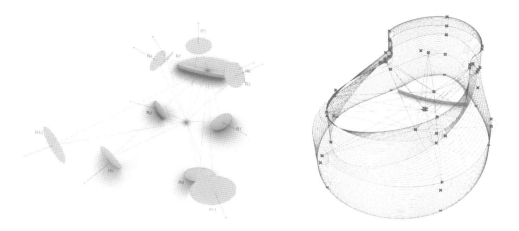

Figure 13.5 Variable reflectors (left) are independently addressable to enable a modulated envelope (right) to be evolved from biometric responses to input stimuli.

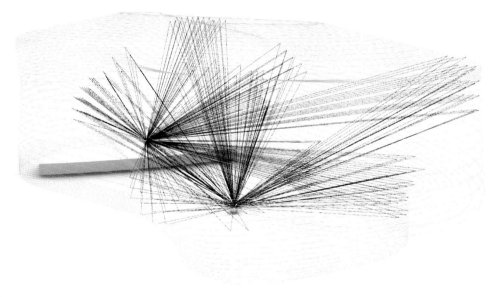

Figure 13.6 Raytracing analysis of a hall evolved to suit Beethoven's 7th Symphony.

either played externally, or in heavily absorbent interiors, the sound propagation being weighted towards purely direct energy, with the internal spatial condition of early and late reflections playing less of a role in the augmentation and perception of the music.[13]

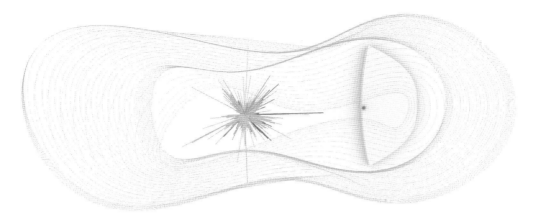

Figure 13.7 Beethoven Hall viewed from the Top with IRIS plot showing acoustic reflection data.

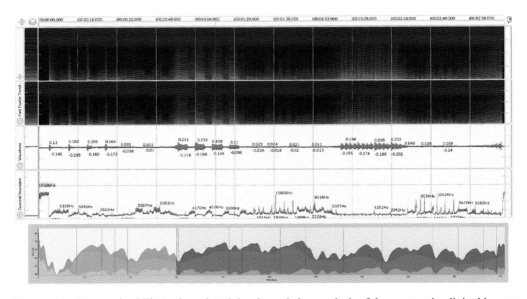

Figure 13.8 Deconvolved Electrodermal Activity shown below analysis of the source stimuli, in this case, a section of Beethoven's 7th Symphony. The red lines on the electrodermal response (EDA) graph represent recorded emotional change. This represents a positive valence, which was triangulated with the post-test questionnaire.

Contextual limitations

Emotional response to music is not just about the music, but the context(s) of the experience. There are many more external conditions that can influence the impact of any given experience.[14] An experimental set-up places an individual in a laboratory environment, which is isolated from external influences and distractions. Public settings are not, and an experience is influenced and modulated by a plethora of distractions, some beneficial and some detrimental, for example:

- Social issues, friends and accompanying guests.
- External moods and contexts; was the listener having a good day anyway?
- Knowledge of the music and priming.
- The sense of occasion and space are contributing factors.
- The energy of the musician.
- The response of other audience members.

Practical limitations

This is a small study undertaken by a single researcher with limited access to funding and resources. However, the work serves as a starting point for a more expansive study with a much larger pool of participants and specialists in their field to support the work.

Sample size

Over the five days of tests that were undertaken, including the pilot tests, and repeat tests we saw 53 persons tested. For the main tests, 23 participants' emotional changes were recorded to the stimuli. This is an adequate number for initial tests, but for the results to be validated, there would need to be significantly more test subjects.

Influence of test site

The results indicate a predilection towards drier environments in the demographic studied. This of course could be related to a response to sound and space as the experiments set out to test, but there is also the possibility that the psychophysiological response was a reaction to the space where the test was actually taking place, and not necessarily the media being tested. The test room was a clearly dry space, intimate, and small. The simulation of large rooms would have been in contradiction to the actual space where the tests took place. It could be that the participants were reacting to the real space and not the virtual. This cannot be ruled out as a possibility and can only be tested by using differing site conditions beyond the scope of this work.

Trends

The experiments documented in this paper were developed as a process to show a new way of looking at subjective analysis of objective acoustic parameters, which is not intended to be seen as a fully realised process of developing new music, architecture and spaces for listening. It is beyond the scope of this work to define a codified, clear, and final answer for biometrically evolved sounds and auditoria, but it is possible to develop a process to see what listeners are feeling during moments of an emotional reaction, and the work turns a critical light on current practice in developing spaces for listening.

The process has shown that when judiciously applied, the technology of evolutionary programming and biometric sensing can be utilised to show trends and a direction of travel as to sound, space, and listening, but it will not be possible to suggest a final and definitive output. Barron states that the concert experience is multi-dimensional.[15] Human nature is fickle, tastes are fleeting and elusive, and change over time; at the heart of the issue of taste is a conflicting series of inputs that are constantly evolving and changing, pushed and pulled by so many moving parts that an attempt to draw a definitive conclusion would be meaningless.

Key observations from the experiments

The success of a series of tests like this where subjectivity is so prevalent is difficult to define. Yet the post-interview questionnaires undertaken showed that a large majority of the participants had a broadly positive experience where a positive outcome was developed as a part of the experience. This feedback was triangulated with the results from the GSR analysis and a meaningful output could be developed.

Each participant had a unique output from each of the tests, and this output went on to be aggregated into mean results that allowed observations in taste to be undertaken. It is remarkable that the elusive chimaera of taste can be studied in such an objective way. Core observations are as follows.

A commonality between the two tests was the predilection towards a drier, high-clarity environment across both tests, generating associated soundscapes and new spaces. This relates to the findings of Barron's 1988 review of British auditoria, Schroeder's 1974 comparative study of European auditoria, and a study on preference ratings of concert hall acoustics using subjective sensory profiles.[16] In all studies, a clear preference towards clarity or reverberant environments emerged. However, there is no data available on the listening habits of the listeners tested. The listeners in the experiments documented here were young – with an average age of 26 – and favoured a more intimate environment. The Fabric nightclub and Barbican foyer were both very dry spaces with high lateral fraction metrics and C80 metrics that were significantly higher than ISO 3382 standards. The spaces evolved in the biometrically evolved space test also generated LF and C80 metrics that exceeded ISO 3382 standards. This commonality is significant, as it questions the trend towards excessively reverberant spaces developed in practice. The listeners all enjoyed music, but mostly listened to contemporary popular music in club environments or on personal hi-fi systems. Some of the participants with a positive valence in the tests expressed a preference for clear articulation of the music, rather than the blurring of the notes traditionally associated with the accepted and established Western canon and typical auditoria design.

Research through design

Whilst significant research has been undertaken to understand how the body reacts to sound and space, no one yet has merged this with evolutionary processes to develop new working methodologies in a design context.

There have been attempts to use biometric sensing as a tool for determining levels of emotional response,[17] and that there have been exercises in reverse-engineering auditoria from impulse responses driven by objective manipulation.[18] There have been attempts at interactive evolutionary strategies to generate music, but none of these methodologies use an unconscious response to sound and site; they all rely on objective and live critique, without addressing the underlying conditions that may be leading the participant to any conclusion.

The major gap in the thinking about new methodologies developing spaces for music can be seen in architecture where, as shown in this chapter, new spaces might be created from the psychophysiological reactions of the body, and new meaning might be found in simulating creative acts as a road map of potential uses for space.

Generating space to fixed music

This work is not the first to reverse-engineer a space from an acoustic signature, as Boning and Bassuet have shown is possible, nor is it the first to use galvanic skin response as an indicator of emotional response to sound and space, as has been shown by Lokki and Pätynen, but this works unites these acoustic interrogations and adds evolutionary programming to develop a wholly new and original working methodology (Figure 13.9).

Future research

If multiple demographics were to be tested in the same conditions and they all responded positively towards conditions that suited the test environment and not the simulated, then it would be possible to draw the conclusion that the test site was a significant influencing factor in the tests, a test in an auditoria or external space. Again, this would only be drawn out in future testing with a larger demographic.

In conjunction with larger sample sizes and wider demographics, a more focused research question should be addressed to limit the search space to a more manageable size and get a much more definitive outcome from any future tests. A much simpler question would have less variability and a smaller chromosome chain, resulting in a more manageable search. A typical example could be the optimisation of fixed absorption in a hall or similar. Smaller, more focused searches would have more precise meaning.

Conclusion: is a democratic space a practical response to new halls?

Generating data sets from occasional music listeners would be a low common denominator to generate new spaces and could be seen as a dictatorship of a majority; if that majority is largely

Figure 13.9 Beethoven Hall overlaid on a digital model of the Musikvereinssaal – note the exaggerated width.

untrained and develops hall shapes to suit mass appeal and only popular repertoire, then the work has turned full circle, and we are back to where we started – a hegemony of popularists.

An alternative would be to obtain data sets from only trained listeners and developers of new spaces, such as the directors and shareholders of orchestras and venues. Again, this may only provide a data set based on economic advantages and commercial imperative.

Inviting professional musicians and acousticians to provide ears and data would certainly be interesting and has the potential to lift the idea above the quotidian and develop something new and novel. There is a risk of elitism, and the output being a series of halls and spaces that only suit a minority, but this would allow a new trend to develop that is outside of the current stream of acoustic thinking.

Most auditoria have the capacity for variability in acoustic performance with deployable absorption and openable coupled volumes to decrease and increase reverberation for example, and future research can input into this adaptability by being able to be tuned to match variations in taste without impacting the visual and spatial aspects of the architecture too detrimentally.

Sound Files

These sounds are available on the Routledge E-Book page: The audio file 'Coldstores Mix', included here, is an anonymous compilation of three evolved soundscapes undertaken during the experiments. The concept of 'music' is deconstructed into a series of acoustic/compositional actions that match existent modes of acoustic understanding. To quote Edgard Varèse, 'Music is organized sound' (Varèse & Wen-Chung1966). Thus 'sounds' were to be generatively optimised in such a way that may reflect a language of musical acoustics, and can be described in qualitative and quantitative terms, just as space is to be generatively optimised to suit the understanding of music in the sister experiment. The outputs presented here come from not a piece of music or composition, but an 'aesthetically potent environment' (Pask 1971) that has agency over its occupation.

Notes

1. Gehry, F. (2001), https://archiscapes.wordpress.com/2015/01/11/frank-owen-gehry-architect/.
2. Blesser, B. and Salter, L.-R. (2009), *Spaces speak, are you listening?: Experiencing aural architecture*, MIT Press.
3. Thompson, E. (2002), *The soundscape of modernity: Architectural acoustics and the culture of listening in America, 1900–1933*, MIT Press.
4. Glanz, J. (2000), 'Art + physics = beautiful music', *The New York Times*, April 18th.
5. Lokki, T. and Pätynen, J. (2018), 'Concert halls should primary please the ear, not the eye', *Proceedings of the Institute of Acoustics*, 40, 378–385.
6. Blesser, B. and Salter, L.-R. (2009), *Spaces speak, are you listening?: Experiencing aural architecture*, MIT Press.
7. Pask, G. writing in the introduction of Frazer, J. (1995), *An evolutionary architecture: Themes VII*, AA Press London.
8. Sato, S.-I., Hayashi, T., Takizawa, A., Tani, A., Kawamura, H. and Ando, Y. (2004), 'Acoustic design of theatres applying genetic algorithms', *Journal of Temporal Design in Architecture and the Environment*, 4.
9. Boning, W. and Bassuet, A. (2014), 'From the sound up: Reverse-engineering room shapes from sound signatures', *The Journal of the Acoustical Society of America*, 136(4), 2218–2218. https://doi.org/10.1121/1.4900050.
10. Fogel, L. J., Owens, A. J. and Walsh, M. J. (1966), 'Intelligent decision making through a simulation of evolution', *Behavioral Science*, 11(4), 253–272. https://doi.org/10.1002/bs.3830110403.

11 Rata, I., Shvartsburg, A. A., Horoi, M., Frauenheim, T., Siu, K. W. M. and Jackson, K. A. (2000), 'Single-parent evolution algorithm and the optimization of Si clusters', *Physical Review Letters*, 85(3), 546–549. https://doi.org/10.1103/PhysRevLett.85.546.
12 Pätynen, J. and Lokki, T. (2016), 'Concert halls with strong and lateral sound increase the emotional impact of orchestra music', *The Journal of the Acoustical Society of America*, 139(3), 1214–1224. https://doi.org/10.1121/1.4944038.
13 Blesser, B. and Salter, L.-R. (2009), *Spaces speak, are you listening?: Experiencing aural architecture*, MIT Press.
14 Juslin, P. N. and Västfjäll, D. (2008), 'Emotional responses to music: The need to consider underlying mechanisms', *Behavioral and Brain Sciences*, 31(5). https://doi.org/10.1017/S0140525X08005293.
15 Barron, M. and Foulkes, T. J. (1994), 'Auditorium acoustics and architectural design', *The Journal of the Acoustical Society of America*, 96. https://doi.org/10.1121/1.410457.
16 Lokki, T., Pätynen, J., Kuusinen, A. and Tervo, S. (2012), 'Disentangling preference ratings of concert hall acoustics using subjective sensory profiles', *The Journal of the Acoustical Society of America*, 132(5), 3148–3161. https://doi.org/10.1121/1.4756826.
17 Pätynen, J. and Lokki, T. (2016), 'Concert halls with strong and lateral sound increase the emotional impact of orchestra music', *The Journal of the Acoustical Society of America*, 139(3), 1214–1224. https://doi.org/10.1121/1.4944038.
18 Boning, W. and Bassuet, A. (2014), 'From the sound up: Reverse-engineering room shapes from sound signatures', *The Journal of the Acoustical Society of America*, 136(4), 2218–2218. https://doi.org/10.1121/1.4900050.

Bibliography

Barron, M. and Foulkes, T. J. (1994), 'Auditorium acoustics and architectural design', *The Journal of the Acoustical Society of America*, 96. https://doi.org/10.1121/1.410457
Blesser, B. and Salter, L.-R. (2009), *Spaces speak, are you listening?: Experiencing aural architecture*, Cambridge: MIT Press.
Boning, W. and Bassuet, A. (2014), 'From the sound up: Reverse-engineering room shapes from sound signatures', *The Journal of the Acoustical Society of America*, 136(4), 2218–2218. https://doi.org/10.1121/1.4900050
Fogel, L. J., Owens, A. J. and Walsh, M. J. (1966), 'Intelligent decision making through a simulation of evolution', *Behavioral Science*, 11(4), 253–272. https://doi.org/10.1002/bs.3830110403
Forsyth, M. (1985), 'Buildings for Music', Cambridge: MIT Press.
Gehry F. (2001), https://archiscapes.wordpress.com/2015/01/11/frank-owen-gehry-architect/
Glanz, J. (2000), 'Art + physics = beautiful music', *The New York Times*, April 18th.
Juslin, P. N. and Västfjäll, D. (2008), 'Emotional responses to music: The need to consider underlying mechanisms', *Behavioral and Brain Sciences*, 31(5). https://doi.org/10.1017/S0140525X08005293
Lokki, T. and Pätynen, J. (2018), 'Concert halls should primary please the ear, not the eye', *Proceedings of the Institute of Acoustics*, 40, 378–385.
Lokki, T., Pätynen, J., Kuusinen, A. and Tervo, S. (2012), 'Disentangling preference ratings of concert hall acoustics using subjective sensory profiles', *The Journal of the Acoustical Society of America*, 132(5), 3148–3161. https://doi.org/10.1121/1.4756826
Meyer J. (1978), 'Raumakustik und Orchesterklang in den Konzertsälen Joseph Haydns', *Acustica*, 41(3), 145–162.
Pask, G (1971), "A Comment, A Case History, and a Plan", in Cybernetic Serendipity, J. Reichardt, (Ed.), Rapp and Carroll, 1970. Reprinted in *Cybernetics, Art and Ideas*, Reichardt, J., (Ed.) Studio Vista, London, 1971, 76–99
Pask, G. writing in the introduction of Frazer, J. (1995), *An evolutionary architecture: Themes VII*, AA Press London.
Pätynen, J. and Lokki, T. (2016), 'Concert halls with strong and lateral sound increase the emotional impact of orchestra music', *The Journal of the Acoustical Society of America*, 139(3), pp. 1214–1224. https://doi.org/10.1121/1.4944038

Rata, I., Shvartsburg, A. A., Horoi, M., Frauenheim, T., Siu, K. W. M. and Jackson, K. A. (2000), 'Single-parent evolution algorithm and the optimization of Si clusters', *Physical Review Letters*, 85(3), 546–549. https://doi.org/10.1103/PhysRevLett.85.546

Sato, S.-I., Hayashi, T., Takizawa, A., Tani, A., Kawamura, H., & Ando, Y. (2004), 'Acoustic Design of Theatres Applying Genetic Algorithms', *Journal of Temporal Design in Architecture and the Environment*, Vol. 4.

Thompson, E. (2002), *The soundscape of modernity: Architectural acoustics and the culture of listening in America, 1900–1933*, MIT Press.

Varèse, E. and Wen-Chung, C., 1966. The liberation of sound. *Perspectives of New Music*, 5(1), 11–19.

14
INFRASTRUCTURES OF INAUDIBILITY
The spatial politics of assistive listening

Jonathan Tyrrell

Introduction: symbol, space, orientation

In April of 1978, a request was made from the UK Department of the Environment to the chairman of the London Transport Executive inquiring as to whether they would consider placing symbols for deaf people throughout their network.[1] While symbols for disability in general were not new, there had been a push from advocacy groups to place specific symbols for the deaf, which in the words of the Department would:

> indicate that there was someone […] who would be ready and willing to take the trouble to help them – not necessarily trained in sign language, but to have an appreciation of deaf peoples' difficulties, to be patient, to speak clearly to help lip reading and to have pencil and paper ready if other communication fails.[2]

The chairman's response was sympathetic but limited. He recognised the need for such symbols, and agreed to run a pilot programme, but could not commit to placing them regularly at ticketing offices or on buses because he could not guarantee consistency in the staff that ran them.

By the early 2000s Transport for London (TFL) had begun mandatory disability training for its employees. Specific material on hearing impairment included deaf awareness and facial expression quizzes, noise levels of common environments, hearing dog provision, and cultural attitudes towards disability.[3] Integrated deafness training for TFL employees looks like progress from the perspective of the late 1970s, however, in the intervening years an important technological addition to deafness infrastructure was introduced: the induction loop (IL). Using the principle of electromagnetic induction these devices were able to transmit sonic information to the receiving telecoils integrated in hearing aids and cochlear implants (CI). They began appearing at operable Help Points on platforms, in Ticketing Windows and dedicated bus seats, and at centralised listening points in major stations, where they relayed public address announcements. Thus, the same pre-millennial period of development that saw mandatory disability training create greater awareness and sensitivity among employees, equipping them with the exact qualities that were considered rare in the 1970s, also laid the foundation for a new approach to deafness; it transformed what was essentially a human resources issue into a technological one. By shifting the mode of communication from the attentive human body (lip reading, patient speech, etc.) to the sonic-spatial figure of the magnetic field produced by coils of copper wire, care was technologised and externalised. While this chapter does question the effectiveness of the IL system, that is not its main goal. Rather it wishes to explore several spatial paradoxes embodied therein, namely the independence of near-field electromagnetic listening from the architecture which contains it, and the idea of a soundscape that is pervasive, yet silent.

The induction loop: sound as spatial solid

The earliest known patent for ILs was granted in the UK in 1941 to Russian-born inventor Joseph Poliakoff.[4] Poliakoff's company, Multitone, became one of the first developers of wearable assistive listening devices which offered both IL and microphone-based listening. It was not until the 1970s however that IL manufacturers overcame technical challenges with amplification and portability, leading eventually to more widespread installation.[5] Many hearing aids and CI today are outfitted with a telecoil ('T-coil') option alongside the microphone and Bluetooth components.

The phenomenon of electromagnetic induction, described both by Michael Faraday and Joseph Henry in the early 19th century, demonstrates how a variable magnetic field can induce a corresponding electrical current in a conductor that intersects spatially with that field. In the case of the IL an audio signal, transduced into electrical current, runs through a coil of wire which creates a field of magnetic flux around the listening point. When another coil of conductive wire (in this case a much smaller one housed in the hearing aid or CI) is placed within this field, the inverse occurs: the oscillations of the magnetic field induce the electrons within the second coil to move in a corresponding pattern. This current is then amplified and transduced into soundwaves through a tiny speaker (in the case of the hearing aid) or as direct stimulations of the cochlea through the electrode of a CI.

Several key points are worth noting for the purposes of this study. Firstly, unlike other forms of information-bearing electromagnetic radiation (television, radio, light) which broadcast over large distances at high frequency, the IL listening point produces a near-field sonic space; one must be in close proximity to hear anything. However, because a magnetic field follows the inverse-square law, the volume intensity increases exponentially the closer one places the receiving wire coil to the source. Secondly, the orientation of the source coil and receiving coil relative to one another produces radical variation in the perceived sound intensity. Owing to the geometry of the magnetic force lines, the maximum amount of induction occurs when the coils are placed perpendicular to one another, and will be virtually cancelled out when placed in parallel. In practical terms this requires the hearing aid, implant, or loop listening device to be oriented in a particular spatial way, which in turn orients the body of the wearer. Lastly, it is worth dwelling on a point that may not be obvious. The magnetic field does not occupy space; in the language of Stephen Connor, it procures space.[6] Sound waves propagating through the air will fill up their containing volume (room, corridor, street), folding back on themselves in successive reverberant reflections that reinscribe the geometry of said volume. The magnetic field of the IL on the other hand has fixed, if somewhat blurry, edges that are defined by the electromagnetic principle in combination with the geometry, scale, and density of the induction coil's winding. That one may experience a sound that does not fill the space of its sounding, and is independent of spatial volumes altogether, is not intuitive. It is also distinct, I suggest, from other non-mechanical transmissions of sound (Bluetooth, radio, Wi-Fi, etc.), which because of their scale have a far-less differentiated experiential field. Insofar as they localise sonic information intended for a particular biopolitical group, and orient the body in so doing, ILs produce pockets of politicised electromagnetic flux that emerge as invisible, yet audible, spatial solids. Before starting into a typological examination of the material artefacts that produce these spatial solids it is useful to explore on a conceptual level what deafness offers sound-space discourse, and why assistive listening infrastructures in the TFL are an appropriate case study.

The work of disability studies operates in two complementary and overlapping modes. First, it is engaged in the real efforts of advocacy; struggling on behalf of (and by) those who identify as disabled and for whom the built environment as it stands offers impediments to fuller participation. This work can take many forms: artistic practices, political activism, interventions, but the common goal is to rebalance spatial justice through practical action. In this mode disability is teleological; it works from what exists towards the twin goals of inclusion and access. But in a second sense, it works backwards from what exists to reveal how practices and products deemed 'normal' have been constructed with ideologies both tacit and explicit. Following Stuart Hall, Jonathan Sterne writes how the primary work of criticism is 'showing that what is obvious is ideological and what is considered to be transcendental is actually contingent; it provides genealogies of things that appear to be given'.[7] The term appearance is important here when thinking about public space, inviting a reflection on a key concept from Hannah Arendt: how can one enact the 'space of appearance' (the space of democratic power), when that very space explicitly prevents the actions necessary to produce it? Moreover, disability politicises spaces that are considered apolitical in Arendt's opposition of the oikos and polis. Public toilets for instance become a contested site under disability frameworks, revealing that even the study areas of political analysis themselves can have shadow spaces that evade the ableist gaze. I wish to reemphasise the overlapping nature of these two modes of operation for disability studies. Practical action and criticism proceed under many guises and there is much slippage between them. However, I find it useful to highlight this bidirectional nature of disability studies as it helps to frame the following study of the TFL network and to address the very valid question of why the object of study for disability is disability infrastructure itself.

Deafness has been a significant and controversial thread within disability studies since its emergence alongside the disability movements of the 1990s. In Enforcing Normalcy: disability, deafness, and the body (1995) Lennard J. Davis interrogates the concept of 'normal' as it was used in the construction of national identity.[8] The press of Gallaudet University (an American research university dedicated to education of deaf and hard-of-hearing students) has been publishing literature on deaf studies since the 1980s. H-Dirksen L. Bauman and Joseph J. Murray, both scholars at Gallaudet, have advanced the concept of 'deaf gain', highlighting the ambivalent place that deafness occupies within disability studies.[9] Deaf studies is also the twin shadow of sound studies, owing to the latter's fixation with technology, in particular, those of early pioneers like Edison (who was hard of hearing) but more importantly of Alexander Graham Bell who remains highly controversial in the deaf community for his practical advocacy of eugenics as a form of cultural and bio-physical erasure. Michele Friedner and Stefan Helmreich (2012) have proposed overlaps between sound and deaf studies, revealing how low-frequency vibrational space literally runs beneath the hearing/non-hearing binary.[10] Recent scholars like Mara Mills[11] and Jaipreet Virdi[12] have published extensively on deafness and assistive listening from a cultural-historical and technological perspective.

However, it is only recently that disability studies has entered architectural discourse. Jos Boys' Doing Disability Differently (2014), and the edited volume Disability, space, architecture (2017) are key indictments of architecture's propensity to begrudgingly enact minimum accessibility standards, relegating them to a tier below 'design'. Understandably, such critiques tend to be directed towards impediments to mobility (armatures of circulation within buildings being often the most contested sites) rather than hearing. Exceptions to this include Hansel Bauman's concept of 'deaf space' which involves rethinking architecture on deaf terms from the ground-up.[13] However, it seems altogether too conspicuous that architectural acoustics is woefully undervalued in design not to draw the conclusion that its minor presence in building-related disability literature is symptomatic of a general indifference, if not ignorance, of the sonic world and its relations to what we build. I wish not however to draw a false distinction between issues of mobility and hearing, on the contrary. What I aim to explore in the following section is just how mobility and hearing are intrinsically linked at the site of assistive listening, most pointedly within the complex navigation and orientation infrastructures that pervade and enable public transport.

Disability and deafness on the TFL

Since at least the early 1900s, disability and public transport have been entwined in policy frameworks. Records from the TFL archives show requests from 1905 by the London County Council Education Department to the London General Omnibus Company for travel of blind, deaf, and 'defective' children.[14] Several internal memos from the 1940s discussing accommodation of blind travellers revolve around an existing policy which barred them from travelling alone on the network without a permit. There are at least three reports of blind passengers in the 1940s actively resisting this regulation, insisting on their right to manage personal risk and bodily autonomy in a publicly owned system.[15] Deafness however was not recognised as an issue until significantly later. The introduction of the Chronically Sick and Disabled Persons Act of 1970 sought equal access for all people to recreational and educational facilities, and by extension the means of transport to access them.[16] Although this signifies a shift from accommodation to right, the mechanism and means for doing so moved slowly, varied between types of disability, and was not without bureaucratic resistance and prejudice.[17] As outlined in the introduction it wasn't until the late 1970s that the TFL began looking at infrastructures for assistive listening (bearing in mind that visual aids were highly limited at that time). A poignant letter from a deaf passenger in November 1984 recounts how her Hearing Dog (guide dog for the deaf), given to her through a charitable organisation set up by the Royal National Institute for the Deaf, was refused free access to a bus, whereas access for blind guide dogs had been well established.[18] It is worth noting that the letter describes how, like guide dogs for the blind, the animal 'replaces the lost sense', fostering an inter-species relationality through hearing which highlights the absurdity of paying for what is effectively an extension of the body. What is also at stake in the letter, and supported by internal correspondence of the TFL around how to deal with the complaint, is the question of visibility; of appearances. Not only was the hearing dog not visibly identifiable (something that would be later remedied by using purple scarves to distinguish them) but in a system that relied on visual and textual codes to enact policy, hearing issues could easily go unnoticed.

Such instances illustrate a further nuance to Arendt's 'space of appearance' which will become important later. For now, it serves to elucidate and underscore the main reasons for choosing the TFL network as a site for the critical examination of assistive listening and what it means for conceptions of sound and space: namely that mobility and orientation are the preconditions for participation in public space. And insofar as this conveyance must be managed through policy frameworks, technologies, data, human resources, bodily autonomy, and personal risk; disability and transport are always already a biopolitical concern. The particular hearing technologies created by ILs represent how these biopolitics are localised in shadow soundscapes that live nested, unheard and mostly unseen, within the larger normative one.

Figure 14.1 Three types of induction loops: Left: TFL Ticketing Window, Centre: Help Point (note the double-negation of the non-functional telecoil symbol), Right: Listening Point (images by author).

IL typologies on the TFL

ILs do have applications for larger spaces (auditoria, places of worship, classrooms), where the differentiation in magnetic field is managed through technical innovation. These instances too have their own spatial-sonic politics, however since the TFL network is furnished almost exclusively with point sources, they will be the typological focus of the chapter. ILs have been installed at Ticketing Windows, Help Points, and on platforms throughout the London Underground and Overground networks (Figure 14.1). They are also available at designated seats on some buses and in Black Cabs (licensed by the TFL). While they all use the same principle of electromagnetic induction, each of these types function quite differently in terms of their position in space, relationship to the source audio, and the nature of their material construction. As this chapter is focussed on critically examining assistive listening infrastructure as a spatial-sonic network I have limited the study to the Overground and Underground train systems. It is important to acknowledge that visual information like signage and digital displays are integral components of deafness infrastructure, and that the inherent risk in examining only the sound-based elements denies what Mara Mills calls the 'always already multimodal' nature of deaf communication.[19] However, given that the ILs do not overlap with any other navigational systems, and are denoted by a unique visual symbolic language (which is furthermore unrecognised by most people) they constitute valid objects of study on their own.

Type 1: Ticketing Windows

As discussed above, the original site for assistive listening infrastructure on the TFL network was the Ticketing Window, the locus of 1:1 communication between passenger and officer. The loop system here consisted of a microphone, amplifier, and desktop-sized loop with the international symbol for telecoils: an ear (outer ear, auricle, or pinnae to be precise), with a line running diagonally through it.[20] Passengers would enable the telecoil on their personal assistive listening device (hearing aid, CI) and, should they wish to verbalise their request to the officer, could receive an answer through the IL. This also relies on some amount of vocalic technique,[21] both on the part of the passenger, who may or may not feel comfortable vocalising, but also on the part of the officer who will not have the same auditory feedback as they might when vocalising into a conventional PA (public address) system. In some ways, the ILs at Ticketing Windows operate the same as all sonic media (stethoscope, microphone, megaphone) in that they erect a barrier between sounder and perceiver that alters the chain of sounding and in doing so produces new power relations. The

Ticketing Windows in which the ILs are situated however constitute thresholds across which all sound encounters resistance. The high visual transmission of an impact-resistant acrylic window shows something of an inverse reduction for sound, resulting in microphones and speakers being employed to bridge the divide. What is therefore meant to be a point of contact, through sound of an electromagnetic nature or otherwise, turns out to be a site of multimodal sonic obfuscation.

Type 2: Help Points

Dotted along platforms on the Overground and Underground are a series of Help Points which have operable buttons for receiving general information, or making emergency contact. Circular, with an approximate 570 mm diameter, they can be found mounted on dedicated tubular steel stands or affixed to columns or walls. They have built-in loudspeakers and microphones for two-way audio communication, and most are fitted with ILs which serve a very different function than those installed at Ticketing Windows. The Help Points are call and response machines; a button is pressed and a call is placed, much like a telephone (complete with standard telephone ring tones). But rather than connecting with an agent on duty, the call is rerouted transnationally much in the same way technical support lines connect locations that are determined more by political economy than geographic relation. Whereas telecoils at the Ticketing Window constitute sonic thresholds that manifest between bodies, the Help Points connect the material 'here' with the immaterial 'elsewhere'. Thus they recall earlier telephonic devices like the 'Theatrophone', a late-19th-century technology in Europe that broadcast cultural events like Opera across telephone lines to listening rooms outfitted with arrays of receiving ear-pieces.[22] The coupling of airborne and electromagnetic sound in the same device also intersects with the history of telephony as it was the electromagnetic fields produced by the telephone handset that were the original impetus for Poliakoff to seek a patent for ILs. Many landline telephones on the market today come equipped with a telecoil function that provides better amplification for hearing-aid/CI users, and before the widespread use of Bluetooth, mobile phone companies also explored integration of telecoils, often in the form of external devices like the now-discontinued Nokia Inductive Loopsets.[23]

Type 3: PA listening points

The last type examined in this paper is the public address Listening Point. Like the Help Points, these can be integrated with the architecture or stand-alone, mounted on steel pedestals. The Listening Point is a device for one-way communication that transduces the audio announcements from the public address system at any given station into a magnetic field. They are, in the majority of cases, contained in a blue square housing around 220 mm × 220 mm that presents on its face no words or visual language other than the international symbol for the telecoil in white, the letter 'T' for telecoil, and a 'CE' to denote its compliance with electrical safety standards. In other words, they are textually mute. Again, the technology and scientific principle of electromagnetic induction remains the same as in the Ticketing Window and Help Points but the content and spatial relations are quite different. If desktop ILs at Ticketing Windows constitute an ambivalent sonic bridge across an otherwise thick threshold, and the Help Points provide telephonic call and response that bridges fixed points to floating space, the listening point, as an extension of the public address system produces a series of spatial and political inversions.

To furnish a conceptual reading of the IL listening point I will invoke two related ideas well known in sound studies: the acousmatic voice and schizophonia. The term 'acousmatic' from the

Greek *akousmatikoi*, traces back to an esoteric pedagogical practice favoured by Pythagoras.[24] Michel Chion describes how certain followers of the Greek philosopher would listen to his teachings from behind a veil or curtain, so as to better absorb the word of the Master: an act of division in visual space producing deeper connection in sonic space. The trope of the authoritative voice emanating from an occluded position is prevalent in sonic-spatial configurations of religious ceremony but is not, as Chion discusses, limited to the archaic. In Freudian Psychoanalysis the Master (in this case the doctor) is seated behind the patient, sequestered from view; and in film, the quintessential modernist artform, there is extensive use of the disembodied voice emanating from off-screen.[25] Pierre Schaeffer, pioneer of musique concrete is credited as reviving the term 'acousmatique' in the 1950s, repackaging it conceptually towards the experience of listening to sounds divorced from their sources. A related concept, 'reduced listening' demands of the listener the near-impossible task of taking electro-acoustically produced sounds at face value, manifested then and there through loudspeakers as sonic matter itself rather than a referent of some recognisable mnemonic or cultural symbol.

Schizophonia, a term coined by one of the originators of soundscape studies, R. Murray Schafer, also describes the spatial separation of sound from source. The resonance with schizophrenia is quite intentional, as Schafer considered the split between sound and source to be cleaved exclusively by the emergence of electroacoustic sound reproduction – itself a menacing appendage of perhaps his greatest spectre: the 20th-century city. Whereas schizophrenia referred to a splitting of individual subjectivity, schizophonia was a correlative symptom of larger maladies in the body politic brought on by modernity. The concept is also wrapped up in the idea of authenticity, casting those sounds produced by loudspeakers as ontologically inferior. In something of a recursive tautology, he writes, 'Originally all sounds were originals'.[26]

The aim here is not to mount a critique of schizophonia specifically, or Schafer more broadly, as this has already been done by Mitchell Akiyama, Dylan Robinson, Mickey Vallee, and many others.[27] And in fact, Schafer's work and soundscape studies have become stronger for this, showing resilience as a concept and methodology while also requiring continual updating, as every good concept should. And so, despite its pathological overtones, and laden as it is with a nostalgic anti-urbanist position, the term still has some value for this study.

In being 'blown up and shot around the world', schizophonic sounds are not only phenomenologically separated from their sources, they are also spatially separated. They are adrift in the cultural aether to be appropriated, reproduced, and reconfigured at will. In this sense it shares much in common with Pierre Schaeffer's acoustmêtre, but with opposite value judgements attached; what R. Murray Schafer saw as a troubling symptom in 1970s Canada, Pierre Schaeffer had seen as a path for aesthetic renewal some 15 years earlier in post-war France. But there is a difference between schizophonia and the original Greek rendering of the akousmatikoi that I wish to dwell on: that of presence. In Schafer's schizophonia and Schaeffer's acousmatic listening the original source of the sound is absent (and in the case of the latter's 'reduced listening' actively negated). However, in the original Greek anecdote Pythagoras is still in the same space, but his presence is validated by sound, not vision. Furthermore, the veil through which he speaks is not simply a framing architecture, it is the material embodiment of his acousmatic voice.

Having established this idea that in the original askmatikoi there lies a spatial and material complexity that is absent from both schizophonia and 'reduced listening' I will now return to the discussion of the IL listening point by focussing on a specific instance in London's Euston station. This example will be discussed through the above concepts, in relation to its specific spatial-sonic context, and how it interrogates what is understood as 'room tone'.

Figure 14.2 Left: Central induction loop at Euston station device occluded from view and blocked from access by a marketing display. Centre: Large Listening Point detail (also obstructed by a construction barrier). Right: Accompanying symbolic denotation mounted further up the column (images by author).

Euston station case study

Below the main notice board in Euston station, affixed to a polished granite-clad rectangular column, is an IL Listening Point of a different order. Measuring approximately 500 mm × 500 mm it is significantly larger than the standard platform Listening Points described above. An aluminium mounting frame surrounds a familiar visual configuration: the international telecoil symbol, the 'T', and 'CE'. Interestingly, this particular case is accompanied by a supplementary symbolic demarcation in the form of a bent-steel sign of matching sedated navy, mounted further up the column (Figure 14.2).

On its opposing faces appear a familiar abstraction of the outer ear but in this case with the 'T' placed towards the centre of the pinna, in rough spatial and physiological correspondence with the entrance to the ear canal (the external auditory meatus). To the left appears a coil of wire and waves of information radiating towards it. This signage appears to be unique among the network, and the only example I have seen where the actual physical apparatus (wire loop) that enables assistive hearing becomes part of the symbolic language. As such it is a diagram of a transductive spatial-sonic process, not, as in the case of the international symbol, a biopolitical identifier, removed from space and time, that negates the sensory organ in question by striking a diagonal line through it.

The IL itself functions the same as the other listening points. It transmits announcements from the public address system that are simultaneously broadcast over loudspeakers into the space, through an amplified magnetic coil. This simultaneity of sound broadcast in air and through electromagnetic radiation is a feature of all listening points. In the Euston station example, however, the discrepancy between the two is radically heightened. The main space is large in scale, with high ceilings, orthogonal walls, and surface materials of predominantly polished granite and glass. It is reverberant enough to obscure basic messages from the PA system even if the space is unoccupied; add the periodic surge of travellers, hard shoes, suitcases, clinking of metal cutlery on ceramic plates, and the occasional improvised piano recital (on a quirky upright below the stairs), and it is near-impossible to decipher the announcements. On the other hand, the audio generated in the IL's near field is entirely free of room tone. This is not to say it is 'noiseless', as IL listening is also prone to some amount of electromagnetic interference, but it is as close to a dry reproduction of the PA announcer's pre-recorded voice as possible. Furthermore, for those using CI-based telecoils, these messages bypass air altogether.

What this points to is a politics of 'room tone': the unique product of an architectural volume's scale, proportion, and material construction animated by the types and rhythms of activities that take place therein. It is the most fundamental 'sound of space'. It is also historically linked with mediated listening as it was the first recording on the phonograph that inevitably captured the background noise as well as the signal. In our case study of the IL at Euston station, the room tone and reverberant signature of a space are so present to in fact become impediments to meaning for the normative ear. Somewhat paradoxically then, the more mediated way of listening through telecoil-enabled hearing aids or implants, puts the listener in a stronger position of privilege relative to the fidelity of messages being broadcast, all the while doing so through a means of listening that is inaudible to the unmediated ear.

Let us return now to schizophonia and acousmatic listening to give more shape to this paradox. The disembodied voice of public address systems is inherently schizophonic in that the source of what is heard, in this case, most often a female voice, has been separated in both time and space from any given moment of sounding. Announcements are everywhere and nowhere simultaneously, issuing forth from loudspeakers most often hidden from view, or camouflaged into ceiling structures. But the non-localisability and disembodiment of the public address system are challenged under the mediated listening practices of ILs. As discussed, electromagnetic listening produces static spatial solids which one must enter to experience sound. This spatial pocket then produces an intimate static voice-field which negates the architecture in which it is housed through the absence of room tone, preferring a spatial encounter where the speaking voice is embodied, in oracular fashion, temporarily in the IL object itself. Much in the same way that I have argued the acousmatic voice of Pythagoras finds temporary embodiment in the veil, so too does the pre-recorded voice of the public address system become temporarily embodied in the acrylic signage, aluminium frames, hidden copper coils and amplification components of the IL listening point. A lateral historical digression to look at John William Waterhouse's Consulting the Oracle from 1884 displays (albeit in a rather sensationalised Pre-Raphaelite way) the temporary fusing of disembodied voices into material form (Figure 14.3).

Figure 14.3 Left: Consulting The Oracle, John William Waterhouse, 1884 (Wikimedia). The oracular voice is embodied in the 'Teraph', a mummified head towards which the priestess strains to listen. Right: Author conducting listening to custom-built induction loop relaying signals from a geophone.

But in this paradoxical condition where powers of listening shift between ability and disability; where those who are commonly understood to have hearing disadvantages ultimately have access to a more direct and undistorted public sound than those are not, the overt symbolism and practical exigencies of contemporary urban life also hold sway. In order for one to make use of the ILs at Ticketing Windows, Help Points, or Listening Points, certain bodily practices must be enacted. The officer at a Ticketing Window must enable the Telecoil device and speak with conscious technique into the microphone. At the Help Points, a conversation with a remote voice is heard through the loudspeaker and the intimate space of the middle or inner ears simultaneously. Lastly, in the case of the Listening Point, a body must be placed within the small pocket of space defined by the inversely squared fall-off pattern of the electromagnetic field. Though this does not require any active identification of one's reliance on mediated hearing, the body is inevitably organised within the visual symbolic space of disability biopolitics. What was difficult to identify in the 1970s because of its lack of visual appearances becomes organised under visual ciphers in the technologised approach to deafness care some 50 years later.

Conclusion: symbolism, ornamentation and the sonic space of appearance

According to Hannah Arendt, true democratic power resides in what she calls, 'the space of appearance'. Independent of governmental form and urban form alike, the space of appearance is contingent; it manifests through the gathering of citizens in speech and action and can dissipate unless continually re-performed. For Arendt, collective action and the senses are closely linked, for it is through the 'common sense', that humans establish the reality of the world. And this is where deafness, considered in combination with transport poses an issue. If the 'common sense', as a confirmation of reality acts as a bulwark against the erosion of democratic power, and if such a sense is established in the space of appearance, then where does this leave those for whom sensory perception does not meet pre-established societal norms? Furthermore, if speech and action are the primary mechanisms of enacting democratic power in public space, how do we account for forms of speaking and listening which are not understood to be sonic (sign language) or are already mediated by technologies that select and filter the soundscape?

As discussed earlier, much of the historical discourse around accommodation of deafness within the TFL hinged on questions of visual appearances; signs of hearing loss are not easily spotted, and through increasing miniaturisation neither are hearing aids and CI's. It is therefore one of the ironies of the IL system that while it is visually and physically pervasive, it is highly underused by those for whom it is intended and is mostly unseen and altogether unheard by those for whom it is not. As such these devices acquire a diminished semantic value, moving from a network of symbols into a system of ornamentation. With their sonic function underused or at best inaudible, what remains is their visual presence as empty signifiers, continually relaying sonic information in public space that falls outside the 'common sense' and runs beneath the 'space of appearance'.

This chapter has discussed IL assistive listening infrastructure on the London public transport network, exploring how the particulars of electromagnetic induction, and the situated installations of these technologies, invite a rethinking of how sound relates to space. How do we talk about the sound of space in the context of electromagnetic listening where sound is independent of the space of its sounding? This raises a challenge to the idea that sound is passively framed by the architecture which houses it. Furthermore, it reveals that what we consider as the soundscape of public space is highly dependent on the biopolitics of listening.

Notes

1. "Signs – Symbol for the Deaf," April 21, 1978, LT000820/044, Transport for London Corporate Archives.
2. "Signs – Symbol for the Deaf."
3. "Signs – Symbol for the Deaf."
4. Joseph Poliakoff and Sneath Oswald Barber, *Method of and Apparatus for the Transmission of Speech and Other Sounds*, United States US2252641A, filed June 30, 1938, and issued August 12, 1941, https://patents.google.com/patent/US2252641A/en?inventor=Joseph+Poliakoff.
5. Noel D. Matkin and Wayne O. Olsen, "Response of Hearing Aids With Induction Loop Amplification Systems," *American Annals of the Deaf* 115, no. 2 (1970): 73–78.
6. Steven Connor, *Dumbstruck: A Cultural History of Ventriloquism* (Oxford: University Press, 2000), 12.
7. Jonathan Sterne, *Diminished Faculties: A Political Phenomenology of Impairment* (Durham, NC: Duke University Press, 2021), 196.
8. Lennard J. Davis, *Enforcing Normalcy: Disability, Deafness, and the Body* (London: Verso, 2014).
9. H-Dirksen L. Bauman and Joseph J. Murray, "Deaf Gain: Raising the Stakes for Human Diversity," H-Dirksen L. Bauman and Joseph J. Murray, Editors. *Foreword by Andrew Solomon; Afterword by Tove Skuttnab-Kangas* (Minneapolis: University of Minnesota Press, 2014).
10. Michele Friedner and Stefan Helmreich, "Sound Studies Meets Deaf Studies," *The Senses and Society* 7, no. 1 (March 1, 2012): 72–86, https://doi.org/10.2752/174589312X13173255802120.
11. Mara Mills, "Do Signals Have Politics? Inscribing Abilities in Cochlear Implants," in *The Oxford Handbook of Sound Studies, Oxford Handbooks* (Oxford, New York: Oxford University Press, 2011).
12. Jaipreet Virdi, *Hearing Happiness: Deafness Cures in History* (Chicago, IL: University of Chicago Press, 2022).
13. Hansel Bauman, "DEAFSPACE: An Architecture toward a More Livable and Sustainable World," in H-Dirksen L. Bauman and Joseph J. Murray, Editors. *Deaf Gain: Raising the Stakes for Human Diversity: Foreword by Andrew Solomon; Afterword by Tove Skuttnab-Kangas* (Minneapolis: University of Minnesota Press, 2014).
14. "London General Omnibus Company Limited Board Minute Book," October 1, 1903, LT000067/029, Transport for London Corporate Archives.
15. "Signs – Symbol for the Deaf."
16. Emily Haves, "Fiftieth Anniversary of the Chronically Sick and Disabled Persons Act 1970," May 7, 2020, https://lordslibrary.parliament.uk/fiftieth-anniversary-of-the-chronically-sick-and-disabled-persons-act-1970/.
17. Several instances of internal communications among TFL administration actually show a sarcastic disdain toward requests for disability accommodation.
18. "Signs – Symbol for the Deaf."
19. Mara Mills, "Deafness," in *Keywords in Sound* (Duke University Press, 2015), 52.
20. Whether this line denotes a negation or bypass of the pinna is ambiguous.
21. This term builds on Jonathan Sterne's concept of 'audile technique' developed in: *The Audible Past: Cultural Origins of Sound Reproduction* (Durham, NC: Duke University Press, 2003).
22. Gascia Ouzounian, *Stereophonica: Sound and Space in Science, Technology, and the Arts* (Cambridge: The MIT Press, 2021), 31.
23. Nokia Oyj, "Inductive Loop for Hearing Aid Users Now Available for More Nokia Mobile Phone Models," *GlobeNewswire News Room*, June 14, 2000, https://www.globenewswire.com/en/news-release/2000/06/14/1845246/0/en/Inductive-loop-for-hearing-aid-users-now-available-for-more-Nokia-mobile-phone-models.html.
24. Michel Chion, *The Voice in Cinema*, ed. and trans. Claudia Gorbman (New York: Columbia University Press, 1999), 19.
25. Chion, 19.
26. R. Murray Schafer, *The Soundscape: Our Sonic Environment and the Tuning of the World* (Rochester, VT: Destiny Books, 1994), 90.
27. See: Mickey Vallee, *Sounding Bodies Sounding Worlds: An Exploration of Embodiments in Sound* (Singapore: Springer, 2020); Dylan Robinson, *Hungry Listening: Resonant Theory for Indigenous Sound Studies, Indigenous Americas* (Minneapolis: University of Minnesota Press, 2020); Mitchell Akiyama, "Transparent Listening: Soundscape Composition's Objects of Study," *RACAR: Revue d'art Canadienne* 35 (January 1, 2010): 54.

Bibliography

Dylan Robinson, *Hungry Listening: Resonant Theory for Indigenous Sound Studies, Indigenous Americas* (Minneapolis: University of Minnesota Press, 2020); Mitchell Akiyama, "Transparent Listening: Soundscape Composition's Objects of Study," *RACAR: Revue d'art Canadienne* 35 (January 1, 2010): 54.

Emily Haves, "Fiftieth Anniversary of the Chronically Sick and Disabled Persons Act 1970," May 7, 2020, https://lordslibrary.parliament.uk/fiftieth-anniversary-of-the-chronically-sick-and-disabled-persons-act-1970/.

Gascia Ouzounian, *Stereophonica: Sound and Space in Science, Technology, and the Arts* (Cambridge: The MIT Press, 2021), 31.

H-Dirksen L. Bauman and Joseph J. Murray, "Deaf Gain: Raising the Stakes for Human Diversity," in H-Dirksen L. Bauman and Joseph J. Murray, Editors. *Foreword by Andrew Solomon; Afterword by Tove Skuttnab-Kangas* (Minneapolis: University of Minnesota Press, 2014).

Hansel Bauman, "DEAFSPACE: An Architecture toward a More Livable and Sustainable World," in Human Diversity. H-Dirksen L. Bauman and Joseph J. Murray, Editors. *Deaf Gain: Raising the Stakes for: Foreword by Andrew Solomon ; Afterword by Tove Skuttnab-Kangas* (Minneapolis: University of Minnesota Press, 2014), 375–401.

Jaipreet Virdi, *Hearing Happiness: Deafness Cures in History* (Chicago, IL: University of Chicago Press, 2022).

John William Waterhouse. *Consulting the Oracle*. Oil paint on canvas. Tate collection N01541, 1884.

Jonathan Sterne, *The Audible Past: Cultural Origins of Sound Reproduction* (Durham, NC: Duke University Press, 2003).

Jonathan Sterne, *Diminished Faculties: A Political Phenomenology of Impairment* (Durham, NC: Duke University Press, 2021), 196.

Jos Boys. *Doing Disability Differently: An Alternative Handbook on Architecture, Dis/Ability and Designing for Everyday Life*. 1st ed. London, England: Routledge, 2014.

Jos Boys. *Disability, Space, Architecture: A Reader*. Edited by Jos Boys. London, England: Routledge, 2017.

Joseph Poliakoff and Sneath Oswald Barber, *Method of and Apparatus for the Transmission of Speech and Other Sounds*, United States US2252641A, filed June 30, 1938, and issued August 12, 1941, https://patents.google.com/patent/US2252641A/en?inventor=Joseph+Poliakoff.

Lennard J. Davis, *Enforcing Normalcy: Disability, Deafness, and the Body* (London: Verso, 2014).

"London General Omnibus Company Limited Board Minute Book," October 1, 1903, LT000067/029, Transport for London Corporate Archives.

Mara Mills, "Deafness," in *Keywords in Sound*, eds. David Novak and Matt Sakakeeny (Durham, North Carolina: Duke University Press, 2015), 52.

Mara Mills, "Do Signals Have Politics? Inscribing Abilities in Cochlear Implants," in *The Oxford Handbook of Sound Studies, Oxford Handbooks*, eds. Trevor Pinch and Karin Bijsterveld (Oxford, New York: Oxford University Press, 2011). 320–346.

Michel Chion, *The Voice in Cinema*, ed. and trans. Claudia Gorbman (New York: Columbia University Press, 1999), 19.

Michele Friedner and Stefan Helmreich, "Sound Studies Meets Deaf Studies," *The Senses and Society* 7, no. 1 (March 1, 2012): 72–86, https://doi.org/10.2752/174589312X13173255802120.

Mickey Vallee, *Sounding Bodies Sounding Worlds: An Exploration of Embodiments in Sound* (Singapore: Springer, 2020).

Noel D. Matkin and Wayne O. Olsen, "Response of Hearing Aids With Induction Loop Amplification Systems," *American Annals of the Deaf* 115, no. 2 (1970): 73–78.

Nokia Oyj, "Inductive Loop for Hearing Aid Users Now Available for More Nokia Mobile Phone Models," *GlobeNewswire News Room*, June 14, 2000, https://www.globenewswire.com/en/news-release/2000/06/14/1845246/0/en/Inductive-loop-for-hearing-aid-users-now-available-for-more-Nokia-mobile-phone-models.html.

R. Murray Schafer, *The Soundscape: Our Sonic Environment and the Tuning of the World* (Rochester, VT: Destiny Books, 1994), 90.

"Signs – Symbol for the Deaf," April 21, 1978, LT000820/044, Transport for London Corporate Archives.

Steven Connor, *Dumbstruck: A Cultural History of Ventriloquism* (Oxford: University Press, 2000), 12.

PART III

Philosophy and politics

Mark Burry

How to link the physical world of sound with the thinking and reasoning that underlies philosophy to the negotiation between differing ideas and interests within society – politics?

At its most visceral we have abundant examples of situations where sound affects us in a variety of social settings ranging from the deeply personal to one of a contrived collective response. In a spiritual setting, for example, sound skidding around a resonant interior will elevate inner reflection whether through oration or music, or song, and similarly in a concert setting. Sound and politics are at play as an expression of collective power expressed as much as through communal chanting at a political rally as by the roar of the crowd ricocheting within an arena as the winning goal is scored just as the whistle blows on full-time.

This section on the philosophy and politics of sound is thus an exploration of how sound helps us give shape to our understanding of space and the implications this has on our shared social, cultural, and political lives. Sounds are created artificially, or at least managed for the purpose of manipulating human experience – one of amplifying one's inner voice such as in a religious or concert setting, or expressing a shared will in a political or sporting context. Clearly, there are a wide range of philosophical implications with the curation of sound in space including ethics, experience, expression, culture, and the nature of the human condition. We become more aware of our responsibilities towards non-human sensitivities and sensibilities given that sound can be used as much to manipulate will as it can be contrived to lift human experience.

Thinking about philosophy and sound, the phenomenology of sound considers how it adds to and affects the way we perceive space and place, or more accurately expressed, how sound helps space become a *place* in terms of human experience. When sound fills an exterior or interior space it alters our sensory experiences augmenting our perception of scale, distance, and presence. This implies sound has cultural significance given that different communities may interpret the significance of sound in quite different ways.

When considering the philosophy and politics of the sound of space we need to take on board the creation of an acoustic ecology. Soundscapes influence the inhabitants of the spaces in which sound is being heard; both artificial and natural sounds contribute to the ecology of the space. The resulting sensorily augmented environment can influence the psychology and behaviour of individuals within the space. In given acoustic ecologies sound has a political dimension when it is used as a tool to promote power and domination including sonic surveillance and crowd control. The eeriness of an air-raid warning has led to a meaningful familiarity over the decades, as has the church bell, and the call to prayer from a minaret. But as much as authority draws on the power of sound to express power itself, so to have protest songs and chants become the media of protest and resistance.

An architect can design spaces with materials to reflect the brilliance of sound to accentuate one set of values – vibrancy within a café for example, or materials that deliberately mute sound such as a funereal setting. In a traditionally ocular-centric profession, there is plenty of room for architects to widen their creative repertoire to design for sound to influence emotions and behaviour. Such manipulation includes consideration of the role of sound to help give shape to the identity of a space and its stimulus to aid personal recall and collective memory. In spaces designed for a rich acoustic experience, sound can evoke certain emotions or pull from the depths obscure memories linked to the place in which the sound is being listened to that might otherwise remain hidden.

The ethics of sound consider how sound is used, who has the right to make a sound in a public space, and who has the authority to demand silence. Noise pollution, sonic rights, and aural privacy impact ethics and individual rights in a rapidly changing technological landscape. Technology is as implicated in new ways to produce annoying sound in public as it is in helping damp it – noise cancellation headphones for instance.

In our post-digital era, new opportunities for the creation and manipulation of virtual soundscapes are constantly evolving as are the implications on how we perceive reality as the technology evolves. This implies a shifting platform for sound as a social practice, and how sound is deployed as part of social interaction and communication. Sound and its capture can be used to monitor and control the population with implications for personal privacy and freedom.

Finally, the philosophy of sound allows us to theorise on the absence of sound – silence. What is communicated through silence and how do we contrast silence to noise? Sound art and political expression provide artists with opportunities to use sound to create an immersive experience, challenge conventional perceptions of space and place, and comment on political issues.

The chapters in this section serve as a foundation for discussing the philosophy and politics of the sound of space from a broad range of perspectives. Ten authors offer diverse pathways to help explore how sound influences our personal and collective experiences and our interactions within various soundscapes and environments. We begin with vocalisation as a communal and performative act. In Reading Aloud: The Vocalisation of Living Space, Paul Carter draws from English Romantic poet Percy Bysshe Shelley and his practice of reading aloud; he explains how this was at the core of his poetic practice, with the voice acting as a bridge between inner and outer worlds. Used in this way Carter argues that the voice can be crucial to the cognitive process, mediating between society and environment.

In From Affordances to Value Chains: Probing the System of Sound, Space and Public, Sven Anderson examines how sound, space, and the public combine to offer unique affordances, and how they influence the way practitioners from diverse disciplines engage with urban environments. He shows how the interaction between legislative, urbanist, architectural, artistic, and activist perspectives helps shape the soundscape of cities, enrich the value chain within urban projects, sway decision-making, and benefit project outcomes.

On Vibrational Architectures by Gascia Ouzounian and Jan St. Werner, they explore an architecture which enhances and makes use of sound's vibrational properties rather than reduces them. They point to the politicisation of architectural materials and assay the extent to which the physical and acoustic properties of materials intersect with their political and historical narratives.

They argue that architecture may be appreciated not only by its permanence but also by its transience and the ephemeral energies that interact with it.

Mark Taylor considers sound perception in domestic spaces in House of Silence, of Stillness, of Solitude, investigating how both internal and external sounds influence the experiences of the inhabitants in any home. He looks at atmospheric silence, conceptualising silence as not merely the absence of sound, but as an atmospheric presence encompassing both stillness and solitude. Memories and the passage of time within domestic environments are evoked through temporal distortion and recollection garnered through sounds.

In Dimensionless Space (With Serrated Edges and Sucking Noises): Intimacy, ASMR, Micro-Magic, Sensory Scholarship and Other Taboos, David Toop investigates the intimate sensory experiences produced by close auditory engagements and their effects on listeners. He looks at sensory anthropology and uses speculative writing to understand the sensory experiences of autonomous sensory meridian response (ASMR) and micro-sound practices, focusing on the social dimensions of these experiences.

Decentring the human subject and recognising the bias in soundscape studies towards the able-bodied, white, European male perspective, Jordan Lacey argues in Posthuman Listening to The More-Than-Human Soundscape for moving towards a more inclusive understanding of soundscapes. He calls for an embrace of the idea that soundscapes are experienced differently by each listener, depending on their unique sociocultural, historical, and biological contexts. We can learn from Indigenous practices that view listening as an active, respectful engagement with both land and community.

In Towards a Topology of Music, Ildar Khannanov argues that the traditional two-dimensional approach to music theory expressed through Cartesian coordinates, misses the mark for post-tonal music. In its place, topological concepts like directionality and continuous functions offer a more fitting framework, leading to what he describes as a 'vagabond geometry', one that moves beyond strict, visually oriented space, aligning more closely with the fluid interpretation of music of the humanities. He introduces a dualistic view of space: the tangible physical world versus music's abstract conceptual realm, contrasting the humanities' interpretive flexibility with scientifically precise quantification, which together offer unique insights for understanding music's temporal and spatial dimensions.

Sound can be seen as an active agent with spatial-material attributes, engaging in various operations, forming circuits responsive to and influences on their environment, as discussed by Raviv Ganchrow in his chapter Sound's Spatial-Material Circuitry. This can offer an appreciation of sound's contextual circuitry, where its role varies in different settings, leading to diverse acoustic narratives. Sonic agency is identified through the patterns and impacts within these sound circuits, while terrestrial environments emphasise sound's role in the broader ecological and material world. He suggests that sound's distinctive properties can contribute to dynamic groupings, shaping and being shaped by their interaction with space and matter.

Jeff Malpas peers at topology and situatedness through a philosophical lens to provide a philosophical-topological approach to the analysis of space and place in architecture, outlined in his chapter Place, Sound, and Architecture. He outlines the entwined relationship between auditory experiences and physical spaces prioritising the essence of being 'in' a space over just being 'at' a space. He delves into how sound can contribute fundamentally to the identity of a place and affect our spatial experiences, enriching the emotional texture of environments, exploring how these sonic interactions shape the character of architectural spaces and influence the boundaries between the private and public, the intimate and communal.

To conclude the section, in Shaping Sounds of Future Environments, Eleni-Ira Panourgia explores the innovative use of sound to mirror and navigate shifting landscape terrains undergoing major environmental changes, such as drought. She examines how such transformations alter the acoustic fabric of spaces, requiring a deeper and more inclusive form of listening enfranchising non-human perspectives. As a narrative, their chapter delves into the fusion of fiction with sonic realities to enhance our understanding of climatic impacts on auditory environments. In this way sound can be framed as a creative tool to probe and portray potential future environmental states, inviting listener engagement.

15
READING ALOUD
The vocalisation of living space

Paul Carter

The idea that voice can be an instrument for sounding out space may appear eccentric. However, certain kinds of voice performance are designed to solicit listening and imply an auditorium adapted to their mode of delivery. One reason why the idea is unfamiliar is the influence of language theories that discount the value of voice (and the sonorous body more generally). The legacy of Structuralist and Functionalist descriptions is a language that walks homeless through the world, stripped of all poetic armature from environmental affordances (for example, caves and their echoes) and mimetic invitations (the entire range of dialogical encounters across differences). A case can be made for saying that contemporary architectural functionalism builds over this silencing of the voice, but the repressed comes back in the form of auditory pandemonium. Against this broad background, the object of this chapter is to present a worked example of voicing the world as a mode of oikos-making. The example is derived from the habit of the Romantic poet Percy Bysshe Shelley of reading aloud his poems-in-draft. Here is another starting point that most audiences may find idiosyncratic. However, this may reflect a cultural prejudice against acknowledging the capacity of poets to think analytically: besides his prolific poetic production, mainly completed in the last four years of life when living in exile in and around Pisa, Shelley developed a sceptical metaphysics that places the poetic vocation at the heart of a relationship – a possibility of communication – between inner and outer worlds. His theory of Voice envisages a harmonisation across scales of existence mediated by echoes and summoned up in the act of reading aloud; reading aloud thus becomes an act of auditory design with nested architectural and environmental analogues.

A history of reading aloud is outside the scope of this essay. Reading aloud is supposed to have been the usual way of reading in the pre-modern era; what may be of relevance here is that the advent of silent reading fostered interiority: 'The words no longer needed to occupy the time required to pronounce them. They could exist in interior space, rushing on or barely begun, fully deciphered or only half-said. Silent reading requires a different cognitive process than oral reading'.[1] With the advent of silent reading, reading aloud occupied a contracted cultural space, where it retained (or acquired) a special relationship with authenticity. In reading aloud, the voice served like the handwritten signature to preserve a ghost presence in the discourse of absence. In the case of poetry, the voice print of the author was perhaps essential to the work's comprehension, and was carried in the rhythm that mediated between the general impulse of the spoken language and the personal impress of the writer's feeling – 'I'm going to read my poems with great emphasis on their rhythm' – thus W.B. Yeats in 1932, explaining what Seamus Heaney called his 'elevated chant'.[2] But that was not all: Yeats sought to revive a living language for the wireless but reading aloud to a live audience involved a temporal contract measured through the time required to pronounce the words. Famously, in S. T. Coleridge's The Ancient Mariner, the narrator has to fix the marriage guest, not simply with his glittering eye, but with the irresistible force of a rhythmic measure that induces identification. As a result, a lingering occurs that opens up a new shared space of audition. Whether in the pre- or post-silent reading era, or in a transitional period,[3] reading aloud had a special meaning for the poet, one invested in the primacy of the voice and the sounding of language. It is reasonable to say – and the evidence about Shelley's compositional habits supports this – that in poetic utterance reading aloud preceded writing a poem down; and in this apparent paradox, with its relocation of inspiration in the exteriority of performance, there opens up the theme of this essay, the sense in which the poet is the architect of an auditorium for echoes.

The reason for focussing on Shelley is that we can trace through his practice an anatomy of this claim: the conviction that a resonance exists between poetic utterance, the inducement of new human community and the perennial harmony of the cosmos informs every aspect of Shelley's poetic method, including his approach to reading aloud. In fact, in mediating between the elemental voices of nature and the voice of the crowd, reading aloud is critical. Joseph Severn's portrait of Shelley composing Prometheus Unbound among the Baths of Caracalla may court a Romantic cliché, but Shelley confirms its accuracy, significantly describing the ruins as presenting 'a scene by which expression is overpowered: which words cannot convey'. Shelley was a plein air poet, opening himself directly to the challenge of vocalising elemental energies whose powers lay beyond words. He was also highly sociable, a theorist and practitioner of 'free love' and fascinated by the theatre. While his ideal theatre was ancient – 'the Greek drama was performed to an audience composed of the entire population, familiar with poetry and the legends it drew on, in masks, with dancing and music, in the open air with a view of wonderful beauty' – modern theatre, despite its limitations ('confined to artificially illuminated interiors'), also had its appeal. The influence of Viganò's mechanically ingenious stage sets on the almost cinematic scenography of Prometheus Unbound has been plausibly argued.[4] In our context, though, it may be another theatrical phenomenon from that period that is most relevant. In 1820–1821, Percy and Mary Shelley made the acquaintance of the great Italian poet, actor improvisateur, Tommaso Sgricci. Percy witnessed two of Sgricci's theatrical appearances, and, after the performance of 22 January 1821, drafted a highly complimentary Italian-language review that ascribed Sgricci's powers of improvisation to divine inspiration, noting its electrifying effect on the audience, and claiming that it 'surpassed perhaps all that Italy has ever known in this kind'.[5] Part of Sgricci's genius was to conjure up a scene without the aid of scenography, to make an audience see what, paradoxically, eluded visual representation.

Sgricci appealed to Shelley because he exemplified the poetic vocation. Shelley saw poetic utterance as forging a link between the cosmos and society. The poet participates in the great story of nature but also amplifies fundamental aspirations in human nature. 'A poet participates in the eternal, the infinite and the one; as far as relates to his conceptions, time and place and number are none'.[6] As far as their mode of expression (language) is concerned, though, they are bound by the limits of human cognition: 'The social sympathies … begin to develop themselves from the moment that two humans coexist'[7] and the communication that follows – words and actions – 'are distinct from that of the objects and the impressions represented by them, all expression being subject to the laws of that from which it proceeds'.[8] The relationship of language to the external world is mediate: it is artificial in the sense that it follows the internal logic of its own rules (syntax, grammar), but because 'there is a "permanent analogy of things," a set of essential "similitudes or relations" that constitute "all knowledge" ("perception") and "human intercourse" ("expression")'[9], 'The linguistic system is analogous (relationally proportional) to the perceptual phenomena it would express'.[10] As Bruhn explains, For Shelley, the developmental problem is not to reconcile different sources of psychological activity ['self and others, present and nonpresent'] in order to formulate a uniform or transpersonal coding for theory of mind, but rather how one learns to symbolically distribute an originally unassigned phenomenon (mental experience) for the purpose of shared representation and the kinds of social interaction, including moral thinking and behaviour, it supports.[11] This problem is the poet's vocation:

> Those in whom [the power of analogy] exists in excess are poets, in the most universal sense of the word; and the pleasure resulting from the manner in which they express the influence of society or nature upon their own minds, communicates itself to others, and gathers a sort of reduplication from that community.

Bruhn focuses on Shelley's theory of mind, or the human domain, but it is clear from Shelley's poetry that he extended the intuition of relational intelligence to the cosmos as a whole and to the realm of Meteora in particular, all that lies between heaven and earth being the scene of Prometheus Unbound. The phenomenon of inspiration proved this point and critically, although this lies outside Bruhn's consideration, inspiration is sonorous – analogy being understood as a harmonisation of vibrations across scales. In the truly inspired poetic act – which, Shelley thought, Sgricci exemplified – there is no before or after in the act of composition. A text is not first written, then read aloud, then (at a distance in time) absorbed into collective consciousness. Ideally, the poetic act is an improvisation whose power to produce pleasure finds instant recognition (duplication) in the listener. This social harmonisation is significant because the origin of the vibration lies outside the poet – in their participation in the eternal, a relation that Shelley, like Milton before him, attributes to the muse Urania.

> If answerable style I can obtain/ Of my celestial patroness, who deigns/ Her nightly visitation unimplored/ And dictates to me slumbering, or inspires/ Easy my unpremeditated verse:/ Since first this subject for heroic song/ Pleased me long choosing and beginning late.

Thus Milton, and Shelley, going a step further, imagines Urania as a link in the great cosmic harmony, opening Adonais, his elegy for the poet John Keats, 'With veilèd eyes/ 'Mid listening Echoes, in her Paradise/ She sate, while one, with soft enamoured breath,/ Rekindled all the fading melodies,/ With which, like flowers that mock the corse beneath,/He had adorned and hid the coming hulk of Death'. Because the language is of its period, we may be inclined to dismiss such lines as rhetorical. In fact, they express elegantly and aptly, a conception of existence defined harmonically: just as echoes emanate from the unheard melody of the universe and are given voice by the poet, so the poet echoing these melodies in language re-enters Paradise in this world, revivifying the primary relation echoically. In such cases, the inner voice originates as an outer voice: the poet ventriloquises or is the echoic mimic of what is dictated. For language is not only syntactically analogous to the order of things but sonorously related, being 'connected with a perception of the order of the relations of thought'; hence 'a certain uniform and harmonious recurrence of sound … is scarcely less indispensable to the communication of [the poet's language] than the words themselves'.[12] Words 'unveil the permanent analogy of things by images which participate in the life of truth', when their expression is 'harmonious and rhythmical' – when they are 'the echo of the eternal music'.[13]

Shelley's theory of cognition, and an understanding of the poetic vocation and the manner and scope of poetic composition, help explain the importance Shelley gave to reading aloud. There are many references to Shelley reading from his works (usually while they were in progress). We know that when he was working on Prometheus Unbound, Shelley read drafts each evening to his cousin, writer and poet Thomas Medwin. Such readings were not afterthoughts or reports of work performed elsewhere: they were performances that invited sympathetic identification. When Prometheus Unbound was sent to the printers in London, Shelley wrote to his friend John Gisborne that he would be better able to revise Prometheus for the press, because 'he heard it recited' Auditory memory might inform editorial direction perhaps because, as Shelley writes in A Defence,

> there is a principle within the human being, and perhaps within all sentient beings, which acts otherwise than a lyre, and produces not melody alone, but harmony, by an internal adjustment of the sounds and motions thus excited to the impressions which excite them.[14]

Producing harmony in practice involved considerable performance aptitude: poem drafts could be regarded as scores, Mulhallen comments that 'The importance Shelley attached to his poetry being read aloud [is] seen by his use of rhetorical punctuation'.[15] Shelley's approach to registering 'the echo of eternal music' was eminently 'outward' – social, imitative and occasional: 'Percy Shelley was known for his ability to improvise oral translations, and there are numerous accounts of him performing extempore translations and interpretations of non-English works for a group of his English friends'.[16]

More radically, the poet might be another Orpheus, finding the resonant frequency to move nature. The circumstances of the composition of Shelley's 110-line verse fragment 'Orpheus' are suggestive.[17] First, the poem exists only as a manuscript in Mary's handwriting, having been taken down at dictation; second, Shelley drafted his Italian review of Sgricci's 1821 improvvisazione on the pages that surround the dictated 'Orpheus' fragment. Shelley's 19th-century editors conjectured that the poetic fragment was either his translation, from memory, of part of a poem that Sgricci improvised on the subject 'Orpheus and Eurydice' – or, intriguingly, that it was Shelley's imitation of Sgricci, 'an improvisation he performed with only his wife as audience'.[18] Whatever the circumstances, the poem offers a precise description of the poetic act in terms of a relationship between environmental sound and analogous language. In the act of reproduction natural sound finds a voice that communicates; a kind of reading aloud occurs that, through the vehicle of poetic language, affects a moral transformation. First, 'There rose to Heaven a sound of angry song'.[19] It is Orpheus abandoned by Eurydice. Next, the 'angry song' is said to be 'as a mighty cataract … that casts itself with horrid roar and din/ A down a steep'. This sound 'from a perennial source/ … ever flows and falls, and breaks the air/ With loud and fierce, but most harmonious roar', but it needs the power of analogy to translate it into poetry, which Shelley describes as follows: 'as it falls casts up a vaporous spray/ Which the sun clothes in hues of Iris light./ Thus the tempestuous torrent of his grief/ Is clothed in sweetest sounds and varying words/ Of poesy'. But the interesting point about this Orphic voicing is that it never dies: 'Unlike all human works,/ It never slackens, and through every change/ Wisdom and beauty and the power divine/ Of mighty poesy together dwell,/Mingling in sweet accord'.[20]

I interpret this as saying that Orphic song or 'Poesy' communicates primarily through something like the meaningful musicalisation of sounds and only secondarily through striking representations.

In seeking to participate in the oneness of nature, poetic language constantly seeks to transcend language, returning perhaps an Adamic time when utterances correlated directly to objects and the harmony this affected had no need of mediation:

> I talk of moon, and wind, and stars, and not/ Of song; but, would I echo his high song,/ Nature must lend me words ne'er used before,/ Or I must borrow from her perfect works,/ To picture forth his perfect attributes.[21]

This is in line with Shelley's sense that the voice has an upward and a downward aspect. Upwardly, 'Verse is the echo of the eternal music', and recapitulates Promethean speech when 'the harmonious mind/ Poured itself forth in all-prophetic song,/ And music lifted up the listening spirit/ Until it walked, exempt from mortal care,/Godlike, o'er the clear billows of sweet sound'.[22] Downward speech is divorced from this echo of the divine. It is the chatter of the oppressed crowd, the 'false, cold, hollow talk/ Which makes the heart deny the yes it breathes/ Yet question that unmeant hypocrisy/ With such a self-mistrust as has no name', as occurs under the tyranny of Prometheus's usurper, Jupiter. According to Vatalaro, Shelley experienced this schism in his own poetic calling: 'Shelley explored the potential of the music of the Other as a means to the end of establishing – and, more to the point, sustaining – an intimacy with the Other'. This fantasy sprang, writes Vatalaro, from Shelley's

> fear of estrangement from his own poetic voice. Ultimately, two things troubled him: first, that a man's poetic utterance, in stark contrast to a woman's singing performance, proceeds from vacancy rather than presence; second, that inscription remains silent and mortal, and that, in the end, written words do no more than constitute a non-speaking subject.[23]

Perhaps, but Shelley had two inter-related responses to hand: listening and reading aloud. Estrangement was a function of not listening to the Other, of failing to acknowledge the primary nature of the echo. Without the echo, no sound can find its voice. It is their attention to the eternal music, which they channel (as the poet channels the wordless voice) into environmental sounds (a kind of wordless music) that mediates our relationship with our surroundings. Echoes are, as it were, already listening out for us to attend to them: as David Levin writes,

> We can begin to realise that our primordial relationship to Being is a relationship through which our hearing is always situated in a field of sonorous energies that have already been gathered for it in a 'layout' that sets the tone and gives it a 'proper' measure.[24]

This sense of primary gathering or ordering corresponds to Shelley's principle of harmony, 'an internal adjustment of the sounds and motions thus excited to the impressions which excite them'.[25] This 'principle' is an auditory one, 'the experience of consonance, harmony and attunement that necessarily underlies the correspondence', and Levin's analysis gives a precise sense to the effect that the subject of 'Constantia, Singing' has on Shelley 'Even though the sounds which were thy voice, which burn/ Between thy lips, are laid to sleep … Her voice is hovering o'er my soul – it lingers …' effecting a double union, with the singer and with voiceless nature: 'I have no life, Constantia, now, but thee,/ Whilst, like the world-surrounding air, thy song/ Flows on, and fills all things with melody. – Now is thy voice a tempest swift and strong … Now 'tis the breath of summer night [which] Lingering, suspends my soul in its voluptuous flight'.[26] In Shelley's auditory ontology, the phenomenon of lingering or fading sound is the trace of the music of the eternal in the temporal: the voice lingers after what it has said, the sonorous body that speaks with the wordless voice of nature lingers on the edge of consciousness. ('The dissolving strain, through every vein,/ Passes into my heart and brain'.)

The embodiment of this protended sound, a sound defined by its dying away, is what Prometheus refers to in Act 1 of Prometheus Unbound as the 'many-voicèd Echoes' of the Elements. Prometheus Unbound is in part the drama of the Echoes' unbinding. For the echo is the auditory mediator between the immortal voice that cannot yet speak its name and the earthly communication between humans (whose voice is currently hollow, empty and ghostlike). It is the evidence of a resonance, a harmonisation and answering sound that the echo provides. In Act II of Prometheus Unbound, Demogorgon, a figure of human collective will, links the impossibility of revolutionary change to the absence of a voice:

> If the abysm/ could vomit forth its secrets ... But a voice/ Is wanting, the deep truth is imageless;/ For what would it avail to bid thee gaze/ On the revolving world? What to bid speak/ Fate, Time, Occasion, Chance, and Change? To these/All things are subject but eternal Love ...[27]

But a clue to the solution has already been given in an earlier scene: the Demogorgon may be the agent of change but how will the supporters of Prometheus (Asia and Panthea) come to him? In a complicated image, Asia informs Panthea of a dream: 'on the shadows of the morning clouds,/ Athwart the purple mountain slope, was written/ FOLLOW, O, FOLLOW'.[28] Then as the Echoes become audible, first mocking 'our voices/ As they were spirit-tongued', then finding voices of their own ('The liquid responses of their aëreal tongues yet sound') the Echoes show the way, explaining to Panthea, 'In the world unknown/ Sleeps a voice unspoken;/ By thy step alone/ Can its rest be broken;/ Child of Ocean'.[29] The Echo is the audible expression of the resonance that enables the poet to mediate between cosmos and community; and echoic mimicry is what the poet performs, their (literal) vocation being to give the Echo human voice.

Shelley's practical theory of poetic composition offers a uniquely well worked-out case study of orientation to the world mediated through voice. The role of echo in this is critical. As Levin notes, the echo 'sets in motion uncountable vibrations of uncertainty; it refuses to be controlled; it cannot be possessed; it makes careful distinctions interpenetrate; it denies the possibility of pure presence; it decentres the ego'.[30] Shelley describes one manifestation of this de-centring when he reflects that in the creative act 'the very mind which directs the hands in formation, is incapable of accounting to itself for the origin, the gradations, or the media of the process'.[31] There is a self-forgetfulness at the heart of the poetic act that is very different from the 'auto-affection' said to be necessary for the voice of consciousness.[32] As a vehicle of echoic mimicry, or self-othering, the poet's inspired utterances propose a shared space whose dimensions are defined by the intertwined acts of hearing, listening and speaking. This insight suggests many fertile lines of enquiry. Recalling that Shelley's poetic practice embodies a highly sophisticated theory of human cognition and its development, one wonders whether a radically different theory of psychological development might be given. Lev Vygotsky's developmental model, in which the child's self-talk evolves into the muted inner speech of adulthood, for example, would yield different results if the earliest phases of private-talk were recognised as echoic mimetic.[33] The psychologist of group behaviour, Erwin Goffman writes that, on the assumption that speech is communicative and self-talk represents 'a threat to intersubjectivity', 'Self-talk is deemed legitimate only when done in private, by children, by people with intellectual disabilities, or in Shakespearean soliloquies'.[34] But this last admission is suggestive. Just as the soliloquy breaks down the representationalist barrier between the inner world of mental illusion and the shared world of common action, so reading aloud as a poetic gesture, channelling what is, as it were, in the air, redefines intersubjectivity as primary, not an after effect of efficient communication.

In architectural circles, the French philosopher of scientific method, Gaston Bachelard, is perhaps best known for his book The Poetics of Space. But, in terms of evoking the oikos that consciousness hollows out of space, its involutions, scalarities and (foundationally) its dependence on analogy, his essay The Poetics of Reverie is equally suggestive. Riffing on the poet Henri Bosco's statement, 'All the being of the world, if it dreams, dreams that it is speaking', Bachelard rhapsodises, 'The words of the world want to make sentences … The poet listens and repeats. The voice of the poet is a voice of the world. In consequence, human and cosmic tonalities reinforce each other'.[35] Could this translate into a spatial poetics? 'To express my attachment to hills, valleys, paths, groves, rocks, and grottos, I would have to write a "non-figurative" geography of names … [a] geography of memories'.[36] How would this work? Following J.J. Gibson (and Shelley), we assume the analogy between 'existence' and 'human cognition' or, in Gibson's terms, the reality of the organism-environment ecosystem: 'hence, the role of key concepts, such as attunement, reciprocity and resonance, and the corresponding perceptual processes of detection, discrimination, recognition and identification'.[37] With regard to attunement, there is evidence that

> Most behaviorally important sounds – animal and human vocalizations in particular, but also musical sounds – have as prominent feature their harmonic spectra with central neural networks of the listener being preferentially attuned to consonant intervals because of their prevalence or biological significance in the environment.[38]

Hence, not all sounds are equal: those

> characterized by aspects of harmonicity and regularity, can easily be set apart from the acoustic background as a whole. As such, they involve an aspect of self-similarity in the sense that they can be recognized as such – sounds as sounds – which makes them apt for processes of differentiation, discrimination and identification.[39]

The notion of 'sounds as sounds', sounds distinguished by certain formal characteristics from the background noise, throws light on the role onomatopoeia plays in language formation. At its simplest, words that name phenomena by imitating the sound they make discern in the source sound a proto-lingual character – thinking of birdcalls – defined by the perceived similarity of the phrase to a phonemic combination, and by the signature consistency of the signal. The next step from casual imitation of the 'Oh-sweet -Canada-Canada' kind[40] is (staying with birds) to attribute to birds a language. According to Palmisciano, when the ancient Greek poet Alcman listened to the birds, he 'found not only the right music but also the right words. The partridge's song has a meaning and the poet is able to understand it'.[41] At the next level of generalisation, in cultures without silent writing systems, onomatopoeically derived names may have a distinct survival value, naming animals after their calls reducing the cognitive effort required to remember a large ethnobotanical vocabulary.[42] But perhaps more suggestive in the context of understanding echoic vocalisation as a form of poetic architecture is a study of Pitjantjatjara sound classification which found that

> categorising and naming according to sound … is consistent with the concept of 'naming' a place by using the correct stylised sung form of the name; with 'singing' a person by

manipulating his behaviour through the medium of song; with taboos on the use of the name of the dead, since their spiritual power can be tapped through the correct sounding of the name.[43]

The authors explain,

> The sounds of the environment in general provide a different form of classification. There seem to be two main aspects. One illustrates value judgements about sounds. The other is an extensive vocabulary of onomatopoeic words which suggest that much of the classification of the surrounding environment is done on the basis of the sounds produced by the thing named.[44]

The writers hypothesise that 'one of the most critical elements in classification by Aboriginal people, probably throughout Australia, is sound, both musical and environmental'. The term meaning 'to sing' in Pitjantjatjara can mean 'the act of causing an event through the power of song'.[45]

The text for reading aloud comes from outside. As a distinctively poetic mode of vocalisation, reading aloud channels a voice that (in my experience) comes from the back of the head and issues an invitation to gather and listen. The poet listens to what the environment is dictating.

> I take great delight in watching the changes of the atmosphere here, and the growth of the thunder showers with which the moon is often overshadowed, and which break and fade away towards evening into flocks of delicate louds. Our fire-flies are fading away fast; but there is the planet Jupiter, who rises majestically over the rift in the forest-covered mountains to the south, and the pale summer lightning which is spread out every night, at intervals over the sky.[46]

I have come to realise that a passage like this is not a picturesque ecphrasis, a representation of an external scene. It is the transcript of an environmental echo that, in the act of communication, imagines a resonance in the reader. Poetic utterance occurs when these environmental cues (thunder, moon, planet, lightning) find their voices as Echoes (see Prometheus Unbound) and, in the poet's performance produce a 'reduplication' in the listener. The drama of the echo is also the drama of love, the sole vocalisation of the eternal that tyranny cannot silence: Love 'fills up that intermediate space between … two classes of beings, so as to bind together, by his own power, the whole universe of things'.[47] Yet, as a socially cohesive bond underwriting a proto-architectural oikos, it is bound to fail. Like the echo and its auditorium, it is always a ruin in prospect: 'though love seeks as a type for its "mirror"-like "antitype"[48], no existing object or social other can perfectly answer the functional demand, and so the "communion" is always incomplete, partial, and still to seek'.[49]

My thanks to Vera Bühlmann and Ludger Hovestadt who encouraged me to improvise a poem for their colloquium, 'Art and Aesthetics of Forgetting', Meteora Academy (Villa Medici, Buti, Tuscany), 13–16 September 2023. This event, supported by a visit to Bagni di Lucca, was an attempt to find out what the theory outlined in this chapter would look (or sound) like in practice.

Notes

1 Alberto Manguel, *A History of Reading*, 1996.
2 https://www.openculture.com/2012/06/rare_1930s_audio_wb_yeats_reads_four_of_his_poems.html.

3 As late as the 1700s, historian Robert Darnton writes, 'For the common people in early modern Europe, reading was a social activity. It took place in workshops, barns, and taverns. It was almost always oral but not necessarily edifying' (Darnton, First Steps Toward a History of Reading, 1990, 168) – more importantly, perhaps, it was always audible, a tribute to vernacular acoustic design. Censorship of the singing voice is a feature of Victorian England (see Paul Carter, *Amplifications*, New York: Bloomsbury, 2018, 40).
4 Angela Esterhammer, 'Improvisational Aesthetics: Byron, the Shelley Circle, and Tommaso Sgricci', *Érudit, Romanticism on the Net*, https://www.erudit.org/en/journals/ron/1900-v1-n1-ron1383/013592ar/ With one fundamental difference. Viganò's 'ideal dramas staged at La Scala, enthusiastically praised by Shelley – 'a combination of a great number of figures grouped with the most picturesque and even poetical effect, and perpetually changing with motions the most harmoniously interwoven and contrasted with great effect' – were spectacles, while Shelley's Prometheus was, in conception and action, a voice drama.
5 P.M.S. Dawson, 'Shelley and the Improvvisatore Sgricci: An Unpublished Review', *Keats-Shelley Memorial Bulletin* 32 (1981): 19–29.
6 Percy Bysshe Shelley, 'A Defence of Poetry', in *The Selected Poetry and Prose of Shelley*, ed. B. Woodcock, London: Wordsworth Editions, 2002, 635–660.
7 Shelley, 'A Defence', 636. James H. Johnson – Listening in Paris, *A Cultural History*, Berkeley: University of California Press, 1995, 82.
8 Shelley, 'A Defence', 636.
9 Mark J. Bruhn, 'Shelley's Theory of Mind: From Radical Empiricism to Cognitive Romanticism', *Poetics Today* 30(3) (Fall 2009): 373–422.
10 Bruhn, 'Shelley's Theory of Mind', 393.
11 Bruhn, 'Shelley's Theory of Mind', 400.
12 Shelley, 'A Defence', 639. James H. Johnson traces the emergence of harmony in the sense that Shelley uses it. 'The revolution of Gluck' encouraged the recognition of 'the orchestral harmonies as legitimately expressive apart from the text'. The new music 'approached "the heart and the imagination and not servile imitation"'. And Johnson traces the correlation between the emergence of this new sensibility and 'a new way of listening' and the phenomenon of the concert audience, listening reverently in silence (*Listening in Paris: A Cultural History*, Berkeley: University of California Press, 1995, 82).
13 Shelley, 'A Defence', 640. In a similar vein, Milton: 'But if our souls were pure, chaste, and white as snow, as was Pythagoras' of old, then indeed our ears would ring and be filled with that exquisite music of the stars in their orbit; then would all things turn back to the Age of Gold' (John Milton, 'Delivered in the Public Schools: On the Harmony of the Spheres', in *Complete Prose Works of John Milton 1*, ed. Don M. Wolfe, New Haven, CT: Yale University Press, 1953, 234–239, 234.
14 Shelley, 'A Defence', 635.
15 Jacqueline Mulhallen, *The Theatre of Shelley*, https://books.openedition.org/obp/766?lang=en), chapter 5, para. 4.
16 Esterhammer, 'Improvisational Aesthetics'.
17 Percy Bysshe Shelley, *Poetical Works*, ed. T. Hutchinson, Oxford: Oxford University Press, 1970, 'Orpheus', 628
18 Esterhammer. *References to* Richard Garnett, ed., *Relics of Shelley*, London: Moxon, 1862, 20; H. Buxton Forman, 'The Improvvisatore Sgricci in Relation to Shelley', *The Gentleman's Magazine* 248 (1880): 115–123.
19 Shelley, *Poetical Works*, 'Orpheus', l.72.
20 'Orpheus', l. 73–84.
21 'Orpheus', ll. 98–103.
22 Shelley, 'Prometheus Unbound', Act II, Scene IV, ll. 75–79.
23 Paul A. Vatalaro, *Shelley's Music, Fantasy, Authority, and the Object Voice*, Farnham: Ashgate, 2009, 2.
24 David Michael Levin, *The Listening Self*, London: Routledge, 1989, 254.
25 Shelley, 'A Defence', 636.
26 Percy Bysshe Shelley, *Poetical Works*, ed. T. Hutchinson, Oxford: Oxford University Press, 1970, 'To Constantia, Singing', 539–540, ll. 4–5, 21, 32–35, 39, 42.
27 Shelley, 'Prometheus Unbound', 204–274, 238, Act II, Scene IV, ll. 114–120.
28 Shelley, 'Prometheus Unbound', Act II, Scene II, ll. 151–153.
29 Shelley, 'Prometheus Unbound', Act II, Scene II, ll. 164–165, 171–172, 190–194.
30 Levin, *The Listening Self*, 237.

31 Shelley, 'A Defence', 657.
32 For Husserl, the ideal order of inner speech is only preserved in outer speech when the hearer repeats immediately in himself the 'hearing oneself speak' of the speaker: there is a chain of 'auto-affection' which somehow remains suspended above the abyss of difference. Derrida criticizes this view (Speech and Phenomena, 86) but perpetuates a silencing of the voice. As Annemarie Jonson observes, 'In impugning the excision of the "worldly" and the "body" from Husserlian expression … Derrida paradoxically appears to effect an annihilation of sounding voice which renders the de-voiced subject immaterial and mute. Annemarie Jonson, Voice and silence: aspects of Derrida's critique of phonocentrism', UTS MA, 1995, 63.
33 For summary, see Nana Ariel, 23 December 2020 at https://psyche.co/ideas/talking-out-loud-to-yourself-is-a-technology-for-thinking.
34 Nana Ariel, 23 December 2020 at https://psyche.co/ideas/talking-out-loud-to-yourself-is-a-technology-for-thinking.
35 Gaston Bachelard, *The Poetics of Reverie: Childhood, Language, and the Cosmos*, trans. D. Russell, Boston: Beacon Press, 1969, 189.
36 Bachelard, *The Poetics of Reverie*, 189, note 33.
37 Mark Reybrouck, 'Music as Environment: An Ecological and Biosemiotic Approach', *Behavioral Sciences* (Basel) 5(1) (2015 Mar): 1–26.
38 Reybrouck, 'Music as Environment'.
39 Reybrouck, 'Music as Environment'.
40 An onomatopoeic rendering of the White-throated Sparrow's call (K. Vella, D. Johnson and P. Roe, 'Describing the Sounds of Nature: Using Onomatopoeia to Classify Bird Calls for Citizen Science', *PLoS One* 16(5) (2021): e0250363, https://doi.org/10.1371/journal.pone.0250363 Editor: George Vousden, Public Library.
41 Riccardo Palmisciano, 'To Speak Like a Bird: Beyond a Literary Topos', in *Rethinking Orality*, eds. A. Ercolani, L. Lulli, Berlin: Walter de Gruyter, 2022, 106. Not only ancient Greeks 'learned to sing by imitating the birds' songs they heard in the most isolated places': likening the skylark to 'a Poet hidden/ In the light of thought', Shelley concludes his famous poem, 'Teach me half the gladness/ That thy brain must know,/ Such harmonious madness/ From my lips would flow/ The world should listen then, as I am listening now' (*Poetical Works*, 602–603, ll. 101–105).
42 Brent Berlin, and John P. O'Neill, 'The Pervasiveness of Onomatopoeia in Aguarana and Huambisa Bird Names', *Journal of Ethnobiology* 1(2) (December 1981): 238–261.
43 Catherine J. Ellis, A.M. Ellis, M. Tur and A. McCardell, 'Classification of Sounds in Pitjantjatjara-Speaking Areas', in *Australian Aboriginal Concepts*, ed. L.R. Hiatt, Canberra: Australian Institute of Aboriginal Studies, 1978, 67–80.
44 Ellis et al. 'Classification of Sounds in Pitjantjatjara-Speaking Areas', (1978).
45 Ellis et al. 'Classification of Sounds in Pitjantjatjara-Speaking Areas', (1978). See also Catherine J. Ellis, *Aboriginal Music*, St Lucia: University of Queensland Press, 1985, 70. In a ritual context, singing has a geographical significance: 'A songline [representing a myth] is a mapped form of a song, each small sung presentation being located at an identifiable place. Each of the series of small songs represents consecutive events in the myth. Pitjantjatjara performers speak about mainkara wanani, "following the way in song"' (59).
46 P.B. Shelley, Letter to John and Maria Gisborne, Bagni di Lucca, July 10, 1818, *The Letters of Percy Bysshe Shelley*, ed. R. Ingpen, London: G. Bell & Sons, 1914, 2 vols, vol. 2, 604.
47 Percy Bysshe Shelley, *The Complete Works of Percy Bysshe Shelley*, eds. Roger Ingpen and Walter E. Peck, 10 vols. New York: Gordian, 1965, vol. 9, 279.
48 Percy Bysshe Shelley, *The Complete Works of Percy Bysshe Shelley*, vol. 6, 202.
49 Bruhn, 'Shelley's Theory of Mind', 415.

Bibliography

Ariel, Nana, 23 December 2020, https://psyche.co/ideas/talking-out-loud-to-yourself-is-a-technology-for-thinking.
Bachelard, Gaston, *The Poetics of Reverie: Childhood, Language, and the Cosmos*, trans. D. Russell, Boston: Beacon Press, 1969.
Berlin, Brent and John P. O'Neill, 'The Pervasiveness of Onomatopoeia in Aguarana and Huambisa Bird Names', *Journal of Ethnobiology* 1(2) (December 1981): 238–261.

Bruhn, Mark J., 'Shelley's Theory of Mind: From Radical Empiricism to Cognitive Romanticism', *Poetics Today* 30(3) (Fall 2009): 373–422.

Carter, Paul, *Amplifications*, New York: Bloomsbury, 2018, 40.

Darnton, Robert, 'First Steps Toward a History of Reading', in *The Kiss of Lamourette: Reflections in Cultural History*. New York: W.W. Norton, 1990, 154–187.

Dawson, P.M.S. 'Shelley and the Improvvisatore Sgricci: An Unpublished Review', *Keats-Shelley Memorial Bulletin* 32 (1981): 19–29.

Derrida (Speech and Phenomena, 86).

Ellis, Catherine J., *Aboriginal Music*, St Lucia: University of Queensland Press, 1985, 70.

Ellis, Catherine J., A.M. Ellis, M. Tur and A. McCardell, 'Classification of Sounds in Pitjantjatjara-Speaking Areas', in *Australian Aboriginal Concepts*, ed. L.R. Hiatt, Canberra: Australian Institute of Aboriginal Studies, 1978, 67–80.

Esterhammer, Angela, 'Improvisational Aesthetics: Byron, the Shelley Circle, and Tommaso Sgricci', Érudit, Romanticism on the Net. https://www.erudit.org/en/journals/ron/1900-v1-n1-ron1383/013592ar/

Forman, H. Buxton, 'The Improvvisatore Sgricci in Relation to Shelley', *The Gentleman's Magazine* 248 (1880): 115–123.

Garnett, Richard ed., *Relics of Shelley*, London: Moxon, 1862, 20.

Jacques Derrida, *Speech and Phenomena, And Other Essays on Husserl's Theory of Signs*. Trans D.B. Allison. Evanston: Northwestern University, 1973.

Johnson, James H., *Listening in Paris: A Cultural History*, Berkeley: University of California Press, 1995, 82.

Jonson, Annemarie, 'Voice and Silence: Aspects of Derrida's Critique of Phonocentrism', University of Technology, Sydney, Master of Arts, 1995, 63. https://opus.lib.uts.edu.au/bitstream/10453/147460/2/02whole.pdf

Kellie Vella, Daniel Johnson, and Paul Roe, 'Describing the Sounds of Nature', *PLoS One* 16(5) (2021): e0250363. https://doi.org/10.1371/journal.pone.0250363. Editor: George Vousden, Public Library.

Levin, David Michael, *The Listening Self: Personal Growth, Social Change and the Closure of Metaphysics*, London: Routledge, 1989, 254.

Manguel, Alberto, *A History of Reading*, New York: Viking, 1996.

Mike Springer, 'Rare 1930s Audio: W.B. Yeats Reads Four of His Poems,' Open Culture, June 13th, 2012. https://www.openculture.com/2012/06/rare_1930s_audio_wb_yeats_reads_four_of_his_poems.html

Milton, John, 'Delivered in the Public Schools: On the Harmony of the Spheres', in *Complete Prose Works of John Milton 1*, ed. Don M. Wolfe, New Haven, CT: Yale University Press, 1953, 234–239.

Mulhallen, Jacqueline, *The Theatre of Shelley*, Open Book Publishers, https://books.openedition.org/obp/766?lang=en, chapter 5, para. 4.

Palmisciano, Riccardo, 'To Speak Like a Bird: Beyond a Literary Topos', in *Rethinking Orality*, eds. A. Ercolani and L. Lulli, Berlin: Walter de Gruyter, 2022, 106.

Reybrouck, Mark, 'Music as Environment: An Ecological and Biosemiotic Approach', *Behavioral Sciences* (Basel) 5(1) (2015 March): 1–26.

Shelley, Percy Bysshe, *The Complete Works of Percy Bysshe Shelley*, eds. Roger Ingpen and Walter E. Peck, 10 vols. New York: Gordian, 1965, vol. 9, 279.

Shelley, Percy Bysshe, 'A Defence of Poetry', in *The Selected Poetry and Prose of Shelley*, ed. B. Woodcock, London: Wordsworth Editions, 2002, 635–660.

Shelley, Percy Bysshe, 'Letter to John and Maria Gisborne, Bagni di Lucca, July 10, 1818', in *The Letters of Percy Bysshe Shelley*, ed. R. Ingpen, 2 vols. London: G. Bell & Sons, 1914, vol. 2, 604.

Shelley, Percy Bysshe, *Poetical Works*, ed. T. Hutchinson, Oxford: Oxford University Press, 1970, 'Orpheus', 628.

Vatalaro, Paul A., *Shelley's Music, Fantasy, Authority, and the Object Voice*, Farnham: Ashgate, 2009.

16
FROM AFFORDANCES TO VALUE CHAINS

Probing the system of sound, space, and public

Sven Anderson

Tuesday, February 26th, 2013

In 2013, I began a public art commission based on the concept of working in the invented position of 'urban sound designer and acoustic planner' within a local authority in Dublin, Ireland, where I have lived since 2001.[1] Following a series of contractual negotiations and introductory meetings with senior personnel in the public art, architecture, and urban planning departments, I remember the first day of fieldwork involved in the commission. It was Tuesday, February 26th, 2013. I took a bus out to Inchicore, where a contact in the housing department had informed me that the last block of an old social housing estate was scheduled for demolition prior to redevelopment.[2] Ready with a bag packed with my Sound Devices 633, DPA 4060's and Canon 5D Mark II, I approached a throng of people gathered at the perimeter of the demolition site expecting to set up and document the event.

Unfortunately, I was running a few minutes late and arrived after the initial impact. Already in motion, the demolition crane repeatedly tore into the abandoned structure, further exposing its frailty with every contact. As I watched, I began to feel daunted by the prospect of the commission I had taken on. I recalled Ultra-red's 2003 project *The Debt*,[3] an expansive body of work exploring the complexities surrounding urban regeneration in the nearby (albeit slightly more peripheral) suburb of Ballymun. I sensed the shadowy network of skeletal building projects paralysed throughout Dublin following the collapse of the Celtic Tiger.[4]

In this instant, the core subjects of my commission (namely, the emergent fields of urban sound design and acoustic planning) felt far removed from the more coherent notions of environmental acoustics and noise control that they are often associated with. Instead, these subjects took the form of a more ephemeral question mark, a pointer that might be oriented towards any number of subjects involving cities and their inhabitants. In the case of the demolition of St. Michael's Estate, this question mark pointed towards an infrastructural rupture as an array of private acoustic memories was torn from its gridded spatial sanctuary into the exposed sonic exteriority of the city.

I left the site several hours later without taking my sound recorder or my camera out of my bag.

The following months of establishing the foundation of this artwork demanded a similar approach, in which not only my field recording equipment but equally all my Max Neuhaus and CRESSON references had to be left deep in my bag.[5] It turned out that the tools I had brought to the job weren't necessarily the right tools for the task at hand.

(Sound + space + public) = ?

Instead of moving towards a resolution, the *MAP* commission focused on keeping that question mark in the foreground. Looking back ten years later, I understand that this question mark is a proposition to place emphasis on a system comprising sound, space and public.[6] This is indeed a system, as these three elements can be understood as at once independent and interdependent and are only brought into temporarily stable configurations via perspectives that seek to reconcile their dynamic and often unstable relationship with one and other.

Within the context of *MAP* and my subsequent public artwork *The Office for Common Sound*,[7] I kept this question mark (or this system) close at hand as I sought to develop opportunities to work with different people to see what form it might take in different circumstances. The more open and less direct I remained, the more my collaborators demonstrated a willingness to follow their instincts and experiences and take the lead within these conversations. When I tried to demarcate the characteristics of this system myself, I encountered resistance. Tactically, I adapted a rhythm within these projects to allow for lengthy discursive phases (mostly premised on listening to others) punctuated by brief moments of accelerated focus and production whenever the system began to crystallise. Indeed, it was in one of these moments of crystallisation when the sound installation *Continuous Drift* – one of the project's more coherent infrastructural outputs – took form (Figure 16.1).[8]

One way to better understand this question mark and how it might perform in the context of urban design and planning contexts is through the concept of affordances. In the context of design, a given object presents different affordances to different users. For many people, the design of a door handle suggests that it can be used for opening the door that it is attached to – but equally, it affords (to a lesser extent and perhaps only to certain people) being prised from the door and integrated with the construction of a rhythmic instrument. Expanding on this, systems can also be seen to present different affordances to those who seek to use them, revealing different meanings, opportunities, and resistances, and allowing themselves to be steered towards particular outcomes.[9]

In the current scenario, the system of sound, space, and public can be seen to have different affordances for architects, artists, developers, planners, acousticians, community groups, and individual members of the public. Although they might at times overlap, these affordances are not equivalent. Consequently, concepts that seek to operate across the gaps between these affordances without acknowledging their plurality are often not the right tools to activate this territory of production.

Advancing an open framework

I developed the project *Sound-Frameworks* as a device to further explore this territory, and to shift this question mark (or this system) into different working contexts.[10] The project was composed with extensive support from the engineering and consulting firm Arup and subsequently positioned within the independent research organisation Theatrum Mundi.[11] The project's conceptual centre was thus somewhere between these two entities, geographically between 13A Clerkenwell Close and 80 Charlotte Street in London. I balanced this centre with a third location by renting Flat 41 John Trundle Court, a small Type F1A studio apartment in the Barbican located on the second floor of a block with a view lancing diagonally across a public garden towards the skyline of the City.

Theatrum Mundi provided a space in which to link my enquiry laterally from sound to other questions around city-making as an intersection between disciplines, with an emphasis on artistic practice as a guiding impulse within these conversations. The research centre's series of *Sonic*

Figure 16.1 Continuous Drift. A public artwork that blurs the boundaries between public sound installation, architectural intervention, and curatorial framework. The project allows the public to use their mobile devices to trigger sounds by different artists that play back from eight loudspeakers integrated in the architecture of Meeting House Square in Dublin, Ireland. Installed in 2015, additional works were added in 2016 and 2017. Developed within the context of the public art commission The Manual for Acoustic Planning and Urban Sound Design, the project is still active in 2024. Photo credit: Ros Kavanagh.

Urbanism publications demonstrate a capacity for the subject of sound and cities to be reconfigured and extended through different prompts, at first as a more general probing of the definition of the term,[12] then through the notion of the city possessing its own voice,[13] and finally by highlighting the role of non-human perspectives in shaping urban sonic enquiry.[14] Equally, Theatrum Mundi's oeuvre outside the subject of sonic urbanism provided hopeful ideas, ranging from the proactive cultural infrastructuralism of *Urban Backstages* to the reflexive choreography of *Navigations: Scoring the Moment* to the private urban retreats collected in *Interior Realms*.[15,16,17]

To engage with the system of sound, space, and public, *Sound-Frameworks* centred around three proposals. First, that a survey conducted with people working with sound in the urban context in different professions would reveal clues about how this field of practice might be better understood. Second, that cues derived from these open exchanges could collectively embody a series of guidelines or prompts that others might take up in their work. And thirdly, that some form of discursive armature or 'design tool' might be crafted in response to these first two proposals, with the intent that it would be ready for others to audition within future projects. Each of these proposals was action-led. For example, the surveys themselves, beyond representing a research modality within this project framework, were equally meaningful as exchanges between myself (as a practitioner) and other practitioners. In this scenario, the role of the interlocutor or researcher is not invisible or unbiased, it is instead highly biased, highly involved, pursuing certain curiosities. Operating through this bias from the outset, the discussions framed within the survey become more directive on both the part of interviewer and interviewee, emphasising a greater mutual agency.

It is the affordances of the system of sound, space, and public that surface through these interviews and conversations. The system affords different things to each practitioner depending on the contexts in which they come to work with it, as well as the methodologies that they are the most comfortable with and the collaborators and partners with whose perspectives they are most attuned. In the context of urban design and other city-making practices, this system is rather unique due to the particular nature of sound in this context. It is often relegated to a somewhat minor status in considerations of urban form and decision-making processes, and yet it is entirely omni-present, a condition that might be accessed in any given project no matter its scale, budget, or objective.

Embedding sound within more expansive value chains

I learned about value chains during one of the first *Sound-Frameworks* interviews in Struer.[18] Struer Kommune (the local authority for the municipality of Struer, Denmark) began to explore its identity through the concept of *Lydens By* (The City of Sound) in 2008, building from a legacy of audio culture stemming from it being the home and headquarters of the high-end electronics company Bang & Olufsen.[19] In a conversation focused on the redevelopment of Struer's harbour-front, the notion that the design concept for this project had to involve sound was seen to present a challenge. After all, within the context of urban redevelopment, sound is generally not something that is taken into consideration. To bring sound into the conversation, it was potentially necessary to find a way to integrate it within a larger value chain, so that value added to the project through sound could be linked to other more legible dimensions of the project's value. In the context of this conversation, this idea of sound being incorporated within a larger value chain was related both to a need to attract developers and a need to follow the directives of the Struer *Lydens By* initiative, which emphasise that sound should be incorporated within every dimension of the city, from urban development to culture, health, and education.

At the outset, the focus on value chains (and perhaps more fundamentally on value itself) didn't sit comfortably with me. As an artist, I am familiar with the task of reconciling the value of my labour; this is demonstrated when an artwork is sold, when a commission is secured, when an exhibition advances the reputation of its creator, when a project generates new relationships or partnerships, and equally – and perhaps most importantly – when it can be demonstrated that artistic labour *does work* on a specific subject. But beyond this, the value of sound as a component of urban conditions, the public realm, and the city is something that I always felt was implicit or given.[20] After all, exploring the system of sound, space, and public in the context of (for instance) a sound study or theoretical text asserts the value of sound as implicitly given, as it is the foundation that supports the argument of the text. When I read *Acoustic Territories* or *Beyond Noise and Silence*,[21,22] the authors have made this value legible from the titles, from the first pages, from the premises of the books. The work that can be achieved in this space moves rapidly, through references to other projects, texts, and situations, and through recounting lived sonic experience.

However, the prompt to consider linking sound within more expansive value chains remained with me. I began to reconsider my own experience prior to *Sound-Frameworks*. From the projects I developed with Dublin City Council, I had shifted my attention between 2016 and 2019 to collaborating with architects, landscape architects, and other designers in the context of proposals for international design competitions. For me, this was a space to explore how to advance sonic considerations in contact with these other disciplines. To a certain extent, what took place within these

Figure 16.2 Proposal for the UK Holocaust Memorial International Design Competition. Honourable Mention. 2016. London (UK). Design team: Heneghan Peng Architects, Sven Anderson, Gustafson Porter + Bowman, Bruce Mao Design, Arup, Event, Bartenbach, BuroHappold, Mamou-Mani, Turner & Townsend, PFB, Andrew Ingham & Associates and LMNB. Competition organised by Malcolm Reading Consultants. Image courtesy of Heneghan Peng Architects.

collaborations was that a small network of people worked to link sound within the value chains that (along with design considerations) defined a series of projects, many of which were characterised by significant (and complex) cultural responsibilities. These negotiations were characterised by a formidable stress-testing of the affordances of the system of sound, space, and public from different perspectives, both in terms of design and use (Figure 16.2).

Tracing the affordances of (sound + space + public) by discipline

This coupling of the concepts of affordances and value chains remained close to me as I proceeded through the interviews that served as *Sound-Frameworks'* foundation. The following sequences begin to trace the affordances of the system of sound, space, and public that manifested through the lens of different disciplinary affiliations. The footnotes for these sequences provide links to a selection of the interviews themselves, as further resources. These sequences only include a limited selection of the perspective (and interviews) encountered through the project, as a more extensive investigation would be outside the scope of the chapter. As they progress, the sequences also introduce the central concerns of the *Sound-Frameworks* design tool that was developed as a publicly accessible beta release within this two-year project, with an intent for expansion in the future.

Artistic practice

For an artist (or a sound artist), the system of sound, space, and public affords a context in which to most explicitly confront and question urban conditions and public life through mediated forms such as performance and installation, whether temporary or permanent. Within this discipline, the significance of this system as a space of continued research and potential production is linked to the emergence of a strong (and vocal) cohort of artists who advocate for work in this domain through the sustained efforts of their practices.[23] Because the provision of art in city-making practices is itself mutable and unstable, the definition of projects in this domain is underpinned by a combination of shared methodologies (which can be observed to arise on a generational level as we move into a third and fourth generation of sound artists) and current directives linked to higher-level cultural objectives (in 2024 these might include focuses such as sustainability, migration, and post-humanism). The affordances of this system in relation to artistic practice are also linked to the prominence of socially engaged artistic practice, as it is often within this genre of work that it is possible to trace solid lines between sound, space, and public through specific projects.[24]

Architecture

In architecture, the affordances of the system are markedly different. Directly addressing sound lies out of architecture's formal remit, although it is implicit in virtually any architectural project through the coordination of its programme, and through considerations of various user activities which carry with them distinct sonic and acoustic dimensions.[25] To activate the system from an architectural perspective therefore acts to extend the discipline and to introduce a new set of design parameters, and equally to reveal a new set of challenges in communicating these parameters with clients. A select grouping of architectural practices achieve this as a recurrent or central element of their work,[26] but more generally, the activation of this system is rarely positioned as central (or as a formal requirement) within the context of typical projects. This consideration helps to understand why this system is only occasionally activated, almost as a reserve methodology within specific design contexts to extend towards subjectivities that are more difficult to address via more conventional architectonic strategies and techniques. This form of extension can be one of responsive (in which architectonic form is itself shaped by sonic concepts or processes), parasitic-symbiotic (in which the system of sound, space, and public actively occupies and repurposes an architectural project or space), or expansive (in which the system expands architecture's intensive visuality to open towards a wider multi-sensorial field). The affordances of the system in the context of architecture are therefore linked to options, not to requirements.

Landscape architecture

The system affords something different again in the context of landscape architecture. As landscape architecture operates in the domain between buildings and within open outdoor spaces, it is more intrinsically aligned with efforts to integrate considerations of sound in the creation of spatial form, as it is within this context that the urban sound environment is often experienced. The rarity of sonic practice in landscape architecture – and the lack of inclusion of sonic practice within related pedagogy – is therefore more notable than in architecture, and when shifted into the foreground seems to demand attention.[27] In other words, because sound (particularly potentially unwanted or unintended sound) is present in the context of the projects identified through the remit of this discipline, landscape architecture is perhaps more poised to be introduced to methodologies involving sound, space, and public experience than architecture.[28] The system affords an intuitive extension in this scenario.

Acoustics and noise control

The affordances of this system in the context of acoustics and noise control are (unsurprisingly) the most clearly defined, as these disciplines have developed identifiable service offerings embedded in both urban design and higher-level legislative processes.[29] The longer-term presence of these service offerings within the composition of multi-discipline built environment projects (and within the value chains that support them) is therefore established; however, this does not indicate that these forms of practice are always well-resourced.[30] Within the context of public realm projects, the tendency for objectives related to acoustics and noise control to be value-engineered out of projects at later stages of development is a frequent concern. Driven by an awareness of this (as well as by the centrality of sound within these disciplines), more future-oriented service offerings advanced particularly within the discipline of acoustics are in rapid flux.[31] At the same time, the more confident position of these disciplines within more complex value chains can potentially deter other practitioners from developing more confident approaches to working with sound through their own experience, vocabularies, and project initiatives. The affordances of the system in this case are complex: on the one hand, they indicate a proliferation of service offerings, while on the other they invite an awareness to listen to other instincts that might advance more expansive approaches to the system of sound, space, and public.

Policy and legislation

From a legislative perspective, addressing the system of sound, space, and public as open, unstable, and dynamic is problematic. It effectively affords a challenge. Strategies to address sound on the level of entire populations and larger urban territories have historically leveraged static definitions of environmental noise premised on quantitative metrics that can be easily articulated, enforced, and leveraged within the context of policy and legislation. Indeed, the interviews conducted within the context of *Sound-Frameworks* reveal a sharp division of perspectives regarding whether static definitions of noise were entirely necessary or extremely problematic. Of equal significance is the prioritisation of measurements focused on thresholds between domestic or private space and more open public spaces. The idea of a mobile, plural, and highly differentiated subject (i.e., not a *population* but a *public*) as a vehicle through which to evaluate noise and upon which to establish policy-level objectives is extraordinarily difficult to imagine from working perspectives rooted within this field. A subset of *Sound-Frameworks* interviewees understood the current evolution of the Soundscape ISO standard as somewhat positioned within this problematic,[32] as a mechanism to advance a more subjective means of evaluating sound within diverse environments within the context of environmental policy and legislation. However, beyond this core set of practitioners, the concepts and methodologies that underlie the establishment of this standard proved to be both unknown and potentially cumbersome when introduced within the context of other practitioners' work. As it gains momentum within the context of policy and legislation, this standard emerges as a singular approach to the system under discussion, and other practitioners' approaches – which do not adhere to its rhetoric – should be prioritised to ensure that a broader field of practice can begin to take form.[33]

Urban development

As noted earlier, the affordances of the system of sound, space, and public are also difficult to decipher within the context of urban development. However, unlike in the context of policy and legislation, there is no default position established to address this void. This gave the *Sound-Frameworks* interviews that addressed this perspective a certain potency, and led to the conjecture that in order for this system to be coherently addressed in the context of urban development, it needed to be possible to first demonstrate its value, and then to link this value to other dimensions of (for example) urban regeneration projects, and to do this in such a way that this sonic, acoustic, or aural dimension could not be simply value-engineered out of the project at a later stage in order to reduce costs.

As this proposition was interjected and recycled within subsequent *Sound-Frameworks* interviews and conversations, the idea of value became to take on more complex characteristics, eventually splintering into the four interrelated concepts of *value*, *narrative*, *legibility*, and *integration*. The centrality of these four concepts led to their being positioned as the core indicators within the *Sound-Frameworks* design tool, where they are defined as follows:

> A project's 'Value' is its perceived usefulness or importance. The project's 'Narrative' is what is told to communicate the project to diverse stakeholders. 'Legibility' measures how much the project's 'Value' can be understood through this 'Narrative'. Finally, 'Integration' indicates how well the project's intended strategy for working with sound is connected with other aspects of the public realm project being developed.[34]

The first objective of the *Sound-Frameworks* design tool is to present and interrogate these four concepts within the context of individual projects from the outset. For a given public realm project to retain considerations of sound from inception to implementation, it is necessary to develop as coherent an approach to articulating the value, narrative, legibility, and integration concerning sound as possible (Figure 16.3).

The second objective of the *Sound-Frameworks* design tool seeks to capture the diverse concerns of different practitioners working in this field. In other words, it is a mechanism for explicitly adjusting the consideration of sound, space, and public encapsulated in the tool to the affordances of this system that a given user (or team of users) perceives as most relevant. By selecting these custom indicators and adjusting their 'importance' and 'rating', users map out the more nuanced dimensions of the project and begin to observe how considerations of sound are related to other components of the more expansive value chains at play (Figure 16.4).

Remaining attentive to peripheral perspectives

Beyond the more prominent perspectives encapsulated in the previous sequences, perspectives rooted in more (disciplinarily) peripheral aspects of city-making and public life propagate a further set of affordances related to the system of sound, space, and public. To venture into this expansive territory, *Sound-Frameworks* pursued interviews with researchers and practitioners involved in the development of water features,[35] the advancement of strategies supporting equitable housing,[36] and the development of entertainment systems for automobiles.[37] Exploring these (and other peripheral) practices introduced a further set of considerations that serve to destabilise and challenge the centrality of artistic, architectural, acoustic, and noise-related practices in the domain of urban sonic research and practice. These practices introduce aspects of urban materiality, experience, domesticity, and mobility that can only be revealed through the pursuit of non-directive, open-ended discourse in which concerns of urban sonic experience are repeatedly positioned in proximity to concerns that are more central to these forms of practice. The affordances of the system of sound, space, and public within each of these contexts are quite specific; they cannot be effectively grouped in relation to more dominant disciplines but require their own vocabularies and objectives to remain intact as they are re-oriented towards urban sonic practice.

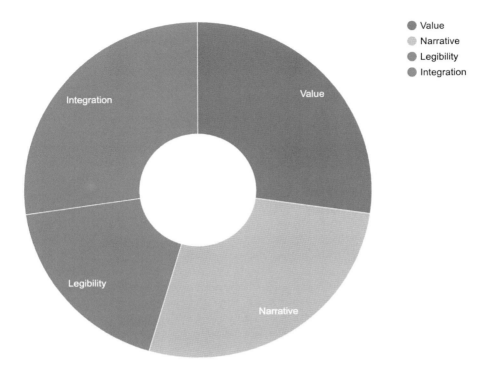

Figure 16.3 The Sound-Frameworks design tool has four core indicators: 'Value', 'Narrative', 'Legibility' and 'Integration'. Users input information concerning the significance and rating of each indicator to develop a visualisation of these parameters in each project. These four indicators receive prominent attention to place emphasis on the importance of communicating about the value of working with sound in specific contexts. In this case, the user has set their 'significance' at 3, 3, 2, and 3 (on a scale of 1–3) and their 'rating' at 1, 2, 3, and 3 (on a scale of 1–3), respectively. Image courtesy of the author.

To accommodate the proliferation of affordances revealed through these perspectives, the *Sound-Frameworks* design tool links the core indicators (value, narrative, legibility, and integration) with the larger list of custom indicators via a diagram demonstrating their connections. Users enter information about these connections based on their analysis of each project, potentially enacting this process as a mode of workshop to sustain dialogue between the project's key stakeholders. The connections diagram that takes form through this collaborative process is a useful asset for quickly assessing which dimensions of a project are the most relevant and require attention and resources (Figure 16.5).

Maintaining a plural approach

The research conducted through *Sound-Frameworks* demonstrated that practitioners working across these different perspectives sense both the opportunity and necessity of advancing projects that explore intersections between sound, space, and public, but that the means of securing such praxis within projects that are fully implemented remains difficult to identify. Thinking through

From affordances to value chains

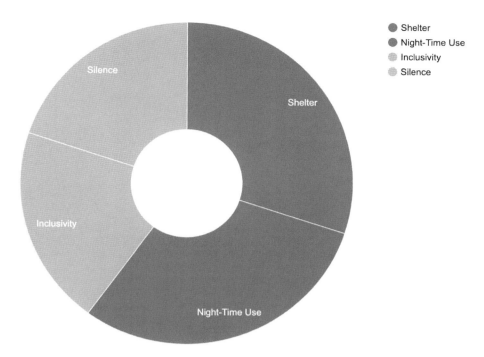

Figure 16.4 The second diagram in the Sound-Frameworks design tool allows users to select and configure up to eight custom indicators selected from a list, including Accessibility, Acoustics, Activism, Ambiance, Architecture, Atmosphere, Biodiversity, Building Codes, Climate Change, Commercial Space, Community, Conflict, Crisis, Culture, Density, Ecology, Education, Environmental Acoustics, Equity, Events, Exclusion, Gentrification, Heritage, information and communication technologies (ICTs), Inclusion, Industrial Space, Landscape, Lighting, Masterplanning, Mobility, Morphology, Music, Nature, Night-Time Use, Noise, Noise Complaints, Noise Regulations, Non-Human Life, Pedestrian Space, Personal Space, Physical Plant, Public Art, Residential Space, Resilience, Redevelopment, Safety, Shelter, Silence, Soundscape, Sports, Sprawl, Suburbs, Sustainability, Tranquillity, Transportation, Vibrance, Waste Management, Water Features, Weather, Zoning. In this case, the user has selected Shelter, Night-Time Use, Inclusivity, and Silence, and set their 'significance' at 3, 3, 2, and 2 (on a scale of 1–3) and their 'rating' at 1, 1, 2, and 2 (on a scale of 1–3), respectively. Image courtesy of the author.

concepts such as value, narrative, legibility, and integration and seeking to position sound-related objectives within more recognised value chains serves as a first step towards securing this form of praxis. The multitude of approaches to exploring the relationship between sound and space explored in this volume sustain this enquiry and can be augmented by these more tactical considerations.

Crucially, it is not only the task of practitioners to work with this awareness of affordances and value in mind. It is important for decision-makers to script opportunities to identify these concepts within project briefs and development requirements. This does not imply a singular, centralised, top-down approach; instead, it indicates a call for a more plural, distributed initiative. For instance, these opportunities can be articulated within the briefs defined for public art commissions, but equally, they might be achieved through action planning related to the Environmental Noise Directive.[38] They can be supported by more prominent architecture firms (that often possess a greater

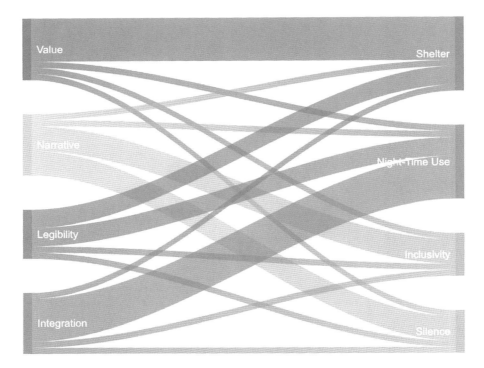

Figure 16.5 The third diagram in the Sound-Frameworks design tool maps the linkages between core indicators and custom indicators, showing how specific aspects of the project contribute to its value, narrative, legibility, and integration. The weights of the connections demonstrate how much each custom indicator contributes to each core indicator. The colours are derived from the settings used in the core indicator diagram and custom indicator diagram. This diagram provides a means of identifying which dimensions of a project require attention to ensure they can progress. Image courtesy of the author.

sense of agency in forming interdisciplinary design teams) who seek out practitioners working with sound in urban contexts to support and extend larger project briefs. They can be implemented by community organisations that flag the potential for sound and listening methodologies within concise local funding applications. The work that goes into this process will remain lateral, tactical, and plural as an awareness of these methodologies – and the projects that they support – advances in both design discourse and public experience.

Related resources

The project *Sound-Frameworks* focused on the development of three central resources, including a survey, a set of design guidelines, and a discursive design tool that can be used by others within the context of future projects and proposals. These resources can be accessed at https://www.soundframeworks.org/ and https://design.soundframeworks.org/. For longer-term preservation, the interviews conducted within *Sound-Frameworks* (along with other selected assets) are available via the open access repository Zenodo at https://www.zenodo.org/communities/soundframeworks/.

Disclaimer and acknowledgements

The opinions presented in this text are my own and do not necessarily reflect the opinions of the partners, collaborators, individual practitioners, and organisations who contributed to the development of *Sound-Frameworks*. I wish to acknowledge the contributions of all of those who participated in *Sound-Frameworks* from its inception at proposal stage to its implementation between 2021 and 2023, as well as towards its future evolution. The project is indebted to its host Theatrum Mundi and its partners Arup, UrbanIdentity, Struer Kommune (and the Sound Art Lab), the University of Oxford Faculty of Music (and the SONCITIES team), and the Sound Studies Lab at the University of Copenhagen.

Sound-Frameworks received funding from the European Union's Horizon 2020 Research and Innovation Programme under the Marie Skłodowska-Curie Grant Agreement No. 101032632.

Notes

1 This public art commission is titled *The Manual for Acoustic Planning and Urban Sound Design*, abbreviated as *MAP*. I proposed this project and carried it out with Dublin City Council commencing in early 2013. *MAP* took part within *Interaction with the City*, the second strand of the Dublin City Public Art Programme and was funded through the Per Cent for Art scheme.
2 The St. Michael's Estate Regeneration Board organised a public event for the demolition of the last remaining tower block of St. Michael's Estate at 11:00 AM on Tuesday, February 26th, 2013, on St. Vincent's Street West, Inchicore, Dublin 8.
3 *The Debt* is a public art commission by the sound art collective Ultra-red developed in 2003. The project evolved as a series of actions uniting residents from the public housing communities of Ballymun in Dublin and Pico Aliso in East Los Angeles to discuss experiences of social housing regeneration.
4 The *Celtic Tiger* is a term that refers to the intensive (and ultimately unsustainable) economic growth of Ireland between the mid-1990s and the late 2000s.
5 Leading into the *MAP* commission, I was researching a multitude of practices related to this field, particularly the discursive praxis of the artist Max Neuhaus and the methodologies developed by Le Centre de Recherche sur l'Espace Sonore et l'Environnement Urbain (CRESSON) in Grenoble.
6 I'm slowly trying to train myself to separate 'public' from 'space', not only to avoid falling back on 'public space' or 'public 'realm' as given terms but also to encourage an implicit inclusion of 'public life'.
7 I developed *The Office for Common Sound* (*OCS*) in 2016 following a realisation that the responsibilities of the *MAP* commission precluded face-to-face interaction with members of the public. *OCS* was initiated in a small shopfront in Bray, County Wicklow over a two-month duration in which I worked in the street-level office full time five days a week with an open-door policy so that anyone could contribute to (or even lead) the directive of the office.
8 More information about *Continuous Drift* and the artists whose work is presented within it is available at https://www.continuousdrift.com (Accessed March 10, 2024), which also serves as the public interface for the installation.
9 I came to question the affordances of systems through the work of the artist Dennis McNulty. Describing his project *Blesh*, McNulty writes: "When designers talk about the things they make in terms of their affordances, they're talking about the way the form of an object can suggest the ways in which it might be used. A good example is the way the design of a saucepan suggests it should be picked up by the handle. Maybe it's possible to think about systems in terms of their affordances?" For more information, see https://blesh.net/origins/ (accessed March 10, 2024).
10 *Sound-Frameworks: Collaborative Frameworks for Integrating Sound Within Urban Design and Planning Processes* is an action-led research project that explores the role of sound in the design and planning of the built environment. I developed the project in London as a two-year research fellowship with Theatrum Mundi between November 2021 and November 2023. The project received funding from the European Union's Horizon 2020 research and innovation programme under the Marie Skłodowska-Curie grant agreement No 101032632.

11 I developed the proposal for *Sound-Frameworks* with input and advice from Arup and positioned the project with Theatrum Mundi as host and beneficiary, drawing from the support and expertise of both organisations.
12 Theatrum Mundi and &beyond (eds.). *Sonic Urbanism* (London: TM Editions, 2019).
13 Theatrum Mundi and &beyond (eds.). *Sonic Urbanism 2: The Political Voice* (London: TM Editions, 2021).
14 Theatrum Mundi and &beyond (eds.). *Sonic Urbanism 3: Listening for Non-Human Life* (London: TM Editions, 2022).
15 Cecily Chua, Labeja Kodua Okullu, and Marta Michalowska (eds.). *Urban Backstages* (London: TM Editions, 2023).
16 Marta Michalowska (ed.). *Navigations: Scoring the Moment* (London: TM Editions, 2022).
17 Andrea Cetrulo and Marta Michalowska (eds.). *Interior Realms* (London: TM Editions, 2021).
18 The second person I interviewed for Sound-Frameworks was Helle Baker, Program Manager of Harbour Development for Struer Kommune (https://zenodo.org/records/10413088). Baker's reference to the concept of value chains introduced me to a new way of interrogating sound in the context of the public realm.
19 More information about the concept and history of *Lydens By* is available at: https://cityofsound.dk/about-the-city-of-sound/ (Accessed March 10, 2024).
20 I grew up a block away from the Massachusetts Turnpike in the suburb of Newton, eight miles outside of Boston. Between the nightly resonance of the highway and days spent exploring the city as a teenager interested in experimental music, the connection between urban form and sonic experience was always present in my imagination.
21 Brandon LaBelle. *Acoustic Territories: Sound Culture and Everyday Life* (London: Continuum, 2010).
22 Salomé Voegelin. *Listening to Noise and Silence: Towards a Philosophy of Sound Art* (New York: Bloomsbury, 2010).
23 The prominence of long-term artistic practice in this field is evidenced in the *Sound-Frameworks* interviews with artist Peter Cusack (https://zenodo.org/records/10419663) and Andres Bosshard (https://zenodo.org/records/10418974).
24 The *Sound-Frameworks* interviews with artists Colm Keady Tabbal (https://zenodo.org/records/10419347), Nicola di Croce (https://zenodo.org/records/10419365), and James Wilkie (https://zenodo.org/records/10419311) reveal the prominence of socially engaged practice within their praxis.
25 The *Sound-Frameworks* interview with architect Shih-Fu Peng of Dublin-based Heneghan Peng Architects (https://zenodo.org/records/10419278) underscores an awareness of the intrinsic sonic considerations latent in architectural practice.
26 The architect Juhani Pallasmaa's *The Eyes of the Skin: Architecture and the Senses* (Hoboken, NJ: John Wiley & Sons, 1996) exemplifies this form of practice.
27 The challenges (and potentials) of introducing sonic methodologies within professional landscape architecture curricula features prominently in the Sound-Frameworks interview with landscape architect Sara El Samman (https://zenodo.org/records/10417508).
28 This curiosity about why tools to explore this system are not more readily at hand for landscape architects to integrate within practice came up in the *Sound-Frameworks* interview with landscape architect and integrated city specialist Dima Zogheib at Arup (https://zenodo.org/records/10417343).
29 The *Sound-Frameworks* interview with acoustician Mhairi Riddet at Arup (https://zenodo.org/records/10417390) centred on expansive considerations of noise assessment procedures required in rural contexts in relation to the emergence of data centres.
30 Amongst many other subjects, the *Sound-Frameworks* interview with acoustician and founder and director of Platform 78 Rachid Abu-Hassan (https://zenodo.org/records/10419113) drew attention to the low portion of overall project resources dedicated to acoustics, even within projects such as concert halls and performance venues in which sonic experience is a central concern.
31 The *Sound-Frameworks* interviews with Arup acousticians Mitchell Allen (https://zenodo.org/records/10413175) and Shane Myrbeck (https://zenodo.org/records/10417306) charted how these services are currently evolving, featuring case studies of projects involving night time safety and the design of outdoor community music venues respectively.
32 The International Organization for Standardization (ISO) defines the Soundscape Standard in three documents: "Soundscape - Part 1: Definition and Conceptual Framework (ISO 12913-1:2014)", "Soundscape - Part 2: Data Collection (ISO/TS 12913-2:2018)", and "Soundscape - Part 3: Data Analysis (ISO/TS 12913-3:2019)".
33 I foregrounded a need to support diverse approaches to this field of practice in my presentation for the Internoise conference in 2022, titled: *Beyond Standards: In Search of Heterogeneous Approaches to Sound in the Design and Planning of the Public Realm.*

34 The *Sound-Frameworks* design tool is online at: https://design.soundframeworks.org/ (Accessed March 10, 2024).
35 The *Sound-Frameworks* interview with Stephane Llorca of JML Water Features (https://zenodo.org/records/10419221) explored the potential of sonic considerations of urban water features, which proved much more extensive than common links between the sound of fountains and masking functions.
36 Researcher Hani Salih's *Sound-Frameworks* interview (https://zenodo.org/records/10419184) probed the role of sound in the work of The Quality of Life Foundation through a case study of a specific housing development project.
37 I met with Gorm Haldor Jørgensen, Senior Director of Innovation for EPIC and Car Audio at Harman International, to discuss how advancements in interior and exterior loudspeaker and microphone arrays in cars contribute to discussion of sound in the context of the public realm (https://zenodo.org/records/10419592).
38 It could be possible to develop a mechanism within the noise action planning subroutines specified by the Environmental Noise Directive (2002/49/EC) that would be similar to the Per Cent for Art schemes employed in many countries to fund public artwork through a percentage of budgets related to larger infrastructural projects. In this case, a similar (small) percentage of the funding allocated to noise modelling and noise mapping processes could be pooled to support artistic practice in this field, which in turn could become a central component of more progressive action plans.

Bibliography

Anderson, Sven. *The Manual for Acoustic Planning and Urban Sound Design*. 2013–2020.
Anderson, Sven. *Continuous Drift*. 2015 – (Ongoing).
Anderson, Sven. *The Office for Common Sound*. 2016 – (Ongoing).
Anderson, Sven. "Beyond Standards: In Search of Heterogeneous Approaches to Sound in the Design and Planning of the Public Realm." Paper presented at the 51st International Congress and Exposition on Noise Control Engineering (Internoise 2022), Glasgow, 21–24 August 2022. https://doi.org/10.5281/zenodo.10465135.
Cetrulo, Andrea and Marta Michalowska, eds. *Interior Realms*. London: TM Editions, 2021.
Chua, Cecily, Labeja Kodua Okullu, and Marta Michalowska, eds. *Urban Backstages*. London: TM Editions, 2023.
The European Union. "The Environmental Noise Directive (END) 2002/49/EC." Accessed March 12, 2024. https://eur-lex.europa.eu/legal-content/EN/TXT/?uri=CELEX:32002L0049.
The International Organization for Standardization. "Soundscape - Part 1: Definition and Conceptual Framework (ISO 12913-1:2014)." Accessed March 12, 2024. https://www.iso.org/standard/52161.html.
The International Organization for Standardization. "Soundscape - Part 2: Data Collection (ISO/TS 12913-2:2018)." https://www.iso.org/standard/75267.html.
The International Organization for Standardization. "Soundscape - Part 3: Data Analysis (ISO/TS 12913-3:2019)." Accessed March 12, 2024. https://www.iso.org/standard/69864.html.
LaBelle, Brandon. *Acoustic Territories: Sound Culture and Everyday Life*. London: Continuum, 2010.
Michalowska, Marta, ed. *Navigations: Scoring the Moment*. London: TM Editions, 2022.
McNulty, Dennis. *Blesh*. 2016.
Pallasmaa, Juhani. *The Eyes of the Skin: Architecture and the Senses*. Hoboken, NJ: John Wiley & Sons, 1996.
Theatrum Mundi and &beyond, eds. *Sonic Urbanism*. London: TM Editions, 2019.
Theatrum Mundi and &beyond, eds. *Sonic Urbanism 2: The Political Voice*. London: TM Editions, 2021.
Theatrum Mundi and &beyond, eds. *Sonic Urbanism 3: Listening for Non-Human Life*. London: TM Editions, 2022.
Ultra-red. *The Debt*. 2003.
Voegelin, Salomé. *Listening to Noise and Silence: Towards a Philosophy of Sound Art*. New York: Bloomsbury, 2010.

17
ON VIBRATIONAL ARCHITECTURES

Gascia Ouzounian (text) and Jan St. Werner (images)

Tremblings

In distinction to the spectacular forms that lined the exhibition grounds of the 13th Venice Biennale of Architecture in 2012, whose theme was 'Common Ground', the Polish Pavilion by Katarzyna Krakowiak was remarkable not in what it portrayed but in what it revealed and transmitted: the sounds and vibrations of the pavilion, its infrastructure, and its surroundings. Vibrations produced by the building were 'explored and amplified'[1]; the capacity of its infrastructure – such as its ventilation system – to funnel, filter, and transmit sound were reinforced; and the sounds and vibrations of neighbouring pavilions were made audible inside it, bringing the building into a resonant dialogue with its environment and thereby crafting through sound, as the biennale judges remarked when awarding the project an Special Mention, a genuine 'common ground'.[2]

When announcing the biennale's theme of 'common ground', the architect David Chipperfield stressed that it was intended to denote the role of architecture in 'defining the common ground of the city' while also evoking 'the ground between buildings, the spaces of the city'.[3] Krakowiak's acoustic reconfiguration of the Polish Pavilion, *Making the walls quake as if they were dilating with the secret knowledge of great powers* (2012), made clear that this common ground could be construed not only metaphorically, as in the shared social and political concerns articulated by architecture, or literally, as in the physical ground shared by buildings, but also in something as immaterial and elusive as the air. It made tangible the idea that buildings are connected through the energies they share, including the sounds and vibrations they produce, mediate, and transmit – vibrations that pass through the ground and through the air; and that buildings themselves have vibratory capacities that, in contrast to longstanding architectural traditions that have sought to diminish them, can be productively enhanced and reinforced.

A critic wrote of the Polish Pavilion that it 'collaborates with neighboring pavilions' by echoing them.[4] Indeed, one could say that the building both resonated with the sounds of other buildings, thus symbolically dissolving the separation between nations; and that it came into a fuller resonance with itself, its materiality and physicality gaining voice through technological, acoustic, and architectural interventions. According to curator Michał Libera, these interventions included reinforcing the pavilion's resonant frequencies and naturally long reverberation time (over six seconds) to the point that speech inside it became unintelligible; installing a temporary wall and a temporary floor and tilting both at slight angles, thereby increasing the reflective capacities of architectural surfaces; and dismantling an artificial ceiling that had been put in place for previous exhibitions, exposing a skylight and opening the pavilion to neighbouring sounds via ventilation pipes and holes. While most of these interventions resulted in increasing the building's reflective capacities, a passage near the entrance of the pavilion was soundproofed to diminish ambient sounds and thus 'prepare' visitors to experience an aural architecture.[5] A final gesture, Libera explained, was to present 'a "live sonification" of the vibrations of the walls of the entire building'. He wrote:

> The trembling of the walls is translated into sounds and made audible in the space of the pavilion, together with the trembling of selected parts of the building. A network of sensors and cables entwines the entire architectural complex, including the façades of the adjacent pavilions, marking the continuity of sound as a phenomenon.[6]

Thus, the pavilion's tremblings – subtle tremors and mechanical oscillations that would not normally produce audible sound – were converted into acoustical energy and amplified; as were the tremblings of adjacent buildings, whose vibrations were similarly made audible and sensible, transmitted via loudspeakers hidden behind the temporary wall and floor in the Polish Pavilion.[7] These 'live sonifications' of architectural tremblings revealed the vibratory nature of architecture, suggesting that a building is not an isolated entity but rather one that operates within a wider ecology of sensible energies. As such, the pavilion was a meditation on architecture as it might be rewritten in the language of vibration, with the transmission of acoustic and vibro-acoustic energies as an organising principle of architectural forms and spaces and a key dimension of architectural experience.

Although Krakowiak's pavilion was centrally concerned with the interrelationships of sound and architecture, the building was not imagined as a sound stage or a structure in which to install sounds. Such gestures have predominated in sound art traditions since the late 1950s, when Le Corbusier, Iannis Xenakis, and Edgard Varèse's pavilion for the 1958 Brussels World's Fair, *Poème électronique*, used a multichannel system to route sounds to hundreds of loudspeakers placed throughout the building, 'engulfing' listeners in what Varèse described as a 'spatial music'.[8] By contrast, the Polish Pavilion was conceived of as a sound-generating entity itself, one whose intrinsic sound-producing, sound-mediating, and sound-transmitting capacities were highlighted and extended. By making audible the tremblings of the building and those of adjacent buildings, the pavilion was reconfigured as a vibrational architecture: an architecture comprised not only of solid forms, fixed structures, and stable materialities, but also of trembling, vibrating, oscillating, quaking energies. By making sensible the vibratory dimension of architecture, the pavilion embodied the idea that architecture can serve not only to 'contain' or reflect acoustic energies, but also to produce, transmit, mediate, filter, combine, and recombine them. It engaged architecture as a dynamic, moving phenomenon, productive of and mediating energies that should be sensed and heard; and it revealed the interconnectedness of seemingly distinct architectures that were nevertheless intertwined in the common realm of vibration.

Vibrational architectures

In considering sound in relation to architecture, a predominating model in Western architectural and scientific thought since 1900 has been that of reverberant space: of architecture as a container for sound and acoustical reflections; and of architectural forms as having acoustical characteristics that derive from their capacity to reflect and absorb sounds that occur inside them. Architects and acousticians routinely describe a room in terms of its reverberation time, the time it takes for sound inside the room to decay over a specific dynamic range (typically 60 dB). The modern scientific discipline of architectural acoustics, generally traced to the work of the American physicist Wallace Clement Sabine – whose equation for calculating the reverberation time of a room was instrumental in establishing acoustics as a scientific enterprise[9] – is rooted in this model of reverberant space, one in which architecture acts as a physical structure that absorbs and reflects sounds, but not necessarily one that produces or transmits sounds and vibratory phenomena, apart from unwanted noises and vibrations that should be reduced and attenuated.[10]

This essay explores the possibilities of a vibrational architecture and its potential to reorient architectural thought and practice. 'Vibrational architecture' here refers to architectures whose energetic capacities are highlighted (not reduced); and to architectural and spatial practices that privilege vibrational phenomena, including audible and inaudible sound. Vibrational architecture can be a way of conceptualising architecture as comprised not only of materialities, but also of energies that modulate and complicate those materialities. Vibrational architectures challenge the perceived fixity and stability of architectural forms and structures, including monumental architectures that are imagined as enduring and immutable. In engaging the possibilities of a vibrational architecture, this essay asks: how would the study of architecture change when architecture is considered not only or primarily in relation to solid, fixed, enduring, material forms but also in relation to energies that are emergent, transient, ephemeral, and unpredictable? How would architectural history change if it were to consider not only what was built and what remains, but also what passes through: those energies that have been produced, transmitted, reflected, and absorbed by architecture?

In engaging with the idea of vibrational architecture, this essay draws on the work of theorists and practitioners who work across sound, architecture, and vibration, notably Maryanne Amacher, Mark Bain, Steve Goodman, Katarzyna Krakowiak, Mendi + Keith Obadike, Jonathan Tyrrell, and Jan St. Werner, among others. In Sonic Warfare, Steve Goodman recalls a scene from the popular Japanese manga series *Patlabor* in which 'the vibrational architecture of the city becomes a weapon'.[11] Goodman suggests that this form of sonic warfare – one that arguably occurs not only in fiction but also in contemporary warfare in which the urban environment is weaponised to transmit harmful vibrations – that 'the city is no longer merely the site of warfare but, as a result of the resonant frequency of the built environment, the very medium of warfare itself. The plot', he remarks, 'tunes into the city as an instrument, not just venue, of terror'.[12]

Goodman is specifically concerned with the capacity of what he calls 'the vibrational architecture of the city' – its vibrational substrate – to be mobilised to participate in an ecology of terror, transmitting fear, dread, and other negatively valanced affect in 'climatic' and 'volatile' ways to a city's population through the spread of 'bad vibes'.[13] In this scheme, the built environment acts as a 'resonating surface' that, tuned to certain frequencies (Goodman is particularly interested in low frequencies and infrasound), transmits bad vibes to a populace, 'sending an immense collective shiver' through it.[14] Goodman's scenario taps into anxieties about the nefarious uses of an invisible medium (sound) by governments and militaries, anxieties that have circulated since early experiments on the effects of infrasound in the 1930s, and that typically map onto wider anxieties about unknown forces at play in the political arena.[15] It is especially salient here in imagining the city as a vibrational topos or network – an interconnected system or, as Goodman puts it, 'vibrational ecology' that can be activated, instrumentalised, and potentially weaponised through sound and vibration.[16]

An aesthetics of material transmission

In his anthropology of ocean science, which considers the underwater sounds that scientists interpret while submerged in submarine vessels, Stefan Helmreich proposes transduction – the conversion of energy from one energetic substrate to another – as an analytic, a frame through which to understand the flows of sound across physical boundaries and media.[17] Transduction, he suggests, can be helpful both in 'thinking about how space, presence and soundscapes are produced', since, for example, underwater sounds must be converted from one form of energy into another to be audible inside submarines; and as a way of 'theorizing against immersion'.[18] The concept of soundscape 'has become haunted by the notion of immersion', Helmreich argues, the idea that listeners are 'at once emplaced in space and, at times, porously continuous with it'.[19]

Indeed, the idea of immersion has haunted not only the notion of soundscape but also the domain of architectural acoustics, which principally considers acoustical environments as ones in which subjects are emplaced, with sounds and acoustical reflections taking place all around them. By contrast, in a vibrational architecture the listening subject is not immersed in sound so much as their body is coupled with acoustic and vibro-acoustic energy, mediating sound and vibration, 'connect[ing] the occupier to that which is occupied'.[20]

In *Architecture's Acoustic Shadow: Unsettling the Sound-Space Relationship*, the architect, artist, and theorist Jonathan Tyrrell suggests that 'an emphasis on reverberation has overlooked how sound operates transversally, moving through bodies and matter, undermining spatial division, and confounding architectural legibility'.[21] Tyrrell asks, 'How [would] a focus on material transmission rather than reverberation change the way space and sound are mutually conceived and experienced?'[22] Tyrrell's incisive question signals what, if adopted more widely, would entail a paradigmatic shift in architectural discourse and practice in relation to sound, as well as conceptions of sound space more broadly. It points to models that challenge the immersion framework articulated by Helmreich and embodied in works such as *Poème électronique*. If Varèse's concept of a spatial music was contingent upon the notion of sounds 'surrounding' and 'engulfing' listeners, and if sound art traditions that have developed from that model have privileged an aesthetics of sonic immersion, what are alternative models that focus on the material transmission of sounds – or, put another way, an aesthetics of material sound transmission – and what are their genealogies?

What Tyrell identifies as a missing focus on material transmission can be traced in the work of practitioners who have emphasised architecture's capacity to transmit and mediate sound and vibration. Tyrell himself points to the work of Maryanne Amacher and Mark Bain, who, he writes, 'have both worked with the direct vibration of architectural matter', drawing attention to Amacher's concept of '"structure-borne sound" (material propagation at high intensities)'.[23]

If Varèse's concept of a spatial music 'engulfing' listeners was foundational in articulating an aesthetics of immersion, Amacher's practice of working with structure-borne sounds can be considered equally foundational in articulating an aesthetics of material transmission. According to Amy Cimini and Bill Dietz, with *Living Sound, Patent Pending* (1980), her first architecturally scaled work, Amacher used both structure-borne and airborne sound transmission to transform the entirety of a vacant mansion in St. Paul, Minnesota, into a sonic architecture, raising 'a host of new technical questions and considerations (e.g., the necessity of idiosyncratic loudspeaker placements so as to transmit and filter sound via physical substrates such as walls)'.[24] In contrast to spatial sound works installed 'in' buildings, with *Living Sound, Patent Pending*, an entire building was reconfigured as a resonating structure – a 'gigantic instrument' – that pulsated with acoustic energy. Amacher remarked:

> A visitor who stepped "off limits" into the kitchen was literally slammed up against the refrigerator by the force of the energy. Others felt themselves pushed, as if by acoustic pressure, out into the garden, where the entire house was heard, sounding, as a gigantic instrument.[25]

With *Living Sound, Patent Pending*, the coupling of sound to building materials magnified sound's physical force and seeming omnipresence. Elsewhere, Amacher spoke to the potential of structure-borne sounds to enhance the expressive potentials of music, saying of her process that:

> It's going to the space, finding these kinds of spots where the sound can sort of take on its life, traveling through the structure, one way or another. […] I never like to work with a lot of speakers, because I would prefer to get the sound alive in the architecture. I now believe that the architecture can make magnifying really expressive dimensions in music in a way that you can't do any other way.[26]

If Amacher sought to 'get sound alive in architecture', Mark Bain has arguably sought the inverse – getting architecture alive in sound. Such an impulse has shaped Bain's vibro-acoustic works which, since the late 1990s, have engaged the conductive and vibrational capacities of architectures and bodies. One of Bain's best-known projects in this vein is *The Live Room: Transducing Resonant Architecture* (1998), originally conceived for a room at Massachusetts Institute of Technology (MIT) that had large concrete isolation pads separated from the rest of the building's foundation to control for external vibrations. For this manifestation of *The Live Room*, Bain mounted 'acoustic-intensifying equipment' directly onto the foundational structures of the room, setting the architecture into vibration by sending intense impulsive energy through it, '[imparting] frequencies into the building, the floor and the persons who were situated in the room'.[27] More specifically, he mounted six rotary-type mechanical oscillators with a frequency range of 1–30 Hz directly onto the floor system, setting it into vibration apart from energetic 'dead spots' where the seven concrete isolation pads were located.[28] He made this vibrational 'topology' visible by placing fine sand on the floor system, revealing vibrational patterns and nodes much in the vein of Ernst Chladni's experiments with vibrating plates in the late 18th century.[29]

Bain's process in *The Live Room* entailed identifying the resonant frequencies of objects and structures; using a series of mechanical oscillators to vibrate and resonate them, and, setting into motion a network of vibrations, triggering vibrational 'feedback' processes that he likened to additive synthesis; and, through mechanically induced vibrations, additionally producing sympathetic vibrations in nearby bodies and objects. *The Live Room* thus operated as a 'vibrational ecology': a complex, dynamic vibrational system or network in which various vibrations reinforced, cancelled out, and produced other vibrations.

Bain conceived of *The Live Room* as a 'vibro-acoustic environment that engages directly the architecture, the room and the people who occupy it'.[30] He treated both the building and visitors' bodies as having specific resonant frequencies that might be excited through vibration, writing:

> Buildings too, along with bodies, have their own particular resonant frequencies. If you locate this frequency, and its associated value of efficient excitation, and through mechanical

reinforcement impart this frequency, it is possible literally to "ring" a material in a manner similar to striking a bell. If, through a feedback system, a phase-aligned addition to this waveform is encouraged, it may become possible for the materials to oscillate out of control.[31]

With *The Live Room* Bain treated buildings, architectural materials, and bodies on equal terms, as objects that have the capacity to conduct vibro-acoustic energy. He noted that architectural materials, in contrast to a gas like air, typically have dense molecular structures and can therefore conduct acoustic energy relatively efficiently; and that not only bodies but also individual body parts – 'organs, bones, and tissue' – have specific resonant frequencies that can be excited through vibration.[32] Here, the listener did not only or primarily 'receive' sound through cochlear audition and interpret it through neurocognitive processes of listening. Instead, the listening subject was arguably reconfigured as an object – a thing that, like other things, participated in vibrational ecology, conducting, mediating, modifying, absorbing, and potentially producing sound and vibration. Such a reconfiguration of the listener recalls Salomé Voegelin's observation that:

> Vibration makes the world appear as an invisible field of connections within which my body *oscillates as a thing amid other things*. Vibration is the inexhaustible condition of this world that … binds me into its texture, not at its center, but in its weave.[33]

Notably, in a vibrational ecology, space is not treated as an empty container but rather as possessing materiality itself. Empty space – or more precisely, air – is a vibrational medium, one that transmits airborne sounds in the form of 'waves of pressure that radiate outwards', compressing, dispersing, and displacing air molecules.[34] A room is not only the setting of a work but an active participant in it, its material elements, volume, form, and connection to other architectural structures all actively shaping the transmission and transformation of energies in and through it.

Aesthetics and politics of material sound transmission

An aesthetics of material sound transmission privileges an engagement with the physical properties – such as the molecular densities and conductive potentials – of things; and an attunement toward how the material properties of things affect the propagation, transmission, and production of acoustic energy. It further invites an engagement with how the combination of various materials can influence the unfolding of energetic processes, for example through feedback or sympathetic vibration; or, to put it another way, the role that materiality plays in dynamic, complex, and potentially chaotic vibrational ecologies.

However, to focus only on the physical properties of materials would be to deny their politics and histories. Such aspects come to the fore in vibrational architectures that contend not only with the conductive capacities of materials, but also their cultural, political, and historical dimensions. The artist duo Mendi + Keith Obadike's *Praise Songs & Installations* (2000–present) – a series of works that pay homage to artists and activists who have transformed social consciousness around issues including race, nation, and power – includes several projects that engage both the sonic and vibrational capacities of architecture and the politics of architectural materials. In their sound art installation *Blues Speaker [for James Baldwin]* (April 1–30, 2015), for example, the Obadikes used 'the glass façade of The New School's University Center as a delivery system for sound, turning the building itself into a speaker'.[35] They transmitted a 12-hour composition that unfolded between 9 AM and 9 PM on each day, transmitting the 24-channel work via three sides of the building's glass façade, 'wrapping' the building in sound which 'emanated' from the glass.[36] *Blues Speaker*, the Obadikes explain, included original blues songs they composed and performed; field recordings of ambient street sounds they recorded in Harlem; tuned sine tones that musicalised Baldwin's writings; and references to Baldwin's writing, including 'an inventory of sound contained in Baldwin's story "Sonny's Blues."'[37]

In a wide-ranging interview with Julie Beth Napolin, Keith Obadike explains that, over the course of the work's 12-hour cycle – a duration chosen to reflect the form of the 12-bar blues – 'there were sound events that happened once a day, moments of silence, dynamic volume swells as well as long looping elements'. He added that the 24 channels of audio were transmitted via three sides of the building:

> The idea was for the sound to wrap the building and emanate from the glass. The sounds chosen for each side of the building depended on how the space was used. For example, the glass wall on the west side of the building was next to a busy stairwell, but it was not next to a seating area like our other zones. This meant listeners would be passing through the area quickly, so we could occasionally do more pithy song-like gestures and low frequency material. In the zones where people sat and lounged we needed to do slower atmospheric moves with brief pauses and soft punctuations.[38]

Thus, the building's architecture, materials, and uses determined the poetics of the installation, which, as Mendi Obadike remarked, engaged the blues 'as a kind of knowledge, a way of processing the sorrowful news and coming out on the other side with information about survival'.[39] The Obadikes described their architectural rendering of the blues as social and political code – as knowledge and as 'feeling' – in both sonic and vibrational terms. Keith Obadike continued:

> We hope one of our contributions to the music through this mix of blues and sound art might be exploring the spatial and architectural possibilities of the blues. Our site for the *Blues Speaker* installation, The New School's University Center, has large glass panels tracing the walkways on all sides of the building. The glass is not just a functional portal, but also a design choice with social implications. We know that glass often represents a kind of utopian gesture in architecture. Some of these ideas came from Paul Scheerbart's *Glass Architecture* (1914). As we studied people moving throughout the building while looking down on a pulsing Manhattan, we wanted to vibrate those portals. We wanted to think about what the blues had to say to this structure; Baldwin gave us the lyrics.[40]

Blues Speaker [for James Baldwin] contended with the utopian modernist fantasy of a glass architecture, imagined by Paul Scheerbart in 1914 as embodying an 'open' culture that would put an end to the 'closed' culture of brick architecture, bringing this utopian impulse into dialogue with the blues – a language, code, and epistemology of survival.[41] Here, the physical capacity of glass to transduce and transmit sound across the surface of a massive structure was inflected by the social implications of a glass building, an architecture that strives for openness and liberation. *Blues Speaker* mapped the liberatory and survival codes of the blues, embedded within Baldwin's writings and the Obadikes' songs and soundscapes onto a building that itself embodies a liberatory impulse in its glass design. Glass functioned both as a physical portal connecting the internal rhythms and flows of the building with those of the pulsating city outside; and as an energetic or metaphysical portal that could be vibrated to evoke shared histories and futures of resistance and survival.

Anti-monumental architectures

The use of sound and acoustic energy to resist the political ideologies manifested in architecture underlined Jan St. Werner's *Space Synthesis* (May 5–July 2, 2023), for which St. Werner transformed Staatliche Kunsthalle Baden-Baden, a monumental neo-Classical building, into a continuously evolving 'sound space' and 'building that speaks'.[42] Maryanne Amacher's practice and her work with structure-borne sounds were important references for the exhibition. As the curator Çağla Ilk wrote, 'St. Werner develops an idea akin to structure-borne sound and takes this idea into the future by developing an innovative choreography that reduces the image and the intensity of a stage setting using a unique composition'.[43]

For *Space Synthesis*, St. Werner emptied the seven massive gallery spaces of the Kunsthalle, an imposing early 20th-century building designed to evoke a Roman temple, of nearly all visual imagery and cues. Various sonic, technological, and spatial interventions – electronic sound pieces projected over custom speakers that triggered a variety of psychoacoustic and spatial effects, producing a sense of shifting dimensionalities and perspectival flux[44]; a skylight mechanism that modulated the flow of natural light into the building, and an artificial light design that emphasised the interplay of light and shadow; and architectural interventions including 'unexpected walls that seem to be falling, raised floors, and a door that moves back and forth through the exhibition space'[45] – combined to produce an anarchic architecture of uncertain, modulating proportions.

Seeking to '[turn] the building into movement',[46] *Space Synthesis* achieved what might be called *structural flow*: the continuous, unpredictable, uncontrollable re-organisation of architectural space through the flux of shifting energies, including acoustic energy as it modulated architectural materials and forms. *Space Synthesis* challenged any sense of architectural stability and fixity, replacing monumental built forms with continuous spatial flux, movement, dissolution, and disintegration. Such a gesture has a political basis (Figures 17.1–17.5). As St. Werner wrote:

> *Space Synthesis* is also a practice against the idea of history as fixed knowledge which often manifests itself in monuments, rigid structures which no sound is strong enough to transform. It sets multi-perspectivity and dynamic interdependency in opposition to singular monumental thought and static histories. *Space Synthesis* is movement and assembly, manifold at its core, and it reflects an understanding of solid structures as porous and borders as transitional.[47]

Sonic materialities and architectural energies

Academic debates on sonic materiality have largely focused on the ontology (nature) of sound and its relationship to the analysis of music and sound art. In a much-cited article from 2011 Christoph Cox suggested that sound art had remained 'profoundly undertheorized' until then because prevailing theoretical models, which privileged textual and visual analysis and were oriented toward signification and representation, 'fail[ed] to capture the nature of the sonic'.[48] Cox proposed a materialist framework that added to addressing sound's material nature, a nature described by Cox in terms of flow and flux, as the 'ceaseless and intense flow' of matter'.[49] Cox pointed to the work of sound artists including John Cage, Alvin Lucier, Christina Kubisch, Toshiya Tsunoda, and others who explore '*the materiality* of sound: its texture and temporal flow, its palpable effect on, and affectation by the materials through and against which it is transmitted'.[50] He suggested that these artists' work revealed that 'the sonic arts are not more *abstract* than the visual but rather more *concrete*, and that they require not a *formalist* analysis but a *materialist* one'.[51]

Figure 17.1 Vibraceptional architecture sequence 1 (2023) by Jan St. Werner. © Jan St Werner. Courtesy of the artist.

Cox's materialist philosophy of sound distinguished between sound art (concrete, materialist) and music (abstract, formalist) to enable a theorisation of sound art adequate to its concerns. While such a materialist framework rightly challenged the formalist leanings of much Western art music theory, it would be limiting, firstly, if we could not also understand music in materialist terms. Analysing music in connection to architecture, for one, invites a materialist approach. Music composed for specific architectural spaces such as churches or salons – whose acoustics are both contingent upon the material properties of those spaces as well as a key *determinant* of musical aesthetics – is a case in point. However, we might also consider music's intrinsic materiality, for example, its inextricability from the bodies and objects that produce and mediate musical sound.

Cox's materialist framework may have relied upon too-blunt distinctions between music and sound art; and scholars including Marie Thompson have raised issue with the 'white auralities' it

On vibrational architectures

Figure 17.2 Vibraceptional architecture sequence 2 (2023) by Jan St. Werner. © Jan St Werner. Courtesy of the artist.

engages in embracing a post-Cagean aesthetics that privileges 'sound-in-itself' or sound beyond representation.[52] Still, it is a generative philosophy that, in considering the ontology of sound in relation to materiality, contends both with sound's intrinsic materiality or nature (its 'texture and flow') as well as its extrinsic materialities, as produced through transmission and mediation ('its palpable effect on, and affectation by the materials through and against which it is transmitted').[53]

We might, however, also consider the possibility that the ontology of sound is uncertain and in flux, residing somewhere between energy and matter, and able to assume either or both 'natures', which are normally understood as distinct states or categories of being. Where does the distinction between energy and matter lie in relation to sound? Douglas Kahn visits cognate territory in *Earth Sound, Earth Signal*, where he suggests that 1960s debates on materiality, immateriality, and the dematerialisation of the art object in the visual art world were inflected by the circulation of the art

Figure 17.3 Vibraceptional architecture sequence 3 (2023) by Jan St. Werner. © Jan St Werner. Courtesy of the artist.

Figure 17.4 Vibraceptional architecture sequence 4 (2023) by Jan St. Werner. © Jan St Werner. Courtesy of the artist.

Figure 17.5 Vibraceptional architecture sequence 5 (2023) by Jan St. Werner. © Jan St Werner. Courtesy of the artist.

object as a commodity in the art market, whereas in music the situation was more fluid. 'Distinctions between matter and energy were never that pressing for music', writes Kahn.[54]

> The object-mission of musical instruments has always been to willingly dissolve between the surface of a page and performance space or to join the voice and vibrate in a complexly audible cosmos. Musical instruments are switching mechanisms, objects to be used at the disposal of energies. They are transductive objects.[55]

Taking a cue from the ontological looseness of music, then, the practice and study of architecture could shift attention from the materiality of the built form to the transient and the ephemeral – the energetic as it modulates, transforms, and reconfigures the material. Vibrational architecture, in which matter and energy intermingle in the realm of the built form, and which reveals architecture as comprised of shifting states of being, is a route toward understanding this.

Acknowledgements

This study was funded by the European Research Council, under the European Union's Horizon 2020 Research and Innovation Programme, as part of the project Sonorous Cities: Towards a Sonic Urbanism (Grant Agreement No. 865032).

All images by Jan St. Werner (2023). All images © Jan St. Werner.

We are grateful to Emma-Kate Matthews, Jane Burry, and Mark Burry for the invitation to contribute to this volume, and for thoughtful and generous communications throughout; and to Katarzyna Krakowiak, Jonathan Tyrrell, and Mendi + Keith Obadike for their insightful and inspiring work and conversations.

Notes

1. Michał Libera, 'Making the Walls Quake as If They Were Dilating with the Secret Knowledge of Great Powers,' accessed July 30, 2023, http://katarzyna-krakowiak.com/making-the-walls-quake/.
2. BiennaleChannel, 'Biennale Architettura 2012 – Michal Libera – Katarzyna Krakowiak,' *Video*, 4: 42 accessed July 30, 2023, https://www.youtube.com/watch?v=u2DK7fci8oA.
3. Karissa Rosenfield, 'David Chipperfield Announces "Common Ground" as the Theme for the 13th International Venice Biennale,' *ArchDaily*, January 17, 2012, https://www.archdaily.com/200806/david-chipperfield-announces-.
4. Irina Vinnitskaya, 'Venice Biennale 2012: Poland Pavilion,' *ArchDaily*, August 19, 2012, https://www.archdaily.com/264557/venice-biennale-2012-poland-pavilion.
5. Michał Libera, 'Making the Walls Quake', https://labiennale.art.pl/en/wystawy/making-the-walls-quake-as-if-they-were-dilating-with-the-secret-knowledge-of-great-powers/.
6. Michał Libera, 'Making the Walls Quake' [5].
7. Both the temporary wall and temporary floor that were installed for *Making the Walls Quake* were tilted at slight angles (though they were at 90 degrees to one another). As Krakowiak mentioned in conversation, this was done in part to encourage visitors to linger in the pavilion for a long time and physically interact with the architecture. Gascia Ouzounian and Jan St. Werner, unpublished conversation with Katarzyna Krakowiak, Zoom, July 18, 2023.
8. As Varèse proclaimed in 1936, when he imagined a 'spatial music' aided by electroacoustics, 'Music will eventually engulf and surround you.'
9. Marshall Long, *Architectural Acoustics* (Burlington, MA: Elsevier Academic Press, 2006), xxv.
10. For clarity, the material transmission of sound is treated in architectural acoustics, including by Sabine (1906); however, this is generally in relation to how sound transmitted through walls and partitions helps in the absorption of sound, and not as a productive or generative capacity of architecture. See Wallace Clement Sabine, 'Architectural Acoustics,' *Proceedings of the Academy of Arts and Sciences* 42, no. 2 (June 1906): 51–84.
11. Goodman, *Sonic Warfare: Sound, Affect, and the Ecology of Fear* (Cambridge: MIT Press, 2010), 75–76.
12. Goodman, *Sonic Warfare*, 76.
13. Goodman, *Sonic Warfare*, 75–76.
14. Goodman, *Sonic Warfare*, 76. In *Sonic Warfare* Goodman is particularly interested in the capacity of infrasound and very low frequencies to behave in such a way. He develops the concept of a 'bass materialism'—'practices of affective engineering through vibrational modulation' (p. 26)—and considers its role in sonic warfare.
15. See Gascia Ouzounian, *Stereophonica: Sound and Space in Science, Technology, and the Arts* (Cambridge: MIT Press, 2021), 83–104.
16. Goodman evokes the concept of a 'vibrational ecology' in relation to Mark Bain's work, writing, 'As opposed to a sound artist, [Bain] describes the sonic effects of his work as side effects, or artifacts, merely an expression of a more fundamental subsonic vibrational ecology.' Goodman, *Sonic Warfare*, 77.
17. Stefan Helmreich, *Sounding the Limits of Life: Essays in the Anthropology of Biology and Beyond* (Princeton, NJ: Princeton University Press, 2015).
18. Stefan Helmreich, 'Listening against Soundscapes,' *Anthropology News* (December 2010): 10.
19. Stefan Helmreich, 'Listening against Soundscapes,' 10.
20. Mark Bain, '*The Live Room*: Transducing resonant architectures,' *Organised Sound* 8, no. 2 (2003): 164.
21. Jonathan Tyrrell, 'Architecture's Acoustic Shadow: Unsettling the Sound-Space Relationship' (PhD diss., University College London, in progress).
22. Jonathan Tyrrell, 'Architecture's Acoustic Shadow.'
23. Jonathan Tyrrell, 'Architecture's Acoustic Shadow.'

24 Amy Cimini and Bill Dietz (eds.), *Maryanne Amacher: Selected Writings and Interviews* (Brooklyn, NY: Blank Forms, 2020), 224.
25 Maryanne Amacher, 'Living Sound (Patent Pending)', in *Maryanne Amacher: Selected Writings and Interviews*, Amy Cimini and Bill Dietz (eds.) (Brooklyn, NY: Blank Forms Editions), 233.
26 Maryanne Amacher, 'Interview with Jeffrey Bartone', in *Maryanne Amacher: Selected Writings and Interviews*, Amy Cimini and Bill Dietz (eds.) (Brooklyn, NY: Blank Forms Editions), 221.
27 Mark Bain, '*The Live Room*,' 163.
28 Mark Bain, '*The Live Room*,' 164.
29 Mark Bain, '*The Live Room*,' 165.
30 Mark Bain, '*The Live Room*,' 164.
31 Mark Bain, '*The Live Room*,' 166.
32 Mark Bain, '*The Live Room*,' 166.
33 Salomé Vogelin makes this observation in connection to Toshiya Tsunoda's *Scenery of Decalcomania* (2004). Salomé Voegelin, 'Sonic Materialism: Hearing the Arche-Sonic,' in *The Oxford Handbook of Sound and Imagination*, Mark Grimshaw-Aagaard et al. (ed.) (Oxford: Oxford University Press, 2019), 566.
34 Ed Yong, *An Immense World: How Animal Senses Reveal the Hidden Realms Around Us* (New York City: Vintage, 2022), 211.
35 Vera List Center for Art and Politics. 'Exhibition. Mendi + Keith Obadike: Blues Speaker [for James Baldwin].', accessed July 31, 2023, https://www.veralistcenter.org/exhibitions/mendi-keith-obadike-blues-speaker-for-james-baldwin.
36 Julie Beth Napolin, 'On *Blues Speaker [for James Baldwin]*: A Conversation with Mendi and Keith Obadike,' *Social Text*, August 21, 2018, https://socialtextjournal.org/on-blues-speaker-for-james-baldwin-a-conversation-with-mendi-and-keith-obadike/?platform=hootsuite.
37 Mendi + Keith Obadike, 'Praise Songs & Installations (2000–present),' accessed July 31, 2023, http://blacksoundart.com/#/praisesongs/.
38 Napolin, 'On *Blues Speaker*.'
39 Napolin, 'On *Blues Speaker*.'
40 Napolin, 'On *Blues Speaker*.'
41 Paul Scheerbart, 'Glass Architecture,' in *Glass! Love!! Perpetual motion!!!: A Paul Scheerbart Reader*, Josiah McElheny and Christine Burgin (eds.) (Chicago, IL: University of Chicago Press, 2014), 20–91.
42 e-flux Announcements, 'Jan St. Werner, *Space Synthesis*,' accessed July 31, 2023, https://www.e-flux.com/announcements/527930/jan-st-wernerspace-synthesis/.
43 Çağla Ilk, 'The Shadow of Ideas,' in *Jan St. Werner: Space Synthesis*, Çağla Ilk and Jan St. Werner (eds.) (Berlin: Hatje Cantz Verlag, 2023), 32.
44 Jan St. Werner, *Space Synthesis* album, Bandcamp, accessed July 31, 2023, https://fiepblatter.bandcamp.com/album/space-synthesis.
45 Ilk, 'The Shadow of Ideas,' 32.
46 See Gascia Ouzounian, 'Unfolding Spaces: A Wayfinding,' in *Jan St. Werner: Space Synthesis*, Çağla Ilk and Jan St. Werner (eds.) (Berlin: Hatje Cantz Verlag, 2023), 90.
47 Jan St. Werner, 'Space Synthesis,' in *Jan St. Werner: Space Synthesis*, Çağla Ilk and Jan St. Werner (eds.) (Berlin: Hatje Cantz Verlag, 2023), 39.
48 Christoph Cox, 'Beyond Representation and Signification: Toward a Sonic Materialism,' *Journal of Visual Culture* 10, no. 2 (2011): 145.
49 Brian Kane, 'Sound Studies Without Auditory Culture: A Critique of the Ontological Turn,' *Sound Studies* 1, no. 1 (2015): 9.
50 Cox, 'Beyond Representation,' 148.
51 Cox, 'Beyond Representation,' 148–149, emphasis in original.
52 Marie Thompson, 'Whiteness and the Ontological Turn in Sound Studies,' *Parallax* 23, no. 3 (2017): 266–282.
53 The capacity of sound to affect materials has shaped debates about its ontology since at least the early modern period. The English natural philosopher Francis Bacon, for example, remarked that '"Audibles" (sounds) had the power to disturb a medium like water or air, whereas "Visibles" (light) did not. In contrast to light, sound could disturb a medium and produce changes in it—in other words, "affect" and "corrupt" it.' See Gascia Ouzounian, 'Becoming Air: On Sonic Spatial Metaphysics,' in *Sound – Space – Sense*, eds. Diedrich Diederichsen, Arno Raffeiner, and Jan St. Werner (Berlin: Haus der Kulturen Welt, Das neue Alphabet/The New Alphabet series, 2023), 17–18.

54 Douglas Kahn, *Earth Sound Earth Signal: Energies and Earth Magnitude in the Arts* (Berkeley: University of California Press, 2013), 218.
55 Kahn, *Earth Sound Earth Signal*, 218.

Bibliography

Amacher, Maryanne. 'Interview with Jeffrey Bartone.' In *Maryanne Amacher: Selected Writings and Interviews*, edited by Amy Cimini and Bill Dietz, 211–222. Brooklyn, NY: Blank Forms, 2020.

Amacher, Maryanne. 'Living Sound (Patent Pending).' In *Maryanne Amacher: Selected Writings and Interviews*, edited by Amy Cimini and Bill Dietz, 231–235. Brooklyn, NY: Blank Forms, 2020.

Bain, Mark. '*The Live Room*: Transducing Resonant Architectures.' *Organised Sound* 8, no. 2 (2003): 163–170.

Bain, Mark. The Live Room: Transducing Resonant Architecture. Temporary site specific sound installation. building N51, room 117, 1998.

BiennaleChannel. 'Biennale Architettura 2012 – Michal Libera – Katarzyna Krakowiak.' *Video*, 4: 42. Accessed July 30, 2023. https://www.youtube.com/watch?v=u2DK7fci8oA.

Cimini, Amy and Dietz, Bill (eds). *Maryanne Amacher: Selected Writings and Interviews*. Brooklyn, NY: Blank Forms, 2020.

Cox, Christoph. 'Beyond Representation and Signification: Toward a Sonic Materialism.' *Journal of Visual Culture* 10, no. 2 (2011): 145–160.

e-flux Announcements. 'Jan St. Werner, *Space Synthesis*.' Accessed July 31, 2023. https://www.e-flux.com/announcements/527930/jan-st-wernerspace-synthesis/.

Goodman, Steve. *Sonic Warfare: Sound, Affect, and the Ecology of Fear*. Cambridge: MIT Press, 2010.

Helmreich, Stefan. 'Listening against Soundscapes.' *Anthropology News* 51, no. 9 (December 2010): 10.

Helmreich, Stefan. *Sounding the Limits of Life: Essays in the Anthropology of Biology and Beyond*. Princeton, NJ: Princeton University Press, 2015.

Ilk, Çağla. 'The Shadow of Ideas.' In *Jan St. Werner: Space Synthesis*, edited by Çağla Ilk and Jan St. Werner, 15–33. Berlin: Hatje Cantz Verlag, 2023.

Kahn, Douglas. *Earth Sound Earth Signal: Energies and Earth Magnitude in the Arts*. Berkeley: University of California Press, 2013.

Kane, Brian. 'Sound Studies Without Auditory Culture: A Critique of the Ontological Turn.' *Sound Studies* 1, no. 1 (2015): 2–21.

Libera, Michał. 'Making the Walls Quake as If They Were Dilating with the Secret Knowledge of Great Powers.' Accessed July 30, 2023. http://katarzyna-krakowiak.com/making-the-walls-quake/.

Long, Marshall. *Architectural Acoustics*. Burlington, MA: Elsevier Academic Press, 2006.

Napolin, Julie Beth. 'On *Blues Speaker [for James Baldwin]*: A Conversation with Mendi and Keith Obadike.' *Social Text*. August 21, 2018. https://socialtextjournal.org/on-blues-speaker-for-james-baldwin-a-conversation-with-mendi-and-keith-obadike/?platform=hootsuite.

Obadike, Mendi + Keith. 'Praise Songs & Installations (2000–present).' Accessed July 31, 2023. http://blacksoundart.com/#/praisesongs/.

Obadike, Mendi + Keith. *Blues Speaker [for James Baldwin]*. Temporary sound installation. Vera List Center for Art and Politics, The New School University Center, 63 Fifth Avenue, New York City, May 23, 2015.

Ouzounian, Gascia. *Stereophonica: Sound and Space in Science, Technology, and the Arts*. Cambridge: MIT Press, 2021.

Ouzounian, Gascia. 'Becoming Air: On Sonic Spatial Metaphysics.' In *Sound – Space – Sense*, edited by Diedrich Diederichsen, Arno Raffeiner, and Jan St. Werner, 13–21. Berlin: Haus der Kulturen Welt, Das neue Alphabet / The New Alphabet series, 2023.

Ouzounian, Gascia. 'Unfolding Spaces: A Wayfinding.' In *Jan St. Werner: Space Synthesis*, edited by Çağla Ilk and Jan St. Werner, 79–95. Berlin: Hatje Cantz Verlag, 2023.

Ouzounian, Gascia and Jan St. Werner. Unpublished Conversation with Katarzyna Krakowiak. Zoom. July 18, 2023.

Rosenfield, Karissa. 'David Chipperfield Announces "Common Ground" as the Theme for the 13th International Venice Biennale.' *ArchDaily*, January 17, 2012. https://www.archdaily.com/200806/david-chipperfield-announces-.

Sabine, Wallace Clement. 'Architectural Acoustics.' *Proceedings of the Academy of Arts and Sciences* 42, no. 2 (June 1906): 51–84.

Scheerbart, Paul. 'Glass Architecture.' In *Glass! Love!! Perpetual motion!!!: A Paul Scheerbart Reader*, edited by Josiah McElheny and Christine Burgin, 20–91. Chicago, IL: University of Chicago Press, 2014.

St. Werner, Jan. *Space Synthesis* album. Bandcamp. Accessed July 31, 2023. https://fiepblatter.bandcamp.com/album/space-synthesis.

St. Werner, Jan. 'Space Synthesis.' In *Jan St. Werner: Space Synthesis*, edited by Çağla Ilk and Jan St. Werner, 38–39. Berlin: Hatje Cantz Verlag, 2023.

Thompson, Marie. 'Whiteness and the Ontological Turn in Sound Studies.' *Parallax* 23, no. 3 (2017): 266–282.

Tsunoda, Toshiya. *Scenery Of Decalcomania*. Musical release. Australia: Naturestrip – NS3003, August 2004.

Tyrrell, Jonathan. 'Architecture's Acoustic Shadow: Unsettling the Sound-Space Relationship.' PhD diss., University College London, in progress.

Vera List Center for Art and Politics. 'Exhibition. Mendi + Keith Obadike: Blues Speaker [for James Baldwin].' Accessed July 31, 2023. https://www.veralistcenter.org/exhibitions/mendi-keith-obadike-blues-speaker-for-james-baldwin.

Vinnitskaya, Irina. 'Venice Biennale 2012: Poland Pavilion.' *ArchDaily*, August 19, 2012. https://www.archdaily.com/264557/venice-biennale-2012-poland-pavilion.

Voegelin, Salomé. 'Sonic Materialism: Hearing the Arche-Sonic.' In *The Oxford Handbook of Sound and Imagination*, edited by Mark Grimshaw-Aagaard et al. Oxford: Oxford University Press, 2019, 558–577.

Yong, Ed. *An Immense World: How Animal Senses Reveal the Hidden Realms Around Us*. New York City: Vintage, 2022.

18
HOUSE OF SILENCE, OF STILLNESS, OF SOLITUDE

Mark Taylor

Introduction

I have often thought about the experiences of people in their homes and places of inhabitation, and to what extent their surroundings and the actions of others affect or impact inhabitants' lives. The lives of individuals have become increasingly important to me, in the wake of generalisations made about the home. Recently this was explored in my coedited book Domesticity under Siege[1] which questioned the domestic interior as a 'safe place'. It challenged the 19th-century notion of 'home as haven', by examining particular situations and moments when the home and its inhabitants are affected by outside agents. That is, to understand that the home is not a benign collection of materials and artefacts, but is a place affected and reactive to both outside and inside agents.

In this chapter, my focus is both on the happenings and occupations of a person in the home and the way sound is sensed and experienced during daily activities. That is, the way an occupant is affected by outside and inside sounds. The intention is not to measure noise levels or to record actual sound, but to explore the various ways sound is described and presented through a narrative setting. To effect this, my methodology is to write the chapter as both a literary narrative and as an academic discussion of sound. The reason for this is to try to write the person and their experience in the chapter. Therefore, I draw upon the novelist's ways of including sound within their writing. In literature, sounds are often described in order to give atmosphere to moments in the narrative. Sometimes they are uttered by various characters, whereas others come from the surrounding environment, and can be locally produced or come from afar. Various adjectives and metaphors are used to help describe the way a sound emanates or is heard by narrators and characters, throughout both day and night.

This methodology of including the novelists' space is because the writer is an active reflector of and contributor to the discourse of architecture and the interior. Writers often capture moments in contemporary society and write them into the novel. Whether the poverty of Britain is reflected in Charles Dicken's work, or more sinister aspects of political change in George Orwell's dystopian view of the world. That is, unlike a real scene perceived by an individual, the literary narrative can offer a more detached or abstract view enabling the absorption of the scene as a whole. By taking in the scene in this manner, the visionary quality of the novelist can reveal a number of relations that normally go unnoticed when the individual is focussed on particular aspects of the unfolding scene.

While writing this chapter, I am reading Virginia Woolf's *Mrs Dalloway* (1925), which like her other novels contains detailed observations and descriptions of her characters and their environments. Her novel accompanies this text as it is set in a day in the life of Clarissa Dalloway from the moment she leaves her house and walks through her local neighbourhood to collect some shopping, to returning to her London home. The novel has been analysed through many perspectives and while the psychological and emotional aspects of the narrative are evident, with Woolf such moments are often predicated on an event, sound, or action. It is this relationship between sound and event that is of interest in this writing. It has been observed that the modern novel is 'saturated with sound—both in content and form' (Frattarola 2010: 9). Leah Toth has commented on how Woolf traces sonic events and provides several examples such as 'the ringing out of Big Ben' 'the backfiring of the car' or 'the whir of an ambulance siren' as they enter the ears of her characters and reverberate through each individual consciousness.[2]

Mrs Dalloway is set in inter-war London at a time when the industrialised city was transforming from the romantic's idyllic haven to a bustling, noisy and somewhat hostile environment. It was a time when machines began to invade the peace of a quieter world. While some of the romantic poets and writers lamented for the days before the din of technology, Woolf understood the reality of the new acoustic reverberations of the city and how such sounds became associative, 'in the same way that natural sounds do'.[3] It is through such sonic devices that aspects and histories of individual characters are revealed. The aural stimulus prompts inner thoughts, what might be thought of as 'an internal re-sounding of a memory or ingrained principle'.[4] We could argue that there is a subjective nature to how we respond to sound stimulus, such that characters respond differently and process aural events relative to their mental or emotional state. For example, one person hearing a particular sound might trigger memories of a difficult past event, whereas for another person it might be uplifting.

For this chapter, the central focus is an elderly woman who is living alone in her London family house. The house is familiar, comfortable and is marked by continuity of time. Based on observation and a shared experience, I will include reflections around the way sound seems to impact and resonate through space, particularly following moments of change. I will discuss the ordinariness of time passing and its relationship to sound where various adjectives are used to highlight an occasion or enhance a transition between events. Critically, what some might regard as the mundane passing of a day and how it measures sound, silence, solitude and stillness.

Outer sound

Sounds made harmonies with premeditation; the spaces between them were as significant as the sounds. A child cried. Rightly far away a horn sounded.[5]

The clink of the milk bottles placed outside the front door beside the Geranium plant pot, the murmur of voices further down the terrace mingled with their footsteps echoing within the narrow street, these sounds together with the smell of blossom from the cherry trees, were familiar and etched into her memories. She had performed this act from the top of the front steps many times before. Today she paused and listened to the street, recalling an article she read in which the author discussed sound and space, particularly how echoes of steps on a paved street have an emotional charge because they bounce around noting that 'sound measures space and makes it comprehensible'.[6] He even suggested that 'the contemporary city had lost its echo', but she thought, the noise of the main road barely reaches into the side street, hence the echo is one of neighbours and passers-by going about their business, walking, talking, opening car doors, emptying rubbish and consoling a child's cry. At the end of the terrace cars and buses passed by as footsteps drew near accompanied by random words spoken into the air. A chain dislodged from a bicycle lock fell to the pavement prompting a dog to bark as if protecting its territory. Far off a dustbin lid banged.

Leaving the house, she closed the gate, noting how much the weeds have grown in the pavement cracks, and carefully avoided the uneven areas. She paused at the kerb, waiting for a car to pass then stepped forward into the crowd. Having lived in the city for over a lifetime, walking this familiar route, she recalled how street sounds have changed. In the past 'the diesel stammer of London taxis, the wheeze of its buses'[7] were signs of the city that she remembered. She glanced across the road to place where many years ago, there used to be a bus terminus. Now there are private apartments. This was where her late husband's father worked when he returned from the trenches after the Great War. It was where his health declined due to gassing in the trenches. The bus engine was his sound, now it is his silence. These are the silences that mark what has gone, such as 'the elevated railway has just about' and he 'misses the sound… the tremor of the thing'.[8] But now there is a babble in the crowd, not the babble of a group of people, talking or drinking outside the corner pub, but a babble that is spoken into the void. Each person connected to another uttering words that fall from the clouds. They are not directed at anyone, but all can receive. Now she knows Nora will be late home, the clicking sound in the car is being fixed, the noisy neighbour Alex has moved out, and the church bells will ring on Sunday. Fragments of conversations across time and space. In the past the sounds of transit invaded her thoughts, now it is other people's dialogues. Perhaps in their speaking, they cannot be heard except by whoever is receiving their chatter – wherever they are in the world. It is a kind of deafness to surrounding sounds and a personal freedom liberated by the Sony Walkman and the iPhone. A loudness that implies no one is listening in, no one is concerned for their conversation. It is no longer the wordless solitude of the individual in a noisy city, but noisy individuals in an increasingly quiet city she mused as the ghostly hush of the electric bus paused at the bus stop. The city offered a soundstage for the drama of modern life, a city 'divided between the engineered and the accidental, between sound and noise'.[9] She looked up into the sky 'the sound of an aeroplane bored ominously into the ears of the crowd. There it was coming over the trees'.[10] High overhead a plane engine whined, as it approached Heathrow. It reminded her of the days when Concorde flew over with a roar that defied its elegance and beauty. It is a sound like the diesel engine of the buses that is no longer heard.

Silence

As a cloud crosses the sun, silence falls on London; and falls on the mind.[11]

The air was chilly and with a slight breeze as she walked back up the steps. Glancing at the Geraniums she picked three and nipped others snapping their dead stalks. Two sparrows perched in the tree sang in voices prolonged and piercing. Turning, and pushing the door open, the hinges gave a little squeak, the heavy door chain bumped the panelled wood and followed the same arc it had always done, adding to the etched surface before banging as the door closed. In this final movement, the momentum of the door caused the brass door knocker to spring into action signalling finality with a clear and sharp tone that cut through the hallway and echoed down the street. Although now inside, the domestic boundary is not impenetrable and despite notions of home as haven, sounds still pass through the domestic screen. Once again, she thought of Pallasmaa's article as he turned to the domestic interior and the harshness of an 'uninhabited and unfurnished house as compared to the affability of a lived home in which sound is refracted and softened by the surfaces of numerous objects of personal life'.[12] She was glad for the familiarity of the hallway and thought maybe he was right that 'every building or space has its characteristic sound of intimacy'.[13]

She was back inside, in a familiar place she had known since her teens, over 70 years ago. It was a modest late-Georgian terraced house with the entry level raised several steps above the street, and an environment that offered a spatial and auditory experience. From the hallway, a spatial sequence revolved around stairs. Beginning outside on the tiled steps the flow extended above to the first floor and below to the lower ground with a mid-level access to the garden. Visually, the decorative ceilings with moulded plasterwork were enhanced by autumn sunlight falling through the landing window. The simplicity of the stair banister with plain square spindles and continuous polished handrail draws the eye to upper level.

Aurally the home breaths. Her coat rustled as she draped it on the hall stand, a fairly unremarkable piece of Victorian furniture, heavy and dark. The floor gave a slight creek as she shifted her position. In this moment the hallway's sounds of intimacy gave way to another image in another modality, one that Pallasmaa had alluded to. Although she could not remember the source he quoted, it was something like 'quiet reigns with nothing to break the silence save the note of the boiling water in the iron kettle'.[14] It is a silence that 'focusses attention on one's very existence'.[15]

Silence. Gernot Böhme confirms that in English the word 'silence' is not strictly related to 'not talking' but has 'a much broader field of application precisely because it is not meant to strictly relate to talking'.[16] He suggests the term relates to something atmospheric and characterises situations. In this moment of silence, the room takes on a shrouding atmosphere, one that is close to the absence of sound, as the mind tunes out exterior noises. Silence therefore is a spatial atmosphere related to a quiet place.

Solitude

quiet descended on her, calm, content, as her needle, drawing the silk smoothly to its gentle pause.[17]

In the quiet of the afternoon, she picked up a patchwork quilt she was making for a grandchild. She continued stitching the pieces together. As her fingers nimbly pulled thread she felt silence was a witness to her actions, not only in terms of Susan Sontag's disclosure that silence in art reveals both a spiritual and provocative aspect,[18] but also silence that falls without drawing attention, without being noticed. It is a silence not composed of the absence of sound, but one between her and the house. It is the sound of solitude, that is revealed and exchanged between inhabitant and house. The tiny noise as a floorboard creaks, or the exhale of breath as the body relaxes. Now as the needle and thread united the various colours and patterns she breathed, becoming one with the house and content in her solitude.

The sewing absorbed her mind as the room with its solid volume and soft materials absorbed sound. In this architectural experience, all external noise was silenced, emphasising solitude. Detached from the present she began to experience 'the slow, firm flow of time and tradition'.[19] Enveloped in the familiarity of her work, the room reflected her life while registering the passing of history. There was a sense of tranquillity, but not that described by Pallasmaa as the aftermath of noisy of construction when 'the building becomes a museum of a waiting, patient silence', as in 'the silence of the pharaohs', or 'the last dying note of a Gregorian chant'.[20] This atmospheric silence cast 'a sense of solitude and sensual opening to the surrounding environment'.[21] Don Ihde has theorised about the body-auditory motion enticed by music, which sometimes enables the body to experience a temporary sense of the 'dissolution' of self-presence. He muses that 'music takes me out of myself in such occurrences' noting how background music, 'floats lazily around me, and I find I can easily retire into my 'thinking self' and allow the perceptual presence to recede from focal awareness'.[22] This moment when an auditory experience enhances inner focus, is also evident when stillness takes over, when the precision of needlework induces its own auditory rhythm.

Atmosphere

The hall of the house was cool as a vault.[23]

Stillness shrouded her. But this was not the kind of silence that is soundlessness. It had something to do with atmosphere and the characteristics of the situation. She knew this from her time nursing her late husband. Although he spoke little, and only to make a joke or recall some past experiences, silence was not strictly related to not talking, but the absence of any noise. It is also a silence experienced by the absence of her husband, whose calm of death is 'definitive silence in the sense of the end of speech and the absence of movement'.[24] In this silent moment, the room is familiar creating a shrouding atmosphere that tunes out exterior noises, creating a spatial atmosphere related to a quiet place. Böhme proposes that atmosphere concerns how 'we feel in surroundings of a particular quality – that is, how we sense these qualities in our own disposition'.[25] She felt 'at home'. There was a familiarity of materials and artefacts combined with the historical depth of the home. Over many years it had been altered, redecorated, and refurnished, a slow process that enabled continuity and a sense of feeling sheltered. Atmosphere is not something that can be conjured through an image 'rather, what is meant by atmosphere is that which is commonplace and self-evident for the inhabitants, and which is constantly produced by their lives'.[26]

The acoustic profile slipped between the background sounds that change daily to the sound events of the clocks, floorboards, pipes and familiar domestic sounds. She had the capacity to create a sound map and understand space through listening – whether the children on the other side of the party wall running up the stairs or piano rehearsal time form the other side. The street, the garden the adjacent neighbours all entered the interior, entered her thoughts and entered her life. Their sounds were as much part of her home as were those generated from within.

Stillness

Within this quietude, the silence-sound is an 'embodied experience' and in some way proffers an authentic experience. The sewing absorbed her. The concentration engendered an acoustic silence, inasmuch as 'intelligible sounds come to the consciousness and senses are fully engaged'.[27] The distant rhythmic tick, tick, tick of the hallway clock, rescued long ago from an overseas post office, spoke unsilently of the past, of centuries, of ancestors, of home. A corner radiator gurgled as it came to life, while once or twice the window rattled as the panes were touched by a passing breeze. These familiar sounds hovered in the background, barely heard or responded to, and never penetrated the 'thinking self'. It was a silence that produced by focussed activity. It was a silence that was not soundless, but was more like 'stillness'. This was tranquillity brought about by stillness, in which her bowed head, focus, and rhythmically moving hand might be mistaken for religious practice.

To the left of the fireplace is a wedding photograph, that now holds a special place in the calm of death. Although her husband's funeral was some months ago, photos and mementoes of the past 70 years abound. But this is not the calm felt at the moment of death but is one extended over time, knowing that person and mentally seeing and hearing that person perform their habitual wanderings in the house. That is not to say her room and home have transformed into a place of rest, but is now inhabited by the memories of another. This is not the silence of the tomb, as the ashes-filled funerary vase might suggest, but is a calm silence in relation to that of laughter, words and music, that once filled the home. This is not the architecture of silence that we feel in a church where the 'atmosphere of silence is therefore an essential element of Christian ritual place',[28] since ecclesiastical spaces are deliberately separated from exterior sounds.

She allowed the perceptual presence to recede from focal awareness, and drifted into a 'thinking self'. Imperceptibly she recalled his funny antics, phrases, and sayings spoken with a London accent. The war, the army, their travels, friends and children came flooding through. She fantasised the conversations between friends and family while recalling the strains of a melody heard long ago. And within the peculiarities of this auditory dimensional space, between the imaginative and the perceptual she remembered the past. This is an inward silence brought about by consistency of environmental sound – the low-intensity constants that help maintain 'inner' focus. In this moment of concentration, as the needle crosses the fabric, 'a sudden noise poses a serious distraction',[29] as it can intrude, distract, and break the stillness.

Intensity

> The violent explosion which made Mrs. Dalloway jump and Miss Pym go to the window and apologise came from a motor car which had drawn to the side of the pavement precisely opposite Mulberry's shop window.[30]

The intensity and volume of a sudden noise pierced the constancy and broke her inner focus. This outer sound of a vehicle in the street disturbed her train of thought. Her body jumped with the shock of the backfire. Now looking through the window the initial noise gave way. Coming to the window with her hands holding thread, she looked out. A lady pushing a pram stopped, a group of children on electric scooters glanced as they silently whizzed by. The engine was still running as she turned, put down her sewing and went into the hallway. The sense of urgency and excitement generated by the sound of traffic gave way to an enclosed silence. Later in the evening, she would sit in the back garden where the sound of birdsong provided a sense of peace and tranquillity. But for now, pausing on the stair, she was aware that her lightness now prevented the floorboards and stair treads from creaking. In her earlier years with the children, they would try to walk the staircase without a sound, in order to not wake grandma, but creeping was never an option, the house had its own voice.

The kettle neared boiling. The water thumped and bashed as bubbles collapsed on themselves. It was a sound intensified by a desire to make some tea. She turned on the radio. She remembered her husband always went up to bed and listened to the radio, some talk-back show or classical music whichever happened to be broadcast. He would often fall asleep as the voices or music became more imperceptible – more like 'white sound', or the way programmed background music, or lift music floats lazily around one. The radio, much like the clock on the wall, and the whistle of the kettle, was just one of the sounds that 'construct geographies of cultural memory'.[31] Each sound is a benchmark of cultural identity and reminder of a place, event or happening. Her husband's tread, his accent and pronunciation of words, and his actions of turning locks, switching off lights and closing the shutters include sounds embedded in cultural repertoire. Sounds are not heard as 'abstract information related to the acoustic properties of given objects',[32] they are perceived as events.

The car backfire outside the window, was simultaneously the sound of gunfire, returning her to a time of warfare, but at the same time was a reminder that we travel and engage with transportation.

Urban sociologist Fran Tonkiss suggests that in such situations sound 'works through metonym, aural fragments that speak of something larger'.[33] The relation of sound to memory Tonkiss argues is 'audibly present in the moment of 'recall', the melding of space, sound, and memory there in the concept of 'resonance'; a movement in the air like sound you can touch'.[34] Hearing has a relationship to recollection and offers a sense of memory, sometimes half-remembered but which is immediate and sensuous. It is recall emanating from the presence of sound. But, as with the car backfire, memory arousal is modified by the parameters of reception (activities, occupations, etc.) and those of source modification (echoing street, foliage, etc.), that result in different perceptions.

Darkness

Dark descends, pours over the outlines of houses and towers; bleak hillsides soften and fall in.[35]

As the daylight bleeds into darkness a small click came from the brass table lamp as it illuminated the corner of the room. The buzz of the city gave way to a quieter murmur. More distant sounds from the football stadium grew louder, as the local cadence diminished. She closed the shutters and drew the curtains escaping the noise of modern life. The television silently lit up the room. This was unlike her old colour television that popped and crackled, and occasionally buzzed and hummed as it came to life. There are now so many more programmes to choose from, she thought flicking through the channels and settled on a 1950s black-and-white film. While tame by today's standards it had an edge of suspense and a little scary, consistent with the early horror genre. She generally disliked such films, but there was something familiar and nostalgic about this one, reminding her of her youth and nights out at the pictures. A 'jump scene' caught her off-guard. Despite the build-up of musical discord, human scream, and door banging, it was still enough to scare. Such sounds were created by Foley artists, but we believe the sound of horse's hooves was created by tapping a pair of coconut shells. Such sounds are often associated with an event or a character – the sound of knife scraping on a blackboard or a wooden leg walking across a cobbled street. But silence is also used to anticipate what follows.

Later she traced the sound of water drops in the darkness outside her bedroom window. She scarcely read a book now, but used to read thrillers and a few 'bodice rippers'. Rustling a page, she reread a paragraph and mused that they seem to take longer to finish. She placed the book on the nightstand and lay in silence. Broken by an occasional cough from a cold that would not leave. Now alone, listening to the patter of rain and without her husband beside her, she awaited sleep in the hushed atmosphere of a room that is both familiar and unfamiliar. But 'late at night the quiet creeps in and it is as though you hear the city sleep. It can make you dream tall'.[36]

Looking back silence is not just the absence of sound defined as an enclosed quiet place. The house of silence, of stillness, and of solitariness concerns sounds, memories, and recollections. She breathed quietly and the house responded, but the silence of the city seemed like a reluctance to communicate. 'She sighed, she snored, not that she was asleep, only drowsy and heavy, drowsy and heavy, like a field of clover in the sunshine this hot June day'[37]

Notes

1 Bloomsbury (2023).
2 Toth (2017: 566).
3 Toth (2017: 569).
4 Toth (2017: 565).

5 Woolf (1925).
 6 Pallasmaa (2011: 43).
 7 Tonkiss (2003: 306).
 8 Tonkiss (2003: 307).
 9 Tonkiss (2003: 304).
10 Woolf (1925).
11 Woolf (1925).
12 Pallasmaa (2011: 43).
13 Pallasmaa (2011: 43).
14 Pallasmaa (2011: 41).
15 Pallasmaa (2011: 43).
16 Böhme (2020: 180).
17 Woolf (1925).
18 Sontag (1969).
19 Pallasmaa (2011: 43).
20 Pallasmaa (2011: 43).
21 Kakalis (2020: 130).
22 Ihde (2003: 62).
23 Woolf (1925).
24 Böhme (2020: 183).
25 Böhme (2014: 50).
26 Böhme (2014: 46–47).
27 Kakalis (2020: 133).
28 Böhme (2020: 186).
29 Ihde (2003: 62).
30 Woolf (1825).
31 Moore (2003: 267).
32 Moore (2003: 267).
33 Tonkiss (2003: 307).
34 Tonkiss (2003: 307).
35 Woolf (1825).
36 Tonkiss (2003: 308).
37 Woolf (1925).

References

Böhme, Gernot. *Architectural Atmosphere: On the Experience and Politics of Architecture*. Basel: Birkhäuser, 2014.

Böhme, Gernot. "Quiet Places – Silent Space: Towards a Phenomenology of Silence." In *The Place of Silence: Architecture, Media, Philosophy*, edited by Mark Dorrian and Christos Kakalis, 179–181. London and New York: Bloomsbury, 2020.

Frattarola, Angela. "The Phonograph and the Modernist Novel." *Mosaic*, 43(1): 146. 2010, 2020.

Ihde, Don. "Auditory Imagination." In *The Auditory Culture Reader*, edited by Michael Bull and Les Back, 61–66. Oxford: Berg, 2003.

Kakalis, Christos. "Silence, Paradox and Religious Topography." In *The Place of Silence: Architecture, Media, Philosophy*, edited by Mark Dorrian and Christos Kakalis, 129–138. London and New York: Bloomsbury, 2020.

Moore, Paul. "Sectarian Sound and Cultural Identity in Northern Ireland." In *The Auditory Culture Reader*, edited by Michael Bull and Les Back, 265–280. Oxford: Berg, 2003.

Pallasmaa, Juhani. "An Architecture of the Seven Senses." In *Toward a New Interior: An Anthology of Interior Design Theory*, edited by Lois Weinthal, 40–49. New York: Princeton Architectural Press, 2011.

Sontag, Susan. "The Aesthetics of Silence." In *Styles of Radical Will*. Edited by Susan Sontag, 3–34. New York: Farrar, Straus and Giroux, 1969.

Taylor, M., G. Downey, and T. Meade. Domesticity under Siege: Threatened Spaces of the Modern Home. Edited by Mark Taylor, Georgina Downey, and Terry Meade. London, England: Bloomsbury Visual Arts, 2023.

Tonkiss, Fran. "Aural Postcards: Sound, Memory and the City." In *The Auditory Culture Reader*, edited by Michael Bull and Les Back, 303–310. Oxford: Berg, 2003.

Toth, Leah. "Re-Listening to Virginia Woolf: Sound Transduction and Private Listening in Mrs. Dalloway." *Criticism*, 59, no. 4 (Fall 2017): 565–586.

Woolf, Virginia. *Mrs Dalloway*. London: Hogarth Press, 1825.

19
DIMENSIONLESS SPACE (WITH SERRATED EDGES AND SUCKING NOISES)

Intimacy, ASMR, micro-magic, sensory scholarship, and other taboos

David Toop

Tissue

> She kept turning pages, the girls fidgeted cautiously, now and then a gurgle came from a hot pipe – the tissue of small sounds that they called silence filled the room to the dome.[1]
> (Elizabeth Bowen, *The Death of the Heart*, 1938)

Elizabeth Bowen, the Irish writer of the thirties and forties whose work is characterised as the link between experimental modernism and mainstream literary fiction, captures here a prickling sensory intimacy generated by (the tissue of) small sounds coalesced as silence in a resonant space. They are little or nothing, yet they fill the room with its dome, and in that allusion to grand architecture – the Pantheon in Rome, the Taj Mahal in Agra, or the Cathedral of Santa Maria del Fiore in Florence – resides the suggestion that silence is a teeming presence whose complex insubstantiality can expand ad infinitum.

Bowen's intentions are unknown but her deployment of the tissue metaphor invokes the familiar expression, a tissue of lies; in other words, a construct so flimsy as to teeter on the edge of disintegration. But tissue paper (simply a name for lightweight papers, so there are variations according to the type) has other significant properties, a latency to be unlocked by the reader as they negotiate fictive space with their inner hearing: translucence, absorption, and unique sound-generating potential. This sound, usually described as crumpling or crackling, generates sound as a result of energy injected into the fibrous structure of the paper. Creases are formed suddenly, releasing energy at wildly contrasting sound levels.

The 1996 paper – Universal power law in the noise from a crumpled elastic sheet – described the phenomenon:

> …elastic sheets with quenched curvature disorder have a variety of glassy characteristics including many distinct energy minima, discrete memory, and stretched exponential relaxation. The sound produced by these sheets is a direct consequence of the complex energy landscape associated with glassy systems.[2]

Kramer and Lobkovsky, the researchers who authored this paper, used Mylar for their experiment but as they point out, their work inhabited an active field of research into crumpled membranes of all types, including thin paper sheets and aluminium foil. To explain the process, they wrote:

> The work done to strain the sheet is stored as elastic energy. When the total work done is enough to overcome the potential barrier to a neighbouring minimum, the sheet buckles and the stored energy is released as heat, vibrations in the material, and the sound of a click.[3]

Esoteric as it is, this research goes some way to demystifying the phenomenon of sound production in sheets of paper and plastic and why this can be both appealing and irritating. On the one hand, these tiny sounds evoke and perhaps inculcate feelings of intimacy; on the other hand, they may remind us of tense situations in the cinema, theatre or a quiet performance of music during which a member of the audience crumples or unwraps food packaging to disturbing, disproportionate effect. Although stored energy can be released as sound at dramatically different dynamic levels (hence, leading to its absorbing qualities as a listening experience and the difficulty of controlling these sounds in a situation that demands restraint), the overall impact falls within the category of microsound.

Microsound is both a specific field of research within computer music and a more general description of an informal minimalist (not to be confused with the minimalist genre) tendency in strategies of exploratory music, including past styles such as lowercase, reductionist improvisation, Japanese onkyo and Berlin-based Echtzeitmusik, and subsequent developments in sound arts that have absorbed the lessons of such strategies and incorporated them into a broader approach. Google 'sound of crinkling paper' however, and something different appears in the search results. There is, for example, on YouTube, ASMR (Autonomous Sensory Meridian Response) Paper Sounds, Old Book, Crinkling (No Talking),[4] in which a person with turquoise nails tentatively ruffles the pages of a George Martin novel, applies pressure to a padded envelope, slowly turns the pages of a worn, large format childrens' book, Cenerentola (better known to the post-Disney world as Cinderella), then presses lightly on folded white tissue paper, applying pressure then raising the hands, allowing the paper's elasticity to return it to its former position with the faintest of rustling sounds. So this episodic sequence continues, reprising the George Martin paperback (a clear sign of demographic intent, surely) and those other materials already familiar to the viewer/listener, exhausting their auditory possibilities over the course of a clip lasting for 21′ 59″. The video was uploaded by Chiara ASMR (1.09 M subscribers) in 2015; by December 2022 it had been experienced by 238,630 people. In the comments section, Neil Barrett speculates, with the hint of a raised eyebrow: 'So, public libraries are actually ASMR nightclubs'.

His point is insightful. The library ideal (if not its 21st-century actuality) is a site of concentration and silence in which breath, necessary movement and the turning of pages hover at a level of audibility somewhere close to silence and its tissue of small sounds. Traditionally, a public library offered a **communal** space open to strangers in which serious research, quests for diversion and shelter from weather and loneliness could combine in the pleasureability of muted existence. Such places anticipated ASMR in the sense that their sensory narrowing intensified the effect of sonic awareness, associating this restraint with detached intimacy, an intimacy with clear boundaries and few responsibilities, other than the capacity to be quiet. Thanks to libraries, comic and authoritarian possibilities still reside in the high-frequency word sound, Shhhh, a whisper in its aggressive form.

Sensory response

What, then, is ASMR, and what distinguishes it from microsound? ASMR (very rarely punctuated) stands for 'autonomous sensory meridian response'. Chained together in this sequence, these words convey very little other than a quasi-scientific impression of bodies responding to stimuli, perhaps in some relationship to Chinese medicine, in which the term 'meridian' denotes a theory of pathways through which qi, or chi, life energy, flows. According to legend, the originator of the ASMR initialism was Jennifer Allen, a woman who experienced euphoric tingling sensations as a reaction to certain stimuli and who sought some explanation for the effect. In 2009 one of her Google searches led her to a post – Weird Sensation Feels Good – on a message board called SteadyHealth. According to a 2019 New York Times article:

> As discourse on the unnamed feeling evolved, users shared accidental triggers found online – a man unlocking a damaged padlock, someone brushing her hair. These videos had a gentleness in common that many of the users found hard to describe. Some spoke of the need for a research group to better understand the sensation.[5]

Allen realised the phenomenon demanded a more convincing name than 'Weird Sensation' to give it legitimacy, visibility and some hope of attracting serious research. In 2010, she invented the ASMR title, then used the SteadyHealth.com discussion forum, an online health community, to announce an ASMR Group she had launched on Facebook. The video-sharing capabilities of YouTube (launched in 2005) proved an ideal environment through which individuals could create ASMR experiences in their own homes, directed at users who wanted to experience the sensation. Without intervention or editorial constraint, they were able to develop the now familiar form and its variants. Typically, the screen frames in close-up a young woman whispering into stereo or binaural microphones while gently activating the textured surfaces of objects by tapping, handling or scratching to produce low-level sounds, though there are many alternative forms, extending out into ASMR cooking, eating, atmospheres, applying make-up and brushing hair, hypnosis and relaxation techniques, pencil drawing, role-playing, storytelling, erotica and pornography, sound healing, lifestyle scenarios and 'accidental' ASMR (scenes from cinema, 'found' video clips or documentary film which inadvertently trigger an ASMR response).

Clearly, the post-2000 confluence of message boards, social media, video sharing, smartphones and the Google search engine created a channel through which this elusive phenomenon could be identified, focussed, created cheaply through many different iterations, uploaded and widely distributed to significant numbers of followers without commercial or academic support. The extent of this activity can be gauged by estimates of the number of ASMR videos uploaded to YouTube since 2013. They range from 5.2 million to 25 million, though it is impossible to assess the accuracy of these figures.

Although the two activities – ASMR and microsound – inhabit entirely different worlds, there is an intriguing crossover in their timing. Curtis Roads' book, Microsound, was first published in 2001, seemingly unaware of concurrent developments in improvised music practice, particularly in late-1990s Berlin, London, Tokyo, and Vienna. Reductionist improvisation, as it came to be known, was characterised by a concentration on microaudial listening, the exploration of quiet, finely detailed textures, individual pitches or sustained tones played in isolation, lengthy silences and slow development. All of these tendencies emerged around the turn of the millennium, though they shared many antecedents, not least the so-called slow movement which encouraged a slower-paced approach to life, as manifested in cooking, cinema, conversation, and so on.

Dimensionless space (with serrated edges and sucking noises)

Rather than restricting the ancestral origins of computer microsound to 20th-century composers such as György Ligeti, Iannis Xenakis and Bernard Parmegiani, Curtis Roads indicates in Microsound a more commonly shared listening experience[6]:

Microsound is ubiquitous in the natural world. Transient events unfold all around in the wild: a bird chirps, a twig breaks, a leaf crinkles. We may not take notice of microacoustical events until they occur en masse, triggering a global statistical percept. We experience the interactions of microsounds in the sound of a spray of water droplets on a rocky shore, the gurgling of a brook, the pitter-patter of rain, the crunching of gravel being walked upon, the snapping of burning embers, the humming of a swarm of bees, the hissing of rice grains poured into a bowl and the crackling of ice melting.

> Microsound is ubiquitous in the natural world. Transient events unfold all around
>   in the wild: a bird chirps, a twig breaks, a leaf crinkles. We may not take notice of
>   microacoustical events until they occur en masse, triggering a global
> statistical   percept. We experience the interactions of microsounds in the
> sound of a spray of   water droplets on a rocky shore, the gurgling of a brook, the
> pitter-patter of rain, the
>   crunching of gravel being walked upon, the snapping of burning embers, the
>   humming of a swarm of bees, the hissing of rice grains poured into a bowl and the
>   crackling of ice melting.[6]

Implicit within this list is a heightened awareness that comes from documentation of such events – non-human sounds, whether bioacoustic, meteorological, machinic or complex auditory sites, or events connected to human activities such as fire making, food preparation and the impact of walking – and the late 20th-century expansion of field recording as a creative sub-genre of experimental music. The practice of field recording has increasingly focussed on the details of small sounds and auditory spaces, intensifying human engagement with sounds at the edge of perception and beyond human perceptual limits. Landmark examples of slow cinema, notably the Ben Rivers film, Two Years At Sea (2011), overlap all of these fields. As if by an anthropologist who asks no questions, an unnamed individual living in a remote setting is unobtrusively observed, slowly and patiently. Details of his personal history are left unexplained, only inferred from photographs which show other people, presumably once part of his life. What is important is the materiality of his existence, its solitary, resolutely independent nature, the textural richness of simple activities and sensory impressions, the unfolding of time regulated only by necessities of survival, and at the end, the dying away of a campfire at night, the subject staring into its fading light, then a black screen, only crackling sound remaining.

The question, then, is why all of these related yet distinct practices emerged into the light during a relatively brief timespan? Technological developments are the most likely catalyst for this convergence. The above-mentioned channels of distribution were one aspect of this flourishing, as if a conduit had been opened, but the tools of creation were also evolving towards a public accessibility model. Digital audio and video recorders, free audio editing software such as Audacity and video editing applications like Apple's iMovie all facilitated a shift away from professional equipment, training, studios and edit suites towards independent, do-it-yourself production.

Linked to these developments but arguably more profound in their impact were two related effects: the sense that time was accelerating or becoming unbearably compressed and excessively managed; and secondly, that space itself was undergoing radical transformation.

Virtuality promised a near-infinite immersive space, an entry into other worlds through the flat mirror/portal of screens, yet it also tended to cocoon the user, potentially isolating them from physical social connectivity and expansive non-screen spaces. With the acceleration of ubiquitous computing and the rapid growth of the internet, alarmist dystopian scenarios inevitably followed. 'Thoroughgoing digital networking and communication have massively amplified the compulsion to conform', philosopher Byung-Chul Han wrote in his 2017 book, Psycho-Politics.[7] While much of this analysis was perceptive and timely, it did little to illuminate detailed on-going effects of virtual space for those who occupied it to various degrees, which is to say much of the world's population.

Night leaves breathing

Speaking autobiographically, my own trajectory mirrored this trend. By the early 2000s, I was able to record and mix at home, work in great detail on individual sound events within the digital domain and move away, albeit temporarily, from the compromises and physical challenges of live performance. In 2008, while deeply engaged in research for my book, Sinister Resonance (published in 2010), I was commissioned to compose a work for the Quiet Music Ensemble, based in Cork, Ireland. Composing to the commission was unfamiliar to me. Since I neither read nor write music notation I had no idea where to begin. The answer came to me one night as I was reading in bed. At that time, I was reading extensively on Dutch 17th-century genre painting, particularly those artists who painted scenes set within interior spaces, often representing acts of sounding and listening. The early works of Nicolaes Maes (1634–1693) were typical of this style, depicting women eavesdropping on mildly scandalous scenarios, working on domestic occupations such as scraping parsnips, sewing, spinning or lacemaking, dozing over account books or the Bible, threading a needle or tending small children.[8] What characterises the majority of these paintings is their atmosphere of quiet and concentration. Imperceptible sounds – a needle passing through fabric or the deep breathing of a sleeper – are implicit elements of the paintings. In all senses, as representations of audio atmospheres at the threshold of hearing and as silent works that invite listening, albeit of the imagination, they can be interpreted as recordings of time-based phenomena existing beyond the limitations of the medium.

As I read one particular book on Dutch genre painting, I became acutely conscious of sounds within the house at night: the turning of book pages, a settling of wooden floorboards, cooling radiators and the snoring of our dog on the bedroom floor. 'But there were leaves as well', I wrote in a programme note for my eventual composition,

> and I was thinking about my eureka moment one night in spring a few years ago. I was sitting in my garden and became conscious of a microscopic sound. As I listened harder, I realised this was the sound of slugs and snails eating the leaves of my plants. Taking the dog for a walk I became well versed in the nuances of leaf sound: underfoot (or should I say under foot and paw); blown by breeze or wind; or simply moving and flexing as atmospheric conditions change within subtle gradations.[9]

For the composition, I decided to record an auditory environment that evoked these entangled experiences. As a starting point, I went into a studio to make a 20′00″ recording of myself reading silently from a large book. Subsequently, I added recordings of the dog snoring, leaves being scrunched, floorboards creaking and electronic atmospheres suggestive of room tone and fluctuations of perceptual awareness. This was my score. The intention was that the five musicians of the

Quiet Music Ensemble would recognise within it their own personal connection to micro-listening and contemplative hearing and improvise in this spirit.[10]

What I found compelling at that time was the potential for these tiny, often unidentifiable nocturnal sounds, many of them produced out of eyesight of the listener, to induce fear. There was an autobiographical element to this, the memory of hearing noises in the house at night when I was a child and fearing intruders, whether human, ghost or monster. Reading authors like Edgar Allan Poe at a young age only intensified these episodes. A residue of threat was evident in performances of night leaves breathing and yet it was also noticeable that the concentration on small sounds gave rise to feelings of comfort, tranquillity and safety. It was as if they fashioned a space of enclosure that echoed the protected bubble of childhood as much as they flickered at the edges of vulnerability, displacement and night terrors. In other words, the space of audition could be in a state of contradiction, intact yet unstable, almost simultaneously.

Intimate space

Poe was not the only writer to recognise the eerie emotional component of microsounds. In Waiting For Godot, Samuel Beckett has Vladimir and Estragon discussing the dead voices, their noise like wings, sand, leaves, rustling, murmuring, whispering:

Vladimir: They make a noise like feathers.
Estragon: Like leaves.
Vladimir: Like ashes.
Estragon: Like leaves.
Long silence.[11]

Beckett returned to these threshold sounds throughout his life. A late work – Ill Seen Ill Said (1981) – describes the existence of a woman living alone in a cabin, evoking suffering and endurance through the near-hallucinatory clarity of co-existing with heightened perceptions:

> All dark in the cabin while she whitens afar. Silence but for the imaginary murmur of flakes beating on the roof. And every now and then a real creak. Her company … Far from it in a corner see suddenly an antique coffer. In its therefore no lesser solitude. It perhaps that creaks.[12]

Beckett's personification of non-human phenomena – the creak that is her company, the antique coffer that may also be lonely and speaks of it in shared language – touches upon the intimacy we share with objects and events entangled in the mystery of private worlds. Perhaps the coffer creaked, the voice of its solitude; maybe the snowflakes were heard or felt as hallucination, despite their silent accumulation. Because of this role as faint voices living within the construction of the self, microsounds address such moments of being in which the body is settled and nurtured within itself yet also open to the liveliness of companion objects.

One clear distinction between ASMR and microsound is that ASMR practitioners seek to isolate and instrumentalise small sounds in order to bring about a calm state, facilitating the emergence and spread of euphoric feelings through the body. ASMR makes no claim to being a form of art, a disruptive process or an experiment in sonic manipulation. Its purpose, measured to some extent by YouTube views, comments and channel subscriptions, is the tingle. The use of microsound in improvisation or electronic composition has a less defined or circumscribed purpose, varying according to individual practitioners. Both deal in intimacy, however, and this raises the question of what intimacy can mean, given emerging, conflicted conceptions of space.

Online intimacy, mediated through video sharing or social media, is inevitably shadowed by a corporate, monetised interpretation of intimacy. The space through which it reaches out for contact is a doubled or mirrored space, flat and lacking physical resonance. There are, however, possibilities for endless replay in multiple regions. As Naomi Smith and Anne-Marie Snider write: 'Bookmarking and "favouriting" videos on YouTube also means the affective experience ASMR videos produce is not temporally constrained but rather exists in a persistent state of "readiness."'[13] The materiality of ASMR could also be described as flat rather than possessing volume, in the sense that it works within a narrow dynamic range, restricting itself largely to upper frequencies and avoiding resonant tones or any hint of reverberant or capacious auditory spaces. Closeness has a higher value than distance. Distance is global, yet highly compressed.

Whispering is the medium of a lover or confidant, a spy or co-conspirator, flat and thin, designed specifically for the ear cup in close proximity. The illusion is a contracted space of two people, operator and receiver: one housed in its flat-screen operating theatre (usually in close-up, sometimes excluding the face, almost always excluding the lower body, with a limited view of the operator's environment); the other in close relation to their own flat screen, perhaps imagining themselves to be sharing the same space. Counteracting this in-built alienation is the yearning for its opposite. According to the catalogue of Weird Sensation Feels Good, an ASMR exhibition at the Design Museum, London:

> ASMR injects the internet with softness, kindness and empathy. As a form of digital intimacy, it offers comfort on demand, standing against the feeling of isolation that constant connectivity can deceptively breed. Anecdotally, ASMR is being used more and more as a form of self-medication against the effects of loneliness, stress, and anxiety.[14]

It should be stressed that ASMR comes in many forms, not solely auditory. The presence of what I call an operator is considered obligatory in most cases, however, though the question of whether they are regarded as healer, guide, companion or some other role will surely vary across so many users. A tendency to anonymity, as in Chiara ASMR's paper crumpling video cited above, is counterbalanced by personality-driven channels in which the operator appears full-face, introduces the theme of each video in direct address to the user and fully engages in all the markers of success typical of YouTube influencers: a 'support this channel by subscribing' button, product advertising, sponsorship messages, and fragments of personal information. Some ASMR channels sexualise their 'content' – in other words, young women dressed (apparently) to infer something more explicit – while others are tangentially linked to digital trends such as guided meditation, livestream cosplay workers, unboxing videos, crafts in progress and sleep (or its opposite, productivity) apps for white noise, pink noise and brown noise.

All of these trends might be considered therapeutic, to some degree. With the emergence of ASMR in an era of prodigious content mining, this potentiality has been explored retrospectively through video clips of cats purring, bespoke tailoring, embroidery and cross-stitch tutorials, consultations with neurologists, opticians and other health practitioners, the Polish accent, Japanese print making, Welsh stone carving, pencil sharpening, even the Victoria and Albert Museum's 'ASMR at the museum' channel, which includes turning the pages of a medieval choirbook, library experience of reading feminist magazines and massaging hands with bath rasps from Iran. Unintentional ASMR extends into cinematic fiction. One example of this, uploaded to YouTube, is a one-hour duration loop of a 4′ 15″ scene from Paul Thomas Anderson's film, Phantom Thread (2017), in which the fictional haute couture dressmaker played by Daniel Day-Lewis measures a woman for a dress. Muted sounds reflect economy of movement, the concentration inherent in the task (in effect, the portrayal of an older man seducing a young woman through a displacement activity) – fabric, pins, shoes on a wooden floor, catches being fastened, hushed speech, the reading out of measurements. What seems to be invoked here are memories of moments, atmospheres, settings and actions, reaching back to childhood, that induce blissful, near-hypnotic sensations of nurture, care, contemplation, a shedding of anxiety, though in this instance they have another subtext, that the routines of work can sublimate another agenda entirely.

As triggers these proliferating scenarios can be assumed to be highly subjective in their efficacy. The phenomenon may be fascinating but my personal response to the sound of many ASMR videos is that they can be harsh, careless, irritating, aimless, uninventive, banal, inane or simply ineffective – judgements I can make both as a gesture towards sensory anthropology and as an expression of my own quirks and boundaries. John Levack Drever, Professor of Acoustic Ecology and Sound Art at Goldsmiths, University of London, has described misophonia, a strong dislike of sounds made by others, as 'ASMR's evil sibling', appropriately, in a social media post.[15] Finger tapping, a popular ASMR trigger, is a common source of misophonic anger. As for sucking, chewing, swallowing, tongue and saliva sounds, also present in a multitude of ASMR uploads, these account for a high proportion of strong misophonic reactions, ranging from rage to disgust. A Dutch study investigating brain activity of emotional responses to audiovisual activity selected mouth sounds such as breathing and lip-smacking for its misophonia-related cues.[16]

Many ASMR eating videos have bewilderingly complex, targeted titles. HunniBee ASMR (8.36 M subscribers), for example, posts 'ASMR clear food, aloe vera drink, frog eggs jelly, frozen honey lip gloss tubes mukbang'. Mukbang is the name given to a type of Korean-origin eating show, in which the host eats while talking to a live audience. Although this culturally specific referent has been internationalised through YouTube ASMR videos, the connection remains by implication. Masked by the context-free, dimensionless space of ASMR, the origins of mukbang are rooted in Korean history and eating styles: family-style communal eating is highly valued, solitary eating is felt to be a lonely experience, poverty is still a strong memory for older Koreans, noisy eating of certain foods (as in China and Japan) is welcomed as a socially acceptable sign of enjoyment. Again, in mukbang ASMR, social auditory space is transformed – made flat yet also vastly expanded through algorithms and likes – into digital space.

Mukbang highlights the misophonic or taboo aspects of ASMR. Ostentatiously noisy eating, or even the sight of other people eating, is as likely to trigger strong reactions of disgust as it is to generate the ASMR tingle. To return to my own practice as a musician and theorist with a strong interest in microsounds, I can relate this polarisation of responses to audiences and their listening boundaries. Present-day listeners who have chosen to attend a concert now tend to be familiarised with music created with small, delicate, quiet, detailed sounds (though this was not always the case). If I bow a leaf with a violin bow, producing tiny variations of sound from the leaf's serrated, gradually abraded edges, I am conscious that the sounds can possess an intense materiality. They have a magical quality, simply because of the complexity of sounds that emerge from such a fragile, humble substance, but the harder I press, the closer I get to those sounds often cited as alarmingly unpleasant: a nail scraping on a chalkboard, the shrill scream of a parrot, a fork scraping against a plate or the friction squeaks of polystyrene blocks rubbing together.

Of course, these latter sound events come unexpectedly, tending to shock the listener into an adverse reaction. My intention is to explore the sensory possibilities of materials and objects – the auditory, visual and ecological characteristics of flat materials such as leaves and other plants, paper and thin wood – transforming these sounds with resonating objects such as cardboard boxes, paper cones or a drum, often at dramatically variable dynamic levels. For me, this contrast between the acoustically unsupported sounds of flat materials and their amplification (perhaps more accurate to say enrichment) through hollow objects acts as a bridge to the larger space of the performance site itself. This is the room – both acoustic and social – in which intimacy can form through close communal attention to minutiae.

As outlined above, there are obvious parallels between ASMR and this approach to music. There are clear differences also. The spaces in which they exist represent two distinct conceptions of sociality, resonance and action. ASMR practitioners aim to reach a specific sensory or therapeutic goal, reducing and refining stimuli and settings to that end and taking care not to disrupt the mood. My practice may use similar techniques and materials but their ordering in time, their emphasis on contrasts (sometimes disruptive) and their cultural context operates in an entirely different sphere. Both inhabit a distinct conception of auditory space – the digital and the physical – and its relationship to the formation of bodies and communities. As such they are complementary. They investigate the constitution of spaces that must be negotiated, to some extent understood, in order for us to exist in the world as it is now.

Notes

1. Elizabeth Bowen, *The Death of the Heart* (London: Penguin Vintage Digital, 2015), Kindle.
2. Eric M. Kramer and Alexander E. Lobkovsky, *Universal Power Law in the Noise from a Crumpled Elastic Sheet*, 1469.
3. Kramer and Lobkovsky, 1467.
4. https://www.youtube/watch?v=wgtbr5oQeQw, Chiara ASMR, 17.3.2015.
5. Jamie Lauren Keiles, https://www.nytimes.com/2019/04/04/magazine/how-asmr-videos-became-a-sensation-youtube.html (New York: *The New York Times Magazine*, April 4, 2019).
6. Curtis Roads, *Microsound* (Cambridge: The MIT Press, 2004), 21.
7. Byung-Chul Han, *Psycho-Politics* (London/New York: Verso, 2017), 82.
8. Ariane van Suchtelen, et al., *Nicolaes Maes: Dutch Master of the Golden Age* (London: National Gallery Company 2020).
9. David Toop, Programme Note Written for Night Leaves Breathing, 2008.
10. David Toop, *Night Leaves Breathing*, The Quiet Music Ensemble, *The Mysteries Beyond Matter* (with Alvin Lucier, Pauline Oliveros, John Godfrey) (Ireland: farpoint recordings CD, 2015).
11. Samuel Beckett, *Waiting for Godot* (London: Faber and Faber, 1965), 62–63.
12. Samuel Beckett, *Ill Seen Ill* Said (London: John Calder, 1997), 34–35.
13. Naomi Smith and Anne-Marie Snider, ASMR, affect and digitally mediated intimacy (*Emotion, Space and Society*, 30, 2019), 44.
14. Weird Sensation Feels Good: The World of ASMR, Exhibition Catalogue (London: The Design Museum, 2022).
15. John Levack Drever, Facebook post, 5.7.2022.
16. Misophonia is associated with altered brain activity in the auditory cortex and salience network (www.nature.com/scientific report, 17.5.2019).

Bibliography

Beckett, Samuel (1965) *Waiting for Godot*. London: Faber and Faber, 62–63.
Beckett, Samuel (1997) *Ill Seen Ill Said*. London: John Calder, 34–35.
Bowen, Elizabeth (2015) *The Death of the Heart*. London: Penguin Vintage Digital, Kindle.
Drever, John Levack (2022) Facebook post, 5.7.2022.
Han, Byung-Chul (2017) *Psycho-Politics*. London/New York: Verso, 82.
https://www.youtube.com/watch?v=wgtbr5oQeQw,, Chiara ASMR, 17.3.2015.

Keiles, Jamie Lauren (2019) https://www.nytimes.com/2019/04/04/magazine/how-asmr-videos-became-a-sensation-youtube.html. New York: The New York Times Magazine, April 4, 2019.

Kramer, Eric M. and Alexander E. Lobkovsky (1996) *Universal Power Law in the Noise from a Crumpled Elastic Sheet*. Chicago, IL: Physical Review E, The James Franck Institute and the Department of Physics, The University of Chicago, Vol. 53, number 2, 1469.

Misophonia is associated with altered brain activity in the auditory cortex and salience network. https://www.nature.com/articles/s41598-019-44084-8 report, 17.5.2019.

Roads, Curtis (2004) *Microsound*. Cambridge: The MIT Press, 21.

Smith, Naomi and Anne-Marie Snider (2019) ASMR, affect and digitally mediated intimacy. *Emotion, Space and Society*, 30, 44.

Toop, David (2008) *Night Leaves Breathing* (musical composition).

Toop, David (2015) *Night Leaves Breathing, The Quiet Music Ensemble, The Mysteries Beyond Matter with Alvin Lucier, Pauline Oliveros, John Godfrey*. Ireland: farpoint recordings CD.

van Suchtelen, Ariane, Bart Cornelis, Marijn Schapelhouman, and Nina Cahill (2020) *Nicolaes Maes: Dutch Master of the Golden Age*. London: National Gallery Company.

Weird Sensation Feels Good: The World of ASMR (2022) *Exhibition Catalogue*. London: The Design Museum.

20
POSTHUMAN LISTENING TO THE MORE-THAN-HUMAN SOUNDSCAPE

Jordan Lacey

Opening remark

This essay presents the concept of the posthuman listener who relates to a more-than-human soundscape. The soundscape term is humanist (*vis-a-vis*, not posthumanist), through and through. It is built on the perspective of the sensing human, with a specific focus given to the human who listens. The soundscape wraps around an individual; its complexities are reduced to the perceptual response of that listener, who it seems is often in some sort of contemplative solitude. This human-centred soundscape is often represented as a circle, with a centre point representing the listener. Every sound heard is mapped into this circle with its circumference acting as a so-called acoustic horizon, marking the very edge of what that individual can hear. What is, is what the 'I' hears; or, what it apprehends. A posthuman approach questions the capacity of a human to fully apprehend the materiality and affective potential of the sound field in which it is immersed, and asks who is the human listening at the centre of this circle? In this way, the circle is reductive. Who is the privileged self at the centre of things? Who gets to decide what the soundscape sounds like, and therefore, how it might be represented and/or transformed? To pose such questions is to decentre human privilege, which is possibly the single most important goal of the posthuman project; therefore, a posthuman approach to the soundscape must consider what a soundscape is when there is no human at its centre. In support of this goal, this essay considers a soundscape to be a multiplicity; it is as diverse as the listeners through which it is perceived. A soundscape is more-than-human insofar as it always exceeds any one person's apprehension, while at the same time being differentiated by each sensing body into a diversity of listening experiences. This essay is comprised of two parts that fully develop this theme. First, it does the work of decentring the human listener within the context of the soundscape, and second, it proposes a new way of understanding relationships between people and sounds by entangling posthuman insights with First Nation deep listening practices. It ends with a proposition for a new soundscape discipline that could be included in future posthumanist pedagogies.

Part 1: Decentring the human listener

Returning to the circle, where might the posthuman be located in a mapping of the soundscape? One way to approach this is to reimagine the circle as a net, or a mesh. Such suggestions have already been put forward by the eco-philosopher Timothy Morton[1] and the anthropologist Tim

Ingold,[2] both of whom, in different ways, explain environments as entangled interconnections of things/objects. The mesh overcomes the problematics of dualities, which separates humans from nature; instead, it puts us in a transversal and interconnected relationship with the environment. Using these concepts, I'd like to propose here that the posthuman (and not just the environment) is more like a mesh than a singular point – the human explodes outwards into a cluster of constituent parts; its social, cultural, historical, political, biological, and temporal intersections, each influencing the reception of sound in different ways. This image can be extended further; each node in the mesh can be considered a posthuman cluster that sits in relationship to every other posthuman cluster; and of course, beyond this, we can consider non-human listeners who also form their own nodes. This image presents a more complex depiction than the circular soundscape described above. The soundscape becomes a multiplicity; a transversal substance expressed differentially through every listener. Readers familiar with Deleuze's thought might recognise this as difference-in-itself.[3] Applied here, we can consider the soundscape to be a pre-personal transversal substance/material, which is expressed divergently and uniquely through each listener. The posthuman listener, as a node in the mesh, brings the soundscape to life in a unique and personal way dependent on its own idiosyncratic clustering. In this depiction there is no one type of soundscape and no one type of listener; rather, the soundscape comprises a multiplicity of listenings that constitute its ever-shifting, amorphous complexity. Before proceeding, I am compelled to present myself as a posthuman cluster. Indeed, it is a necessary act of posthumanism (or at least should be) to declare one's own positionality. How do I hear, this 49-year-old white cis-male employed by an institution as a sound study academic? I will answer this through three reflections that, in keeping with the mesh concept, extend beyond the reflexive towards a curious probing of the environment with which I am presently interconnected. This positioning, I present as three posthuman principles concerning soundscapes.

Three posthuman principles concerning soundscapes

1 Entangled soundscapes: Firstly, my hearing range is probably something in the order of 20 Hz–15 kHz (20 kHz for those with excellent hearing). That means I am only hearing certain sounds that fall within that frequency range. Outside my window are a flock of rainbow lorikeets. Wonderfully playful and frenetic birds. The hearing range of a parrot is approximately 200 Hz–8.5 kHz.[4] They have no ear lobes; their ear openings are covered by feathers that act as a shield during flight. Consequently, their hearing is less sensitive than my human ears, and so they can screech and shout in close proximity. These are volumes I find overwhelming, but to them no doubt, it is a comfortable vocal exchange. Across the road is an occasionally barking dog. The average hearing range of a dog is 65 Hz–45 kHz.[5] The dog is living in an entirely different soundscape to me. It is hearing a whole layer of sounds that I can never experientially access (without technological tricks, like transposition, etc.). High-pitched sounds generated by machinery or the call of other animals beyond my upper 15 kHz range are sonic worlds I can only imagine. And then there are the earthworms in my garden, only metres from where I am sitting. Worms hear through their skin.[6] Ground-borne vibrations (or air-borne when surfaced) transfer through the skin of the worm; thus, they feel the vibrations of the world around them. Recent research suggests the worm can feel airborne vibrations in the range of 100 Hz–5 kHz.[7] with the worm's body acting as a whole-body cochlea – a whole-body ear, so to speak. Skin as ear, felt as vibration. And so, while I could list for the reader the sounds I am hearing right now, I will not privilege my own human ears. Instead, I ask the reader to consider the lorikeets, the dogs and the earthworms I am sharing this moment with – what are they hearing? Four

soundscapes for four species. An entangled relational complexity – each contributing sounds and hearing sounds in a co-constitutive soundscape.

2. Noise can be natural: Secondly, I have a favouritism towards 'natural' sounds. So, I am not too happy with the busy road about a block away from my desk. Now this presents an immediate problem. Each of those vehicles producing noise from their rubber tyres rolling over a tarmac road is being operated by a human. Some going to work, some driving their children to school, some going to shops, etc. In its totality, all of those human drivers create a continuous sound that penetrates my home. My bias tells me this is not natural sound, because it is technological. And yet, the very humans that create this flow of sound are as natural as I am; insofar as they are primarily constituted of biological material. And this presents yet another problem! What is this biological material that I refer to? Is it the many autonomous organs that make up a human's biological system that work together to ensure a body can breathe and eat and function? Is it the myriad microbes that have made a home in my body and/or on the surface of my skin? My 30 trillion human cells are matched by an approximately equal number of bacterial cells.[8] Physically speaking, in which group of cells am I located – my human cells, or my bacterial cells? The 'I' that is 'natural' is its own complex ecosystem housing an extraordinary array of life, all of which will die when the 'I' that I claim I am dies as well. All of those fellow humans generating traffic noise comprise a similar complex biological ecosystem as I do; and indeed, each of those vehicles is comprised of resources extracted from the very earth we share (metal, petroleum, etc.). This begs the question, what exactly is natural sound? Each one of us perceives sound uniquely; after all, the experience of noise is perceptual and not an objective reality. There is only sound (materially speaking). Personally, I abhor traffic noise; I find it incredibly intrusive. But this is my bias. I remember a surprising conversation with a friend's aunt in Sydney. She moved from the city to the seaside. She used to live by a busy highway and was now living by the ocean. She complained how disturbing the ocean sounds were which encouraged her to move back to the city so she could be close to the sounds of traffic. I asked why she preferred the sound of traffic to which she replied, it makes me feel connected with other people. By her own admission, it was a symptom of her loneliness. The ocean reminded her she was alone, while the sounds of traffic made her feel connected with the activities of her fellow humans. On the same trip, I met the owner of Zen Music Studios in Sydney, located directly next to the centrally located Sydney airport. As a Boeing 747 with its wheels lowered passed overhead, while multiple punk and rock bands played in adjacent rehearsal rooms, he exclaimed with a smile on his face, this is the loudest place in Sydney! He wasn't complaining. He genuinely loved it – noise for him was a joy. My love of natural sounds is simply my own bias; my preferred soundscape would be of little interest to the aunt or the music studio manager. Soundscapes exceed bias.

3. Hearing identity: Thirdly, as a six-foot-tall white cis-male with an academic position, I am afforded certain privileges in this life, even if I am not from a privileged background (which, as the offspring of working-class parents, I don't deem myself to be). It took a soundwalk with Mexican artist Amanda Guiterrez, in Montreal, for me to realise the reality of my privilege, that I possess through my identity. She took me through the darkened streets of urban areas in Montreal and asked me to imagine what this experience might be for a woman, or for a person of colour or for a transgender person. It was a very direct way to bring to my attention the inherent bias of the actions of the flaneur. The flaneur is a romanticised figure, the poetic soul who wanders through cities unconcerned by the directions and dictates of urban planners. I have long identified with this figure. I love getting lost in new cities – it is one of the joys of my life, and indeed, being able to relax and absorb myself into the sounds of new places is a great

pleasure. But after my walk with Guiterrez, I came to realise how privileged I am to have the body shape and height, and the gender, that probably keeps me safe in certain situations that may not feel safe for others. The sounds of approaching footsteps, dark corners, or the cries of revellers I may have dispassionately perceived, but these could be experienced as a danger to others. Our identity directly affects our sonic experience. Recent sound study scholarship is bringing this to our attention. Johnathon Sterne in Diminished Faculties presents an impairment phenomenology recognising that our bodies are always changing, consequently, our bodies (we) always experience the world differently. John Drever's Aural Diversity brings attention to the diverse ways in which people hear according to (but not limited to) biological, cultural and gender factors. While neither scholar identifies as a posthuman thinker (as far as I am aware) their work brings attention to a posthuman approach insofar as any notion of a single type of human listening or listener is replaced with a diversity of possible listeners and listening experiences. How we hear the soundscape depends on the specific posthuman cluster that forms our unique identity.

This short tripartite reflexive experiment demonstrates just how complicated a soundscape is. As a posthuman thinker, I must question what it means to be human, and this begins with a questioning of myself. Am I human, and if not, what am I? Through this questioning the posthuman thinker attempts to de-centre human privilege; accordingly, they complicate the role of the listening human in their definitions of what is and isn't soundscape. By questioning my own presumptions of what it means to be human, I become the posthuman listener through my awareness of the diversity of other possible listening conditions to which I have no access. As such, the posthuman listener does not locate at the centre of a circle of established knowledge, but coexists in a mesh of endlessly connecting nodes knowing its knowings are only one of many possible knowings.

Towards a posthumanist understanding of soundscape

According to the much-cited posthuman theorist and feminist intellectual Rosi Braidotti, we must be cognizant that when we refer to the human, we are often referring to a white European male who has historically been apotheosised (mostly by himself) as a superior human – superior to women, people of colour, people with psycho-social disabilities, etc. This is problematic as the humanities have been written by these same powerful white men and are taught from their perspective, meaning the voices and perspectives of those considered to be inferior are omitted from the discussion. The answer, according to Bradiotti,[9] is the establishment of a posthumanities that is able to respond to the changing conditions of the world in which we live. The soundscape term, as it has become popularised, seems to fit this critique. The term is rooted in the concerns of the white, male academic and composer, R. Murray Schafer (though the term was in fact invented by Michael Southworth in a separate study; see Sterne, p. 13 for further details). Schafer was no fan of the urban and was troubled by its deteriorating environmental sound conditions (as he perceived it). He desired to reverse this situation with the creation of a generation of acoustic designers – inspired by his work and that of the World Soundscape Project (WSP) – who could improve the urban soundscape. His extensive pedagogical oeuvre is largely concerned with teaching people how to listen so that their ears might be 'cleaned' in preparation for the design of acoustically desirable urban soundscapes. In the context of Braidotti's desire to establish a posthumanities that might overcome the pedagogical limitations of the humanities, Schafer's soundscape seems to fit her critique. His definition of a healthy (or hi-fi) soundscape is the one he hoped acoustic designers would seek to recreate; that is, the centre of his circle becomes the definition of soundscape to which other listeners should defer.

Schafer, in a lesser-known book Temples of Silence (1993), includes the image of the enraged musician from the famed painting by William Hogarth (1741). It presents a musician enraged by the disturbing noises occurring outside his window. This enragement can be read as a desiring silence (the shadow of desiring silence being the silencing of the other). The people outside his window are working people (or an underclass) surviving and making culture through activity. Our wealthy musician is positioned above them in a well-to-do home, clearly aggrieved that he cannot continue hearing the pure notes of his instrument; this is privilege par excellence! Schafer, himself a famed composer, had similar grievances; indeed, he eventually relocated to the countryside, escaping the unpleasant electrical soundscapes of the city (and as a lover of nature sounds, I am sympathetic to his decision). While the need for silence is very real and important, the silencing of the other presents a troubling side rarely confronted by soundscape researchers. When the world is silenced, typically by the desires of the middle classes, livelihoods can be destroyed. There are notable research examples that address these concerns. Gascia Ouzounian presents an important discussion of the silencing of outdoor African-American Go-Go music cultures, caused by inner-city gentrification.[10] Emily Thompson (2004) brings similar issues to light in her historical work The Soundscapes of Modernity. Thompson writes about the silencing of the culture of Coney Island (in 1907) when police put a ban on megaphones: 'noise abaters typically identified relatively powerless targets, noisemakers who impeded, in ways not just acoustical, the middle-class vision of a well-ordered city (which) foretold of far more ambitious efforts to regulate and harmonize the sonic disorder of urban life'.[11] Similar scholarship was undertaken by Derek Valiant, who writes about the silencing of working-class cultures at the beginning of the 20th century: '…control over sound constituted an important domain of refined public and private behaviour in the view of the bourgeoise (…) Cultivating a discerning ear permitted the individual to better appreciate or make discriminating judgements about the character of others'.[12] The aforementioned discussions demonstrate that the silencing of acoustic cultures by middle-class desires for quietude remains an active social force. It is fair to say that Schafer's concerns were mostly directed at the rise of the electrical soundscape rather than the cultural expressions of people; and yet, it is difficult to discriminate his concerns from those of the enraged musician. This anti-noise sentiment, so concomitant with the soundscape term, has an ongoing cultural bias, the dark side of which is to silence the other in favour of the preferred living conditions of those who demand silence.

In my own work, I have tried to demonstrate the fact that there has always been a counter-narrative to anti-noise sentiments. Sound artists, in particular, have long presented another narrative. Max Neuhaus (a former percussionist who turned to the public sphere, and it must be said was no fan of the term Sound Art) had a much more nuanced appreciation of urban noises, which enabled him to create sound works that transformed city spaces.[13] Times Square is Neuhaus' most enduring work. In it, he placed a synthesiser and speaker below a grate in New York's Times Square. It propagates a sound that is strangely familiar and yet unique within the immediate environment; through his appreciation of the complex character of urban sound he was able to create a place of sonic intrigue in one of the nosiest – or at least busiest – urban places on Earth. This is a remarkable achievement, the possibilities of which cities have yet to properly understand in relationship to urban soundscape design (despite it being in place since 1977, with a ten-year break from 1992 to 2002). In my own work as an urban sound artist who works with, rather than against, the sounds of the city I have always sought the positive rather than the negative in the urban experience, even if, in all honesty, I am often annoyed by unwanted sonic intrusions. But there is no point in being oppositional to urban noise when pluralistic transformations (by practice) can achieve effective outcomes. Neuhaus demonstrated this with an acumen that outstripped his contemporary, R. Murray Schafer, in the acoustic design of public spaces. To be fair, Schafer's position would never have allowed success in these efforts, as the need to remove noise as a first step is extremely difficult to achieve. The closest he came to this was his description of the 'Soniferous Garden' in his book The Soundscape which proposed the use of earth mounds to block unwanted traffic noises intruding into urban green spaces.[14] However, Neuhaus was able to bypass Schafer's problem by taking (what I would consider) a posthuman approach to urban sounds – he embraced and played with their materiality (artistically) instead of falling into the trap of judging them from a reified position; a.k.a. trying to concretise personal bias instead of interconnecting with the real/material.

The essay will now turn to a discussion of posthuman listening in relationship to more-than-human soundscapes. I intend to do this by focussing on First Nation/Indigenous scholarship. Transversal listening practices to the more-than-human are ancient and continuous. They precede the humanities and are now re-emerging (in the context of Western academic institutions) as the posthumanities.

Part 2: Learning from First Nation deep listening practices

Posthuman theorist Francesca Ferrando argues that humans have always been posthuman.[15] She comes to this position through her interest in spirituality as practised historically by many cultures, especially Eastern cultures (i.e. Hinduism, Buddhism, Tantra). Spirituality she argues, as an 'all-encompassing signifier', complements posthumanism's 'post-dualistic perspective'. This is a transversal approach to understanding nature and culture that applies equal value to all living and non-living things, thereby eschewing divisions between humans and animals, culture and nature, men and women, etc. She goes on to argue that posthumanism takes these traditional spiritualities in a new direction, by applying it as a shared 'technology of existence' that is free from the anthropocentric foci of organised religions. I introduce the second part of this essay with this reference as it illuminates two important points regarding both the posthuman term and the soundscape term.

On the posthuman term: we should be careful not to get too caught up in the meaning of the prefix, post-, which means 'after'. As argued by Ferrando, there is no after-human here, but rather an exceeding of what it means to be human. Understood as such, it may be that the more-than-human term is a better substitute for the posthuman term. It approaches more closely the driving premise of posthumanism, which decentres human privilege by placing us in a healthier and more respectful relationship to each other, other animals, to the Earth and to spiritualities (a.k.a. forces exceeding human understanding). However, I have no intention of trying to overturn the posthuman term in this essay, which I embrace. Rather, my intention is to bring to attention that posthumanism can mean many things: it asks for a rethinking of the humanities; it asks for a rethinking of what it means to be human, it calls for a more respectful relationship with other animals, and the world in which we live. Of course, there are many more explanations than that, but what is of interest here is the insight that posthumanism does not mean a progression (as it might for transhumanism, as we will see below) but rather a re-engagement with practices that transversally connect humans with the land and with more-than-human forces (spiritualities). As we shall see, posthumanism connects Western thought with thought systems and practises from the East and from First Nations that, combined with technological developments, could lead to creative futures.

On the soundscape term: by connecting the word spirituality with soundscape, we can use Ferrando's argument to develop a different understanding of the soundscape term. As discussed, understandings of soundscape are most often human-centred. Soundscape studies privilege human hearing; particularly the able-bodied, rational, Western listener (such as the enraged musician). However, by concentrating on First Nations deep listening practices a different image of the soundscape emerges. Not one of anger and judgement towards unwanted sounds, but an engagement, via 'deep' listening with the land. In his recent work Hungry Listening, Dylan Robinson does the work of decolonising listening practices by bringing awareness to indigenous listening practices. The book has many things to say, but of interest here is the way in which he brings attention to the fact that certain songs sung in Indigenous cultures are alive; that is, a song is a living thing (in the way a human is a living thing). To disturb a song while it is being sung is akin to removing a body part; it is as Robinson explains, akin to an act of violence. This is particularly relevant to songs in media form that might be played back in institutions such as museums. Typically, these would be thought of as non-living media fragments. It is common practice to cut, sample, and manipulate digital media in endless ways. But this is anathema to indigenous societies where certain songs (as sung) are steeped in cultural significance. That a song is alive extends from connections with land and society; vis-a-vis the transversal continuum of non-dualistic nature-culture, from which those indigenous songs emerge. Indeed, Robinson suggests in his book that perhaps we should be (re)turning our attention to a listening to the land (rather than just our media); for in the land, there is meaning and connection to be found. What would it mean for us to listen to the land – and our cities – in such a way? We might speculate that this is what Neuhaus (and others) are trying to achieve when creating works that entangle installation sounds with the noises of the city; in so doing they create new listening relationships between human society and the land. This speculation resonates with Ferrando's provocation that we have always been posthuman: if humanism has been accompanied by a slow withdrawal from our connection with the land as we transfer to the semiotics of everyday urban life, then perhaps posthumanism can be recognised in those transversal practices that are attempting to reconnect listener and land. Therefore, to turn our listening to the land is to reconnect with those meaningful relationships that connect human and terra (urban or otherwise).

Taking these points into account the proposition of this essay can now be properly framed. To be posthuman is to connect with the more-than-human; that is, the posthuman listener interconnects with the more-than-human soundscape. This removes the bias of the human-centred, privileged listener, and instead puts us in a more respectful, transversal relationship with the land. Soundscapes are not subservient to humans but instead are environmental expressions alive with more-than-human forces that a posthuman listening acknowledges and reflects through practice. These relationships are referred to as deep listening practices in Australia.[16,17] Deep Listening means much more than the tuning into the sounds of instruments, and/or the sounds of the environment; an approach attached to the compositional and educational oeuvre of Pauline Oliveros. While Oliveros' approach directs us towards a listening that is interconnected, it continues a duality in so far as the listener is asked to connect with something with which they are typically not attuned. Indigenous deep listening practices are different. Firstly, a deep listening practice means quietly and respectfully hearing the words and narratives of your interlocutor and/or group. To listen deeply to what is being said ensures a respectful sharing of knowledge[18]. Secondly, and of most interest to this essay, deep listening is a tuning in of listener with land. This is not listening in a dualistic sense; that is, the land speaks and we hear it. Rather it is suggestive of a resonant relationship in which bodily connection and expressive land are co-constitutive of nature-culture (and not culture in opposition to nature). This resonance creates an interconnected positionality that is transversal, and without a centre. This resonance can be profound; for instance, in Indigenous cultures, it can include methods[19] and experiences that enable connections with ancestors and other forces. This is a position that a human-centred (humanist) perspective finds difficult to accept (there is no reality beyond that which can be perceived by the human); however, it is one which posthuman philosophy attempts to accommodate in its attempts to establish connections between the paradigms of indigenous research and posthumanist theory.[20,21] It is at this point that words risk losing their meaning; they become redundant signifiers trying to grasp experience. The only option for the reader, as I am right now, is to plant their feet on the ground and to feel the resonance of the very Earth on which every other living and non-living thing resides. Consider the feet to be like the skin of the worm, feeling vibrations of earth and air. Becoming-worm! It presents an entirely new way of listening to the world and connecting with the mesh of life with which we are networked. Listening with feet is simply one example of how a posthuman listening decentres human privilege by connecting with the ways in which other animals might hear, and how we might resonate with those more-than-human forces swirling through the land.

It is important to point out that a deep listening is not intended as an anti-technology, back-to-nature ideal. In fact, posthumanism embraces sociocultural technological transformations; not as something to be feared, but as something that leads to new possible definitions of what it means to be human. This misconception is rarely applied to the closely related transhumanism. Transhumanism investigates how humanity will transform as it becomes integrated with technology: becoming-cyborg, as such.[22] There are already transhuman listeners; for instance, cochlear implants, hearing aids, and access to audio-technologies all enhance the human capacity for hearing. These biotech developments are only going to increase, and with them, the advent of new human types. The difference between these two philosophical vantage points is that some approaches to transhumanism tend towards an anti-humanism by desiring the overturning of human fragility/weakness via our transmogrification into cyborgs; however, in so doing, it risks ignoring important ethical considerations, which a posthumanities would hope to investigate and (re)define. On the other hand, posthumanism is a desired transformation of the humanities that responds to these technological changes. It asks, what does it mean to be human when these technological changes so fundamentally change what we thought it meant to be human? For instance, a posthuman approach desires an understanding of First Nation insights into the becoming of technological futures whereas transhumanism, perhaps unwittingly, dismisses the past as frail in comparison to the promise of future technological transformations. The possible charge that transhumanism is a form of colonialism is a very small step to take from here, and for this reason, the subject should be approached with great care and not without the insights of posthuman theorists that compel us to position privilege in any discussion regarding the meaning of being human.

A final proposition

Sound-art and soundscape design practices are in an excellent position to join the emergence of a new type of practice-led posthumanist pedagogy. Can we create cities that speak to, and with, the posthuman listener? Technological options abound, including real-time field recording, sound system cultures, biophilic design, etc. A particularly powerful development is the emergent technologies of artificial intelligence (AI). With the use of AI, we can create soundscape generators that respond to the ever-morphing conditions of soundscapes. AI playback of sounds in response to real-time soundscape monitoring could lead to new possibilities in the creation of soundscapes at multiple scales. This might resolve issues of trying to understand what types of soundscapes we should design, as AI has the potential to cycle through multiple soundscape types carefully calibrating them to existing conditions. Additionally, AI could be trained to respond to local communities, who could use their own hand-held technologies to provide feedback about those sounds desired or undesired. The approaching potentials of these technologies seem limitless. Perhaps this continues the slow journey of reconnecting with the land – not by wishing noises away, but by reimagining their possibilities through interfaces that transform environments into something as yet unimagined. Indeed, this could present a new stage of Neuhaus' vision. Using his oeuvre as a precedent, noises could be transformed in spectacular ways. This is not to lose site of the importance of quietude, but to demonstrate that where noises are inevitable, so are the creative possibilities of their transformation.

It would, of course, be ludicrous and disrespectful to try and use these techniques to recreate indigenous deep listening to the land given that these emerge from culturally specific practices. Indeed, indigenous groups are doing their own important work in the creation of places of cultural significance in cities. For instance, Canadian Indigenous leader Lewis Cardinal worked with local government to locate a Kihciy askiy (meaning 'scared land') in a park in the heart of Alberta's capital, as a means to connect local urban indigenous people with their cultural history especially those 'seeking healing from addictions, abuse, or other trauma'.[23] Embracing indigenous knowledge should never ignore the struggles these communities face, having survived the shocking power abuses of colonialism. But part of a respectful healing means listening to and learning from these ancient cultures, and the wisdoms so generously shared in their scholarship and other forums[24]. Posthuman approaches to the more-than-human soundscape can try to create listening places that acknowledge and learn from Indigenous wisdoms by practicing an intention of evoking those more-than-human forces that circulate within the land; not as sociocultural construct, but as material realities expressed through nature-culture practices. This should be the ambition of a posthuman listening approach. How can we use a combination of technology, design and biology to create listening practices/places able to generate affective interconnections between land and listener? Such a practice-based discipline, part of the new posthumanities, would aim to produce and enact knowledge through experimentation, play and co-design (between land/technology/people), with a deep respect for land and indigenous cultures at the center of these practices.

I presented a related argument in Sonic Rupture. It applied affect theory to imagine ways to diversify listening experiences across cities; the final chapter Dreamings was a nod to Australian Aboriginal deep listening practices. I followed this book up with Urban Roar that postulates the existence of more-than-human forces in the land, called autonomous affectivities, with which artists can connect to imagine new cities into being. I continue the discussion here, by arguing that the theory and practice of soundscape must transmogrify in accordance with the decentring politics of a posthuman approach. This requires a leap of the imagination, that simultaneously embraces technology and the theories of posthumanism; imagination not as fancy, but as a type conceptual technology that is unafraid to explore new possibilities in city design. I conclude this essay with a proposition: a course of study for the posthumanities called posthuman listening to more-than-human soundscapes could be a new practice-driven discipline that encourages the decentring of human privilege by foregrounding our transversal interconnections with terra, the land or earth we share with other species. It reimagines the humanist dream of a silent soundscape controlled by a cultural elite; instead, city soundscapes imagine themselves into being through the land, technologies, peoples, and cultures through which they are expressed.

Notes

1 Morton, *The Ecological Thought*.
2 Ingold, *Bringing Things to Life: Creative Entanglements in a World of Materials*.
3 Deleuze, *Difference and Repetition*, 28–69.
4 Wright et al., "Hearing and vocalizations in the orange-fronted conure", 257–87.
5 Meek et al., "Camera traps can be heard and seen by animals", 2004.
6 Iliff et al., "The nematode C. elegans senses airborne sound", 3633–3646.
7 Iliff et al., "The nematode C. elegans senses airborne sound", 3633–3646.
8 Abbott, "Scientists bust myth that our bodies have more bacteria than human cells".
9 Braidotti, *The Posthuman*.

10 Ouzounian, "Urban sonic cartographies: The sonic counter-mapping of cities".
11 Thompson, *The Soundscape of Modernity: Architectural Acoustics and the Culture of Listening in America*, 123.
12 Vaillant, "Peddling noise: Contesting the civic soundscape of Chicago, 1890–1913", 260.
13 Lacey, "Sound installations for the production of atmosphere as a limited field of sounds", 315–324.
14 Schafer, *The Soundscape: Our Sonic Environment and the Tuning of the World*, 246–252.
15 Ferrando, "Humans have always been posthuman: A spiritual genealogy of posthumanism", 243–256.
16 Ungemerr, "To be listened to in her teaching: Dadirri: Inner deep listening and quiet still awareness", 14–15.
17 Couzens and Eira, "Meeting point: Parameters for the study of revival languages", 313–344.
18 Wilson, *Research Is Ceremony: Indigenous Research Methods*.
19 Yunkaporta, *Sand Talk: How Indigenous Thinking Can Save the World*, 155.
20 Ferrnando, "Humans have always been posthuman: A spiritual genealogy of posthumanism", 243–256.
21 Lacey, *Urban Roar: A Psychophysical Approach to the Design of Affective Environments*, 15–85.
22 Sorgner, *We Have Always Been Cyborgs: Digital Data, Gene Technologies, and an Ethics of Transhumanism*.
23 Rendell-Watson, "'Era of reconciliation': Building kihciy askiy in Edmonton", 2022.
24 See for example; Kimmerer, *Braiding Sweetgrass*; and, Wilson, *Research is Ceremony*.

Bibliography

Abbott, Alison. "Scientists bust myth that our bodies have more bacteria than human cells." *Nature*, 2016. https://rdcu.be/duLQx

Braidotti, Rosi. *The Posthuman*. Cambridge and Malden, MA: Polity Press, 2013.

Couzens, Vicki and Christina Eira. "Meeting point: Parameters for the study of revival languages." In *Endangered Languages: Beliefs and Ideologies in Language Documentation and Revitalization*, edited by Peter K. Austin and Julia Sallabank, 313–344. Oxford, U.K.

Deleuze, Gilles. *Difference and Repetition*. New York: Columbia University Press, 1994.

Drever, John and Andrew Hugill (editors). *Aural Diversity*. New York: Routledge, 2022.

Ferrando, Francesca. "Humans have always been posthuman: A spiritual genealogy of posthumanism." In *Critical Posthumanism and Planetary Futures*, edited by Debashish Banerji and Makarand R. Paranjape, 243–256. New Delhi: Springer Link, 2016.

Iliff Adam. J., Can Wang, Elizabeth A. Ronan, Alison E. Hake, Yuling Guo, Xia Li, Xinxing Zhang, Maohua Zheng, Jianfeng Liu, Karl Grosh, R. Keith Duncan, X.Z. Shawn Xu. "The nematode C. elegans senses airborne sound." *Neuron* 109, no. 22 (2021): 3633–3646.

Ingold, Tim. "Bringing things to life: Creative entanglements in a world of materials." *Horizontes antropológicos* 18, no. 37 (2010): 25–44.

Hogarth, William, The Enraged Musician, The Tate, London, image of engraving last accessed 9 August 2024 at https://www.tate.org.uk/art/artworks/hogarth-the-enraged-musician-t01800

Kimmerer, Robin Wall. *Braiding Sweetgrass: Indigenous Wisdom, Scientific Knowledge and the Teaching of Plants*. Canada: Milkweed Editions, 2013.

Lacey, Jordan. "Sonic placemaking: Three approaches and ten attributes for the creation of enduring urban sound art installations." *Organised Sound* 21 (2016): 147–159.

Lacey, Jordan. *Sonic Rupture: A Practice-led Approach to Urban Soundscape Design*. London and New York: Bloomsbury Academic, 2016.

Lacey, Jordan. "Sound installations for the production of atmosphere as a limited field of sounds." In *The Bloomsbury Handbook of Sonic Methodologies*, edited by Michael Bull and Marcel Cobussen, 315–324. London and New York: Bloomsbury Academic, 2020.

Lacey, Jordan. *Urban Roar: A Psychophysical Approach to the Design of Affective Environments*. London and New York: Bloomsbury Academic, 2022.

Meek, Paul D., Guy-Anthony Ballard, Peter J. S. Fleming, Michael Schaefer, Warwick Williams, Greg Falzon and Zhigang Jiang. "Camera traps can be heard and seen by animals." *PLoS One* 9, no. 10 (2014). https://doi.org/10.1371/journal.pone.0110832

Morton, Timothy. *The Ecological Thought*. Cambridge, MA and London: Harvard University Press, 2010.

Ouzounian, Gascia. "Urban sonic cartographies: The sonic counter-mapping of cities." *Presented at More-Than-Sound Symposium 2021*. Accessed, December 30, 2023, https://www.youtube.com/watch?v=YgGa9tTIDBY&t=32s

Rendell-Watson, Emily. "'Era of reconciliation': Building kihciy askiy in Edmonton." 2022. Accessed, December 30, 2023, https://parkpeople.ca/blog/era-of-reconciliation-building-kihciy-askiy-in-edmonton/

Robinson, Dylan. *Hungry Listening: Resonant Theory for Indigenous Sound Studies*. Minneapolis: University of Minnesota Press, 2020.

Schafer, Murray. *The Soundscape: Our Sonic Environment and the Tuning of the World*. Rochester, VT: Destiny Books, 1977.

Schafer, Murray. *Voices of Tyranny Temples of Silence*. Toronto: Arcana Editions, 1993.

Sorgner, Stefan Lorenz. *We Have Always Been Cyborgs: Digital Data, Gene Technologies, and an Ethics of Transhumanism*. Bristol: Bristol University Press, 2021.

Sterne, Johnathon. *Diminished Faculties: A Political Phenomenology of Impairment*. Durham, NC: Duke University Press, 2022.

Thompson, Emily. *The Soundscape of Modernity: Architectural Acoustics and the Culture of Listening in America, 1900–1933*. London: MIT Press, 2004.

Truax, Barry. *Acoustic Communication*. Westport, CT: Ablex Communication, 2001.

Ungunmerr, Miriam-Rose. "To be listened to in her teaching: Dadirri: Inner deep listening and quiet still awareness." *EarthSong Journal: Perspectives in Ecology, Spirituality and Education* 3, no. 4 (2017): 14–15.

Vaillant, Derek. "Peddling noise: Contesting the civic soundscape of Chicago, 1890–1913." *Journal of the Illinois State Historical Society (1998-)* 96, no. 3 (2003): 257–287.

Wilson, Shawn. *Research Is Ceremony: Indigenous Research Methods*. Nova Scotia: Fernwood Publishing, 2008.

Wright, Timothy F., Kathryn Cortopassi, Jack Bradbury and Robert Dooling. "Hearing and vocalizations in the orange-fronted conure (Aratinga canicularis)." *Journal of Comparative Psychology* 117, no. 1 (2003): 87–95.

Yunkaporta, Tyson. *Sand Talk: How Indigenous Thinking Can Save the World*. Melbourne: Text Publishing, 2019.

21
TOWARD A TOPOLOGY OF MUSIC

Ildar Khannanov

Introduction. (Musical) space as a problem

Space is extension, *écart* (French), Dichtung (German), an interval, widening, opening. All these terms presuppose real existing space. What is it? Is it something that is pertinent in any discussion? It perfectly suits the natural scientific language, but does it work in humanities?

Space is filled with things – *re*. *Res extensa* and *res cogitans*, according to Descartes, are two fundamentally different categories.[1] Spinoza objected by referring to the existence of God, but this binary opposition remained standing.[2] He simply added that God – an ultimate *res cogitans* – although not present in space as such permeates all real things.

There is the *outer space* – that which lies outside of our individual inner spaces and is given to our eyes in the form of an image (the core of Plato's theory of ideas). It is filled with things – tangible objects, and without these things in it, space would lose points of reference and would cease to be *extensive*. Extension, widening, and opening of the space is thus contingent upon things, *re*. The very existence in the form of extension, signified by the verb to be, depends on the presence of things. Only *res* makes existence *real*. Therefore, it is the real *space*.

It is difficult to prove that *res cogitans* can exist without any references to *res extensa*. Yet their connection is murky: only the circumstantial evidence – albeit sufficiently true – allows us to connect the two.

Martin Heidegger, among very few, dared to inquire into this dark hole of thought, into the links and service mechanisms of language that users normally do not even notice. Heidegger asked the questions that stop the natural flow of thought, such as: What is a thing? What is the jug?[3] What is a pair of peasant shoes?[4]

If we start pondering the thingness of the thing, it will bring us to the verge of the visible and intelligible Universe. The thought does not seem to have resources to go beyond that question despite the calls of phenomenologists 'to go back to things'.

A scientific approach, i.e., the strategy of conquering and harnessing Nature, relies upon the unshakable arguments of truth. These are based upon tangibility (even if it is about the dark matter or black holes), repeatability of the experiment, and the falsification principle (according to Karl Popper[5]). Scientists established that the thing exists as truth not because it is such but because it is impossible to falsify. There is a caveat though: one imperative condition for such truth is the complete exclusion of the thinking subject.

On the other side of the Universe, isolated by a kind of Great Void,[6] there is what is called human condition. This domain, addressed by humanities, implies the imperative of the presence of the subject. The work of art demands the author; even in folk art, the subject is present in the body of a song or a poem. Its essence is not so much that which dwells in the artwork as it is the soul – the ψυχή of the one who creates it. If for a scientist it is an annoying circumstance, for the art theorist it is the core element.

There are things – real things; and there are the things only in metaphorical sense. That is, they cannot be seen, but there is a small hidden passage in our mind that, by the power of bricolage,[7] or by the virtue of the symbolic, of two heterogeneous things thrown together (literally, συμ-βάλλουν[8]), allows us to fetch something after the event, post festum, and enables us to elaborate on it using other real things, their images, and real names as tools. One of the things of such nature is music. From any angle, music is not a thing; we call it a thing habitually and for the mundane purposes, just as we name Heideggerian jugs and shoes. Even to memorise music we need to translate it into the names that are related to real objects. Notker Balbulus did not discover the essence of music; he figured out how to memorise the tunes, to fend off his musical aphasia.[9] It would not be an exaggeration to claim that musical aphasia is not an anomaly; it is a norm. Notker came up with a gimmick. Later, Guido of Arezzo stumbled upon another gimmick – the music notation on the staff. This was a major step toward the progress in music understanding (and a devil's advocate may add – toward the demise of Western music, together with later harmful inventions, such as equal temperament, sic. Philip Gosset or Ross Duffin[10]). In contrast with that, in previous neumatic notation, the musical lines were not harnessed to reality with such violence as in this new line notation. Neumes alluded to musical essence by following the traces – the ash paths – of the melody that has sounded and already disappeared. In stark contrast with that, the line notation placed music on the Euclidian two-dimensional surface, pinned it to the plane of reality by breaking it into segments, in which the movement was cancelled (arrested) by points. The composers and theorists of ars nova added to that a two-dimensional binary blueprint of rhythm. Music has been thus enslaved by Logos; it was brought down from heavens to the level of real things. Something similar took place in the architecture. Sigfried Giedeion in his *Time, Space and Architecture* writes about Florentine model, with a very similar strategy of harnessing space and time:

> At the start of the 15th century in Florence, a new [conception of space] was translated into artistic terms through the discovery of perspective. Throughout the following five centuries perspective has become a canon to which all artists had to conform. In linear perspective—etymologically "clear seeing"—objects are depicted upon a plane surface in conformity with the way they are seen, without reference to their shapes or absolute relations.[11]

This account of the logocentric revolution in visual arts and architecture of the Renaissance fits quite literally the description of the similar outbreak in music. Indeed, all the finesse of the individual space – *topos* – was sacrificed for the unity of the Euclidian (Cartesian) plane and the individual shapes yielded their status to the universal regulations. Since that time – and, indeed, for five centuries to follow – even the ultimate, extreme cases of musical expression, such as ekmelic, lament, sobbing, screaming – had to be notated, like the opening of Igor Stravinsky's *Les Noces*, in which the desperate cry of the bride whose hair is violently combed, torn and weaved into a braid is meticulously measured, calculated, and pinned to the diatonic pitches (points) with the grace notes.

Humanities, including music theory, is pragmatic. Perhaps, too pragmatic to matter for scientists. There were the music theory teachers in the second half of the 19th century who could instruct the students in the harmonisation of the unfigured melody and the aural identities of chord progressions on an unprecedented level. Yet, the conclusions of many contemporary scholars regarding their concepts and categories are mostly negative: for example, Hugo Riemann, according to Daniel Harrison, did not provide a scholarly definition of tonal-harmonic function.[12] The Italian teachers of partimento were even less preoccupied with formal notions. Giorgio Sanguinetti insists that partimento techniques were non-verbal. Noteworthy is the reception of the theory of Jean-Philippe Rameau by his contemporaries. The main accusation went along the lines of the inapplicability of natural scientific ideas to musical art. The most eloquent critique of this sort can be found in François-Joseph Fétis's *La musique mise à la portée de tout le monde*.[13]

Thus, tonal music appears rather imperfect to the eye of a classical scientist. New music – that which came to existence after World War II and managed to cancel some of these imperfections – has been approved and fits well into logocentric worldview. Yet only the tonal paradigm carries a distinct alternative to real space conquered by physics and mathematics. Tonal music offers its own teleology, directionality, and continuity.

It is evident that the paths of art and sciences diverge. This difference is fundamental and needs clarification. Edmund Husserl, in his lectures of the 1930s, marked this difference as the 'crisis of European sciences'.[14] While pushing the project of phenomenology, he has done great service to humanities as a whole by carving out a space for the languages of art theory proper. One of the most important points of his disagreement with natural sciences – what he called classical metaphysics – relates to the perception of space and time. After Husserl's Copernican Turn, the true language of humanities has received its sovereignty and independence from the metalanguage of the sciences.

So, there is a living space and a time of living. The one who lives does not fit well into abstract space and time – the environment that is eternal and unchanging (essences do not change and do not multiply). Living space and time of living are inscribed into the horizon of human life[15] (in terms of Husserl). This dichotomy is seen in the history of 20th-century architecture. Le Corbusier suggested the idea of dwelling as the machine à vivre of illusive irregular shapes, while Mies van der Rohe confronted it with a lined-up and rectified segmented three-dimensional Cartesian (Euclidian) real space of glass-and-concrete. Architecture ever since has been divided: machine à vivre is left for the residential realm, while the Cartesian mechanism of symmetry and Florentine lined-up perspectival constructions are saved for commercial designs. Indeed, humans live in the houses (including churches – God's machine à vivre!), while the work, obligation, fight for survival (lute pour vie, according to Emmanuel Levinas) are allocated for the few hours spent in the concrete towers and glass pyramids of modern downtowns. In general, architecture constantly struggles with space. Sigfried Giedion calls for the 'organization of the outer space' with the purpose to create 'architecture as an organism'. He suggests that with efforts and time 'we encounter the examples of [our] ability to bring separate and often already existent elements into splendid, coherent, and surprising unity'.[16]

These two worldviews, each locked airtight in its circle of beliefs and perceptions, consider what is generic in regard to space, differently. For a scientist, the generic form of all-round objects is an ideal circle. In contrast with that, human space does not offer ideal shapes and colours. The strategy of straightening the lines and bringing the naturally available shapes to a certain kind (*eidos*) invites an exercise in psychoanalysis, or what Foucault termed as archaeology of ideas. The ideal shape is the exigency of the system of views, established by Plato. The operation on naturally available shapes in order to bring them to the order of the ideal circle is not necessitated by the study of natural phenomena: it simply fits into the rules of the game, offered by Plato. Many abstract theoretical concepts that followed Plato's presupposed exactly that: a kind of orthopedy of ideas with the goal to comply with the rules of a single concept. In Plato's case, this procedure is also contingent upon translation of the data from the non-visual domains into the visual. Needless to say, it can be very forced and intrusive. Therefore, following Gilles Deleuze, we can assume that for the human space the oval (any round shape) is generic, while the ideal circle is a peculiarity. Such perception unites the views of Husserl, Deleuze, and Derrida. Husserl subjected traditional geometric postulates to phenomenological reduction. Deleuze, after reading Husserl, came up with a new subfield – vagabond geometry – as the reflection of his concept of the nomadic.[17] The difference is eloquent: while a geometer arms himself or herself with the ruler and the compass in order to create a precise Cartesian two-dimensional plane, controlled by the axes of abscissa and ordinate, the nomadic geometer gallops on the uneven plane with random placement of hills and ditches in seemingly random direction and order. One may imagine an animal (the beast – Tier, in Husserlian terms) or an insect, creating loops and twisted lines on the surface of the steppes. These lines are not controlled from a single position of a viewer outside the plane. Sedentary culture, without a doubt, will label it as esoteric. However, behind and beneath all sedentary achievements lies a surface of the steppes with nomads galloping at the fast speed (or, without having points of reference, one can say: standing still) in different directions at the same time (again, without the points of arrival – in no directions, or in all of them simultaneously!). For example, the music of Frederic Chopin is both sedentary and nomadic, in this sense.[18] The moment some fixed real structure is established, as a State Apparatus, it is crushed by the emotional outburst of the Machine of War (using Deleuzian metaphors).

Mathematical topology applied to music

The conclusion to the discussion above can be formulated as follows: musical space is a misnomer. Space defined by things (real space) is appropriate for classical scientific research; it presents an obstacle to the study of music. There is a need for a different term and different category. Fortunately, there is no necessity to reject mathematics, geometry, and physics: scientific discourse in its latest post-classical form offers such an alternative. If the subject exceeds the ordinary parameters that are tangible and countable in positive whole numbers, mathematicians move on to a higher level, that of topology. In such situation, they question the very calculability of number – its ability to calculate and measure such complex object – and introduce the methods of topology that proved to be able to solve this issue. It is much less problematic to borrow musical ideas from this highly sophisticated subfield of mathematics. Therefore, here and further in this chapter, the major distinction will be made: there is no musical space; there is musical topology.

There had been a prolonged affair of music theorists with the mathematical category of set. However, the way set is discussed and used in mathematics differs significantly from its musical appropriation. The set has attracted the attention of mathematicians after the publication of the *Theory of Transfinite Numbers* by Georg Cantor in 1915. He avoided the intrinsic limitations of the finite sets – the unordered state of their elements – by operating exclusively with the transfinite sets – those described by the infinite cardinal number aleph. At that level, there is no possibility to count the elements of the transfinite sets in positive whole numbers, to measure their sizes and compare them. The very possibility of mathematics that does not operate with digits and does not measure shapes in digitised segments has been a significant step forward – toward mathematical topology. The comparison of infinite or transfinite sets, suggested by Cantor, offers a complete overhaul of the idea of calculus.

This mathematical apparatus – a topological alternative to quantitative methods – has been developing for several centuries. One may begin with Leonard Euler's *Solutio problematis ad geometriam situs pertinentis* (1736) (that includes the problem of 'Seven bridges of Königsberg'), or with Nikolai Lobachevsky's 'On the Origin of Geometry' (1829). Bernhard Riemann continued the work on the non-Euclidian geometry with his ideas of manifolds and tensors. Carl Weierstrass's works on irrational numbers and Georg Cantor's development of uncountable sets completed the doctrine. The final block to the edifice was added by the group of French mathematicians under the alias Nicolas Boubaki. The doubts and concerns of music theorists about the non-quantifiable elements of musical structure can be easily dispelled by these solid contributions to contemporary mathematics by the representatives of analysis situs.

Thus, Bourbaki explains the difference between, on the one hand, the algebraic interpretation of limit, continuity, and neighbourhood through the experimental determination or measurements in whole numbers, and, on the other hand, the topological method, which excludes the necessity to operate with the whole numbers and measurements. The algebraic definition says: 'A subset will be a neighborhood of a if it contains all elements whose distance from a is less than some preassigned strictly positive number'. Bourbaki adds to this: 'Of course, we cannot expect to develop an interesting theory from this definition unless we impose certain conditions or axioms on the "distance"'.[19] (Notice, how the Heideggerian question of 'distance' reemerges in this new mathematical context.)

> The general concept of topology … does not depend on any preliminary theory of rational numbers. We shall say that a set E carries a topological structure whenever we have associated with each element E, by some means or other, a family of subsets E which is called neighborhoods of this element—provided of course that these neighborhoods satisfy certain conditions (the axioms of topological structures).[20]

In other words, contemporary mathematics is able to discern the phenomena which exist beyond their measurement in rational numbers. This fact justifies some anti-Pythagorean aspects of the theories of Aristoxenus of Tarentum and Ernst Kurth by suggesting that the musical interval can be understood through the discussion of intensities and continuities. It also allows music theory to deal with the idea of musical space in non-quantifiable topological terms.

The examples of non-quantifiable spatial and temporal elements in music are so numerous that it is impossible to ignore them. The most obvious is the idea of musical pitch fixed in space. The problem of tuning has not been resolved by the introduction of logarithmic division of an octave: equal temperament 'ruined harmony'.[21] Music – and especially, the music of the continents beside and beyond Europe – is filled with steps that are not fixed in pitch, in principle, e.g., Turkish 'spinning' tones. A major logocentric error of Western music theory has been made in an attempt to calculate and measure the position of each tone of the mode. Turkish music allows for 53 increments within an octave and the maquam is non-octavic. In performance practice, the outer notes of the tetrachord are fixed; the notes that fill in this topology are spinning and gliding. They are mixed with the elements of vibrato and small embellishments. Such was the situation in ancient Greek music as well. In Aristoxenus' terms, diatessaron was the combination of tones that can 'sound together' (σύμφωνα), while the intervals smaller than diatessaron did not sound well together and could only follow one another, passing through the musical topology (διάφωνα). In performance practice and in cognition, even today, in contrast, and disregard of prescription of notation, tones, and intervals are perceived this way. As Juan Roederer describes it:

> Time distribution of neural pulses is not utilized in the perception of the pitch of a pure tone[22] [] When we slightly increase frequency F, we continue hearing one single tone, but of slightly higher pitch, corresponding to average frequency. [] When the frequency difference surpasses the so-called limit of frequency discrimination, we suddenly distinguish two separate tones.[23]

Measuring and counting the position of all tones of the mode 'makes their positions clear'. It also arrests the movement that these imperfect tones and, in Zarlino's terms imperfect intervals realise. Counting, in general, stops movement. There is a curious detail in Greek etymology: rhythm is derived from the verb *rheo*, flow (ῥέω – ῥυθμός). By allusion and assonance, it is related to the word 'number', ἀριθμός, with alpha privative suggesting its annihilation. It is possible to assume that in Greek terms the number stops the flow.

From the musical note, the next step taken in the obviously wrong direction was a line perceived as a simple geometric object ('by analogy' with graphic lines). Musical topology does not project the exact positions of the notes on the screen (of the score, or the screen of perception, in terms of platonic theory of knowledge). Instead of the 'clear picture', topology of musical linearity is perceived as vague (vagabond) sensations of continuity, discontinuity, and directionality. Two topological spaces, neighbourhoods of two tones, may be linked together by means of 'continuous uniform function'. It can be absent despite the fact that the two tones appear to be adjacent in real space.[24] The traditional definition of a straight line, as the closest distance between the two geometrical points, is principally incompatible with topology. Instead of depending on the row of the rational numbers, topology offers a category of the 'rational line', which is a 'fundamental group of neighborhoods'.[25] It is the 'indiscrete space'.[26] Such a line consists of 'a set Q and the additive group topology'.[27]

An example that may clarify this difference is clearly seen in comparison of topographic and topological maps. The former presents a strict point-to-point projection of the real space onto the space of the map. The topological map – a kind of treasure map – suggests only the rough comparisons of shapes and distinctions of directionality. What is of value for the study of music and musical space here is the possibility of developing a rigorous mathematical language of description for the study of non-extensive uncountable and non-measurable objects. Analysis situs, as Euler labelled it, allows for the rational description of *res cogitans* in general, and musical events, in particular.

The real space of atonal music and the human topology of tonal music

In the decade after World War II, sciences and technology breached the traditional barrier that isolated them from humanities and social sciences. There were objective reasons for that. One was the development of military technology and logistics (discussed, among others, by Paul Virilio in his Vitesse et politique (1977)). This dramatic paradigm shift brought in a new model of music-making. One of the leaders of post-WWII avant-garde, Edgard Varèse suggested that 'scientists are the poets of today. "Art" means keeping up with the speed of light'.[28] The conceptual approach replaced the old pragmatic one. With that music was introduced into the general scientific discourse, including the concepts of time and space promoted by physics and geometry. Karlheinz Stockhausen wrote music that utilised the idea of space directly. For example, *Kreuzspiel* is a composition for the grand piano and a set of instruments. The piano is not intended for playing; it is used as the space, into which all instruments direct their sounds. Le carré employs four orchestras that create spatial fluctuations of sound, comparable to the fluctuations of the noises of four engines of an airplane, according to Stockhausen. This is the use of real space. Highly theoretical, conceptual explanations follow all of Stockhausen's compositions; his music cannot exist on its own without such commentary. Similar in intention are the compositions by Tōru Takemitsu and Iannis Xenakis[29] for the exhibitions in Japan and in Netherlands. The electronic composition harnessed with the avant-garde techniques and technologies – the very term avant-garde belongs to military vocabulary – was intended to conquer the real space by the force of music.

In the second decade of the second millennium, tonal music seems to survive the age of atomic bombs and the space of tonal music remains actual as it was in the 19th century. In contrast with the real space of musical avant-garde, the space of tonal music is invisible. One way to interpret it is to label it as transcendental, referring to the category of theology of Thomas Aquinas and Philip the Chancellor. It may as well be interpreted as immanent. One way or another, it is not real. It is not contingent upon the placement of speakers in the room or on orchestral seating diagrams. Its aural markers do not translate directly into visual domain. It exists in the form of tension and relaxation that result from the fluctuation of the affects, the inner movements of the psyche, ranging from asthenic to sthenic. That musical tone and musical interval are the products of tension was known to Greeks – to Aristoxenus of Tarentum in his Harmonikon stoikheion, in particular. It has been developed further by Ernst Kurth in *Romantische Harmonik und ihre Krise in Wagners 'Tristan'* (1920). A good analogy is the experience of pain. This commonly known and ubiquitous phenomenon is measured in size. There is small pain, and large pain. Natural language avoids these direct translations by using, instead, mild and acute. Musical experiences seem to belong to the same category. In most of the traditions beyond the Western-European canon, musical sounds are distinguished not as high and low, but as, say, ὀξύς καὶ βαρύς (Greek) [literally, 'shrieking and wet'], or ashe and kalun (Tatar) [literally, 'acidic and thick'], etc. Remarkably, these oppositions are non-binary: their two terms are non-correlative. On this background, the Western idea of measuring tones in height appears to be rather peculiar (just as peculiar is an ideal circle in the universe of ovals). It takes the work of a detective, or, in Foucault terms, a geologist, to unearth the real reasons that led Europeans to establish the musical ladder – scala – with the steps – gradus, as the locations of tones in space, from low to high. Not without the general influence of the Christian doctrine, musical space has become oriented vertically. Compare, for that matter, the shape of the pre-Christian Parthenon with that of Strasbourg Cathedral. The overall shape of the Parthenon is a flat rectangle, in which the height is not emphasised. The Strasbourg Cathedral's proportions offer the overwhelming dominance of its height. On the drawings of the time of completion of its construction, the city of Strasbourg is dwarfed by this skyscraper of the 15th century. Greeks did not treat preferentially the higher grounds; even Mount Olympus was not an unreachable mountain peak. Rather, it was a hill that was slightly elevated, just like the Acropolis. The lives of Olympian inhabitants were rather normal – oriented horizontally. It was not before the appearance of the metaphor of Jacob's Ladder and, say, the *Celestial Hierarchy*[30] of Pseudo-Dionysius the Areopagite the bottom-up orientation was established as a norm. In this sense, the space of a tonal interval is not a real space, and it is not characterised by trivial parameters of width or height. Yet, anybody who understands and appreciates music would confirm that it moves – hence it exists in a space. Perhaps, such space is, indeed, transcendental (non-visual and non-tangible).

An easy – and unfortunately completely misleading – way of explaining musical interval is based upon the Pythagorean (Ptolemeian) idea of interval as a distance measured in digits. Thus, a perfect fifth must be an interval that is wider than perfect fourth, and major seventh is, according

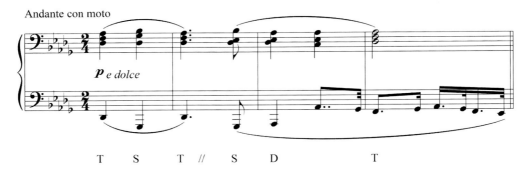

Figure 21.1 Beethoven, Piano Sonata op.57, Appassionata, Slow movement.

to this logic, is the largest interval within an octave. The actual behaviour of musical space may completely disorient such calculations. The tones a perfect fifth above the one and perfect fifth below it, form the closest circle – the home base, the bird's nest, the triangle that ensures the stability of the universe, as shown in Example 1 (Figure 21.1).

The dominant tone creates the opening – écart, Dichtung – and it fills it with the energy of τόνος. The drama of harmony that unfolds (silently for an untrained ear and unbearably loudly for the trained one) in this space is happening between tonic and dominant. Music travels from tonic to dominant – within most of Classical forms the arrival at the large dominant is marked with the double bar line; then music comes back to tonic. There is another pole – and nobody has been able to locate it in tonal space with any degree of precision yet. It is the subdominant – a tone perfect fifth down from tonic, or perfect fourth up from it (indeed, tonal space is akin to Moebius strip – it is circular, centralised, and glued at fringes, connecting sharps and flats, ascending direction with descending, hot and cold, dry and wet, sharp and soft). The subdominant is that grain of sand that sets off the mechanism, puts the system off balance, and gives an impetus for its forward movement. Otherwise, tonal space would simply oscillate between tonic and dominant. The subdominant generates a point in an asymmetric triangle and adds a temporal dimension to the otherwise hard static structure of dominant to tonic relationship. The triangle in Example 1 is asymmetric: it allows to triangulate Tonic, to find the place of the topos-oikos by vague comparisons of directionalities, similar to following the arrows on a treasure map.

Thus, the perfect fifth and the perfect fourth are the smallest intervals in musical topologic space (they are not such in real topographic space). The thirds and sixths are not related to the fifth linearly. They seem to belong to a different plane. The thirds and sixths are located between the primary notes of tonality (tonic, dominant and subdominant). The major and minor seconds have two substantial definitions in musical topology: they are intervals and steps. The interval that exists in the outer space – according to Boleslav Yavorsky – is the tritone.[31]

In harsh competition between Pythagoras and Heraclitus, the former took a logocentric direction – that which has led to the problem of compatibility of science and art that is discussed in this chapter. The latter, rather paradoxically, took an entirely different path. Heraclitus established what can be called philosophy minor (by analogy with literature minor). In a nutshell, harmony is an explosive, tragic development that is based on economy of tension (ὁ τόνος). Harmony is not a correlation of eternal numbers but a continuous war, conflict, or tumult. It is irreplicable on the blueprint of theory of knowledge. It is excessive and defiant. The relationship of lyre with bow-and-arrow describes harmony, which is also hiding from sight – a serious blow to Platonic theory of ideas. Nothing in harmony is affixed to preexistent timeline and lines of space.

Rather, harmony is a transcendental invisible process; only traces of ash[32] of its explosive existence are left in real time and space. These lines are the routes of escape from a given space, the paths of transfiguration of the subject. Musical time is segmented not by the points on the line but by the events of *ekpurosis*. Aeons are not measurable in time units. In addition, harmony likes to hide – just like nature ('Η φύσις κρύπτεσθαι φιλεῖ).

Such hiding brings in another idea: tonal musical space, the space of harmony, is not equal and even everywhere and does not dwell in the points in preexisting infinite space. Rather, there is a place, location, τόπος, unrelated to other places. If harmony is the economy of tension, then the economy should be viewed not as global monetary exchange but, as the original Greek meaning suggest, an ὄικος – a bird's nest or the hearth. (An animal keeps lurking on the background of this discussion, animal as such, represented by body, and not by the concept). A place, home, borrow, refuge: not existence in general, but existence here and now (alluding, perhaps, to Dasein of Heidegger). Musical events are localised and hidden; a circle around the fireplace constitutes and guards the subject – the main figure of humanities. Human space – the place that is enveloped in silence and darkness, while the fire from the fireplace lights up a sphere, in which the subject is cocooned and saved from the elements. In other words, a song – a ritournel – that keeps a child safe and confident in a dangerous walk at night. This place is known to all musicians – it is tonic (from both Greek tonos and Latin tenere – to hold, to keep, to maintain with force).

All this explains the modest sizes and dimensions of tonal music in comparison with the atonal experiments. Instead of the full audible spectrum from 80 speakers (as in Takemitsu's opus mentioned earlier), tonal music commonly presents a few notes of a tune in the middle of the vocal range. The most impressive and remarkable in their movement melodies contain, surprisingly, just a few notes. They generate an extreme sensation of motion, often exceeding anything that can be represented by real space. Bach's 'Erbarme Dich' and Tchaikovsky's theme of the finale of Symphonie Pathétique, as well as his Pas de deux from the Nutcracker come to mind. All these sweeping melodies are limited to an octave of tones.

The invention of the chord was a quantum leap from an interval. In general, the hierarchy of tonal musical structures was built in the process of irreversible evolution. It spans from the tone, to interval, to chord and to function; it is of ascending, emergence type. With every higher level, the system acquires an entirely new quality (quale novum), irreducible to that of the lower level. Chord function is the last achievement in this evolution; it is the most advanced component of musical speech. When all these elements come together, they generate a function as quale novum. It is not culturally predetermined; it is an evolutionary phenomenon.

Jean-Philippe Rameau suggested that melody is a chant of one voice: 'they say that music is melodious when each voice reflects the beauty of harmony'.[33] Thus, if melody is folded onto itself and closed for communication with the outside worlds, harmony allows to unfold it and expose its inner space and motion.

Something else is happening in harmonic progression. It often appears as the jolt or a sudden change of spatial mode of existence of melody. There comes a time when harmony, instead of gathering everything around tonic (one of the meanings of Greek word λόγος), gets on a path of destruction. The sides of the triangle that were working so well together are being detached from each other, the triangle becomes fragmented, the listener – disoriented in inner tonal space. It is labelled by the perceiving subject as opposition of keys, digressions, or modulation. The dynamic, process-like existence of tonal music has been reflected upon by many theorists. Composers of tonal music perceived musical form as modulation.

Such is the topology of musical space. It is twisted, crooked, uneven, and unstable, yet there is always a possibility of a local place – topos – for tonicity of tonic that generates the sense of

Figure 21.2 Chopin, Mazurka op. 68 no. 4, F minor.

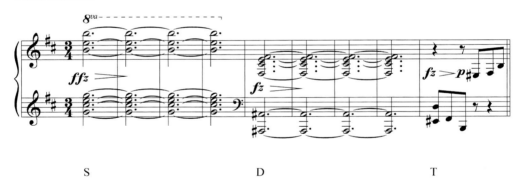

Figure 21.3 Chopin, Scherzo op. 38, B minor.

tonality. Tonic is the point; tonality is its neighbourhood. Remarkably poignant question of Martin Heidegger – that about nearness and distantlessness – applies directly to tonal topology. What we normally perceive as near, in music appears the most distant, and vice versa. The following example of what music theorists habitually call 'linear chromatic progression' reveals exactly that discrepancy between topography (what we see in the score as 'a smooth line' and topology of musical events that create a rugged line or not line whatsoever (Figure 21.2).

The opposite – visibly large gestures with actually small musical movement – is present in the introduction to Chopin's Scherzo No. 1, op. 38 (Figure 21.3).

Here, the hands fly from one chord to another over the three octaves. Not only that it creates an illusion of great journey – moving, from ii4/3 to V6/5 breaks all the rules of smooth voice leading (the seventh remains unresolved, dissonances – taken by leaps). It should be ii4/3 to V7, by the book. And yet, the fact that the first chord is Subdominant, the second – the Dominant, and the last one – Tonic offers the ultimately smooth connection within the peaceful and reasonable continuous space (continuous function).

Conclusion

The history of European thought of the new times (modernity, from Enlightenment to the middle of the 20th century) manifests the overwhelming achievements of the scientific worldview. In this celebration of the victory of humans over the forces of nature, the attempts of those who are not included in the scientific pantheon are easily brushed away as esoteric, impertinent, and impractical. In this continuum of objective truth, there is almost no room for humanities. More precisely, the objects produced by artists are commonly visited by scientists with easy and ready explanations, using the meta-scientific language of description. In the 20th century, after World War II, in the wake of the technological revolution, composers started leaning toward scientists and abandoned the specific orientation, directionality, and the flux of tension, offered by tonal music.

Yet, although some of the elements of tonal music submit to segmentation and orthopedy of the visual translation, its essential aspects elude the well-established methodology. Melody, rhythm, harmony, form, and genre exist in their own domain that is not directly linked to the real space. Any direct application of calculation and measurement proves to be too crude and ineffective in grasping the spirit of music.

Moreover: sciences experienced internal crisis in the 20th century at the time when the objects of study had become infinitely complex. After reaching the threshold of calculability of number, mathematicians retreated to the more general discussion of the analysis situs. Here, rather paradoxically, their paths crossed the paths of musicians, artists, poets, and architects. This new approach allowed music theory to retain its original terminology and to deal with the musical space in non-quantifiable topological terms.

Notes

1 Réne Descartes, *Principia Philisophiae* (Online resource: https://www.gutenberg.org/cache/epub/4391/pg4391.html).
2 Benedict Spinoza, *Ethics*, Latin text 2006; English translation 2017 (Online resource: https://www.gutenberg.org/files/3800/3800-h/3800-h.htm), 23.
3 Martin Heidegger, *Poetry, Language, Thought* (New York: Harper and Row, 1971), 163–164.
4 Martin Heidegger, *The Origin of the Work of Art* (New York: Harper & Row Publishers), 1971.
5 This famous Popper's Falsification Principle was introduced in his book Logik der Forschung in 1935.
6 Robert P. Kirshner, Oemler, Augustus, Jr., Schechter, Paul L., Shectman, Stephen A., "A Survey of the Bootes Void." *Astrophysical Journal* (Part 1. 314, March 1, 1987), 493–506.
7 Literally, "to play by bounce." Used in billiard or tennis. Levi-Strauss used it to describe the improvisatory approach to creation of myths and artworks. Claude Levi-Strauss, *The Savage Mind* (Chicago, IL: University of Chicago Press, 1966).
8 Greek verb συμ-βάλλω means "to throw together, to collide." Hence, the symbol: συμβολή—confluence, coincidence, or merging.
9 As it is seen from his last name, Notker suffered from stutter (hence his name in English is "Notker the Stammerer"). His contribution to early notation was the Liber Hymnorum (circa 890) that contains short poems with acronyms for memorization of the parts of chant.
10 See, Ross Duffin, *How Equal Temperament Ruined Harmony* (New York: W. W. Norton Company 2006).
11 Siegfried Giedeion, *Time, Space and Architecture* (Cambridge: Harvard University Press, 1959), 30.
12 Daniel Harrison, *Harmonic Function in Chromatic Music* (Chicago, IL: University of Chicago Press), 36–37.

13 François-Joseph Fétis, *La musique mise à la portée de tout le monde* (Paris: Alexander Mesnier, 1830), 95.
14 Edmund Husserl, *Die Krisis der europäischen Wissenschaften und die transzendentale Phänomenologie* (The Hague: Springer Verlag, 1993).
15 See, Edmund Husserl, *Cartesian Meditations* (The Hague: Martinus Nijhoff, 1960).
16 Siegfried Giedeion, *Time, Space and Architecture* (Cambridge: Harvard University Press, 1959), 24.
17 Gilles Deleuze, *Mille Plateau*, 454–455, quoted by Ildar Khannanov, "Line, Surface, Speed: Nomadic Features of Melody," *Sounding the Virtual. Gilles Deleuze and the Theory and Philosophy of Music* (London: Ashgate Publishers, 2010), 254.
18 See, Ildar Khannanov, "Line, Surface, Speed: Nomadic Features of Melody," *Sounding the Virtual. Gilles Deleuze and the Theory and Philosophy of Music* (London: Ashgate Publishers, 2010).
19 Ildar Khannanov, "Line, Surface, Speed: Nomadic Features of Melody," *Sounding the Virtual. Gilles Deleuze and the Theory and Philosophy of Music* (London: Ashgate Publishers, 2010).
20 Op. cit., 12.
21 Many have voiced disagreement with equal temperament (for example, already mentioned Ross Duf and Philipp Gossett); even more encounter difficulties in singing and using unpitched instruments in ensembles.
22 Juan Roederer, *Introduction to the Physics and Psychophysics of Music* (New York: Springer-Verlag, 1975), 50.
23 Op. cit., 28.
24 Here and further the term "real space" is used as the category of classical sciences, based upon the cartesian definition of the two-dimensional plane or three-dimensional place for real things, *res extensa*. It is the space of classical physics, mechanics, and geometry. It is distinct from the topology.
25 Nicolas Bourbaki, *General Topology* (Don Mills: Addison-Wesley Publishing Company, 1966), 330.
26 Op. cit., 331.
27 Op. cit., 330.
28 Edgar Varese quoted in: Karlheinz Stockhausen, *Stockhausen on Music. Lectures and Interviews*, Compiled by Robin Maconie (London and New York: Marion Boyars, 1989), 169.
29 Subtitled "Program of Steel Pavilion at Expo 70." "Hibiki Hana Ma" is an electronic-music composition for 12 channels and 800 speakers composed especially for the Japan Iron and Steel Federation's Pavilion at Expo '70, and is dedicated to Kuniharu Akiyama, Seiji Ozawa, Yuji Takahashi, and Toru Takemitsu.
30 Dionysius the Areopagite, *On Heavenly Hierarchy* (electronic resource: https://www.tertullian.org/fathers/areopagite_13_heavenly_hierarchy.htm).
31 See Boleslav Yavorsky, *The Design of Musical Speech* (Moscow: Kompozitor, 2022).
32 This is an allusion to Derrida's terms trace and ash. The event (in this case, the flux of harmonic syntax, the arsis-thesis of tension-resolution) leaves after itself only traces and ashes (cinders).
33 Jean-Philippe Rameau, *Traité de l'harmonie* (Paris: J. B. C. Ballard, 1722), xiii–xv.

Bibliography

Areopagite, Dionysius the. *On Heavenly Hierarchy*. Online resource: https://www.tertullian.org/fathers/areopagite_13_heavenly_hierarchy.htm.
Bourbaki, Nicolas. *General Topology*. Don Mills: Addison-Wesley Publishing Company, 1966.
Deleuze, Gilles. *Mille Plateau*, University of Minnesota Press, Minneapolis, 1987, 454–455.
Descartes, Réne. *Principia Philisophiae*. Online resource: https://www.gutenberg.org/cache/epub/4391/pg4391.html.
Duffin, Ross. *How Equal Temperament Ruined Harmony*. New York: W. W. Norton Company, 2006.
Euler, Leonard, *Solutio problematis ad geometriam situs pertinentis (The solution of a problem relating to the geometry of position)* 1736 University of the Pacific Scholarly Commons last accessed on 9 August 2024: https://scholarlycommons.pacific.edu/cgi/viewcontent.cgi?article=1052&context=euler-works
Fétis, François-Joseph. *La musique mise à la portée de tout le monde, expose succinct de tout ce qui est nécessaire pour juger de cet art, et pour en parler sans l'avoir étudié*. Paris: Alexander Mesnier, 1830.
Garbuzov, Nikolai. *Zonal Nature of Pitch Perception* [*Zonnaya Priroda Zvukovusotnogo Slukha*]. Moscow: Academy of Sciences, 1948.
Giedeon, Siegfried. *Space, Time and Architecture*. Cambridge: Harvard University Press, 1959. Online resource: https://archive.org/details/spacetimearchite00gied.
Daniel Harrison, *Harmonic Function in Chromatic Music* (Chicago, IL: University of Chicago Press), 1994, 36–37.

Heidegger, Martin. *The Origin of the Work of Art*. New York: Harper & Row Publishers, 1971.
Heidegger, Martin. *Poetry, Language, Thought*. New York: Harper and Row, 1971.
Husserl, Edmund. *Cartesian Meditations*. The Hague: Martinus Nijhoff, 1960.
Husserl, Edmund. *Die Krisis der europäischen Wissenschaften und die transzendentale Phänomenologie*. The Hague: Springer Verlag, 1993.
Khannanov, Ildar. "Extension and Directionality. A Sketch for Musical Topology." *Music and Space. Theoretical and Analytical Perspectives.* Ed. Ivana Ilić, Jelena Mihailović-Marković and Miloš Zatkalik. Belgrade: SANA, 2021, pp. 15–37.
Khannanov, Ildar. "Line, Surface, Speed: Nomadic Features of Melody." *Sounding the Virtual. Gilles Deleuze and the Theory and Philosophy of Music*. Ed. Brian Hulse and Nick Nesbitt. pp. 249–267.
Kirshner, Robert P., Oemler, Augustus, Jr., Schechter, Paul L., Shectman, Stephen A., "A Survey of the Bootes Void." *Astrophysical Journal* (Part 1. 314, March 1, 1987), 493–506.
Kurth, Ernst, *Romantische Harmonik und ihre Krise in Wagners "Tristan." 1920* [Leather Bound]. Generic, 2018.
Levi-Strauss, Claude. *The Savage Mind*. Chicago, IL: University of Chicago Press, 1966.
Lobačevskij, Nikolai Ivanovich. Berlin and Paris: Mayer et Müller, 1840. The main idea was first formulated in Russian language: Лобачевский Н. И. "О началах геометрии." *Казанский вестник*. — Казань: Императорский Казанский университет, 1829—1830. — № 25—29.
Rameau, Jean-Philippe. *Traité de l'harmonie réduite à ces principes naturels*. Paris: J. B. C. Ballard, 1722.
Roederer, Juan G. *Introduction to the Physics and Psychophysics of Music*. New York: Springer-Verlag, 1975.
Spinoza, Benedict. *Ethics*. Online resource: https://www.gutenberg.org/files/3800/3800-h/3800-h.htm.
Stockhausen, Karlheinz. *Stockhausen on Music. Lectures and Interviews*. Compiled by Robin Maconie. London and New York: Marion Boyars, 1989.
Virilio, Paul. Vitesse et politique. Paris, Galilée 1977.
Yavorsky, Boleslav. *The Design of Musical Speech*. Moscow: Kompozitor, 2022.

22

SOUND'S SPATIAL-MATERIAL CIRCUITRY

Raviv Ganchrow

Sound operations

Every sound already manifests meandering interlocking trajectories of spatial agency and contextual dynamics.[1] Curiously though, those itineraries are not readily apparent in the sounds. In order to observe sound's activities requires sounding its *qualities*, taking its manifestations as *sites* and observing its operations as *circuits*. Sound circuits provide access to semblance domains where the dynamics of entities-in-the-making can be observed.

Sound's spatial-material appearances reveal its *situated operations*. By focusing on the contextual circuitry of sound an entirely different dynamics opens up, far exceeding the motions constrained to its waves. To observe sound actions require converging with its flows, following diverse travelogues of context.

Operational sounds are circuitries of the extant.[2] Each instance presents its own unique case with quirks and dazzles that cannot, and should not, be generalised. On the other hand, there are tendencies of action and behaviours of dynamics that recur in sound circuits. Closer observation of the mechanisms and infrastructures of sonic eventfulness provides guidance for reading sound's particularities. This text attempts to open up some traits of sonic agency and in doing so provide tools towards more robust, detailed and inclusive accounts of sound's being. Unpacking sound's *operational spaces, contextual agency* and *terrestrial circuitry* are paramount to that approach. The list is far from exhaustive, but it aims to provide some foothold into sound's operational expanse.[3]

In the current epoch, there is a growing urgency of attending to interlinked, interdependent, porous, fluid-energetic, reciprocal conditions of terrestrial milieus. Observing spatial-material qualities of sound and tuning into its contextual infrastructures provide crucial interlinks.

Sound circuitry

When attempting to qualify a sound, one shouldn't ask what it is but rather what it *does*. Sound operates. Its operations involve contextual circuits. These circuits are integral to the processual extant. Sounds are movements with qualities that persist, seemingly escaping the motions bringing them about. This is not a contradiction but rather a central *mechanism* through which sonic spatial-material specificity comes about.

DOI: 10.4324/9781003347149-26

Every sound already arrives with distinctive spatial-material features. Its attributes are conditional of its appearance. Sound circulates agencies through qualitative specificity. Sonic qualities are expressive of circuit operations where its shifting qualities perform dynamic assemblages of context. Qualities operate. They reflect intricacies of the circuit and sense their operations. Identities, qualities, spatial characteristics observed in sounds are real, though their properties remain tightly bound with more expansive interlinked movements. Situated qualities are also calls to action, ensuing further operations in the expanse. Nothing escapes this ongoing dynamics. Anything and everything may participate in sound's circuitry from forces and signals to materials, artefacts, subjects, places and knowledge – just as long as they circulate along interlinking pathways that give rise to subsequent movements, and their configurations are productive of discernible effects.

Attention circuits foster spaces, materials and times through ever-changing currents, augmenting agencies of flux. Appearances that mediate also participate in the characteristics they dynamically bring about. Their forms perform. Materiality is a habitat for actions. Not all of those materialisations are accessible to the incidental observer, and sentience is not always the addressee. Yet some of the ongoing exchanges escape their primary spaces of action resurfacing in incidental resonances or in the general radiance of a locale.

The collective movements of events, energies, vibrations, sounds, impressions, ideas, and residues are heterogeneous territories of shifting, interweaving dynamics perpetually active in the extant. Observing sound's spaces of action and the related contextually circulating attentions provide glimpses into that dynamics.

Spaces of sound

There are multiple spaces in an auditory event: in the relations between subjects and objects, between vibrating air molecules, between sequences in an oscillation. All these spaces somehow manage to wrestle themselves loose of their initial agitations, yet each example seemingly resides within its own independent, mutually exclusive, relational space. However, there is no common 'vibrational space' from which sonic dimensionality emanates and to which sounds eventually return. Sounds materialise and propagate through a broad range of spaces – besides acoustic territories – such as audio mechanically transduced onto vinyl or travelling over copper wires and fibre optic cables. Some sounds even propagate silently, through written text, without ever budging from the space of a screen or page. Written sounds also sound, though in manners that cannot be monitored on audio technologies.

Commonplace definitions of sound as a series of longitudinal pressure waves propagating through compressible media (e.g. air, liquids, solids), no doubt present a space where vibrations can reside, though the rise to prominence of *that* specific sound-space has its historicity. Calculable wave-space is tied to universalised presumptions of matter, scaled up through technical and tactical industrial production that should be approached today with a measure of caution. Otherwise, how are we to understand the multitude of other interlinked movements, materially and anecdotally involved in adjoining spaces of sound, such as redistributed human-made minerals in earth's crust directly tied with the universalised wave-space modelling of sound? Minerals such as piezoelectric crystals, ferrite, and rare-earth neodymium magnets, are central for production of vibration-coupling transducers such as sensors, loudspeakers and microphones that today more than likely outnumber the sum total of human mouths and ears on earth. The cumulative force of abstract universalised wave-space may in fact be empirically measureable through a stratigraphy of anthropogenic transduction deposits across worldwide landfills.

A key to sound's ontological pluralism is that it always *specifically* appears (haecceity) and furthermore that such specificity is *always* nested in implicit material, spatial and possibly temporal confines. Diversities of sound's manners of appearance also hint at its myriad ways of being. Variations in perceived extensity of sound reveal differing spatial ontologies of fluctuation. Sounds open into polyphonic, at times discordant spaces reflecting concurrent and potentially incommensurable worlds. Space in general – and sonic spaces in particular – are emergent *properties* of relations they involve. And those relations can be many in sound. Complicating things even further, once emplaced, sound's forces continue diverting elsewhere, on alternate pathways, through alternate agencies manifesting in other movements of disparate domains.

Sound-space context

Sonic-spatialities are heterogeneous, intermittent, and *contextually* constituted materialisations of sound. Context is sound's circuitry. Far from being inert backdrops, contexts produce dynamic fields of agency actively present in every moment of hearing. Contexts *conduct tendencies* of movements in the expanse, providing space for sounds' operations. Each contextual dynamic spawns slightly divergent spatial characteristics. For example, the space of *acoustic waves* versus the space of sound's *perceptual* identities present two contrasting spaces sharing an energetic expanse. Context is hard to pin down in the abstract as it can only be gauged in relation to its operations. When entangled with observations, its clustering contexts often cut across diverse subjects and domains.

Sound-space scale

Spatiotemporal materialities of sound organise their operations. In other words, every qualification of sonic materiality is already spatialised within the limits of its operational scope, keyed on actions in its corresponding surrounds. The catch is: sound's spatial determinates are themselves indeterminate due to the instability of the processes through which they arise. As contexts cluster and recombine – an essential property of sound's becoming – determinates determining 'determined space' continually evolve, especially in the social milieu. So do relations amongst, and presence of, various subjects active in a circuit.

Operations provide spatial *scales* and agency in context. To understand a scale of a given sound-space, one must first grasp the scope of its context. Refracting sounds back at their features provides hints of their formative contexts. Sound-spaces do not express linear scale progressions common to geometric or quantised space (e.g. small-to-large, slow-to-fast) but rather scale to their situated configurations. Temporality, too, likely synchronises with agencies of a context, beating time in rhythm with native domains. Scales and rates of sound-spaces mirror their territorial responsiveness.

Relational spatiality

Sounds are inherently relational, they have influence on, and are shaped by, the contexts within which they operate. Sounds provide armatures for endured relations: of the frictions providing them structure; of the mediums they traverse and waymarks encountered along the way; of the entities they relationally bind; of substances they activate and attentions they cluster. Sounds never occur in isolation, and when they appear as such it's due to constraining factors of their circuits (such as propagating through anechoic chambers). In other words, sound is not ever a thing in itself but are rather roadmaps of situational orientations and dynamic configurations.

Acoustic sounds exhibit relational spaces through the correlation of frequencies, distances and environments. The propagation of acoustic waves is closely related to the physical dimensions of the frequency, which are influenced by the properties of the propagation medium and environmental conditions. For instance, a single frequency cycle of the note 'A' (the sixth note of the C major diatonic scale, oscillating 440 times per second) is 78 cm long when measured at sound's conventional speed in air of 343 m/s.[4] However, in seawater, this note expands more than four times to a length of 341 cm due to the increased propagation speed of sound in the ocean.

In fact, even in the open air, the note 'A' is not typically encountered at its standard 78 cm textbook definition due to fluctuations in temperature, humidity, wind, and elevation, which are inevitable outdoors. The speed of sound in a medium is dependent on the density of the material, which in turn depends on the temperature, all of which are deeply relational to terrestrial constraints, which are again relational to the speed and position of an observer, when determining a resultant pitch. Furthermore, hearing pitch as a primary category in sound entails learned practices that are often socially and culturally sustained.

Spatial qualities

Sonic qualities emerge from eventful in-situ processes. Space is one such quality. Space denotes directions and operation. The spaces and materials heard in sound are domains of situated relations. Sound is relational to the extent that it entails interactions among various components of a circuit. To 'hear' is to participate in relations and in doing so also assist in making them other. Hearing engages movements and also partially alters their courses of action. Conditions of sound endowed with material qualities, spatial properties and temporal dynamics indicate diverse relations as well as ascribed territories of action. Contextual circuit operations inform sonic properties, providing sounds with signature qualities. Those qualities in turn perform agencies that feed back into movements of the extant, potentially giving rise to other contexts and other spatial-material emanations. Sensory attentions sustain contextual flows. They are vortices in vast circulatory configurations, such that when attention shifts so do the contextual assemblages. As energies traverse terrestrial expanse, encountering interfaces of *refraction-transduction*, their sounds multiply and with them the spaces and contexts of operation.

Sounds do not reside *in* space; sounds *are* spaces taking place. Sounds *produce* space dynamically. Sound-spaces are characteristically tactile, malleable, fluctuant and occasionally pulsed. There is no vacant sound-space, when a sound ceases so does its space. Spatial configurations, orientations and extents reveal sound's *operations*. There is no containing space for sounds, no singular native 'acoustic space' where vibrations dwell from which the rest of sonic spatiality can be derived. Instead, there is a rich heterogeneous agglomeration of overlapping, nested, complementary or contradictory sonic spaces reflecting equally dazzling contexts and sensations at play.

Fluctuant emplacement

Locational properties of sound arise from in-situ interactions; there is no generalising sound's emplaced formations, no guiding principles to the inexhaustible manifestation of sound-space. Space outlines sound's possible relations and potentials in its surroundings suggesting fields of activity, courses of action and manners of movement. Just as other spaces, sound-space imposes relational logic embedded in tectonics of its formations, permitting certain movement logics to take precedence over others. Sounds provide a wide range of spatial typologies, spanning from quasi-Newtonian object-retaining spaces to turbulently elastic shape-shifting wonders.

Some of that spatial diversity can be corporeally inhabited. When listening to an echo, sounds appear distinctively separate. Sound-things travelling about *inside* seemingly enclosed empty space. On other occasions, when listening to resonance, structural features of discontinuous objects seemingly melt into one another, producing an interlocking presence challenging their disconnected optical confines. Imagine a distant airplane rattling a nearby windowpane. Occasionally, space itself seems to emanate from within undulations of ongoing oscillations as experienced when walking through the palpable, pulsating, fluid architecture of standing wave interference patterns taking shape inside buildings. Each configuration responds to differing spatial logics, providing varieties of movement and sustains.

Participating in sound's spatial diversity – observing echo, merging with resonance or swaying with oscillations – transforms perceptions of an event's locale while also subtly reconfiguring subject/object divides. Attending to sound's spatial logic provides partial mapping of its operational extent as well as a roadmap towards possible modifications.

Attention circuits

Perception often has been mistakenly portrayed as a one-way flow, where energy passes on from stimulus to receptor to the perceptual apparatus, the final destination, within which impressions then arise. If, on the other hand, perception is taken as a *circuit of relations*, linking subjects to surrounds, then an entirely different impression of perception arises, understood as a system or *environment* of reciprocal exchange. Sensed sounds are not an endpoint of perceptual representation, barricaded within an isolated subjective space, but rather essential components of subsequent mobile and lively interactions. Impressions seemingly belonging to an object are in fact neither fully contained *in* the object nor sealed off in impressions of the mind. Their qualities are rather *relational agencies* of *contextual conductivity* native to a given attentive circuit.

Features that attract attention are nascent action-functions in search of conductive pathways through their native milieus. Attentive qualities are endowed with agency that invites future motions. Qualities function as connective relays, providing situational conductivities and spatial orientations in activated circuits. Physiological structures in the organism, electrochemical biological pathways, conscious and unconscious meanings, environmental features and affordances are all part of a continual, malleable interconnected *terrestrial circuitry*.

Modes of hearing

Repeating attentions engrain and recursively enhances anticipated perceptions. Attentions tuned to resonate are akin to resonance modes of physical systems, though perception's modes are malleable. Modes of hearing express repeating attentions often adopted in shared social domains. Modes facilitate categorisations and grouping of sounds based on perceptual consensus pertaining to discerned commonalities. Shared modes demonstrate cohabitations of contextual sound circuits, activating rehearsed attentions (conscious and unconscious), patched through selective meanings, modulated by expectations and often ballasted by dedicated technologies and techniques. Configured modal attentions can include listening practices, historical moments, technological and geographical constraints as well as subjective topes, coalescing in distinctive manners of hearing sounds in categories of meaning. Configurations in a mode of hearing are conditional to their circumstantial appearances. Like sounds themselves, modes are essentially situational and inexhaustible. Hearing tends to cling to familiar modes in the same manner that habits cling to routines. Breaking a hearing habit opens paths to discovery of sounds unheard. Hearing in multiple simultaneous modes provides numerous vantages onto fluctuant events, opening up differing aspects of eventful surrounds.

Urban modes of hearing

Modes appear in conscious as well as unconscious states of attention, resonating nascent attributes so that they stand out from other respective qualities. Some modes are carefully fostered, as for example, diagnostics modes practiced amongst physicians and car mechanics. Other modes appear haphazardly, as if the environment itself practices its inhabitants. Vibration-shaping landscapes, resounding geographies as well as built environments produce complex *aural agencies*. Cities are expansive *audio technologies*, their calibrations manifest incidental resonances, at times appearing unexpectedly, notably in musical practices.

Ambient selective attention

Sentient attentions are borne in tangles of mechano-tactile and electrochemical nervous circuits, interlinking organisms and expanse. Functions underlying electrochemical and transduction circuits of the body can also be found in more general dynamics of the extant. Electrochemical interactions of geobatteries or mechano-tactile piezoelectric mineral transductions suggest that aspects of animal attentions may overlap or coincide with less apparent inorganic or geological attentions.[5] Sensory techniques employ mechanisms native to elemental physical/chemical activities, possibly borrowed from inorganic domains. The use of tools modifies connectivity, repatching linkages of the energetic expanse.

Figure 22.1 Device for measuring radio wavelengths with a copper coil tapped at various points in the windings. Invented by Adolf Slaby developed at Telefunken, Berlin c. 1902–1904 (Photo by author).

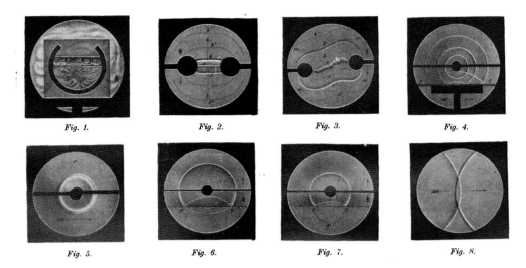

Figure 22.2 Pressure–wave interactions from spark-gap discharges in various reflection/refraction configurations. Images were produced by Schlieren photography, a method developed by the German physicist August Toepler to visualise fluid dynamics in mediums such as air. Sequence from A. Toepler's book *Beobachtungen bach der schlierenmethode* (1906). Ostwalds Klassiker Nr. 158. P117.

Sensation clusters (organic/inorganic)

Inorganic sensation

Sensations are conditions through which qualities arise. If qualities present the features and specificities of an event, then sensations denote underlying *connective configurations* producing particular effects.[6] Sensitivity is a fundamental principle of the extant, possibly more general than motion. Embracing a broadened organic–inorganic approach, sensation processes are understood as *tendencies* of the extant, emerging where fluctuant interchanges abound.

Interference patterns found in vibrational phased space (the locale of acoustic waves) may be examples of pre-auditory sensation complexes, through the manners in which molecules engage and imprint traces of traversing flux. Schlieren photography opens spaces into that remote fluid dynamics. What are Chladni patterns if not snapshots of qualitative states, organically inert, yet intricately featured unfolding sensation complexes?[7]

Aspects of perception share commonalities with other features of the extant by providing temporary habitats for energies and agencies traversing the expanse. As vibrations carry-over they accumulate incidental sensations, joining in on the travel. The terrestrial expanse is saturated with crisscrossing negotiations escaping their locales, all nesting within one another. Shared sensations transform attention, eventually altering features of the extant. Co-evolution is a prime example that performs aspects of expansive sensations intermingling.

Hearing long-distance

When attending to sounds from great distances – rumbling infrasound transmitting halfway around the globe, or, the tangle of earth's telluric currents – their signals arrive saturated with additional presences. Distant sounds interleave multiple encounters cohabiting the same expanse. The remote listener, often unknowingly, shares senses with other places and times, enacting immediacy from the nonlocal and diffuse. Hearing takes part in those ongoing conglomerations of intermingled being. The paramount question is what's kept in attention? And how does attention corroborate with the distant and remote?

Noisy signals

Noisy signals are quintessential domains for hearing ambiguity. They demonstrate pluralist milieus. By attending to their multiplicities, qualitative outlines keep shifting presence, skipping from meaning to meaning. Ambiguity imparts relational interactions occurring simultaneously in overlapping contextual circuits. Attending to convoluted sounds, especially noisy signals from activity-saturated environments or signals from afar, explores the many interlocking, multi-spatial presences a single stream sustains. On closer scrutiny, when honing into any one of the presences, the details tend to fray and resolve into corresponding motions. Repeating loops and eddies piling up in patterned resonances, their internal movements laced with the contradictions and paradoxes of appearing.

The annals of communication could be viewed as a sustained attempt to *suppress* the cacophony. Insisting on the determinacy of messages-in-transit. Communication models unambiguously, and effectively, relegate comingling environmental noises to the status of disturbance. Environments agency downgraded to signal distortion. Ironically, the tensions between sensation (what *could* be heard in a signal) and perception (what *is* heard in a signal) seem to increase symmetrically with lengths of communication cables spanning oceans and land.

Plotting frequency waveforms for both words – 'attention' and 'sensation' – recounts a more nuanced narrative. Use of both terms, in English language printed matter, swells in the 19th century corresponding to trends in the natural sciences at the time, with sensation clearly in the lead. By the early 20th century, interest wanes and trends reverse. Both waveforms intersect only once, in the early 1940s, at the time when Harvard's anechoic chambers were tactically active. In the padded chambers, human sensations were tested for functionality as integral components in hybrid bio-electro-mechanical combat circuits (Ganchrow 2016). Since that time the primacy of *perception* has been experiencing a steady incline, on its way to current battlegrounds of *perceptual meanings* in post-truth politics (Ngram 2023) (Figure 22.3).

Even when senses are functionalised and signals seemingly de-cluttered, folded into themselves and piped-through insular channels, their transmissive propagations exert extraneous pressures, inducing haphazard sensation along their paths of propagation. Inorganic sensations are omnipresent in conjoining flows. Radio signals picked up in conductive minerals of geologic formations or oscillations from national power-grids intermingling with native telluric ground currents demonstrate the degrees to which anthropic expressions permeate the surrounds.

The birth of radio also signals the domestication of lightning. The crude form of emission in spark-gap transmitters (the origin of radio transmission techniques) has audible characteristics that exceed the semiotic function ascribed to it by wireless telegraphy. In its most elemental appearance, the spark is more of a feature of matter than a proper signal. Its ambiguous status in the circuit pathway, as a natural rather than synthetic phenomenon, draws kinship with earth's other circuits. It also points out an in-built *terrestrial* component in naturalised circuits such as electrical signalling in plants, electrocommunication in honeybees, or within the human nervous system. The co-evolution of minerals and life on Earth may also be that of electrical and sensory circuits.[8]

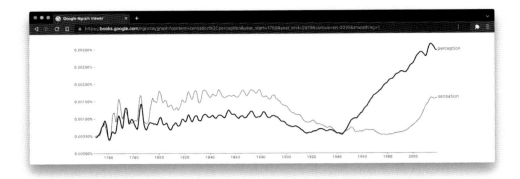

Figure 22.3 Waveforms plotting the frequency of 'perception' versus 'sensation' (ngrams) on Google's corpus of English language books (1750–2019). Graph illustrating the frequency of occurrence over time (x-axis publication years and y-axis frequency of occurrence).

Source: Google Books Ngram Viewer, Mar. 26, 2023.

Physicality of sense

Energetic behaviours of the expanse combine and collide, refract and transduce in a variety of manners giving shapes and contours to unfolding sensation complexes. Imprinting sensations sometimes appear on the surface of things, like patterns in the ocean. Combining currents of wind and water are *sensed* in undulating surface tensions through perturbations of conjoining ocean–atmosphere relations, or in tangles of dendritic electrical relays signalling through minerals in rock and soil. Seismic motions in the earth's crust induce electricity into ground-based telluric currents by way of micro-fracturing piezoelectric rocks, especially along tectonic faults. Do mountains, participating in seismic fault lines, sense a pending quake? That's not to endow geology with anthropomorphic wonders, but rather the reverse, to propose that aspects of perceptions themselves may be borrowed from pre-existing proven configurations of the extant.

If organic consciousness is situated stuff, stratifying aspects of physical matter, a converse statement could also be proposed: that the physical universe *itself* be considered a form of expansive consciousness.[9] Strong nuclear force, clumping together the contexts of physical matter would then present altogether different *magnitudes of attention*. But the term 'consciousness' is too vague and broad a category for examining materialisations of sound. Its potential anthropomorphism is equally troublesome. Suffice to say that every portion of existence bisects many endured actions of which sounds are important constituents. Sounds are *tendencies* in those ceaseless ongoing processes. Every sonic spatiotemporal materiality is an eddy in complex torrents of relational interactions. Observable, differentiated qualities materially express traces of situated affordances and contextual actions simultaneously *taking place* as well as *making sense*.

Vibrant subjectivity

Sound enacts the delays between occurrence and sensation. It resonates relation between attention and context, and in doing so provides partial contours to the *spaces of being*. Subjects undulate and co-evolve with surrounds through turbulent tangles of organic–inorganic sensations productive of myriad agencies and formations. Attentions are not merely functions of an organism's survival but rather coupled interdependencies in complex planetary conductivities, productive of countless effects. To hear is first and foremost to partake in an earthbound ebb and flow of actions and contribute to the intermingling.[10] Demarcations of intensity are recursively pulse-patterned yet also importantly emergent and dynamic. Subjects arise through conditions of sense. Shared percepts delineate continuities and discontinuities between selves and surrounds in perpetual reconfiguration. Adopting listening as a conscious practice exercises those relations.[11]

Distributed subjectivity

Observations close situational circuits, circulating both distinctive object qualities as well as a subjective stance. In the example of sonic experience, hearing becomes a site whereby attention, surrounding, and subjectivity are mutually conductive. Acts of listening emplace subjectivity and situate eventfulness. To hear a 'place' implies internalising oscillations and negotiating limits of the self. Subjectivity breathes in tandem with shifts in the limits of its objects. No form is invariable. Redistributions of subject are reminders that the subjectivity was never fully contained in the first place. The elusiveness of subjectivity is inherent in distributions of self that is as much inside the body as they are spread over its *territories of action*. The organism is a crucial node in that distribution as it is where the possibilities of immediate action are concentrated.[12] Making sense involves dynamic extensions of conductive traces through elements of an attentive circuit. Refractions glean out from events those aspects the conductive medium is capable of influencing. However, since actions *of* the body are never fully *in* the body, corporeality is always scattered in orbits of its subject. Distributed subjectivity includes cumulative attention circuits, conscious and unconscious, active at any given moment in various overlapping domains.

Observer-observed subjects, traced back through their action-temperaments, disperse and are rediscovered within curious formations and incidental conditions structuring the quotidian and geological expanse. This complex dynamics, crossing human/non-human and organic/inorganic divides, opens up numerous relational interconnections, reacquainting the subject with terrestrial heritage at large. A geological subjectivity actively seeks manners of sensing relational reciprocities. An awareness of being at once located and distributed in multiple places and times, finding aspects of the subject in agencies of the surrounding and vice versa, finding terrestrial dimensions in corporeal milieus.

Transverse flows

Just before lightning strikes, intense electrical build-up and ionisation of air cause human hairs to stand on end, as if the hairs themselves reach out to participate in conductivity of a nascent bolt. Extreme energetic situations induce polarisations where the subject becomes a composite of minerals and chemicals forming conduits of exchange in circuits exceeding human domains.[13]

Postures of reaching-out-opening-in are expressions of acute connectivity that are also temporary paralysis, where the intensity of sensations short-circuits comprehension. Excesses of energy at times over-intensify sensations in manners that bodies can no longer sustain. Extreme conditions of terrestrial energetics in which mineral provides support often also spell the organism's demise. Conductive properties of extending hairs seem to *welcome* such corporeal transgression, like the sloshing of calcium carbonate otoliths (ear-stones), trapped in the inner ear, longing for ocean returns. Occasionally fossilised otoliths and tympanic bulla wash ashore, dredged up from sediments of extinct whales swimming in palaeo-oceans (Figure 22.4).

Raviv Ganchrow

Figure 22.4 Fossilised whale inner ear (tympanic bulla), c. 18 million years, South Carolina (Photo by author).

Sound, earth-bound

Interlinking sounds

Sonic attentions, the circuits they maintain and materialities they sustain are integral to dynamics of the extant. Sounds are only ever determinate in relation to contextual conductivities to which they correspond. But circuits never occur in isolation. Circulations are inevitably tied into countless other movements, of other spaces and times and *their* related agencies and interlocking worlds.

By placing the human sensorium in circuit with other terrestrial processes, hearing along with the spaces ballasted in sounds, endures a cascade of relations, blurring distinctions between cultural/natural, bisecting other times, locales and processes that encroach on the immediate chambers of hearing. Once in-circuit, the spaces of sound appear formatted with multi-layered, heterogeneous, poly-temporal *agencies* incorporating diffuse and discontinuous fields of multi-scaled relational interactions. Sound's diverse material presences and spatial logics reflect that interlinked multiplicity.

Attention sediment

Vibration sensing and acoustic sounding features of biota (trembling vocal chords, resonating cavities, undulating membranes, swaying cilia) are physical sediments of vibrant co-evolutions that integrate substances and forces native to developments of the planet. The term paleohistoricity can be coined to describe deep-time components embedded in sensory organs. Sensory morphologies are shapes that are also manners of operation, slowly yet persistently recalibrating their limits and sensitivities in ongoing phylogenic processes. If historicity describes contextual constraints of a sensory circuit, paleohistoricity scopes the geological temporality of sense calibrations transitioning intermittently and across multiple generations.

Organism physiology is also an existential technique; morphogenesis expresses the coalescing of actions *and* energetic attentions combining with surrounds. The natural history of substances and structures involved in sensory circuits cannot be decoupled from the fluctuations they endure and sustain. Spaces and places come in tow of multifarious processes.

Organisms, mineralogy and geography collectively evolve in periodicities of milieu. Subterranean organisms evolved in tactile contact with perpetual substrate vibrations and their various sensing and sounding mechanisms are shaped by, and contribute to, categories of ground motions. Plant roots respond to, and grow towards, vibrations from moving subterranean water.[14] At the same time, dendritic root structures dissipate sounds on sloped terrain by knitting topsoil in place, alleviating rumbling landslides. In turn, vibrational properties of substrates alter with life and mineral compounds inhabiting their domains.

Other terrestrial forces, unrelated to vibrations, also imprint in morphologies of ears. Vestibular systems, present in the inner ears of all vertebrates, perpetually tether organism moments to earth's gravitational pull. The system, most commonly comprised of three semicircular canals oriented in Cartesian XYZ-axis, coordinate body movement with balance. Palaeo-ear evolution in hominins sets the rise to two-legged locomotion – unique to humans among living primates – at approximately 1.8 million years ago, mirrored in the shapes of vestibular canals.[15]

Environmental frequency

The ground is permanently traversed by 54-minute-long waves – the strongest signature in earth's resonant modes – continually ringing with microseism oscillating at around 0.0003 Hz.[16] Steep topography such as alpine ranges, have their own signature resonances. It was recently discovered that Switzerland's distinguished Matterhorn Mountain has a fundamental frequency of 0.42 Hz due to structural properties of the massif.[17] On larger timescales, terrestrial sonority potentially demonstrates vast recursive feedbacks, inaccessible to sounds accumulating in the lifetime of a single listener, of sounds informing structures imprinting sounds informing structures imprinting, et cetera.

Substrate vibration sensing abilities through soil and foliage, widespread in animals, evolved several times independently in various locales. Snakes developed ground-conductive hearing, sensing feeble soil vibrations.[18] Mole-rats employ it for burrow-to-burrow signalling.[19] Conversely, general tendencies of ground vibrations are conditioned by animal presence, most recently noted in an abrupt drop of 'background noise' on seismograph readings during global pandemic lockdowns.[20] At planetary scale, tendencies for earthquake frequency and magnitudes themselves seem to be coupled with cycles of life on Earth, with ocean sediments lubricating Earth's plate subductions in a feedback circuit of ecosystems, climate and movements of tectonic plates.[21]

It's been recently discovered that much of terrestrial life happens underground, where the collection of archaea and bacteria is estimated at nearly twice the volume of all Earth's oceans.[22] Some of those microorganisms live for millennia, barely moving except with shifts in tectonic plates, their lived worlds progressing in cycles of geological time. Surface-dwelling organisms as ourselves, cannot grasp sensations of the depths. Microorganisms of the substrate, on the other hand, may be the best qualified to comment on the experiences canyons endure.

Poly-temporal

Energetic events refract and transduce haphazardly through the extant. Their circulations follow interweaving principles of refraction and shape-shifting transductions. Refraction entails territorial contortions altering orientations and rates of change as well as relational *transformations* within a given terrain. As splayed and warping rates meet up with one another, they express their time-slip, thickening the present tense. Every locale contains multiple rates unfolding simultaneously, only some of which seemingly address the clocks of biological time.

Temporality is an emergent *quality* of processes, as is space. Time variance appears with transformations in quality, like the red-shift in light travelling over great distances or Doppler pitch shift of relative motions. Polychronicity is the accumulation of multiple times packed into the now, but also importantly times *shifting* and *shearing* in relation to one another. Delay is a persistence of the past that carries-over into the present just as pre-delay is a nascent state of the future also active in present, operating simultaneously with differing rates of becoming. Every sound circuit synchs and scales to operations in its context, presenting its own temporal manners and ways of becoming.

Dynamics future-tensed

Spatiotemporal qualities of sound are features of agency. Sonic qualities aggregate and condense eventful pasts as much as they facilitate follow-up states. The vibrancy in material presence is also a call for future actions. Sonic spaces remediate events, preserving and prolong motions to facilitating further interactions. If the past persists in the present through refractive transductions, tendencies of the future also potentially manifest.

Fragments of future-states can also be found in anomalous wave behaviours of the extant, as the tensioned 'anticipation' air experiences, becoming nearly a vacuum just moments before a shockwave passes through or the reverse reverb that ends in a crescendo of ocean echoes traversing the deep sound channel.[23] Likewise, the shoreline slowly recedes, sometimes for several minutes, before the arrival of a tsunami as approaching waves temporarily pull water away from the shore. Both wave retreats, and the reverse echo playback, are possibly forms of inorganic recoil, pre-delays anticipating future impacts.

Refraction-transduction

Temporal shapes

When approaching dynamics of the expanse, principles of *transduction* and *refraction* provide useful guidelines, especially in accounting for transformations of orientation, spread and temporality integral to circuitries of sound. Both terms are applied in a broadened sense, denoting *processes* and *movements*, typical of energetic behaviours of processual actuality. Both refraction and transduction have pronounced *spatial* traits in their movements exemplified in refractive *fields* and *arborescent* transductions respectively. Many terrestrial oscillations can be traced back to ur-transductions of chlorophyll (photosynthesis), approximately 3.4 billion years ago, that send radiant sunlight circulating in a host of emergent chemical, gaseous, organismic and tectonic pulsations.[24]

Transduction is useful for observing circuit propagations weaving through and bridging over material distinctions, where transference of energy is not only in type (standard transduction) but also potentially transference between *systems* (context transduction). An example of *standard transduction* is acoustic energy transferring to electrical energy when passing through the ear or

Sound's spatial-material circuitry

a loudspeaker. An auditory event transducing into descriptive words and gestures could be an example of *context transduction* (movements transferring from a milieu of gasses to movements in the social milieu). The frayed meander of sounds, transducing along media pathways, is possibly more instructive of context transductions.

Waveforms

Storage mediums of sound (and other energetic phenomena) function as refractive-transductive delays. Media containers alter an event's rates of becoming. Recordings on vinyl, invisible patterns on magnetic tape, waveform diagrams, spectral charts, bits encoded in silicone chips are all still part of the agencies of energetic waves, bending, changing directions and velocities when subjected to transduction. Their waves are prolonged, not halted, through sedimentation in media delays. A sound wave embedded in a wax cylinder is first and foremost a physical transduction, in other words, an actual trace of eventfulness still actively performing in other manners what acoustics behaved in air. Although their waves are seemingly inert and captured, they still participate in dynamics of the expanse through their presence as substances, or when closing sensory circuits with unforeseen observers. Knowledge sediments attention, storing latent configurations for future interaction; information stored in media is distinctly context-sensitive in its modes of operation. From the many examples of transduced fluctuations, waveforms, in particular, have taken on enhanced fields of affect in the late modern era, especially in their activation capacities across diverse contexts when movements render in visual artefacts.

One of the longest terrestrial waveforms charted to date spans 590 million years of sea-level fluctuations, published in 1977 by geologists in the petrochemical industry. The Exxon/Vail curve waveform can be grasped in its entirety at a glance (Figure 22.5). When loaded to an audio buffer

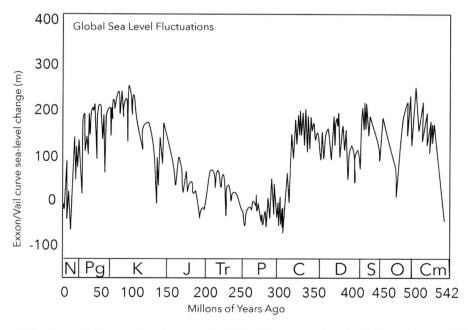

Figure 22.5 Exxon/Vail curve. Waveform charting 540 million years of sea-level fluctuations was initially published in 1977 by geologists from the petrochemical industry (Chart by author).

and played back billions of times faster it can even be perceived as an acoustic sound. But the oscillations from which those waveforms project can only really be sensed within the time scales inherent to their movements, within the ground and rocks to which that particular sloshing relates. A durational accumulation from the last quarter of the waveform is the stuff holding up the white cliffs of Dover, which, incidentally, are also almost entirely composed of calcium carbonate from sinking phytoplankton.

Biominerals such as calcium carbonate (calcite) found in phytoplankton and oyster shells, coral reefs and the eyes of Trilobites, prolong aspects of their agency into processes of the present. Stripped of organic actions their activities persist in follow-up mineral interactions, ballasting the Dover cliffs, participating in zoomorphic limestone cave formations, or contributing to an increased potential force of future earthquakes off the coast of New Zealand.[25] Animal bones too refract their pasts, prolonging them into the present. Calcium phosphate is their delay line. Carbon-14 isotopes from plants and animals (applied in radiocarbon dating) measure the time-slip when biological clocks roll back into geological time.

Discussions pitting linear against cyclic time fall short when observing poly-temporal dynamics of the extant. Rates of change warp and shift continually in physical domains. Ocean floors and glacial mass endure geophysical time; their sediments express intermittent and warping states of becoming. Mountain ranges reveal delays in layered-rock formations. On some occasions, stones exhibit waveforms that can even be read like prehistoric seismograms.[26]

Organisms tune to faster rates, dipping in and out of phase with predominant environmental cycles, synching to earth spin, sun cycles, phases of the moon, tidal motions and seasonal modulations. Predominant geophysical cycles calibrate temporal perceptions that in the case of humans are also potentially clocked to heartbeats.[27] More hidden, yet widespread oscillations have been discovered in core structures of mammalian and plant cells, with collective emergent behaviours, coordinating rhythms of organism processes in resonance with rhythms of the surrounds.[28] Diurnal cycles in organisms were observed linking up with 'master clocks' (housed in neurons), regulating local rhythms of 'slave oscillators' housed in countless peripheral cells and organs.[29]

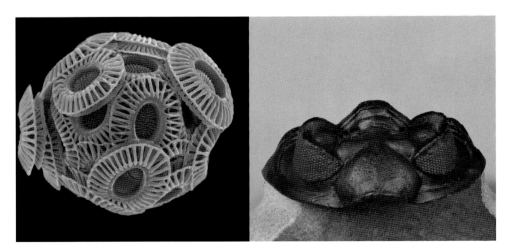

Figure 22.6 Left: Coccolithophores (phytoplankton) capable of forming calcium carbonate (calcite) structures called coccoliths in the process of biomineralisation. Right: Compound eyes of a trilobite (genus Coltraneia), made of the rigid mineral calcite lenses. Early Devonian trilobite from Hamar Laghdad, Morocco (Photo by R. Wilber, CC BY 4.0).

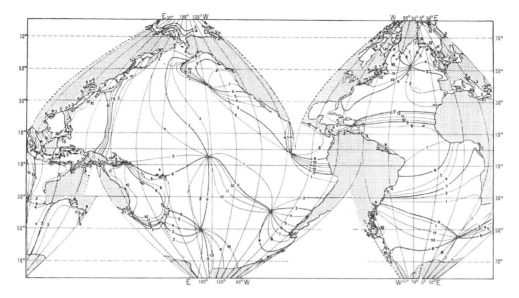

Figure 22.7 Map of cotidal lines along which tides meet, constituting still nodes of oscillation convergence, at each given hour of diurnal oceanic rhythms (Map after A. Defant, 'Physical Oceanography, vol. 2', 1961).

The mechanisms are notably illustrated in the literature with the electronic oscillator graphic symbol, transposed from electrical circuitry. As is the master/slave jargon that electrical engineering itself absentmindedly transduced from dreadful colonial domains.[30] Importantly, cellular clocks, that generate spontaneous oscillations, are nonlinear dynamic systems radically open to environmental energetic exchange.[31] Clocks synch subjects and surrounds, at time skipping between couplings or re-setting phase in cyclic circadian patterns that according to Arthur Winfree can be accounted for in a peculiar topological space likened to 'crystals of living time'.[32]

Shared hearing

Senses inhabit movements without necessarily taking command. Their ethics, politics and aesthetic evolve in-situ, in sway with the contexts at hand. There is an in-built inclusivity to energetic movements indiscriminately bridging sensations, bodies and the milieu. That intrusiveness of un-invited experiences is also potentially its greatest asset. Care stems from taking notice of diverse reciprocities, acknowledging participation and granting access and provisional trust to unfamiliar sways, possibly synching to certain rhythms and following through with consequences of their motions. Hearing diversely is possible, but never all-encompassing, because hearing always synthesises selectively. Listening, on the other hand, opens the circuit up for unexpected encounters. Experimental modes of hearing can practice and share the discoveries. *Hearing differently is the beginning of becoming elsewise.*

An ethics of listening tests degrees of response-ability conditional to sound's spaces of action, turning affordances into questions that perpetuate discovery. Experimentalising the ecologies of hearing test what's at stake in shared modes of attention. Every emplacement also entails distinctive orientations and subject distributions. Even the most mundane, quotidian act of hearing already contains understated metaphysical and, by extension, aesthetic and political vectors. When coordinated, they can empower collective force, such as the sounds emanating from public protest, or the growing sonority of thunderstorms, wildfires, flash floods and landslides in the current epoch. Hearing unfamiliar spaces has the potential of redistributing subjectivity towards other courses of action and other ways of being. Geological listening awakens deep-time awareness of ongoing involvements with past and future terrestrial dynamics. Coalescing with sensory impressions and participating in diverse presences foster an unending procession of spaces and formations with unforeseen agencies and actions.

Acknowledgments

I would like to thank the editors for the invitation to contribute to this volume and Emma-Kate Matthews for helpful comments on the initial draft. This text abridges, and attempts to synthesise numerous inklings and intuitions, accumulated over two decades of sonic practice. Spending time in sway with complex oscillations has taught me much of what I know about sound. This text is dedicated to sound's unending polyvalent splendor. Special thanks to Sila for her profound insights into poly-temporal materiality, hatched across shared explorations. Work on this chapter was partially made possible with a grant from The Royal Conservatoire, University of the Arts The Hague.

Notes

1. Throughout the text, the plural 'dynamics' is used to highlight the diverse relational interactions that give rise to a variety of dynamic categories associated with qualities of the extant. The singular 'dynamic' implies a single overriding principle, which is not the case in mechanisms of the manifest. Instead, the text argues that the diverse formations of sounds occur in the draw of contextual constraints, exhibiting a variety of arrangements with differing dynamic traits.
2. The term 'extant' aims to provide a synonym for existence that emphasises the dynamics *in* presence, hence the noun-ing of the adjective.
3. The word 'expanse' denotes the full reach of intermixed movements in processual actuality, with emphasis on its expanded and emergent spatiality.
4. 440 Hz corresponding with a 78 cm length is calculated at sea level with an ambient temperature of 20°C.
5. A brief passage in Henry Bergson's book *Matter and Memory* (1896) discusses attention in terms of its choosing capacities, a 'force that provokes [specific] reactions', readily observed in plants absorbing selective nutrients from the ground. A similar force was recognised in abilities of hydrochloric acid to 'selectively' pick out the base when combined with carbonate of lime, regardless if it's in marble or chalk (Bergson 1988, 159, orig. ed. 1896). Another example is the attention a copper spool maintains towards electromagnetic fluctuation. The coil's selective attention to frequency nests in its wire and windings. Copper coils tuned frequencies by gliding a conductor along the length of a coil, tapping the windings to dials in a station's carrier frequency corresponding to the coil's resonance properties. Mach argues more general equivalences for physical processes and perceptual states 'our hunger is not so essentially different from the tendency of sulphuric acid for zinc, and our will [is] not so greatly different from the pressure of a stone' (Mach 1919, 464, orig. ed. 1883).
6. Ernst Mach briefly brings up the question of inorganic sensations but then diverts by stating that it is a question without scientific significance (Mach 1914, 243–244). Arguably, if such questions *were* asked, then other manners of empiricism could take shape.
7. For a more detailed discussion of acoustic phased space as a category of inorganic sensations see (Ganchrow 2010, 188).
8. For an explication of radio's entanglements with the domestication of lightning see Ganchrow's radio work *Knallfunken* (2018) and the contextual compendium https://ravivganchrow.com/page/spark-gap/knallfunken-text.html.
9. Exponents of panpsychism – a theory currently gaining traction – hold such statements to be true, as did prominent figures in early experimental psychology (or psychophysics) such as Gustav Fechner and

William James. As a young adolescent, the physicist Ernst Mach experienced the surrounding itself as an expansive *sense formation* ensnared with his own immediate impressions (Mach 1914, 30). Henry Bergson came to a similar conclusion regarding 'pure perception' as a component of the physical universe in his meditations on matter and human memory (Bergson 1988, orig. ed. 1896).

10 Jane Bennett likens agency to an oscillating, tensional property of the expansive dynamics in which there is a 'pulsating, conative dimension of agency […] engaged in a system of pulses, in an assemblage that links them and forms circuits of intensities' (Bennett 2010, 80–81).
11 For a discussion of the psychological, ethical and philosophical implications of listening as a practice of self see Kleinberg-Levin's *The Listening Self* (Kleinberg-Levin, 1989).
12 There is an essential pragmatic underpinning to sensations that are always oriented towards operations. Bergson recognised that aspect in the organism's poise to actions (Bergson 1988, orig. ed. 1896). J. J. Gibson's ambient psychology of 'affordances' is another example.
13 An account from the vicinity of a bolt strike at Morro Rock (Sequoia National Park, August 20, 1975) describes altered spatiotemporal perceptions including the expansion of time and temporary weightlessness (McQuilken 2013).
14 Gagliano et al. (2017, 151–160).
15 Spoor, Wood, and Zonneveld (1994, 645–648).
16 Park et al. (2005, 1139–1144).
17 Weber et al. (2022).
18 Friedel, Young, and van Hemmen (2008).
19 Narins et al. (1992).
20 Gibney (2020).
21 Behr and Becker (2018).
22 Watts (2018).
23 Hamilton (1949).
24 Rosing et al. (2006), Kiang (2008).
25 Boulton et al. (2022).
26 Mach (1914, 236).
27 Sadeghi et al. (2023).
28 Golden and Strayer (2001), Schibler and Sassone-Corsi (2002).
29 Reppert and Weaver (2002).
30 Ellis (2020).
31 Kruse and Jülicher (2005).
32 Winfree (1987).

Bibliography

Behr, Whitney M., and Thorsten W. Becker. 2018. "Sediment Control on Subduction Plate Speeds." *Earth and Planetary Science Letters* 502 (November): 166–173. https://doi.org/10.1016/j.epsl.2018.08.057.
Bennett, Jane. 2010. *Vibrant Matter: A Political Ecology of Things*. Durham, NC: Duke University Press.
Bergson, Henri. 1988. *Matter and Memory*. New York: Zone Books.
Boulton, Carolyn, Marcel Mizera, André R. Niemeijer, Timothy A. Little, Inigo A. Müller, Martin Ziegler, and Maartje F. Hamers. 2022. "Observational and Theoretical Evidence for Frictional-Viscous Flow at Shallow Crustal Levels." *Lithos* 428–429 (November): 106831. https://doi.org/10.1016/j.lithos.2022.106831.
Ellis, Leonard. 2020. "It's Time for IEEE to Retire 'Master/Slave.'" *EE Times* (blog). June 19, 2020. https://www.eetimes.com/its-time-for-ieee-to-retire-master-slave/.
Friedel, Paul, Bruce A. Young, and J. Leo van Hemmen. 2008. "Auditory Localization of Ground-Borne Vibrations in Snakes." *Physical Review Letters* 100 (4): 048701. https://doi.org/10.1103/PhysRevLett.100.048701.
Gagliano, Monica, Mavra Grimonprez, Martial Depczynski, and Michael Renton. 2017. "Tuned in: Plant Roots Use Sound to Locate Water." *Oecologia* 184 (1): 151–160. https://doi.org/10.1007/s00442-017-3862-z.
Ganchrow, Raviv. 2010. "Phased Space." In *Chrono-Topologies: Hybrid Spatialities and Multiple Temporalities*, edited by Leslie Jaye Kavanaugh, 179–193. Critical Studies, vol. 32. Amsterdam; New York: Rodopi.
Ganchrow, Raviv. 2016. "Padded Sounds: S. S. Stevens Conducting Speech Intelligibility Tests, Harvard University, Circa 1943." *Sound & Science: Digital Histories*. https://soundandscience.de/node/1231.

Ganchrow, Raviv. 2018. „Knallfunken" – Klangkunst Über Drahtlose Kommunikation." https://www.hoerspielundfeature.de/hoerstueck-ueber-drahtlose-kommunikation-knallfunken-100.html.

Gibney, Elizabeth. 2020. "Coronavirus Lockdowns Have Changed the Way Earth Moves." *Nature* 580 (7802): 176–177. https://doi.org/10.1038/d41586-020-00965-x.

Golden, Susan S., and Carl Strayer. 2001. "Time for Plants. Progress in Plant Chronobiology." *Plant Physiology* 125 (1): 98–101. https://doi.org/10.1104/pp.125.1.98.

Hamilton, Andrew. 1949. "SOFAR: The Navy's Lost and Found." *Popular Mechanics*, 92(06): 166–169 & 246–248.

Kiang, Nancy. 2008. "Timeline of Photosynthesis on Earth." *Scientific American*. https://www.scientificamerican.com/article/timeline-of-photosynthesis-on-earth/.

Kleinberg-Levin, David Michael. 1989. *The Listening Self: Personal Growth, Social Change, and the Closure of Metaphysics*. London; New York: Routledge.

Kruse, Karsten, and Frank Jülicher. 2005. "Oscillations in Cell Biology." *Current Opinion in Cell Biology* 17 (1): 20–26. https://doi.org/10.1016/j.ceb.2004.12.007.

Mach, Ernst. 1914. *The Analysis of Sensations, and the Relation of the Physical to the Psychical*. Chicago, IL; London: Open Court Publishing Co.

Mach, Ernst. 1919. *The Science of Mechanics*. Translated by Thomas J. McCormack. 4th ed. London: Open Court Publishing Co.

McQuilken, Michael. 2013. "A Hair Raising Experience on Moro Rock." *Social Positive*. https://web.archive.org/web/20151025093819/http://www.socialpositive.com/c/stories/A-Hair-Raising-Experience-on-Moro-Rock-133301.

Narins, Peter M., O. J. Reichman, Jennifer U. M. Jarvis, and Edwin R. Lewis. 1992. "Seismic Signal Transmission between Burrows of the Cape Mole-Rat, Georychus Capensis." *Journal of Comparative Physiology A* 170 (1): 13–21. https://doi.org/10.1007/BF00190397.

Ngram. 2023. "Frequency of Sensation and Perception in Printed Matter, 1750–2019." *Google Books Ngram Viewer*. https://books.google.com/ngrams/graph?content=sensation%2Cperception&year_start=1750&year_end=2019&corpus=en-2019&smoothing=1.

Park, Jeffrey, Teh-Ru Alex Song, Jeroen Tromp, Emile Okal, Seth Stein, Genevieve Roult, Eric Clevede, et al. 2005. "Earth's Free Oscillations Excited by the 26 December 2004 Sumatra-Andaman Earthquake." *Science* 308 (5725): 1139–1144. https://doi.org/10.1126/science.1112305.

Reppert, Steven M., and David R. Weaver. "Coordination of Circadian Timing in Mammals." *Nature* 418, no. 6901 (August 2002): 935–41. https://doi.org/10.1038/nature00965.

Rosing, Minik T., Dennis K. Bird, Norman H. Sleep, William Glassley, and Francis Albarede. 2006. "The Rise of Continents—An Essay on the Geologic Consequences of Photosynthesis." *Paleogeography, Palaeoclimatology, Palaeoecology* 232 (2–4): 99–113. https://doi.org/10.1016/j.palaeo.2006.01.007.

Sadeghi, Saeedeh, Marc Wittmann, Eve De Rosa, and Adam K. Anderson. 2023. "Wrinkles in Subsecond Time Perception Are Synchronized to the Heart." *Psychophysiology* 00 (e14270). https://doi.org/10.1111/psyp.14270.

Schibler, Ueli, and Paolo Sassone-Corsi. 2002. "A Web of Circadian Pacemakers." *Cell* 111 (7): 919–922. https://doi.org/10.1016/S0092-8674(02)01225-4.

Spoor, Fred, Bernard Wood, and Frans Zonneveld. 1994. "Implications of Early Hominid Labyrinthine Morphology for Evolution of Human Bipedal Locomotion." *Nature* 369 (6482): 645–648. https://doi.org/10.1038/369645a0.

Watts, Jonathan. 2018. "Scientists Identify Vast Underground Ecosystem Containing Billions of Micro-Organisms." *The Guardian*, December 10, 2018, sec. Science. https://www.theguardian.com/science/2018/dec/10/tread-softly-because-you-tread-on-23bn-tonnes-of-micro-organisms.

Weber, Samuel, Jan Beutel, Mauro Häusler, Paul R. Geimer, Donat Fäh, and Jeffrey R. Moore. 2022. "Spectral Amplification of Ground Motion Linked to Resonance of Large-Scale Mountain Landforms." *Earth and Planetary Science Letters* 578 (January): 117295. https://doi.org/10.1016/j.epsl.2021.117295.

Winfree, Arthur T. 1987. *The Timing of Biological Clocks*. Scientific American Library Series, no. 19. New York: Scientific American Library: Distributed by W.H. Freeman.

23
PLACE, SOUND, AND ARCHITECTURE

Jeff Malpas

Introduction

Every space belongs to a place. The only exceptions to this rule are the abstracted spaces of formal analysis – and even these hypothetical spaces belong to places indirectly, their exceptionality being purely notional. Sound belongs to space in virtue of this prior belonging of space to place, since sound does not arise in any mere space, but originally in place. Sound is always itself 'placed' before it is 'spaced'. Indeed, were there such a thing as 'pure' space, outside of formal analysis, a space apart from place, then it would be as soundless as it would also be empty.[1] Sound, understood experientially or phenomenologically, always presupposes one who listens or hears (even if the relation to the one who hears is in terms of the absence of sound – of a lack of hearing), and so the appearance of sound also presupposes the being-placed of both the sound, as well as that which sounds, and the one who hears. The way sound is tied to place, and thereby to space, is also indicative of the fact that, contrary to what is sometimes assumed,[2] sound does not lack a relation to spatiality, nor is it primarily or solely temporal in character (although it exhibits a complex temporal dynamic).[3] Belonging to the structure of place and being-in-place, sound is part of a structure that combines spatiality and temporality in a single topology – space and time thus emerge out of the complex unity of place and can be properly understood only in relation to it.

Place, as it is relevant here, does not refer to mere 'location'. The latter is ambiguous between spatial position and genuine topological situatedness.[4] Spatial position is simply the location of a thing insofar as this is related to its movement or movement in relation to it, or as it enables the thing to be positioned in relation to other such things. Spatial position is essentially what can be captured using a plan, diagram, or map, and even, in verbal rather than strictly pictorial form, by means of a set of spoken or written directions, and that can be specified, so long as an appropriate framework has already been established, using a set of spatial coordinates. The notion of space that such location or positionality implies and on which it depends is essentially that of homogenous extensionality – an extensionality that has no intrinsic limit or bound (this being the sense of space at work in abstracted formal analysis) and that is often represented in terms of a Cartesian plane or intersecting Cartesian planes. On this basis, both spatial position and the notion of space associated with it, are amenable to a purely geometrical or mathematical, and so also a quantitative or numerical, specification. Such specifications are also readily amenable to forms of visual representation.

In summary terms, spatial position can be understood as a form of 'being-at'. Topological situatedness, in contrast, is a 'being-in'. As such, topological situatedness is qualitative rather than quantitative, and whilst there is a mode of spatiality that belongs to it, the spatiality of topological situatedness, which is to say the spatiality of place (or topos) is a differentiated or heterogenous spatiality. It is a spatiality that is structured in terms of near and far, interior and exterior, centre and periphery. What characterises topological situatedness above all else is an openness within bounds, although the boundedness at issue is horizonal and so, like the horizon of the visual field, not capable of precise determination (which does not mean that it cannot be 'determined' at all, but that there can be no unique determination).

Such topological situatedness, which always involves a relation of both subjectivity (individual and collective) and objectivity (thing and 'world'), implicates a complex structure that encompasses the perceptual, cognitive, and behavioural capacities of subjects as well as the properties and features of things and environments. Such situatedness is never restricted to a single mode of apprehension or presentation, operating across different modalities of appearance, perception, and activity (it also has an important affective character – an issue to which I shall return towards the end of this discussion). Similarly, the structure of such situatedness involves spatial positionality – as place involves space – but cannot be restricted to such positionality alone any more than it can be restricted to single types of properties of things (say, surface geometry), a single mode of perceptual presentation, or even a single type of activity. For the architect, spatial position cannot be overlooked – architectural design depends on being able to plan and map out spatial positions and structures in ways that are indeed tied to forms of quantitative specification (this applies both to the design of individual buildings and to the design of larger urban and landscape forms, although much of the focus of the discussion of this chapter will be on the former[5]). But from an architectural design perspective, it is also important that spatial position be understood in relation to topological situatedness since it is the latter that is the primary concern of design as it is addressed to specific built or constructed forms – forms realised materially – intended for use and occupation by human beings.

Since situatedness does indeed encompass both subjective and objective elements, and, from the perspective of the subject, the entirety of perceptual, cognitive, and behavioural capacities and engagements, such situatedness cannot be restricted to any one mode of appearance of things or any single mode of perceptual acquaintance. It thus encompasses sound or hearing no less than vision. In contrast, however, spatial positionality, and especially the way such positionality is grasped, tends to have a much narrower range and focus. Moreover, although spatial positionality can be grasped acoustically as well as visually (the use of echolocation being an obvious example here), the focus on positionality nevertheless encourages a bias away from the acoustic or auditory. Indeed, when it comes to representations of such positionality that are readily communicable, it may be argued that the acoustic must inevitably give way to the visual and tactual, and especially the visual. This is a consequence both of the fact that visual and tactual representations can more easily be given in stable, persisting forms (texts, maps, plans, pictures, and so on) that are available on multiple occasions by multiple users as well as the greater facility offered by vision and touch for the simultaneous presentation of co-existing particulars – and, in respect of the latter, vision has certain advantages even over touch.

The tendency for positionality to win out over situatedness in architectural design contexts, in particular, and especially in building design, is reinforced by the fact that situatedness, because of its more complex character (which is partly a function of its irreducibly qualitative character and its resistance to purely metrical analysis), is much less amenable to manipulation and control. And this is a special problem in the case of sound. Unlike light, which can more readily be excluded or directed, sound is highly penetrative and difficult to contain (a direct consequence of the physical character of sound, as discussed further below). However, the difficulty in the management of sound is reduced with respect to enclosed or interior spaces, and so with respect to such spaces, the amenability of sound to design is greater. It is also with respect to enclosed or interior spaces – almost exclusively so – that the management of sound arises as a significant topic in architectural practice.

A real concern with sound, even in interior spaces, is, however, the exception rather than the rule. For the most part, sound is an issue only in those buildings and spaces in which sound, and acoustic experience, is already identified as a key element.[6] Such buildings and spaces include concert and lecture halls, theatres, assembly and debating rooms, churches, mosques, temples and other places of worship, celebration, and contemplation.[7] Here the focus is indeed primarily on interior spaces (although in some cases there is also a need to accommodate sound as it might be projected outwards, as in the case of bell towers or minarets). The acoustic character of a space is largely a matter of the reverberative properties of a space – of the way sound is transmitted within that space. The reverberative properties of a space derive from a combination of factors but are primarily determined by the conjunction of surface materiality and spatial configuration. Large spaces whose walls, floors and ceiling are made of acoustically reflective and non-penetrative materials will typically be more sonorous, more acoustically 'alive', than smaller spaces whose interior surfaces are acoustically non-reflective or absorbent, with the latter being acoustically 'dead'. The positioning of walls, screens, or other architectural features which may block or reflect sound waves back into a space also operate to alter and complicate acoustic performance.

Since different activities, events, or purposes require different acoustics, the spaces associated with those events, activities and purposes must be designed with regard to their different reverberative properties. A concert hall for classical music, for instance, requires different properties as against a Protestant church. This is not merely because one is a space intended for musical performance and another for religious worship. Some religious practices include an emphasis on music whereas others (including many traditional Protestant practices) give priority to the spoken word, and such differences make for differences in the sort of spaces appropriate to those practices, and so to different religious traditions and denominations.

The focus, in these cases, on the reverberative character of interior spaces, although quite specific, is nevertheless entirely in keeping with the primary function of the sorts of buildings and spaces at issue, where the ability to hear certain sounds, and to hear in certain ways, is paramount. The specificity of focus thus reflects the specific nature of the building, of the space, and so also, of the design task – the acoustic character of the building being effectively set by the building's function and socio-cultural context in a way that the designer cannot afford to ignore. One might say that the particularity of the function and socio-cultural context, and the consequent requirement to attend to a more complex spatial structure that includes acoustic or auditory as well as visual and other elements, forces greater attentiveness to topological situatedness rather than mere spatial position. The culturally charged character of the sorts of buildings in which acoustic and auditory considerations are prominent undoubtedly also brings with it, if sometimes implicitly, greater attentiveness to such situatedness. It is hard to engage adequately with the design of a religious building, for instance, if one does not attend to the complex set of elements, and the mode of orientation to the world, that religious thought, feeling, and practice encompass.

Yet the attention given to sound, and to acoustic issues generally, in most design projects, whether residential or non-residential, public, or private, is typically less concentrated and much less sophisticated, if it arises in any significant way at all, than in the sorts of cases just considered.

Most often, where acoustic issues are given attention, it is in relation solely to the management of sound penetration: for instance, the control of external sound from a nearby highway as it might intrude into an apartment or of sound as it might issue from a private meeting room into a more public foyer. Moreover, in many cases, such problems are completely overlooked in the design phase and are only recognised or addressed later.

The increased employment of open plan designs in many institutional workplaces is driven by the widespread desire for cost reduction above all else (increased managerial control and surveillance is sometimes also at issue), with inadequate attention, if any, to acoustic considerations.[8] Nevertheless, such designs frequently give rise to problems of acoustic penetration and interference within spaces in ways that almost always have a negative impact on the usability of those spaces (such problems arise not only as a result of open-plan office design, but also as a result of circulation and other public spaces that are constructed with acoustically reflective materials and are insufficiently separated from other areas, especially from areas where reduced levels of noise might otherwise be required). Once again, these sorts of problems are frequently given attention only after the building has been occupied and users begin to voice complaints. The acoustic issues are then addressed, if at all and then usually inadequately, through the addition, for instance, of baffles and screens, sometimes the insertion into open spaces of acoustically insulated 'pods' (or retrofitting of separate rooms), and even such ad hoc solutions, in some workspaces, as the distribution of noise-reducing headphones.

Of course, sound is neglected in both interior and exterior spaces. And it is arguably even more neglected when it comes to exterior spaces than with respect to interiors. There are some prima facie exceptions, most notably in the case of stadiums and arenas that have a degree of exteriority in being open to the elements (though some are entirely enclosed or capable of being so) and with respect to which sound projection is frequently a key issue in virtue of the size and openness of the space and the tendency for acoustic dispersion and distortion. Yet partly because the distinction between exterior and interior is correlative rather than absolute, and partly because such spaces do possess an interiority of their own, stadiums and arenas do not present a straightforward case of acoustic design as addressed to exterior spaces. Moreover, the acoustic design issues that tend to dominate in such cases are almost entirely those related to sound amplification and projection, and are typically addressed, not as an integrated element in building design, but through the installation of separate sound and public address systems and with little attention to the more complex interaction of sound with the structure of the building.

More relevant to the issue of sound in exterior spaces is the fact that the sound generated in stadiums and arenas, as a result of the nature of both the sound and the structure, means that the sound produced within them readily extends beyond the boundaries of the structure and its site. In these cases, the penetrative or expansive character of sound gives rise, not to problems concerning the exclusion from within of sounds from without, but of the exclusion from without of sounds from within – although it is seldom that this figures as a problem in stadium and arena design as such, being addressed at what might be viewed as another, more general level, namely that of planning and urban design.

The issue of sound as it relates to the spaces exterior to built structures, and not just stadiums or arenas but across the entire spectrum of architectural design, concerns the larger acoustic or auditory landscape, the soundscape, to use the term coined by R. Murray Schafer,[9] to which architecture contributes. The contribution at issue here includes the sound generated by buildings through the projection of amplified sound into the surrounding locale and through the sound produced by components in building infrastructure (generators, ventilation, heating and colling systems, opening and closing mechanisms, and so forth) along with the activities those buildings enable (which includes the sound produced by building occupants and users and the infrastructure that enables building access, especially as related to transportation systems). However, it also includes, though this is less frequently given explicit attention, the way the exterior surfaces of buildings, no less than their interior surfaces, give rise to acoustic effects of their own.[10] The general acoustic issues that appear here sometimes enter into urban design thinking, but this is most often the case in relation to the generation of specific forms of sound of the sort associated with the stadium or arena as well as other buildings, whether entertainment venues or commercial and industrial facilities and including the sound indirectly generated by urban structures through specific elements of their associated infrastructure (most notably, transportation). Relatively little attention is given to the many other acoustic effects to which buildings give rise, and the tendency is often to focus on reducing the level of sound produced rather than a broader design concern with the overall acoustic environment.[11]

It would be easy to say the general treatment of sound in design, and the tendency, except in certain special cases, to give it only secondary consideration, if any at all, provides evidence, in addition to that which is commonly cited in much of the literature, of the supposedly problematic ocularcentrism of architecture and design more broadly (the charge is often directed at European thought in general). Sound, like touch and smell, appears as secondary to vision, often addressed as an afterthought, and usually as something peripheral. It is certainly true that vision does have a priority over sound and hearing in relation to spatial perception, and this is reflected in the centrality given to space as visually experienced and understood. This is not, however, an arbitrary prioritisation or one that lacks foundation. Rather it is grounded in certain facts about the character of human beings and differences in the ways the visual and acoustic, as well as other sensory modalities, relate to spatiality.

Sound, as already noted, is not nonspatial, nor is it temporal rather than spatial, and as was also indicated earlier, neither does it allow for the same sort of spatial apprehension and discrimination that is possible with respect to vision. What gives vision, and visual representation, an advantage here is the fact that it allows, in especially effective fashion, the simultaneous perception of multiple objects within a single, extended 'field'. This is not impossible within a purely acoustic or auditory frame,[12] and that it is so indicates something of the spatial character that belongs to sound. But such simultaneity of presentation is more limited when given in acoustic or auditory terms alone, lacking the same degree of complexity and precision that is possible through the involvement of vision. The importance of vision in spatial cognition, in general, is confirmed by studies that show significant differences in spatial capacities on the part of sighted individuals compared to individuals blind from birth.[13] Moreover, the importance of vision here is not some contingent consequence of the way the human perceptual system happens to be configured – a result of some evolutionary accident – but instead relates to the very nature of sound, vision, and spatiality. It is thus, one might say, ontological rather than merely biological.

That sound and vision are different is an obvious truism, but the exact nature of the differences and their implications is less clear or straightforward. This remains true even though many of the differences between sound and vision are a consequence of the differences in the physical properties of light and sound waves. Sound is mechanical, being caused by vibrations in objects that give rise to vibrations in the surrounding medium which move out in waves parallel to their direction of travel (sound waves are thus longitudinal). Light is electromagnetic, therefore requiring no medium, and moving in waves perpendicular to the direction of travel (light waves are transverse). Yet although these physical differences do indeed make for certain basic differences in hearing and vision, the way the latter actually function, and so also the way sound and vision are experienced, is complicated by the way the different sensory modalities interact, by the way, they engage with higher level cognition and with agency, and, when it comes to space, in particular, by the way, all sensory modalities already depend upon, and so effectively encode, similar spatial and topological elements and structures.

One of the features of sound that is often taken as a key point of contrast with vision is its encompassing or environing character. Thus, Don Idhe writes that

> [the] auditory field ... is not isomorphic with visual field and focus, it is omnidirectional ... Were it to be modelled spatially, the auditory field would have to be conceived of as a 'sphere' within which I am positioned, but whose 'extent' remains indefinite as it reaches outward toward a horizon ... as a field, the auditory field-shape is that of a surrounding shape.... The auditory field surrounds the listener, and surroundability is an essential feature of the field-shape of sound.[14]

It is true that the auditory and visual fields are differently configured – they are not isomorphic – and this is tied to the difference in the longitudinal and transverse character of sound and light waves. But the difference at issue here is not properly one of omnidirectionality in the case of sound – sound and hearing both have a directional character with the acoustic and auditory fields also being directional or anisotropic, and not only does vision alter with the orientation of the body, but so too does hearing (as anyone with hearing impairment can attest). And although one can hear and respond to a sound directly behind even if one's auditory attention is elsewhere, whereas this generally cannot happen in the case of vision (although a light source placed behind the head may still be partially discernible), there is still a degree of attenuation in the case of audition from the centre of auditory attention to the periphery. Moreover, the sense of being surrounded or encompassed by sound can also occur in the case of vision – light can be experienced as encompassing, sometimes as penetrating, even if this is experienced in a different fashion than in the case of sound.

After quoting from the first (1976) edition of Idhe's work, R. Murray Schafer asserts that the supposedly environing and omnidirectional character of sound 'is one reason why aural societies are "unprogressive" – they don't look ahead; their world is not streamlined, as the "visionary" would make it'.[15] Whether Schafer's assertion can be substantiated seems questionable,[16] but it exemplifies a tendency to overestimate the differences between sound and vision, to underestimate the similarities, and to take the supposed differences as giving rise to significant philosophical and cultural divergences that, insofar as they exist, depend on a much wider array of factors. The emphasis on difference in Schafer's account, and in many similar accounts, is itself reinforced by talk of acoustic or auditory as opposed to visual space – as if there were indeed two spaces here rather than one. Yet the truth is that space, and the structure that belongs to it, must be common to and underlie both hearing and vision (just as must the structure of place also), even though the way that space is apprehended differs between the sensory modalities at issue.[17] Talk of different spaces is thus best understood as a shorthand way of talking about different apprehensions of space. As noted earlier, the differentiated character of space as it is tied to place and situatedness, no matter the sensory modalities by which it is apprehended, will always be structured in terms of relations of near and far, interior and exterior, centre and periphery.

Understood as critical rather than descriptive, the charge of ocularcentrism, at least as it is usually expressed, is not well-founded despite being widespread. And this is especially so when questions of space and spatiality are at issue.[18] It is not so much the centrality of vision in relation to spatiality, taken on its own, that is problematic in design and elsewhere, but rather the almost exclusive reliance on certain forms of visual representation and modes of visualised spatiality, and the associated neglect of other sensory modalities and the broader topology in which they are embedded. This is most obviously so in the emphasis on the plan and section in architectural design (notwithstanding their importance as elements in the design process) and on the photograph or rendered perspective as the primary means by which architectural design is presented.

Sound can certainly be represented even when it cannot be heard (as music can be written and read without being played) – it can be represented discursively (through verbal or written description), through diagram or notation, and through static and dynamic images (sometimes through otherwise silent images that already have acoustic associations or connotations). That the representation of sound should often depend on visual representations, as in the case of diagrammatic and notational forms as well as written discourse, should not itself be seen as already implying any illegitimate bias. There is thus no barrier to the incorporation of sound in design even in represented form nor should the treatment of sound as it given representationally or purely conceptually be assumed itself to be necessarily problematic. Yet neither sound nor representations of sound, nor indeed acoustic ideas or concepts generally, are commonly found as elements in design and architectural presentations or proposals (though there are exceptions), and they seldom figure in the often-hyperbolic narratives with which so many contemporary architects like to clothe their projects. Instead, those narratives tend to focus on a limited range of visual elements (including conceptual elements as these are said to be visually represented) and from which references to sound or acoustics, as well as other modes of perceptual presentation, are generally absent (the exceptions mostly concerning sound as it is supposed to indicate the busy-ness, 'buzz', or sociality of a space).

The underlying problem here, and what the relative neglect of sound really indicates, is a tendency to ignore topological situatedness in favour of what is effectively a form of spatial positionality. In other words, the real connection of space to place is lost and the way space is itself derivative of place is forgotten. It is this that is manifest in the tendency to focus on certain forms of visual presentation or representation, like the photograph or plan, and to neglect sound and the other sensory modalities. Consequently, although the priority of vision in spatial perception and cognition undoubtedly contributes to the relative neglect of sound and acoustics in design, such neglect is not a necessary consequence of that priority. The neglect of sound reflects factors within the contemporary practice of design, of which the emphasis on specific visual-spatial forms and representations is one aspect, as well as arising out of broader tendency within modernity for the dominance of a narrowly spatialised understanding that tends to take space, understood as an extended field lacking any intrinsic limit, as having priority over place, the latter reduced to little more than spatial location or position within such a field. Such spatialisation underpins both the emphasis on specific visual and representational forms within design, but it is also the basis for the relative neglect of sound and audition in favour of the visual.

The emphasis on certain forms of visuality in contemporary architectural design is particularly evident in the prominence, particularly when it comes to non-residential design, given to exterior forms over interior spaces (often the two are dealt with separately in the design process, with the exterior being designed first, usually by a different architect, and the interior design having therefore to accommodate to an already determined exterior). Architectural presentation and discussion are thus dominated by visual images of the exterior appearance of the building, whilst the style of individual architects or architectural firms is most often a matter of exterior form and composition rather than any interior character. Again, this can be taken to be closely related to the emphasis on specific types of visuality and visualisation: the 'iconic' image is almost always an exterior rather than interior image, and interiors are much harder to present, in a way that conveys any distinctive character, in the photograph or the rendered perspective (or even in the sketch). This is partly because the focus of the exterior image is typically at some distance from the viewer – the object viewed, in this case, the building, can thus be seen more or less in its entirety, and so apprehended as a single object – at the same time as it is restricted to a single aspect of what is a single mode of perceptual presentation. Images of interiors, however, are, by their very nature, already inside, close to rather than at a distance from. And it is much harder, if not impossible, to gain an overall sense of the character of an interior by reference to any single mode of perceptual presentation or a single aspect of such presentation, whether visual, acoustic, or otherwise, taken on its own.

The difference that appears here between the 'exterior' and 'interior' as they may be amenable to presentation in the form of a single perceptual representation or 'image' echoes the difference noted earlier between 'being-at' and 'being-in'.[19] The exterior presentation can thus be understood as that which is associated primarily with being-at a location rather than being-in a situation (and this remains so notwithstanding fact that most such presentations will include, as part of the visual representation, elements of the surrounding environment). It is such exterior presentation that is typically given in the common forms of visual representation (across many different domains and not only in architecture) that are exemplified, as already noted, in the photographic or rendered perspectival image, in the plan and the 3-D walkthrough, in the diagram, the sketch, and the map. Such being-at is more difficult to capture in acoustic or auditory terms, despite the fact that sound can deliver considerable information about objects including but not restricted to location or position, and purely acoustic or auditory representations (which must be distinguished from visual representations of acoustic or auditory information) can also be more difficult to use – vision generally provides a more efficient means of representing, and so also storing and communicating, spatial information, and this is one of the factors that underlies its importance in spatial cognition.

Not just visual representation, but any single perceptual representation, including the acoustic or auditory, is much less well suited to enabling a grasp of being-in. And this applies equally to exterior and interior presentations – being-in is simply not amenable to adequate presentation by means of a single perceptual image or set of images – even though one might argue that it is indeed more of a problem in the case of interiors. One of the main reasons for this difficulty in representation overall is that being-in is tied to active engagement in place rather than mere perceptual acquaintance. Contemporary immersive technologies provide a closer facsimile to being-in just inasmuch as they also enable capacities for bodily action and agency. However, the sense of 'being-in' made available by such technologies, in their current form, is always compromised and limited, partly because of the restricted perceptual engagement they allow as well as the disjunction that remains between the active and perceptual possibilities in the immersive space as they sit alongside the persisting active and perceptual possibilities of the encompassing 'real world' space.[20] Genuinely immersive space would require complete active and perceptual immersion so that it would no longer be possible to speak of a mere 'representation'. The 'immersive' space would then have to be understood as a space constituted in relation to a place that was as 'real' as any other.

When sound appears as a focus for design, it is most often as already noted, in relation to the design of interior built spaces. And it arises because of the need for design to attend to the activities, and the perceptual capacities associated with those activities, that the spaces at issue are intended to enable and support. The neglect of sound in design can thus be seen as not merely a neglect of sound alone nor a result of a simple prioritisation of the visual. Instead, it can be said to arise out of a neglect of interiority, a neglect of being-in in favour of being-at, which is to say that it is also a neglect of the way spaces and places are constituted in terms of activity at the same time as they encompass a complex of perceptual engagement that includes both sound and vision.[21] Moreover, the relative neglect of sound is also tied to another issue that deserve additional discussion, namely, the affective quality of designed spaces and places, and the role of sound in that affectivity.

Affectivity is an important element in the character of topological situatedness or being-in (as was briefly alluded to near the beginning of this discussion), such situatedness arising out of the interplay between sensory modalities as well as encompassing the perceptual, cognitive, and agential. Even though sound may often be neglected in design, both the acoustic properties of a space and the sounds that are able to penetrate into that space or are generated within it contribute significantly to the overall character of the space, to its 'atmosphere' or overall affectivity. Even the experience of silence can depend on the way sound is experienced in a space, rather than being an experience of the simple absence of sound. Silence is something heard, and often heard in the gaps between sounds or in the acoustic space that remains even against a background of other sounds. The role of sound in the affective character of space, and so also of place, means that sound can sometimes be seen to be taken up in design, even when it is not directly referred to or acknowledged, through considerations of affectivity, and of atmosphere, mood and 'sense of place'.

However, the way sound contributes to the affective character of spaces and places is also tied to another related issue, namely the connectedness of spaces and places, and especially of interiors and exteriors. Because of its penetrative character, sound often provides one of the most important means by which interior spaces connect to the exterior environment. That connectedness can be a significant part of the experience of a space and the character that belongs to it, and so of the affective or atmospheric character of that space. Hearing rain on a roof, the sound of birds, the bells of a church or the call to prayer of the muzzein, even the chatter of voices from a street, can all contribute to the way an interior space is experienced, and to the sense of that space as an interior and of connectedness to an exterior – such sounds thereby contribute to the sense of being indeed in a place that is not just the space or place of the room or building (the interconnectedness of places being itself a central element in the character of place).

Initially, one might suppose that such connectedness is important primarily in relation to private and residential buildings, but that need not always be the case. Wherever the experience or atmosphere of a place is deemed significant and wherever value is given to a sense of situatedness in a larger environment or landscape (to a genuine sense of being somewhere) attention must be given to the contribution of sound to such affectivity and situatedness through the sense of connectedness that it enables. Arguably, connectedness is an important element in all spaces and places – being as important to what are conventionally thought of as exterior spaces and places no less than conventional interiors – and so must also be recognised as an important element in all buildings (with the possible exception only of the most narrowly utilitarian). Consequently, wherever design attempts to engage with space and place, there will always be questions to be asked as to the role and contribution of sound to the sense of spatial and topological connectedness, and how that role and contribution will be encompassed by the design process as well as its outcomes.

In those cases, in which sound is made an explicit concern in existing design practice, the focus is almost invariably on the management of sound within the enclosed interior. Sometimes, as was pointed out in the discussion above, this is about ensuring the acoustic performance of specialised spaces in accord with the activities those spaces are intended to support and enable, but most often what is at issue is the management of sound penetration – the aim typically being to reduce or prevent such penetration from the exterior to the interior. In this way, acoustic concerns readily lead to the acoustic isolation and insulation of interior spaces, sometimes from one another, and very often from the surrounding environment in which the building is embedded. In some locations, such acoustic separation of interior and exterior goes hand-in-hand with climatic separation – with the use of materials, mechanisms, and structures to ensure efficient management of airflow and temperature. Such climatic considerations are obviously important, but even where they do not require such a high level of separation of interior from exterior concern (for instance, in warmer climates or seasons), acoustic separation often remains an issue. And it is so largely because of the acoustic problems that attend on the wider environmental context – problems already alluded to earlier and to which aspects of building design, as well as the activities and infrastructure associated with built form, contribute.

The tendency towards the acoustic separation of building interiors may be seen as one manifestation of a broader tendency for separation between interior and exterior in many built forms, especially in urban situations. It is also evident in the separation between exterior and interior design both in its academic treatment and, at least with respect to most large design projects, in the very practice of design.[22] The exterior design of many buildings is thus undertaken separately, and typically by a different architect, from the design of the interior, with the exterior design usually being given precedence. But such separation, whether as evident in design discourse or architectural practice, or as it may be realised in built form, should not be taken for granted and it certainly ought not be assumed to be a basic starting point in building design. Where sound represents a problem, then clearly the acoustic separation of spaces can be an important design imperative. But it also needs to be recognised that such separation may not always be beneficial, and that there can also be virtues in enabling the sort of connection and communication between interior spaces, between interior and exterior spaces, and between buildings and their environments, that sound makes possible. The tendency to favour acoustic separation is often a result of a building's situation amidst a problematic acoustic landscape or soundscape. But such soundscapes are, as already noted, themselves the product of design and other choices at the civic and urban level as well as in relation to individual buildings. Inasmuch as it represents a kind of solution to the problem presented by those soundscapes, then acoustic separation can also be seen as effectively contributing, if indirectly, to their continuation.

If sound is taken seriously in design, then not only the acoustic environments provided by built interiors need to be considered, but also the exterior acoustic environments in which buildings are themselves situated. But this suggests that taking greater account of sound in design, especially as sound relates to space and place, must involve designers in attending to more than just the management of acoustic performance in specialised built spaces or the management of sound penetration. Sound is, no less than vision, an integral element in the way spaces and places are constituted and experienced, and so in the way they affect us and in the way we respond to them. Moreover, because of its highly penetrative and dispersive character, sound also represents a particularly important aspect of spatial and topological communication and connection. Greater attention to sound in design ought thus to lead to greater attention to both the connection and separation of spaces, to their overall affectivity and the importance of such affectivity to those spaces and places, and to the larger environmental context in which built forms are situated. In short, greater attention to sound requires a greater attentiveness to the placed character of built form and of the human interaction with it.

I would like to thank Randall Lindstrom for his critical but always helpful comments and suggestions on earlier drafts of this chapter.

Notes

1 Such a space is not the same as the 'space' that is referred to in talk of the interplanetary and interstellar regions that lie beyond the sphere of the earth. However, as a matter of fact, even those regions or 'spaces' are not entirely soundless, sound waves being generated by stars and other bodies, and transmitted by ionised gases or plasma, as well as taking the form of gravitational waves that require no other medium that the fabric of space-time. Well beyond the frequencies available to normal human hearing, such sounds are nevertheless detectable by scientific instruments.
2 See, for example, Strawson (1959), chapt.2.

3 Similarly, vision is not primarily or purely spatial, possessing its own spatiotemporal structure that is also dynamic. Even the simultaneous presentation of particulars within a single visual field is not atemporal, since the grasp of something as a particular or of a manifold of particulars depends on a set of consistent presentations over time as well as in space.
4 See Malpas (2018: 25–26).
5 As is discussed further below, sound most often arises as an issue in architectural design in relation to individual buildings, and especially in relation to building interiors. However, even the individual building is situated in a larger landscape, and even the acoustic treatment of individual buildings and their interiors ought not to prescind completely from the larger acoustic landscape or 'soundscape'. This reflects the character of architecture, more generally, as properly oriented towards place rather than merely on specific built forms apart from place, despite the tendency for architectural practice to adopt a much narrower focus on the latter. And although the focus of this chapter is more specifically on the relation between place, space, and sound in architectural design, the discussion is underpinned by a more general claim concerning architecture as fundamentally topological in character – see especially, Malpas (2021).
6 Although there are certainly prominent figures in the history of architecture who have attempted to include acoustic elements in their designs – including, for instance, Louis Kahn and Frank Lloyd Wright. As Randall Lindstrom has pointed out to me, Wright designed one of his most famous buildings, Fallingwater, so that the scenic waterfall that is such a prominent feature of the site would be heard rather than seen. One might argue that many of the best practitioners have always been attentive to sound even when not explicitly so, but the best practitioners are not always representative of the practice as a whole.
7 See, for example, the case studies of Raj Patel (2000). Patel's book is notable, however, for including case studies that are not restricted simply to those indicated above, but also include a wide range of spaces including residences and workplaces as well as retail and research facilities. It is intended, as its title suggests, to provide a basic guide to architectural acoustics in general. For another such work, see also Marshall Long (2005).
8 In the past, it was common for the acoustic issues to be partially addressed through the use of partitions made of acoustically absorbent materials that also allow for a degree of spatial segregation and visual separation (sometimes white noise was also introduced into office spaces to reduce background sound). Even such measures as these are largely absent from most contemporary open-plan workspaces.
9 See Schafer (1977).
10 See, for example, Yildirim and Arefi (2023).
11 That sound in exterior spaces (and its intrusion into interior spaces) is a significant concern and has been recognised since at least the 1970s. Understood as a problem of 'noise' and primarily in relation to its effects on health, it was the subject of a World Health Organisation report published in 1980 (WHO Task Group on Environmental Criteria for Noise, 1980), as well as other studies (see Floyd, 1973; Rosen 1974), and the issue has since been the subject of many international and national reports (see, e.g., Peris, 2020), also giving rise to various legislative measures around the world.
12 See Malpas (2018).
13 See, for example, Homma (2000), Ungar (2000), Pasqualotto and Proulx (2012) and Gori et al. (2014).
14 Idhe (2007: 75–76).
15 Schafer (1985: 94–95).
16 The comments also manifest an ambiguity in Schafer's discussion between the contrast of acoustic and visual, on the one hand, and of the aural and verbal (though the latter contrast is never explicitly invoked).
17 In defence of Schafer, his work has been enormously important in drawing attention to the importance of sound and hearing and in giving impetus to work on acoustic and auditory issues, especially from an aesthetic, philosophical, and broadly socio-cultural perspective. The same is also true of the work of Idhe's phenomenological inquiries into hearing and listening which have pioneered a path that many others have since followed. Both Schafer and Idhe can also be seen to be reacting against the indisputable tendency for acoustic and auditory issues to be widely neglected.
18 Although even with respect to conceptuality and language, vision retains a necessary centrality, in part because of the connection between vision and spatiality – see Jay (1993: 8–9).
19 The latter being the fundamental mode of interiority – see Malpas (2021: 167–179).
20 On the relation between 'virtual' and 'real' spaces and places more generally, see Malpas (2009).
21 Interestingly, the dominance of exteriority, of positionality, of the narrowly spatial, and so also of certain forms of visuality, can be seen as reflected in a widespread tendency to neglect individual variations in sensory capacity – including hearing.

22 It should also be noted that a similar separation occurs with respect to the treatment of sound in design – issues of acoustic performance usually being handled by an acoustic design specialist or acoustician. There is no doubt that specialist knowledge and expertise are often needed, but the importance of such specialization should not be allowed to over-ride the need for an integrated approach to design. Just as sound ought to be an important consideration in design from the outset, so too, in those cases where specialist acoustic knowledge and expertise are necessary, that knowledge and expertise ought to be closely involved from the very beginning. Similarly, in academic and educational discourse, acoustic considerations ought to be encompassed as part of any general approach to design at the same time as they are also given specialist treatment. The fact that a topic may be deserving of specialization does not mean that it can simply be left to the specialist alone.

Bibliography

Floyd, Mary K. (1973). *A Bibliography of Noise 1965–1970* (Troy, NY: Whitston).
Gori, Monica, Giulio Sandini, Cristina Martinoli, David C. Burr (2014). Impairment of Auditory Spatial Localization in Congenitally Blind Human Subjects. *Brain* 137: 288–293.
Homma, Akinobu (2000). Spatial Cognition of Blind and Visually Impaired People in Their Daily Living Space. *Geographical Review of Japan* 73: 802–816.
Idhe, Don (2002). *Listening and Voice: Phenomenologies of Sound* (Albany, NY: SUNY Press).
Jay, Martin (1993). *Downcast Eyes: The Denigration of Vision in Twentieth-Century French Thought* (Berkeley: University of California Press).
Long, Marshall (2005). *Architectural Acoustics* (Amsterdam: Elsevier).
Malpas, Jeff (2009). The Nonautonomy of the Virtual. *Convergence: The International Journal of Research into New Media Technologies* 15: 135–139.
Malpas, Jeff (2018). *Place and Experience: A Philosophical Topography* (London: Routledge, 2nd edn).
Malpas, Jeff (2021). *Rethinking Dwelling: Heidegger, Architecture, Place* (London: Bloomsbury).
Pasqualotto, Achille and Michael J. Proulx (2012). The Role of Visual Experience for the Neural Basis of Spatial Cognition. *Neuroscience and Biobehavioral Reviews* 36: 1179–1187.
Patel, Raj (2000). *Architectural Acoustics: A Guide to Integrated Thinking* (London: RBA).
Peris, Eulalia, et al. (2020). *Environmental noise in Europe–2020* (Copenhagen: European Environment Agency).
Rosen, George (1974). A Backward Glance at Noise Pollution. *American Journal of Public Health* 64: 514–517.
Schafer, R. Murray (1977). *The Tuning of the World* (New York: Knopf).
Schafer, R. Murray (1985). Acoustic Space. In David Seamon and Robert Mugerauer (eds.), *Dwelling, Place and Environment* (New York: Columbia University Press): 87–98.
Strawson, Peter (1959), *Individuals: An Essay in Descriptive Metaphysics* (London: Methuen).
Ungar, Simon (2000). Cognitive Mapping Without Visual Experience. In Rob Kitchin and Scott Freundschuh (eds), *Cognitive Mapping: Past Present and Future* (London: Routledge): 221–248.
WHO (World Health Organization) (2011). *Burden of Disease from Environmental Noise – Quantification of Healthy Life Years Lost in Europe* (Copenhagen: WHO Regional Office for Europe).
WHO (World Health Organization) Task Group on Environmental Criteria for Noise (1980). *Environmental Health Criteria for Noise* (Geneva: WHO).
Yildirim, Yildiray and Mahyar Arefi (2023). Seeking the Nexus between Building Acoustics and Urban Form: A Systematic Review. *Current Pollution Reports* 13: 1–15.

24
SHAPING SOUNDS OF FUTURE ENVIRONMENTS

Eleni-Ira Panourgia

Introduction

Environmental change affects the propagation of sound and shapes acoustic environments.[1] In this chapter, I focus on sound in relation to landscapes and their future environmental states affected by drought. I explore imaginary dimensions of environmental sounds and the way their acoustic qualities transform spaces. My research involves incorporating sensory and embodied forms of knowing landscapes and non-human bodies such as those of plants, animals, water and soil by embodying their frequencies.[2] Listening in this manner provides another dimension for engaging with the possibilities and manifestations of the environmental sounds and the bodies they relate to. Guided by my listening experience, I realise field recordings of airborne, underwater and surface vibrations in landscapes of Potsdam, the state capital of Brandenburg, Germany whose sandy soils make them especially susceptible to drought. I seek to form an interplay between the listening experience and elements otherwise unreachable to our human ears through the development of a speculative sonic process for transforming environmental sounds as creative responses to drought scenarios for the region.

Based on my sound work Water-Drought Patterns (2023), I reflect on the proposed speculative sonic process and the perception of change through non-human perspectives. In working between real and fictional dimensions of landscapes, I propose environmental sonic states as a creative strategy and conceptual framework for speculating about future situations through the processing of field recordings in terms of texture, frequency, temporal and acoustic characteristics. Speculative sonic processes are used on the one hand to structure the sonic experience through fictional elements, and on the other hand act as motivating factors for imagining and relating to the environment differently, through alternative aural relationships. Such stories can be told through the perspective of an environment responding to change. This invites us to navigate unknown and unexplored areas of our perception through listening. In developing a speculative sonic process, my intention is to explore and produce experiences of imaginary states and future conditions, as ongoing and changing manifestations of landscapes and environments.

Attuning to changing relationships

[…] attunement speaks to subtle, affective modulations in the relations between different bodies.[3]

The sounds of elements and weather are interconnected to those produced by organisms. Their interaction shifts as climatic effects shape the ways they inhabit a given environment.[4] Such changing patterns influence sound propagation and our perception of environmental sounds. As Jérôme Sueur, Bernie Krause and Almo Farina mention, 'temperature directly governs the acoustic behavior of terrestrial, aquatic, and marine animals through chemical and physiological regulation'.[5] In addition, sound propagation is influenced by temperature and materials, as according to Garth Paine, 'sound travels further through denser material, such as cold air, than through warm summer air'.[6] Change in the density of plants and forest foliage is another factor that characterises the acoustic qualities of landscapes as '[r]educed plant density will change the balance between absorptive surfaces, such as leaves, and reflective surfaces such as rocks and buildings [...] [which] will increase reverberation and make sound environments more harsh'.[7] Environmental acoustic qualities and soundings are therefore, deeply affected by changes in climatic conditions.

To explore the sensory dimensions of change in acoustic environments, we need to direct our listening towards the relationships and interactions between multiple soundings. Drawing from Pauline Oliveros's deep listening practice,[8] Yolande Harris highlights that listening 'demands our attentiveness to other beings', and the importance of 'enter[ing] into a relationship with' sound.[9] For being attentive to 'the multitude of soundscapes' layers and rhythms [...] is to remain open to a possibility of being continuously recomposed by them', as Jacek Smolicki mentions.[10] Considering acoustic environments through inseparable relationships demands listening to operate at the nexus of their interactions rather than independent or single phenomena.

The sound of the wind is, for example, shaped through such interactions in landscapes.[11] The wind is according to Tim Ingold, 'the tree's double: there can no more be a living tree without wind than a living body without breath. We cannot see the breath of trees, but we can hear it in the rustling of leaves as the wind wraps around them'.[12] Listening to the trees together with their doubles is paying attention to their relationships with their surrounding settings, to 'the dynamics of ecological processes, interactions, and interrelationships, and [...] the principles of the interconnected ecosystem', as stated by Jonathan Gilmurray.[13] Natasha Myers refers to the importance of attuning to plants for 'learn[ing] how to sense alongside them'.[14] In dealing with the expression and perception of environmental change, attunement can propose alternative processes for relating differently, '[n]ot only through aural listening. Attunement means to tune into. To listen and to make adjustments to build harmonies or relationships between things', as AM Kanngieser suggests.[15]

Sensing through bodies and landscapes

Shifting frequencies of environmental sounds increase the complexity in perceiving their changing state; they also transform existing spaces and landscapes as well as ones that are emerging or unknown. For Brandon LaBelle 'encounters with unknown energies or foreign bodies' characterise spatial and temporal experiences 'and that extends our listening [...] upon the borders that work to delimit bodies and subjects, and acts of recognition'.[16] It is through such 'relational intensities' between bodies, noticeable and unrecognisable activities that environmental sound manifestations can offer new insight and ask about how these encounters might generate other experiences of the spaces to which they relate.[17] The sensory dimensions of the process of recognising and exchanging with other bodies through sound draws in this sense from the characteristics of acoustics spaces in which they interact. Astrida Neimanis and Jennifer Mae Hamilton consider bodies and landscapes in terms of 'weathering', as 'pay[ing] attention to how bodies and places respond to weather-worlds which they are also making'.[18] Weathering can offer more profound understandings of the intensities and temporalities involved in climate change impacts such as drought on bodies. How can we express such intensities and temporalities of changing bodies and landscapes through sound?

In AM Kanngieser's and Polly Stanton's radiophonic narrative And Then the Sea Came Back (2016)[19] we can listen to the ways the weather shapes bodies and places through the voice of the geolinguist, a fictional figure coming from Ursula Le Guin's work.[20] The geolinguist narrates moments before the tsunami hit the province of Aceh in Indonesia in 2004 through the earth's frequencies and vibrations allowing listeners to sense through them. Alex Wand proposes in Court of Cicada (2021) a mode of perspective listening that prompts us to listen from the perspective of non-humans, which can be enacted as follows:

> Begin by listening to all sounds. Your ears are an open book. After several minutes begin to focus attention on a sound that stands out to you. Then begin to listen from the perspective of where the sound source is coming from. If you're hearing a bird, listen as the bird [...].[21]

In employing our imagination as we listen, we engage with new ways of being in the environment and sensing through other bodies. Both the narration of the geolinguist and perspective listening create opportunities not only for attuning to environmental sounds and relationships, but also for furthering our imagination to shift our focus beyond the human perspective.

Speculating with sound and listening

> [...] learning to feel, think and listen to the slow effects of climate change is an orientation that is possible through speculative practices [...] a productive mode of thinking, feeling and acting that responds to problems that seem intractable. Such responses are experimental, emergent and creative.[22]

Kaya Barry, Michelle Duffy and Michele Lobo argue that listening is paying 'attention to that which cannot be easily grasped or not accessible to the human (individual) alone that strengthens our focus on listening with the planet'.[23] They propose speculative listening 'for reconceptualising our relations with the world, even as the planetary scale of the Anthropocene can exceed our comprehension'.[24] From a similar standpoint, Ivo Louro et al. call for 'more poetic methods, such as storytelling and sonic fictioning [...] [to] address dimensions of the spectrum that fall beyond the human hearing range while alluring us to sense how other-than-human entities and diverse modes of existence might hear'.[25] Speculating on future environments regards focussing on the ongoing change that is already effective, but also on the imaginative potential of frequencies and relationships through ways of listening that we might not be able to physically manage as humans.

Imagination is required for activating such speculative propositions. Anthony Dunne and Fiona Raby refer to design users as 'imaginers' to underline the importance of shifting perspectives for fully perceiving the speculative experience.[26] Drawing from Dunne and Raby, we suggest with Guillaume Dupetit the term 'listeners-imaginers'[27] to adapt this approach to sound for perceiving and further acting upon speculative sonic processes. For James Auger 'a design speculation requires a bridge to exist between the audience's perception of their world and the fictional element of the concept'.[28] Fiction as a method cultivates then the imaginary potential and reinforces our ability to speculate on future or unknown situations. It enables us to 'turn toward these unexplored, under-explored, and often denigrated territories of thinking and awareness […] moving us beyond the impasses of the present, in opening to the radically new, embracing or reinvigorating the incoming future', as Jon K. Shaw and Theo Reeves-Evison affirm.[29]

Listening through fictional elements plays a central role in questioning existing ways of perceiving. For Simon O'Sullivan, '[f]ictioning inserts itself into the real in this sense – into the world as-it-is (indeed, it collapses the so-called real and the fictional), but, in so doing, it necessarily changes our reality'.[30] With the term fictioning, Burrows and O'Sullivan indicate 'the writing, imaging, performing or other material instantiation of worlds or social bodies that mark out trajectories different to those engendered by the dominant organisations of life currently in existence'.[31] My interest lies in the action that is contained in the term fictioning; how to act upon environmental sounds and their realities to produce other audible perspectives. This takes a step further from listening to the existing sound sources as in perspective listening, to constructing alternative realities that demand from us to relate differently. To do so, embodied and technological dimensions combined can act as 'imaginary devices', as we discuss with Carla J. Maier.[32] In this respect, speculating with sound 'can be activated and perceived through the processing of sound material'.[33]

Shaping sounds of future environments

In attempting to express drought effects through audible speculative forms, I engage with environmental sounds and investigate ways of sensing through non-human bodies. In the following part, I expand on my listening and field recording practice, my inquiry about drought scenarios in the Brandenburg region, and the development of my speculative sonic process and its employment in my sound work Water-Drought Patterns.

Listening and field recording in Potsdam

I start by tuning in to landscapes of Potsdam in the areas of Innenstadt, Nördliche Vorstädte and Babelsberg, and I immerse myself in sound. Listening and physically interacting with landscapes by moving with them, offers me ways of knowing the environment. In listening I connect with landscapes, the Havel River and lakes, and explore plant and animal, soil and watery vibrations at different times and seasons. I familiarise myself with the acoustic characteristics of landscapes and their patterns by focusing on the frequencies and textures of interactions among their noticeable sounds. I also engage with their subtler and unnoticeable frequencies by taking the perspective of the bodies of other beings and imagining the ways in which they respond in sound, as they undergo change. Using microphones as tools for exploration, I concentrate on non-human beings and climate, and how their relationships can be grasped and reimagined. Recordings are undertaken with different types of microphones including omnidirectional, cardioid, hydrophone and contact microphones to pick up a variety of different signals ranging from ambiences to specific sound sources, to underwater acoustic signals and surface vibrations on trees. This configuration allows focussing across levels of relationships from the more distant ones involving the overall acoustic environment to more distinct soundings and frequencies inaudible without technological means.

I reflect on my listening experience and field recordings through notetaking and photographing while in the field, and then through the analysis of my recordings in the studio. I ask, in what ways can the relationships heard be emphasised and composed as alternative structures in space? How can they propose other readings of existing acoustic spaces and lead us to imagine changed environments? Building on the sensory and embodied qualities of my experience in the landscape,

I experiment in the following parts with the processing of field recordings using granular synthesis[34] as a main tool. My intention is to delve into the microcosm, textures and rhythms that emerge from the recordings and create more complex forms that articulate drought effects through non-human bodies and perspectives.

Drought scenarios in the Brandenburg region

I draw from climate models and scenarios for the region of Brandenburg, where Potsdam is located. The soil morphology of the region of Brandenburg makes it particularly vulnerable to drought. The drought severeness in Brandenburg in terms of total soil can be further seen in the drought monitor map based on soil moisture data in Germany by the Helmholtz Centre for Environmental Research (UFZ).[35] Ulrich Cubasch and Christopher Kadow indicate that the temperature in Berlin-Brandenburg region will increase by three to three and a half degrees Celsius by 2100 based on the Intergovernmental Panel on Climate Change (IPCC) scenario A1B.[36] Mirschel et al. stress the dramatic future impacts of water deficit due to climate change and the low absorbing sandy soils of Brandenburg leading to problems in agricultural production. The region will suffer 'serious limitations in plant-available water during the main growing period of agricultural crops'.[37]

In response to my inquiry about climate change impacts, models and scenarios in the region of Brandenburg, Claas Nendel, Co-Head of Research Platform Data Analysis and Simulation, Head of WG Ecosystem Modelling and Landscape Modelling at The Leibniz Centre for Agricultural Landscape Research (ZALF) and Professor of Landscape System Analysis at the University of Potsdam, shared information about the increasing weather extremes and the way these affect landscapes and ecosystems. In our online conversation on November 28, 2022, we engaged with the question of expressing the impacts of a transition to drier conditions leading to drought through sound. For example, how can sound express plant water supply? What kind of textures, frequencies and rhythms can be employed to manifest dry plant bodies moving and interacting with elements, forces and weather? Based on Nendel's experience, people tend to connect easier to sounds that they are familiar with. For instance, relating sounds of insects that only appear on hot summer days or the sounds of walking on dry sand paths to the feeling of high temperature and aridity, while sounds of creaking snow and other dampened sounds are immediately associated with snow-covered landscapes and cold temperatures.

As possible directions, we considered the creative transformation of field recordings in terms of liquid and solid or wet and dry states, and their impacts on soil and plants. Following Nendel's observations, the more the creative transformations of the sounds include the essence of such previously experienced elements including impact sounds on materials that reflect dry conditions, the more likely they will cause the intended association. Furthermore, we thought about how plants that do not grow well could produce different sounds. Trees suffering drought conditions can become less flexible, and tree movement can sound different if water supply is inadequate. Similarly, barley fields might sound different over the change of seasons and conditions such as when they are green or dry. This resonates with recent research by Itzhak Khait et al. whose recordings and analysis of ultrasonic airborne sound of plants shows that 'plants emit sounds, and that stressed plants – both drought-stressed […] and cut plants […] – emit significantly more sounds'.[38] In their recordings, we can listen to rhythmic-like patterns of emitted plant signals that respond to such change.

Our discussion about drought also led us to think about expressing different seasons through sound as shifting weather conditions affect biological cycles in landscapes. We listened to one of my field recordings realised at Nuthepark in Potsdam (Figure 24.1) in October 2022 near the

Figure 24.1 Landscape in Nuthepark, Potsdam. Photograph by Eleni-Ira Panourgia.

Havel River featuring mainly sources of water, ducks and human activity. We then listened to a transformed version of the same recording using granulation, pitch shifting and a high pass filter to produce an environment of higher frequency, drier impacts and additional sources. The transformed sounds suggest a change in biodiversity with more bird species and insects, and drier plants populating the landscape. This reminded us of an environment that we would encounter in that same location in July. Sound can contribute to enhancing our perception of transitions between environmental states and the effects these bring to landscapes and bodies.

Water-drought patterns

Drawing from the discussion with Claas Nendel about the drought scenarios presented above and the knowledge that I have developed about the landscape through my listening and field recording practice, I think about sound in relation to water in the landscape and the effect of its presence or absence. Water-Drought Patterns[39] focuses on the transition of a landscape close to water in Potsdam such as the one in Figure 24.1 to drought conditions. I think of my creative process through sonic states that operate between environmental and sonic qualities to express drought effects. I author environmental sonic states informed by my listening experience in the landscape with an emphasis on the wateriness of interconnected bodies.[40] This reflects the listening perspectives of non-human bodies and their connection to other bodies including water and soil, and weather. Considered as a both conceptual and creative framework, environmental sonic states introduce fictional elements that guide the transformation (or fictioning) of field recordings. I apply the environmental sonic states to transform the recorded material in direct response to drought and the ways in which it weathers such bodies. I guide my explorations across four states: *listening perspectives*, *shifting frequencies*, *other temporalities*, and *states of matter*.

Shaping sounds of future environments

The environmental sonic state 'listening perspectives' portrays ways of listening from within non-human bodies such as plants as they respond to change. These are considered through textural events, listening positions in space and esoteric activities and functions of bodies. In the state 'shifting frequencies', new versions of sounds and their frequency content are explored to express change in the soundings heard; pitch variations and polyphony speculate about other seasons, or the different species present in the new version of the landscape. The state 'other temporalities' plays with the temporal scale of the recordings and of non-human bodies; these effects focussing on changed rhythms such as stretched and accelerated time, micro and reversed rhythms seeking to emphasise change and propose other or future temporalities as in different seasons or temporal perception of non-humans. Lastly, the environmental sonic state 'states of matter' places the focus on transitions between wet and dry conditions, and liquid and solid states; it is about paying attention to effects of change in non-human agency through tactile dimensions of sound and acoustic characteristics.

The environmental sonic states feed into each other and are applied in a semi-improvisational manner to shape and transform my field recordings near water and underwater in the Havel River, on surfaces and tree barks (Figure 24.2). I perform and record live granular synthesis manipulations of the recordings digitally, to create connections between the microcosm of the sounds, and more complex overviews of sonic layers. Drawing from the concept and directions that each environmental sonic state offers, I scan through my recordings to explore micro events, textures and rhythmic patterns in different positions of their waveforms. I further manipulate the selected sounds in terms of grain size, spray, velocity, pitch and envelope parameters to express non-human perspectives and changing states of matter. Scanning through the sounds changes the time of the

Figure 24.2 Recording underwater and surface vibrations on a tree bark in Nuthepark. Photographs by Eleni-Ira Panourgia.

recordings through reverse, forward, speed up and slowed down time or multiple temporal moments heard at once. This transformed time indicates non-human temporal scales in play that challenge our human perception of duration or linearity in the perception of environmental change. All four environmental sonic states suggest a fictional dimension of non-human sensing. These encourage us to imagine what our perception of drought effects could be like from the perspective of a plant, speculating about the future auditory experience in climate-changed landscapes.

Water-Drought Patterns presents multiple listening perspectives such as listening from within plants and wandering around them, and multiple temporal levels of accelerated, reversed and stretched time that communicate the rhythm of change of plants as they are adapting, struggling or behaving differently. The water sounds in the beginning of the work move from a soothing and healthy effect to a point of uncomfortable contrasts highlighted through dry sonic textures that become more pronounced and tactile as the drought effects gain intensity. Sound impacts on surfaces and plants, rattling and clicking sounds and rhythmic patterns create an interplay between higher-pitched textures and gestures.[41] Drying plants moving differently to adapt are expressed through their esoteric activity and exteriority of the landscape sharing their condition with other bodies such as those of animals, changing water levels and supply, and soil condition.[42] The more synthetic sounds and sustained materials give space for new conversations between textures and grains, stretched over time to generate a different version of changing state and considering the overall landscape.[43]

The listening experience reveals a changing environment that requires us to continuously refocus on the acoustic characteristics and relationships situated in-between changes. The transformed sounds give way to more abstract interventions reflecting the perception of change, its spatial and temporal connotations. Starting from a sound that has a fixed relationship with its source towards processed versions, we challenge at the same time the question of recognisability through a sense of intensity and interruption of familiar sounds and spaces.[44] While environmental sounds characterise places and indicate their acoustic qualities, they can also act as a reference point for discovering change through sensory encounters. The proposed speculative sonic process produces experiences of change by embodying other organisms. Imagining sound from within other bodies contributes to creating alternative spaces that have the capacity to accommodate transformed sounds and promote speculative thinking. The speculative sonic process allows us therefore, not only to become more aware of non-humans, rather expanding our sensorium to feel as, and through the bodies of others; to forget our human perspective and become immersed into other scales and temporalities.

Expanding environmental sonic states

In the beginning of this chapter, I discussed the potential of listening, attunement and sensory engagement in response to the need for paying closer attention to environmental sounds through the relationships formed between non-human beings and climatic effects. I suggested that in employing fictional elements in the transformation of environmental sounds besides guiding listeners through the auditory experience, also operates as a way to bridge climate scenarios with acoustic environments. The transformation of field recordings through the proposed speculative sonic process and the listening experience resulting from Water-Drought Patterns prompts us to reposition ourselves within the resulting environmental soundings by listening from non-human perspectives, while imagining alternative relationships in changing environments. The acoustic spaces that emerge from the transformed relationships expand the limits of bodies[45] facing drought effects.

Attuning to and thinking through those soundings is a first step towards sensory understandings of change. This further demands that we act with them. The environmental sonic states explored constitute renderings of environmental sounds as the means for expressing climatic effects related to drought. In switching bodies, scales and perspectives, new spaces are constructed in which speculative events unfold. There is more to be done towards dynamic and collective considerations of this speculative sonic process. Beyond the radiophonic and installation formats of the work Water-Drought Patterns, the next steps involve incorporating the site-specific, sensory and embodied experience of listening and the transformation of sounds based on the environmental sonic states to offer listeners the possibility to construct their own speculative versions and relate differently to landscapes and non-humans. Imagining how to feel and sense through other bodies is hoped to make us more familiar with change in our acoustic environments and help us envisage alternative relationships between spaces and bodies in future landscapes.

Acknowledgements

The work presented in this chapter is part of the research project Listening to Climate Change (2022–2024) funded by the Postdoc Network Brandenburg and undertaken at Film University Babelsberg KONRAD WOLF. I would like to thank Angela Brennecke for her mentorship and support in this project. I am also grateful to Claas Nendel for his valuable input and expertise.

Notes

1 Jérôme Sueur, Bernie Krause, and Almo Farina, "Climate Change Is Breaking Earth's Beat," *Trends in Ecology & Evolution (Amsterdam)* 34, no. 11 (2019): 971. https://doi.org/10.1016/j.tree.2019.07.014; Garth Paine, "Listening to Nature: How Sound Can Help Us Understand Environmental Change," *The Conversation*, December 2018. https://theconversation.com/listening-to-nature-how-sound-can-help-us-understand-environmental-change-105794
2 Such form of embodiment draws from Astrida Neimanis's 'watery embodiment' in which "[w]ater extends embodiment in time – body, to body, to body. Water in this sense is facilitative and directed towards the becoming of other bodies. Our own embodiment, as already noted, is never really autonomous." Astrida Neimanis, *Bodies of Water: Posthuman Feminist Phenomenology* (London: Bloomsbury Academic, 2016), 3. https://doi.org/10.5040/9781474275415
3 Julian Brigstocke and Tehseen Noorani, "Posthuman Attunements: Aesthetics, Authority and the Arts of Creative Listening," *GeoHumanities* 2, no. 1 (2016): 2. https://doi.org/10.1080/2373566X.2016.1167618
4 Jérôme Sueur, Bernie Krause, and Almo Farina, "Climate Change Is Breaking Earth's Beat," *Trends in Ecology & Evolution (Amsterdam)* 34, no. 11 (2019): 971. https://doi.org/10.1016/j.tree.2019.07.014
5 Jérôme Sueur, Bernie Krause, and Almo Farina, "Climate Change Is Breaking Earth's Beat," *Trends in Ecology & Evolution (Amsterdam)* 34, no. 11 (2019): 971. https://doi.org/10.1016/j.tree.2019.07.014
6 Garth Paine, "Listening to Nature: How Sound Can Help Us Understand Environmental Change," *The Conversation*, December 2018. https://theconversation.com/listening-to-nature-how-sound-can-help-us-understand-environmental-change-105794
7 Garth Paine, "Listening to Nature: How Sound Can Help Us Understand Environmental Change," *The Conversation*, December 2018. https://theconversation.com/listening-to-nature-how-sound-can-help-us-understand-environmental-change-105794
8 Pauline Oliveros, *Deep Listening: A Composer's Sound Practice* (New York: IUniverse, Inc., 2005).
9 Yolande Harris, "Melt Me into the Ocean: Sounds from Submarine Spaces," in *The Bloomsbury Handbook of Sonic Methodologies*, ed. Michael Bull, and Marcel Cobussen (New York: Bloomsbury Academic, 2021), 473. https://doi.org/10.5040/9781501338786.ch-030
10 Jacek Smolicki, "Composing, Recomposing, and Decomposing with Soundscapes," in *Soundwalking: Through Time, Space, and Technologies*, ed. Jacek Smolicki (London: Focal Press, 2022), 183. https://doi.org/10.4324/9781003193135
11 R. Murray Schafer, *The Soundscape: Our Sonic Environment and the Tuning of the World* (Rochester, VT: Destiny Books, 1994).
12 Tim Ingold, *Correspondences* (Cambridge: Polity Press, 2021), 34.

13 Jonathan Gilmurray, "Ecological Sound Art," In *The Bloomsbury Handbook of Sonic Methodologies*, ed. Michael Bull and Marcel Cobussen (New York: Bloomsbury Academic, 2021), 456. https://doi.org/10.5040/9781501338786.ch-028

14 Natasha Myers, "Becoming Sensor in Sentient Worlds: A More-Than-Natural History of a Black Oak Savannah," in *Between Matter and Method: Encounters in Anthropology and Art*, ed. Gretchen Bakke and Marina Peterson (London: Bloomsbury Academic, 2017), 76.

15 Anja Kanngieser, "Listening to Ecocide," Filmed February 2020 at Sonic Acts Academy, De Brakke Grond, Amsterdam, The Netherlands. Video, 47:37. https://www.youtube.com/watch?v=dh1Cs4G7mxk

16 Brandon LaBelle, "Minor Acoustics: Sound Art, Relationality, and Poetic Listening." In *The Oxford Handbook of Sound Art*, ed. Jane Grant, John Matthias, and David Prior (Oxford: Oxford University Press, 2021), 455. https://doi.org/10.1093/oxfordhb/9780190274054.013.30

17 LaBelle discusses relational intensity in acoustic spaces through the field recording work of Jana Winderen mentioning that "sounds are worked through as a relational matter, specifically intensifying contact or a sense of encounter with what is out there […] give way to an intensity of presences found in habitats different from one's own, allowing for the distant to become intensely proximate, to interrupt us." Brandon LaBelle, "Minor Acoustics: Sound Art, Relationality, and Poetic Listening." In *The Oxford Handbook of Sound Art*, ed. Jane Grant, John Matthias, and David Prior (Oxford: Oxford University Press, 2021), 456. https://doi.org/10.1093/oxfordhb/9780190274054.013.30.

18 Astrida Neimanis and Jennifer Mae Hamilton, "Open Space Weathering," *Feminist Review* 118, (2018): 81. https://doi.org/10.1057/s41305-018-0097-8.

19 AM Kanngieser and Polly Stanton, "And Then the Sea Came Back," 2016, accessed March 31, 2023. https://amkanngieser.com/posts/and-then-the-sea-came-back

20 Ursula Le Guin, "Keynote Speech" (*Keynote presentation, Arts of Living on a Damaged Planet Conference*, Aarhus University, Denmark, May 8, 2014). https://anthropocene.au.dk/conferences/arts-of-living-on-a-damaged-planet-may-2014

21 Alex Wand, "Court of Cicada," *Mapping Sonic Futurities*, 2021, Accessed April 10, 2023. https://mappingsonicfuturities.com/court-of-cicada

22 Kaya Barry, Michelle Duffy, and Michele Lobo. "Speculative Listening: Melting Sea Ice and New Methods of Listening with the Planet," *Global Discourse* 11, no. 1–2 (2021): 116. https://doi.org/10.1332/204378920X16032963659726

23 Kaya Barry, Michelle Duffy, and Michele Lobo. "Speculative Listening: Melting Sea Ice and New Methods of Listening with the Planet," *Global Discourse* 11, no. 1–2 (2021): 117. https://doi.org/10.1332/204378920X16032963659726

24 Kaya Barry, Michelle Duffy, and Michele Lobo. "Speculative Listening: Melting Sea Ice and New Methods of Listening with the Planet," *Global Discourse* 11, no. 1–2 (2021): 121. https://doi.org/10.1332/204378920X16032963659726

25 Ivo Louro, Margarida Mendes, Daniel Paiva, and Iñigo Sánchez-Fuarros, "A Sonic Anthropocene: Sound Practices in a Changing Environment," *Cadernos de Arte e Antropologia* 10, no. 1 (2021): 11. https://doi.org/10.4000/cadernosaa.3377

26 Anthony Dunne and Fiona Raby. *Speculative Everything: Design, Fiction, and Social Dreaming* (Cambridge: The MIT Press, 2013), 93.

27 Guillaume Dupetit and Eleni-Ira Panourgia, "Field Recording et Fictions Soniques: Vers Une Représentation des Espaces Urbains," *Filigrane: Musique, esthétique, sciences, société* no. 26 (December 2021). https://dx.doi.org/10.56698/filigrane.1109

28 James Auger, "Speculative Design: Crafting the Speculation," *Digital Creativity* 24, no. 1 (2013): 12. https://doi.org/10.1080/14626268.2013.767276

29 Jon K. Shaw and Theo Reeves-Evison, "Introduction," in *Fiction as Method*, ed. Jon K. Shaw and Theo Reeves-Evison (Berlin: Sternberg Press, 2017), 8.

30 Simon D. O'Sullivan, "Art Practice as Fictioning (or, Myth-Science)," (*Paper Presented at the Arts of Existence: Artistic Practices, Aesthetics and Techniques Seminar*, Unit of Play, Goldsmiths, United Kingdom. May 14, 2014), 6.

31 David Burrows and Simon O'Sullivan, *Fictioning: The Myth-Functions of Contemporary Art and Philosophy* (Edinburgh: Edinburgh University Press, 2019), 1. https://doi.org/10.1515/9781474432412

32 Eleni-Ira Panourgia, and Carla J. Maier, "Imaginary Devices for Listening to Water: Environmental Change, Wateriness and the Partiality of Perception," (*Paper Presented at RGS-IBG Annual International*

Conference, Royal Geographical Society and Imperial College London, United Kingdom. August 29–September 1, 2023).
33 Guillaume Dupetit and Eleni-Ira Panourgia, "Field Recording et Fictions Soniques: Vers Une Représentation des Espaces Urbains," *Filigrane: Musique, esthétique, sciences, société* no. 26 (December 2021). https://dx.doi.org/10.56698/filigrane.1109. Own translation.
34 Granular synthesis allows to work on small segments or grains of field recordings, which "serve as […] building block[s] for sound objects. By combining thousands of grains over time, we can create animated sonic atmospheres." Grains can be manipulated across "two perceptual dimensions: time-domain information (starting time, duration, envelope shape) and frequency-domain information (the pitch of the waveform within the grain and the spectrum of the grain)." Curtis Roads, *Microsound* (Cambridge: MIT Press, 2001), 87.
35 Helmholtz Centre for Environmental Research, "Drought Monitor Germany", accessed March 31, 2023, https://www.ufz.de/index.php?en=37937.
36 Ulrich Cubasch, and Christopher Kadow, "Global Climate Change and Aspects of Regional Climate Change in the Berlin-Brandenburg Region," *DIE ERDE – Journal of the Geographical Society of Berlin* 142, no. 1–2 (2011): 3–20. https://www.die-erde.org/index.php/die-erde/article/view/40; Intergovernmental Panel on Climate Change (IPCC), *Climate Change 2007: The Physical Science Basis. Contribution of Working Group I to the Fourth Assessment Report of the Intergovernmental Panel on Climate Change*, ed. Susan Solomon et al. (Cambridge: Cambridge University Press, 2007).
37 Wilfried Mirschel, Ralf Wieland, Karin Luzi, and Karin Groth, "Model-Based Estimation of Irrigation Water Demand for Different Agricultural Crops Under Climate Change, Presented for the Federal State of Brandenburg, Germany," in *Landscape Modelling and Decision Support. Innovations in Landscape Research*, ed. Wilfried Mirschel, Vitaly V. Terleev, and Karl-Otto Wenkel (Cham: Springer, 2020), 312. https://doi.org/10.1007/978-3-030-37421-1_16
38 Itzhak Khait, Ohad Lewin-Epstein, Raz Sharon, Kfir Saban, Revital Goldstein, Yehuda Anikster, Yarden Zeron, Chen Agassy, Shaked Nizan, Gayl Sharabi, Ran Perelman, Arjan Boonman, Nir Sade, Yossi Yovel, and Lilach Hadany, "Sounds Emitted by Plants Under Stress Are Airborne and Informative," *Cell* 186, no. 7 (2023): 1330. https://doi.org/10.1016/j.cell.2023.03.009
39 Water-Drought Patterns was originally presented on the radio. A 12-minute version of the work was premiered on Radio Otherwise with Eleni-Ira Panourgia, Archipel Stations / Radio Otherwise, CTM Vorspiel, 31 January 2023, broadcasted online and on FM in Berlin 88,4 and Potsdam 90,7. A 6′44″ version of the work was presented in Radio Utopia/News from the World installation and online catalogue as part of the French Pavilion for the XVIII Venice International Architecture Biennale 2023, 1–5 August 2023. https://cressound.grenoble.archi.fr/son/2023_VENEZIA/Catalog_NewsFromTheWorld.html
40 Astrida Neimanis, *Bodies of Water: Posthuman Feminist Phenomenology* (London: Bloomsbury Academic, 2016). https://doi.org/10.5040/9781474275415
41 Sound excerpt 1: Water-drought patterns_Panourgia_excerpt 1.wav.
42 Sound excerpt 2: Water-drought patterns_Panourgia_excerpt 2.wav.
43 Sound excerpt 3: Water-drought patterns_Panourgia_excerpt 3.wav.
44 Brandon LaBelle, "Minor Acoustics: Sound Art, Relationality, and Poetic Listening," in *The Oxford Handbook of Sound Art*, ed. Jane Grant, John Matthias, and David Prior (Oxford: Oxford University Press, 2021), 453–470. https://doi.org/10.1093/oxfordhb/9780190274054.013.30
45 Brandon LaBelle, "Minor Acoustics: Sound Art, Relationality, and Poetic Listening," In *The Oxford Handbook of Sound Art*, ed. Jane Grant, John Matthias, and David Prior (Oxford: Oxford University Press, 2021), 453–470. https://doi.org/10.1093/oxfordhb/9780190274054.013.30

Bibliography

Auger, James. "Speculative Design: Crafting the Speculation." *Digital Creativity* 24, no. 1 (2013): 11–35. https://doi.org/10.1080/14626268.2013.767276

Barry, Kaya, Michelle Duffy, and Michele Lobo. "Speculative Listening: Melting Sea Ice and New Methods of Listening with the Planet." *Global Discourse* 11, no. 1–2 (2021): 115–129. https://doi.org/10.1332/204378920X16032963659726

Brigstocke, Julian, and Noorani, Tehseen. "Posthuman Attunements: Aesthetics, Authority and the Arts of Creative Listening." *GeoHumanities* 2, no. 1 (2016): 1–7. https://doi.org/10.1080/2373566X.2016.1167618

Burrows, David, and Simon O'Sullivan. *Fictioning: The Myth-Functions of Contemporary Art and Philosophy*. Edinburgh: Edinburgh University Press, 2019. https://doi.org/10.1515/9781474432412

Cubasch, Ulrich, and Christopher Kadow. "Global Climate Change and Aspects of Regional Climate Change in the Berlin-Brandenburg Region." *DIE ERDE – Journal of the Geographical Society of Berlin* 142, no. 1–2 (2011): 3–20. https://www.die-erde.org/index.php/die-erde/article/view/40

Dunne, Anthony, and Fiona Raby. *Speculative Everything: Design, Fiction, and Social Dreaming*. Cambridge: The MIT Press, 2013.

Dupetit, Guillaume, and Eleni-Ira Panourgia, "Field Recording et Fictions Soniques: Vers Une Représentation des Espaces Urbains." *Filigrane: Musique, esthétique, sciences, société* no. 26 (December 2021). https://revues.mshparisnord.fr/filigrane/index.php?id=1109

Gilmurray, Jonathan. "Ecological Sound Art." In *The Bloomsbury Handbook of Sonic Methodologies*, edited by Michael Bull, and Marcel Cobussen, 449–458. New York: Bloomsbury Academic, 2021. https://doi.org/10.5040/9781501338786.ch-028

Harris, Yolande. "Melt Me into the Ocean: Sounds from Submarine Spaces." In *The Bloomsbury Handbook of Sonic Methodologies*, edited by Michael Bull, and Marcel Cobussen, 469–480. New York: Bloomsbury Academic, 2021. https://doi.org/10.5040/9781501338786.ch-030

Helmholtz Centre for Environmental Research. "Drought Monitor Germany." Accessed March 31, 2023. https://www.ufz.de/index.php?en=37937

Ingold, Tim. *Correspondences*. Cambridge: Polity Press, 2021.

Intergovernmental Panel on Climate Change (IPCC). *Climate Change 2007: The Physical Science Basis. Contribution of Working Group I to the Fourth Assessment Report of the Intergovernmental Panel on Climate Change*, edited by Susan Solomon, Dahe Qin, Martin Manning, Zhenlin Chen, Melinda Marquis, Kristen Averyt, Melinda M.B. Tignor, and Henry LeRoy Miller, Jr. Cambridge: Cambridge University Press, 2007.

Kanngieser, Anja. "Listening to Ecocide." Filmed February 2020 at Sonic Acts Academy, De Brakke Grond, Amsterdam, The Netherlands. Video, 47:37. https://www.youtube.com/watch?v=dh1Cs4G7mxk

Kanngieser, AM, and Polly Stanton. "And Then the Sea Came Back." 2016. Accessed March 31, 2023. https://amkanngieser.com/posts/and-then-the-sea-came-back

Khait, Itzhak, Ohad Lewin-Epstein, Raz Sharon, Kfir Saban, Revital Goldstein, Yehuda Anikster, Yarden Zeron, Chen Agassy, Shaked Nizan, Gayl Sharabi, Ran Perelman, Arjan Boonman, Nir Sade, Yossi Yovel, and Lilach Hadany. "Sounds Emitted by Plants Under Stress Are Airborne and Informative." *Cell* 186, no. 7 (2023): 1328–1336. https://doi.org/10.1016/j.cell.2023.03.009

LaBelle, Brandon. "Minor Acoustics: Sound Art, Relationality, and Poetic Listening." In *The Oxford Handbook of Sound Art*, edited by Jane Grant, John Matthias, and David Prior, 453–470. Oxford: Oxford University Press, 2021. https://doi.org/10.1093/oxfordhb/9780190274054.013.30

Le Guin, Ursula. "Keynote Speech." *Keynote Presentation at the Arts of Living on a Damaged Planet Conference*, Aarhus University, Denmark, May 8, 2014. https://anthropocene.au.dk/conferences/arts-of-living-on-a-damaged-planet-may-2014

Louro, Ivo, Margarida Mendes, Daniel Paiva, and Iñigo Sánchez-Fuarros. "A Sonic Anthropocene: Sound Practices in a Changing Environment." *Cadernos de Arte e Antropologia* 10, no. 1 (2021): 3–17. https://doi.org/10.4000/cadernosaa.3377

Mirschel, Wilfried, Ralf Wieland, Karin Luzi, and Karin Groth. "Model-Based Estimation of Irrigation Water Demand for Different Agricultural Crops Under Climate Change, Presented for the Federal State of Brandenburg, Germany." In *Landscape Modelling and Decision Support. Innovations in Landscape Research*, edited by Wilfried Mirschel, Vitaly V. Terleev, and Karl-Otto Wenkel, 311–327. Cham: Springer, 2020. https://doi.org/10.1007/978-3-030-37421-1_16

Myers, Natasha. "Becoming Sensor in Sentient Worlds: A More-Than-Natural History of a Black Oak Savannah." In *Between Matter and Method: Encounters in Anthropology and Art*, edited by Gretchen Bakke and Marina Peterson, 73–96. London: Bloomsbury Academic, 2017.

Neimanis, Astrida. *Bodies of Water: Posthuman Feminist Phenomenology*. London: Bloomsbury Academic, 2016. https://doi.org/10.5040/9781474275415

Neimanis, Astrida, and Jennifer Mae Hamilton. "Open Space Weathering." *Feminist Review* 118 (2018): 80–84. https://doi.org/10.1057/s41305-018-0097-8

O'Sullivan, Simon D. "Art Practice as Fictioning (or, Myth-Science)." *Paper presented at the Arts of Existence: Artistic Practices, Aesthetics and Techniques Seminar*. Unit of Play, Goldsmiths, United Kingdom. May 14, 2014.

Oliveros, Pauline. *Deep Listening: A Composer's Sound Practice*. New York: IUniverse, Inc., 2005.

Paine, Garth. "Listening to Nature: How Sound Can Help Us Understand Environmental Change." *The Conversation*, December 2018. https://theconversation.com/listening-to-nature-how-sound-can-help-us-understand-environmental-change-105794

Panourgia, Eleni-Ira, and Carla J. Maier. "Imaginary Devices for Listening to Water: Environmental Change, Wateriness and the Partiality of Perception." *Paper Presented at RGS-IBG Annual International Conference*, Royal Geographical Society and Imperial College London, United Kingdom. August 29–September 1, 2023.

Roads, Curtis. *Microsound*. Cambridge: MIT Press, 2001.

Schafer, R. Murray. *The Soundscape: Our Sonic Environment and the Tuning of the World*. Rochester, VT: Destiny Books, 1994.

Shaw, Jon K., and Theo Reeves-Evison. "Introduction." In *Fiction as Method*, edited by Jon K. Shaw and Theo Reeves-Evison, 5–72. Berlin: Sternberg Press, 2017.

Smolicki, Jacek, "Composing, Recomposing, and Decomposing with Soundscapes." In *Soundwalking: Through Time, Space, and Technologies*, edited by Jacek Smolicki, 181–199. London: Focal Press, 2022. https://doi.org/10.4324/9781003193135

Sueur, Jérôme, Bernie Krause, and Almo Farina, "Climate Change Is Breaking Earth's Beat." *Trends in Ecology & Evolution (Amsterdam)* 34, no. 11 (2019): 971–973. https://doi.org/10.1016/j.tree.2019.07.014

Wand, Alex. "Court of Cicada." *Mapping Sonic Futurities*, 2021. Accessed April 10, 2023. https://mappingsonicfuturities.com/court-of-cicada

PART IV

Sound art and music

Emma-Kate Matthews

The final section of this book explores how sound in the form of 'music and sound art' interact with a wide variety of spatial contexts, from remote physical environments to intimate, performative, and virtual realms. This section examines how practitioners and researchers are able to cross disciplinary boundaries to challenge the roles of architecture, technology, and pedagogy in creating sonic experiences and how sound can shape these elements in return. The chapters presented here encompass diverse approaches to music and sound art, from collaborative compositions and performances in iconic architectural spaces to the development of digital archives that preserve global acoustic heritage. These chapters underline the richness and complexity inherent in the many ways in which sound and space can interact through sound-based creative expressions in their various forms. Throughout 11 chapters, music and sound art are presented as versatile and critical sonic mediums, capable of interrogating, inhabiting and defining built and theoretical spaces. Sound can be used to express, react, perform, question, analyse, interrogate, define, and enact space, amongst many other functions. Many of the chapters in this section discuss project-led research and practice-based examples drawing from live projects to glean insights and observations from works within both virtual and physical spatial contexts. In these chapters, the diverse spatial categories such as extreme, exterior, domestic, interior, performative, public, socio-political, virtual, and technological are intricately interwoven, often making them inseparable. This mirrors the multidimensional nature of the works discussed, as they frequently engage with several types of space simultaneously.

The first two chapters inquire into the numerous reciprocal relationships between architecture and music. This work not only questions the role of architecture in music performance and composition but also explores how music can transcend its traditional role as a tool for expression. It investigates how music also has a capacity to serve as a medium to understand the nuances of architectural character beyond what is immediately visible or geometrically tangible. Architect and composer Emma-Kate Matthews, also the editor of this book, reflects on a project which challenges disciplinary conventions by writing and performing music in collaboration with architecture rather than merely within or for it. Her exploration of extreme acoustic environments like the interior of the Basílica de la Sagrada Família allows her to scrutinise relationships between sound and space, where their interactions become especially pronounced, as if 'under a microscope'.

Similarly, musician Fabricio Mattos challenges conventional notions of music performance within fixed architectural layouts. In his work, he advocates for the embrace of 'empty spaces' that defy predetermined configurations, thereby reshaping models for audience–performer relationships as set by the traditional concert hall.

Continuing the exploration of sound and music as dynamic tools for unveiling architectural insights, acoustician Pedro Novo and opera director Rosalind Parker share their cross-disciplinary endeavour in crafting an opera tailored for the unique acoustics of an Edwardian Bathhouse. Through a blend of scientific acoustic analysis and collaborative engagement with the local community and musicians, they uncover the acoustic intricacies of the space, shaping dramaturgical decisions based on clarity in musical and spoken elements. Following this, interdisciplinary artist Angela McArthur and acousmatic composer Emma Margetson explore the role of spatial audio technology in revealing spatial dynamics. Their inquiry canters on the icosahedral loudspeaker (IKO) as an instrument for uncovering architectural characteristics whilst concurrently forging connections between sonic content, spaces, and audiences through carefully crafted sonic compositions.

The subsequent trio of chapters share a common focus on listening practices. In addition to scrutinising the material and geometric attributes of spaces conducive to sound performance and reception, composer and sound designer Lawrence Harvey examines sound as a material entity. Harvey employs the term 'intimate', this time not to describe an acoustic condition, but to characterise a direct, tactile approach to comprehending and manipulating sound as a material which can be manipulated and designed. Harvey's exploration extends to the practical application of sound in design education and practice, with a notable mention of Royal Melbourne Institute of Technology University's SIAL Sound Studios. Artist Ben McDonnell further investigates the intersection between sound and education, particularly focusing on the pivotal role of listening and transcription in shaping pedagogical strategies within design and art education, notably during student critiques ('crits'). Drawing parallels between the structure of musical compositions and the dynamic exchange between educators and students, McDonnell explores how transcribing these interactions, including ambient sounds like the sound of a chair scraping across the floor, unveils insights into pedagogical methodologies within visually oriented disciplines. Expanding on the theme of listening, curator and musician Sasha Elina explores the importance and development of music curation in performance art and sound-based practices. Elina examines various ways in which architectural structures influence our experience and expectations of performed sound and highlights the importance of the curator as liaising between performances and performance space. Drawing on examples such as London's pleasure gardens and various other public outdoor venues, Elina clarifies how curatorial choices shape social interactions, influence audience expectations, and affect the overall auditory experience.

The theme of 'outside' spaces continues with a chapter by sound artist Philip Samartzis. In this context, 'outside' refers to remote sites in Australia and Australian Antarctic Territory. His artistic practice focuses on the creation of sound art against the unpredictability of spaces 'in extremis' and examines the various tensions between the untouched natural world and the intrusion of human actions and technology. This chapter speculates on ways in which sound art practices can be used as mediums to observe and document and navigate the technical and cultural complexities inherent in such tensions.

The final three chapters discuss the role of technology in shaping our relationship with sound and space, as well as how sound and space interact with each other. This topic is particularly pertinent given the opportunities presented by emerging spatial audio technologies and the improved accessibility to mixed reality tools and environments. These advancements enable us to transcend or navigate seamlessly between virtual and physical realms. Composer and sound artist Gerriet K. Sharma discusses the creative potential of ambisonic audio technologies. He explains how these tools have a unique capacity to shape sound within space and create their own spatial constructs, transforming our perception of and interaction with our sonic environment. Sharma examines the ways in which these tools also facilitate practitioners and audiences to examine their own relationships to space and sound, encouraging dialogue and collaboration across diverse disciplinary and cultural backgrounds. Finally, we conclude with a chapter by interdisciplinary artist Cobi van Tonder. Transitioning from the poetic to the highly technical, she extrapolates from the mythological figure of 'Echo' towards the 'Acoustic Atlas'. This project depicts the world as 'an orchestra of echoes'. This online, open-access resource gathers impulse responses to create a digital archive of acoustic fingerprints and heritage from around the globe. This allows visitors to convolve and transplant their own sounds, enabling them to 'imagine' how they would sound in a range of far-flung locations.

25
SPATIOSONIC DIALOGUES
Exploring architecture's role in music composition and performance

Emma-Kate Matthews

Introduction

This chapter explores a project titled 'Construction 002: Tracing / Occupying' (C002)[1] composed for and performed in two extremely different acoustic and spatial environments: an acoustically dry anechoic chamber and the interior of the Basílica de la Sagrada Família, which has an average reverberation time of nearly 12 seconds.[2] The project aims to understand the current state of symbiosis between architecture and music, specifically focusing on the role of architecture in music composition and performance, where architecture may be considered as either an instrument or a performer. Additionally, the project examines how music can reveal non-visually apparent characteristics of buildings – an aspect often overlooked by designers during the design stage, as these qualities cannot be easily drawn or quantified. In this context, architecture is examined not merely as a container for events, but as an active component in music composition and performance. The project adopts a multidisciplinary approach, using tools, methods, and vocabularies commonly used in the design of spaces for music, as compositional tools. Such an approach is referred to as 'Spatiosonic'[3] throughout this chapter.

The motivation for this research comes from a fascination with a quote from physicist Hope Bagenall in which he recalls:

> A legend exists that a Mass by Fairfax – a mediaeval organist of St. Albans – was composed with a fourth part supplied by the church. Even if this was no more than a legend it shows that the building was recognised as an instrument.[4]

Whether real or fictional, Bagenall's account of this spatial, resonant phenomenon sparked the idea of music being composed and performed WITH a building, rather than just happening TO a building. This research explores the practical and conceptual resonances between architecture and music as a means of identifying creative opportunities and moving away from the problem/solution-based approach that currently dominates these fields.

C002 was initially performed in an anechoic chamber at University College London, where the room's acoustic response was completely dry. By contrast, it was also performed in the highly reverberant space of the Basílica de la Sagrada Família (BSF). In quantifiable terms, the vast interior of the BSF has an average reverberation time of almost 9.1 seconds and an early decay time of 11.9 seconds.[5] This is highly problematic for achieving speech intelligibility in the delivery of spoken-word sermons. Many engineers and designers have attempted to fix such a problem by proposing the retrofitting of acoustically absorbent surfaces. However, in acknowledgment of the desires of the building, C002 takes full advantage of this unique condition, by locating musicians in spatially diverse positions around the space, whilst also employing the building's high levels of reverberation to blend their soloistic parts into a series of tonally undulating and spatially immersive harmonic events. Whilst acoustic response is not the only crucial factor in the creation of Spatiosonic works, it significantly influences the aesthetic of a musical performance, especially when deliberately harnessed. By capitalising on the inherent acoustic and spatial characteristics of a performance space, this composition challenges the traditional role of architecture in shaping musical experiences. It also seeks to uncover insights beyond what is immediately visible or quantifiable, using music – a medium that must be experienced at a 1:1 scale, in real time. This approach allows us to perceive architecture over time, rather than merely observing it instantaneously, as we often do with photographs and drawings.

Context

Architectural space has historically influenced music, as seen in the 'spatialised choirs' of Dutch renaissance composer Adrian Willaert, and in the spatialised compositions of 20th-century American composer Henry Brant. Both of these examples explore how aspects of physical space (particularly the parameters of distance and direction), can be as compositionally active as the explicitly musical elements of pitch, rhythm and timbre.[6] Conversely, architects and engineers have long recognised the desires of music and sound in space. Many of us are familiar with Vitruvius' attempts to improve theatre acoustics by embedding 'echea', or resonant pots, within the walls of performance spaces.[7] More recently, collaborations between architects and composers, such as the project 'Prometeo' by composer Luigi Nono and architect Renzo Piano, have seen compositions and buildings conceived in tandem.[8] Despite these prominent examples, only a small handful of sonic and spatial practitioners have dedicated their research and practice to the rigorous exploration of creative, interactive parallels between the acts of making space and making sound in their work.

This is perhaps surprising, given that architecture, acoustical engineering, music composition, and performance all offer ways to develop interactions between sound and space – whether these interactions are explicitly calculated in advance or emerge from empirical and intuitive exploration and experimentation. Whilst the nature of these interactions is potentially rich and varied, current practice relies on highly reductive abstractions and representations, in order to anticipate, record and recall both architectural and compositional works. The trend of formalising and quantifying these relationships seemingly began when physicist Wallace Clement Sabine developed a method for predicting the time it takes for sound to decay in a space (based on room volume, surface area and absorption coefficient of materials which constitute the room).[9] Although predicting reverberation time is useful in building design as it offers an easily quantifiable description of (otherwise invisible) acoustic phenomena, it only discusses a very limited set of interactions between sound and space. As such, this approach overlooks more poetic dimensions, leaving plenty of unexplored potential at the intersection of spatial and sonic practices. This is not a new observation – it is a frustration shared by other practitioners and researchers. Architecture is rarely interrogated as anything beyond a mere container for music. The project explored in this chapter attempts to address this by questioning if architecture can itself offer compositional content.

Acoustic response as a compositional parameter

The composition that forms the focus of this chapter, is part of a larger series of site-responsive compositions which examine a series of practical and conceptual resonances between architecture and music. C002 explores both the spatialisation of performers and the acoustic response of the performance space as deliberate compositional parameters. The lengthy reverberation times in the BSF have caused significant issues with speech intelligibility, particularly during sermons.[10] Teams of engineers and designers have suggested remedying this issue by retrofitting the interior with absorbent acoustic treatments. However, from a music composition perspective, this extreme acoustic environment presents a unique creative opportunity for a highly site-specific and acoustically aware musical project.

Music composition often serves as a medium for expressing ideas. However, in the case of C002, it poses a question: 'How might architecture offer conceptual and phenomenological content for music composition?'. To explore this question in greater depth, C002 evaluates the same composition in two contrasting and extreme acoustic environments, focusing on the acoustic phenomenon of reverberation – one of the most intuitively understood components of acoustic behavior discussed in both architectural and musical contexts.

The notion of 'extreme' – derived from the Latin *extrēmus*, meaning 'situated at the end'[11] – is defined here as a relational condition, necessary for comparing two vastly different architectural conditions. The anechoic chamber and the BSF differ not only in their reverberant behaviour, but also in their capacity to spatialise sound sources (in this case, the performers). In the BSF, performers can be positioned at much greater distances than in a typical stage setting or the anechoic chamber (up to 20 m vertically and 60 m horizontally). This capacity enables the exploration of relationships between a performer's position (relative to each other and to the listener, or microphone) and the acoustic response of the space. Crucially, it provides insight into the role of reverberation and performer spatialisation in shaping and defining musical logics.

C002, a duet for violin and cello, also exemplifies an experimental approach that refrains from explicitly defining the parameters of the acoustic condition within the BSF through quantitative analysis, as these definitions have already been established elsewhere.[12] Instead, its aim is to explore the atmospheric and poetic potentials within this acoustic environment. Architecture, being a largely visually focused profession, often neglects phenomena that cannot be quantified or visually represented. Through music composition and performance, this project simultaneously articulates spatial concepts and reveals architectural character. This approach, referred to as 'Spatiosonic practice' in the wider context of this research, could serve to broaden architectural discourse and suggest methods for furthering such an interdisciplinary practice. The development of Spatiosonic practice relies not only on the performance of projects but also on the methods and tools used in their composition. Given this, it is worth noting that C002 is a multidisciplinary endeavour and was created using methods and tools typically used by architects and engineers to design spaces for music, thereby forming direct practical bridges between sonic and spatial practices. Computer programmes such as Pachyderm for Rhino,[13] Evaluation of Acoustics using Ray-tracing (E.A.R.) for Blender[14] and Computer-Aided Theatre Technique (CATT) acoustic[15] were used to rehearse musical ideas ahead of their performance. Specifically, these tools were used to predict the effects of varying performer positions and to understand how this spatialisation interacts with the building's acoustic response.

Architecture as a performer

When designing buildings, architects often discuss 'performance' in terms of attributes like thermal efficiency and fire safety. However, we rarely describe buildings as artistic 'performers'. This project explores the notion that architecture, when activated by music, is a dynamic entity capable of embodying and expressing spatially dependent musical parameters. In essence, it challenges the traditional boundaries between architecture and performance, envisioning buildings not just as static structures but as active participants in the artistic realm of music composition and performance. This work then elevates the role of architecture beyond just being a container for sound. It suggests that architecture can provide compositional content, equal to traditional musical elements like pitch, time, and timbre. This idea derives from Brant's approach, which viewed space as the fourth element of music.[16] This in turn prompts the question: if architecture is an essential part of a composition, does a change in performance location, especially when transitioning between environments as architecturally distinct as the anechoic chamber and the BSF, result in a variation of the musical piece, comparable to the alteration of other fundamental musical characteristics? This issue inevitably raises significant and subjective questions about the nature of 'the piece'. This is especially relevant in the era of AI-generated music,[17] where content is typically based on libraries of existing examples and outcomes may closely resemble the original set of examples. There are many varying theories on what determines the portability of a musical work or idea.

Whilst a detailed discussion of 'work identity' is beyond the scope of this chapter, it is worth acknowledging the common practical notion that performances of musical compositions always contain deviations from the score. These deviations can be errors, embellishments, or omissions, but they do not necessarily change the fundamentals of the piece altogether. According to philosopher Nelson Goodman, the score (and in the case of C002, also the drawing) clarifies compositional intent, which is crucial for interpreting and performing a work, thereby forming its identity.[18]

Goodman's theory surmises that 'a performance is a performance of a work only if it corresponds precisely to those elements of a score that are capable of precise definition'.[19] Whilst this theory has been critiqued for its lack of adaptability to real-world scenarios,[20] the principle of the identity of the piece as being held in the documents that facilitate its performance can be usefully applied to the questions underlying this research, provided the key variables, including the architecture, are acknowledged and articulated in some way. From an aesthetic perspective, the music played in the anechoic chamber differs significantly from that played in the BSF. Despite using the same scores, the duration and presentation of musical content varies noticeably. The performance in the anechoic chamber has a distinctly melodic sound, whilst the BFS performances appear more drone-like, steering our attention towards harmonic over melodic concepts. This highlights the role of architecture as being implicit in the performance of compositional intent, where differences in performance space could be more deliberately considered at the point of composition. It subsequently prompts the designers of buildings for music to consider 'what does music want'? Specifically, new music.

Along with the score and drawings for C002, the project's title significantly contributes to the understanding of the work. The full title, 'Construction 002: Tracing/Occupying' (C002) serves multiple purposes. It not only situates the work within a sequence of other Spatiosonic 'constructions',[21] but also expresses the intention for the music to expose (or trace) the otherwise hidden sonic potential of architecture as a musical performer. Moreover, it outlines the compositional processes. The geometric and acoustic boundaries of the performance space (the vast interior of the BSF) were initially 'traced' from drawings and photographs to create a 3D digital model, which was used to run ray-tracing simulations that demonstrate how reflected sound 'occupies' the space. The simulation results revealed the spatial and acoustic capabilities of the architecture, and guided the positioning of performers to capitalise on the unique reverberation profile of the BSF. In this research, composition is a mechanism for inquiry rather than a means of expressing known things. Thus, acknowledging the compositional process is a crucial part of the work. Like the others in the series, C002 is not intended to be a 'finished' product. It learns with every performance and serves as a tool for fostering creative resonances between the simultaneous practices of making space and sound in architecture and music, respectively.

Rehearsing and simulating space

The project was composed at a physical distance from both the anechoic chamber and the BSF. With only a distant memory of a visit to the BSF as a tourist several years prior to this project, it became necessary to simulate its acoustic as it was not feasible to revisit the space to conduct tests and gather information before starting the composition. Access to the BSF for the eventual performance of C002 was granted through an Australian Research Council-funded project which quantifiably examined the acoustic condition of the space.[22] Consequently, engineer Dr. Jim Barbour was able to provide several previously recorded impulse responses (IRs)[23] from the space as an acoustic reference during the composition process.

Whilst useful for providing the acoustic 'essence' of the space, the IRs captured have certain limitations which need to be considered. Firstly, they are limited to the locations where they were physically captured. Also, since the BSF interior was under construction during the capture, the IRs only represent a snapshot of the acoustic condition at that time. However, the IRs allowed for the simulation of the sound of the space during music composition. This was done by convolving the IRs with a series of dry Musical Instrument Digital Interface (MIDI) outputs from notation software, Sibelius.[24] This method was particularly useful in the early stages to understand how different instruments might sound in the space and to identify the appropriate timbral and tonal ranges that would capitalise on the unique reflection profile of the space. By visualising the IRs in a spectral frequency display,[25] it was also possible to visually verify which frequencies were the most present in the reflections, and for how long.[26]

To explore performer locations more flexibly, a digital model of the space was used for ray-traced acoustic simulations, akin to the auditory equivalent of an architectural render or physical model.[27] These simulations also generated IRs which helped to auralise different spatialisation options. With Pachyderm, it was also possible to visualise early reflections, identifying areas of intensity, especially in the more intimate peripheral spaces of the BSF where reflection distances and times are shorter due to smaller distances between surfaces. The different computer programmes (listed earlier) each had their own advantages and limitations. In purely subjective terms CATT Acoustic was arguably the most accurate in terms of sounding the most like the eventual performance. Although EAR for Blender was the least audibly similar to the eventual real performance (and unfortunately is no longer supported), it proved invaluable in the sculptural and iterative process of quickly moving sound sources around virtually to sketch out different options. This process is perhaps the digital equivalent of what musicians and composers did centuries ago, where they operated through trial and error, working directly and empirically within the space. This was before acoustics became a specialised field, and before it was possible to make calculations remotely from the physical site. Since the resulting auralisations were subtly yet noticeably different, they were not treated as an absolute measure of the sound in the actual space. Instead, they aided in constructing a relational picture, for comparing how the sound might change between performer positions within each space, rather than providing reliability for the certainty of the eventual sound.

Typically, such tools are used in the design of buildings for music, rather than in the composition of the music itself. At this point, it is worth casting a critical ear across these tools. With non-visual media, it is easy to forget that the outputs are merely representations, and not a replacement for reality. Researcher Jonathan Sterne helpfully acknowledges that acoustic behaviours are complex, stating that 'so many things are happening in so many different ways that they cannot be calculated or captured by any modern computing device....'[28] Sound does not neatly travel in straight lines and many simulation tools do not account for complex wave behaviours,[29] so these tools can only ever provide an impression of the sound in the space. This cannot substitute for the richness and complexity of the real experience. C002 presented an opportunity to use and critique acoustic simulation technology from the viewpoint of a composer with architecturally sensitive expectations, as opposed to a building designer or client seeking an audible representation of a yet unrealised space, which is the typical use of these simulation tools.

The composition process relied less on the numerical data output from simulations, such as early reflection and reverberation times, and more on a sculptural approach. This meant that the content was developed based on how it sounded, using auralisations and spectral frequency displays, rather than checking exact pitches or durations. For playback of simulated content, stereo headphones were primarily used. However, for higher resolution and spatially immersive playback, the SoundSpace[30] at Max Fordham was utilised. This was particularly helpful in understanding how the sound changes with differences in the performers' positions. Modeling these variations before the eventual performance enabled a more thorough and informed phase of experimentation and iteration during composition, revealing, for example, that some performer positions produced more dramatic results than others.

Articulating space

From studying Gaudi's models and drawings, researchers have concluded that Gaudi always intended the BSF to play a role in the projection and propagation of music, particularly the sound of the tubular bells that he also designed. One particular study proposes that the geometry of the louvred windows, in relation to Gaudi's tubular bell design and their sound radiation, would allow sound to propagate throughout the city, therefore turning the BSF into a giant instrument.[31] Inside this instrument, the vast space, and reflective materials – in relation to both sound and light – enhance and prolong the presence of sound, surrounding the listener and articulating space as if another architectural material. The architecture invites music to express its presence in much the same way that the physically tangible materials of stone and glass define architectural space. This creates a reciprocity between space and sound, where the architecture shapes the music whilst the music activates the architecture. The building might therefore be considered not only as an instrument but also as a performer, activated by compositional intent and potentially embodying both roles simultaneously.

As mentioned earlier, the choice of instrumentation was partially guided by the need to align the tonal range of the performed music with the frequencies that the building is most reflective to (as confirmed by the IRs). The timbral characteristics of the instruments were also important: The Violin and Cello have the capacity to sound both similar to, or distinct from each other, based on aspects like playing technique and tonal range. The capitalisation of their timbral similarities when coupled with a highly reflective acoustic condition, has the effect of 'filling up' the space. A principle that draws on Brant's experimentation with the spatialisation of performers.[32]

On the other hand, distinguishing their timbral differences separates the two voices and aids in examining the audible clarity and potential musical value of sound localisation in highly reverberant environments. In the BSF, the effect of spatialisation is particularly complex to understand as there is a directional ambiguity, partially attributed to the long early reflections as a result of the large distances between acoustically reflective surfaces.

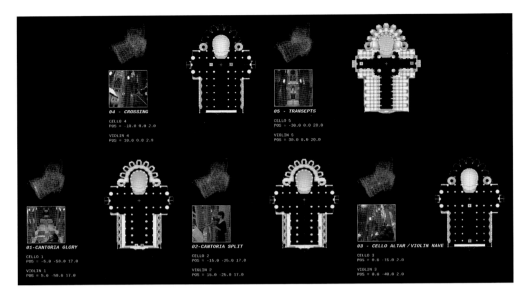

Figure 25.1 BSF performance layout diagrams (Emma-Kate Matthews, 2017).

Two crucial parameters for articulating space through sound are the distance and direction between sound sources or performers, and between the sound sources and the listener or microphone. In the case of the BSF, locating the direction of the sound source is difficult due to the high level of acoustic reflection. However, the effect of distance is audibly clear, especially in the way that larger distances between the performers and the microphone seem to produce a longer attack, even for shorter notes. For example, within the acoustically dry space of the anechoic chamber, the notes in bars 60–63 were heard sequentially, and distinct from each other, forming a perceivable melodic line. However, when the same passage was played in the BSF, the attack of the individual notes was noticeably softer. It was also partially obscured by the decay of the preceding notes, making the passage sound more like an overlay of notes from uncertain spatial and temporal origins. The resulting sound therefore had a more harmonic than melodic aesthetic, influenced by a combination of the distance between the performers and the microphone, and the reflectivity of the space.

The positioning of the performers was first rehearsed in the digital models, allowing for the exploration of several different locations for comparison. The eventual performer locations were instructed as follows (Figure 25.1):

First, the cello and violin were positioned 10 m apart horizontally and 17 m high on a balcony. Next, they remained at the same height but were moved 50 m apart from each other. Then, both instruments were moved to the ground floor, with the cello near the altar and the violin at the other end of the nave. After this, they were placed opposite each other on the ground floor, only 20 m apart and closer to the microphone. Finally, the instruments were separated again, positioned 20 m high and slightly over 60 m apart.

The relationship between performer positions and the corresponding acoustic reflections resulted in distinct differences between each position. These differences were more significant than the simulations predicted. The large distances also affected the musicians' approach to the music, particularly when the distances were greater. For example, the tempo heard in the Sagrada Familia was noticeably slower than in the anechoic chamber. In the Anechoic chamber at 80 BPM, the piece was completed within 5 minutes, in line with the notation. However, in the BSF each performance lasted around 6 minutes and 20 seconds, indicating a tempo closer to 60 BPM. This discrepancy could be due to the musicians waiting for the acoustic reflections to conclude, or perhaps they were simply playing more cautiously as they could not see each other for visual verification of each other's starting tempo. This difference could also result from the delay caused by distance.[33] When asked to reflect on these temporal differences, both musicians said they had not perceived a difference in tempo between the Anechoic chamber and the BSF performances, and they did not make a conscious decision to slow down. The violinist also compared playing in the BSF to 'walking underwater', noting that the sound sometimes behaved unpredictably. The simulations that were ran beforehand failed to detect such nuances, highlighting the limitations of these tools in predicting human behavior or more complex wave interactions.

Despite the value placed on the physical immediacy of live performance and the challenges in capturing a detailed spatial image in sound recording, the performances were still recorded to provide a retrospective reference for the variations between each performance. These recordings were made using a microphone engineered by engineer Dr. Jim Barbour. Barbour's research focusses on capturing the vertical dimension of space in ambisonic recordings. To achieve this, he has devised a method using a tall stack of 12 microphones, arranged radially in plan, to capture the verticality of the BSF under one of the highest parts of its interior (Figure 25.2).

The ambisonic microphone array was always kept static, near the altar, where the original IR files were created. By this rationale, the microphone's position could be considered the 'optimum' listening point as it is the theoretical 'standing point' from which the music was constructed. This is akin to a measured perspective – a familiar concept to architects and visual artists – where a horizon line and vanishing points are set, and a standing point is defined as a 'viewer' (or in this case, the listener). When viewing a perspectival drawing or listening to a sound recording, the viewer or listener is inevitably implicated in the scene. This suggests there might also be an 'optimum' position for viewing or listening in spatialised musical performances. This concept is particularly emphasised in anamorphic art, where visual content only assembles,

Figure 25.2 Left to right: Dr. Jim Barbour with ambisonic microphone array'; view of microphone array from balcony; microphone against the altar (Emma-Kate Matthews & Finneas Catling, 2017).

or makes sense from a certain viewing position. Spatial composer Brant maintained that in his spatialised compositions, where the performers' locations are as meticulously determined as the music they perform, no single listening, or performer position is considered optimal.[34] In this research, the aim of placing performers in different locations within the same space is not to find a single 'optimum' arrangement. Instead, the goal is to understand how the interplay between musical content and performer placement can 'activate' the space. This process can tease out both conceptual and physical resonances between sound and space, using music composition and performance as the medium.

Activating space

There are many ways to explore creative relationships between sound and space, especially when aspects such as architectural acoustics are considered on both poetic and technical terms. C002 offers just a few of countless other approaches that similarly have the capacity to reveal alternative resonances. The compositional methods employed in this research combine both calculative and intuitive processes. The tacit knowledge gained from the performances in physical space is empirical, making it difficult to explain explicitly through notation or numerical data. Nevertheless, it is valuable to consider certain specific compositional decisions and their impact on charging the music with the ability to form a dialogue with the space. To examine the piece, section by section:

C002 begins with a short, *fff* chord, immediately followed by a pause. This initial gesture is like a question to the architecture, somewhat similar to an IR, soliciting an acoustic response. This gesture recurs throughout the piece, repeating the question in subtly different ways. It also serves as a reference point for the performers to synchronise, as there is no conductor, due to the inability to rely on clear visual cues.[35] The notes chosen for this chord, Db2 (69.3 Hz)[36] on the cello and G3 (196 Hz), F4 (349.23 Hz) on the violin, were selected based on the frequency range most prevalent in the IRs. This was also balanced with the ease of playing double stops on the instrument. Open or near-open strings were favoured to maximise the instrument's resonant capacity and enhance the interaction between the sound and the architecture. Section A sets the tonal range for the entire piece, influenced more by the acoustic behaviour of the space than by conventional logics of Western music theory. It primarily features long drones that swell and recede in dynamic, dovetailing with the decay of the initial shorter 'stabs'. Throughout this section, the violin maintains a timbral distance from the cello by playing in a different octave and alternating between sul tasto and sul pont[37] bowing positions. The section culminates in a longer chord which increases in dynamic from *ppp* to *fff* followed by another 'stab' similar to the opening gesture, but with different notes in the cello. The slight change in note selection, now including C2 (65.41 Hz) and Db3 (138.59 Hz), aims to investigate the 'range' of frequencies to which the building is most responsive, instead of trying to pinpoint specific pitches as narrowly defined entities.

The intentional use of 'dissonant' intervals, such as a minor 9th in this case, is aimed at generating as much sonic material as possible during these IR-like moments. Choosing dissonant combinations reduces the potential for tonal ambiguity, as there are fewer shared partials between notes. This makes it easier to pinpoint which musical inputs are being affected by the architecture, how and to what extent.

Section B further explores the effect of attack and dynamic variation on the space. Starting at bar 14, this section features a shorter series of dynamic swells on various notes within the established range from the first section. This aids in understanding the role of reverberation in binding harmonic content. Altering the length of the note tests the point at which individual notes become audibly distinguishable, amidst the 'blending' effect of the acoustic reflections.

In Section C, at Bar 29, the cello starts playing in the same pitch range as the violin. It uses an open C string to create a continuous 'background' or 'horizon' line, occasionally accentuated by a harmonic on the open string. This section is deliberately more tonally ambiguous. The two voices partially merge, testing the capacity of the spatial separation to preserve the clarity of individual voices. The aim here is to determine the threshold at which the acoustic response begins to obscure this separation.

Starting at bar 39, Section D introduces intermittent, short-note melodies in the violin. These are deliberately 'choppy' to contrast with the low-pitch continuity provided by the cello. Occasionally, the violin reinforces this continuity with drone-like material between the intermittent melodies. This creates an oscillation between the merging and separation of voices, testing the temporal thresholds at which the two voices become audibly distinct.

At section E, bar 60, the content established in earlier sections is fragmented and thrown back and forth between the violin and cello. This highlights the relationship between spatial separation, which emphasises each instrument's voice, and the acoustic response of the space, which in contrast, blends these voices together.

In section F at bar 64, the alternation of material continues, marked by frequent swells in dynamic that mark a gradual lengthening of the notes again, and marking the transition into section G at bar 82. Here, there is a return to the drone-like content like that was played at the beginning, but with both instruments playing in the same octave this time. The piece concludes with much more tonally ambiguous material, where it is intentionally difficult to distinguish whether the articulation of the music is produced by the musicians or the building. For instance, it becomes difficult to determine if the slow attack is caused by the musician's articulation, their distance from the microphone, the building's acoustic response, or a combination of all these things. At this stage, the music becomes aligned with the space, and the building performs in unison with the musicians.

Whilst it was possible to examine these sections as separate entities, a conscious decision was made to organise them into a cohesive narrative within a 'piece' of music. This decision was driven by the nature of its analysis, which revolves around comparing performances in different spaces and configurations. Maintaining a consistent sequence in the structure of each rendition offers a more straightforward framework for making meaningful comparisons.

Observing Spationsonic relationships

The development of these musical strategies was influenced by a combination of feedback from digital simulations, intuitive insights gained from past projects, and an examination of Brant's work. It also draws on scientific research examining how acoustic response influences both the behavior of sound in the space and the performers. It has been observed that performers of string instruments frequently adjust aspects of their playing, such as tempo, loudness, and bowing pressure, in response to changes in room response.[38] Such aspects were not anticipated in the digital simulations. Following performances in both the anechoic chamber and BSF, many empirical observations were made that exceeded or were not highlighted by the digital 'rehearsals'. To summarise a few of the most notable:

A strong connection was observed between performer distances and the tempo of their performance, where greater distances corresponded to a slower pace. This phenomenon can be reasonably attributed to the inherent delay in sound propagation over large distances between the musicians. To illustrate, when the performers were positioned 64 m apart, assuming sound travels at a rate of 343 m/s, the resulting delay amounts to approximately 0.18 seconds. Given a tempo of 80 beats per minute, where a quaver lasts about 0.75 seconds, the delay caused by the distance equates to roughly a semiquaver. This delay proves consequential enough to perceptibly influence temporal precision in the performances. This effect is particularly prominent during passages that contain more 'melodic' content from section C onwards, where the shorter notes appear to exaggerate the impact of delay due to distance.

The large distances between interior surfaces also contribute to longer early reflections. At the scale of the BSF, this appears to noticeably soften the attack of each sounding note making individual notes in melodic phrases less distinct and harder to locate in time and space. The resulting sound is blended, immersive, and therefore harmonically rich. This challenges Brant's observation that 'spatial separation destroys harmony and ensemble'.[39] In the case of C002, the extreme acoustic condition of BSF's interior disrupts the effect of physically separating sound sources, meaning that the effect of spatial separation is not distinct from the building's acoustic condition, as Brant's theory suggests. The relationship C002 examines between spatial separation and acoustic response therefore builds on Brant's work. Whilst Brant neither formally studied architectural acoustics nor explicitly acknowledged acoustics in his work, he recognised the lack of cross-pollination between sonic and spatial practices. He suggested starting with pedagogy:

> …it would be worthwhile if you could find the time sometime to get a book on acoustics, musical acoustics, and read about these things, which every musician should learn in music school, and none of them know about it…[40]

This observation is key in enhancing the exchange of knowledge, tools, and skills between spatial and sonic disciplines. A shared grasp of both a technical and poetic acoustic vocabulary would facilitate this process.

A more nuanced observation concerning the effect of reverberation is in the behaviour and presence of partials[41] after the initial attack of some of the more frequency-rich notes (typically those played on open strings). Some partials decayed quickly, whilst others appeared to continue reflecting long after the fundamental tone had faded. Once again, the spectral frequency display of the recordings was helpful in visually verifying the way that these partials were behaving in the space as the ephemerality of sonic phenomena often makes it difficult to comprehend more subtle details of what is being heard. After the first chord (which includes a Db2 (69.3 Hz), G3 (196 Hz), and F4 (349.23 Hz)), these lingering partials can be heard. Whilst only the G is played on an open string, the remaining notes are played low on the fingerboard, with a firmly pressed finger and long length of vibrating string, thus enhancing the presence of partials.[42] Following this first 'stab', partials at frequencies of C3 (130.81 Hz), G4 (392 Hz), and D5 (587.33 Hz) remain audible, where others reliably decay more swiftly. This is consistent with IRs which indicate that the space is particularly reflective to frequencies within this range. This may explain why the fundamentals outside of this range were decaying before their upper partials. These enduring 'after images' are reminiscent of the anecdote by Bagenall that inspired this research, suggesting that the building contributes its own voice to the music.

Whilst this research does not explore the scientific specifics of these phenomena, it benefits from the tools, methods, and vocabulary used in both acoustic and architectural practice to gain a holistic understanding of what the space is doing to the music (and the musicians who play it). It seems that aspects as fundamental to music as tempo, timbre, and even pitch (due to the later decay of partials over the fundamentals) change significantly between the anechoic chamber and the BSF. The music was specifically composed to tease out these differences. Therefore, the musical logics that play out in this composition are governed by architectural logics over those governed by traditional western music theory. The resulting musical aesthetic – a popular topic in musical discourse – arises from a combination of spatially driven musical logics and the audible acoustic and spatial traits of the architecture in the eventual performance. This composition, along with others in the series, collaborates with architecture to make music, rather than simply performing for or at it. Therefore, this music not only projects outwardly during its performance, but also learns, discovers, and gains knowledge in return.

Conclusions

When we empower architecture to become an active participant in the composition and performance of music, rather than merely serving as a container, we engage in more nuanced explorations of the potential range of interactions available between sound and space. These interactions sometimes transcend quantifiable metrics, inviting a subjective and sometimes uncomfortable dialogue amongst architectural, musical, and acoustic practices. This approach opens up new avenues for creative expression and fosters the development of innovative practical and conceptual methods for how architecture and music can interrelate and enhance each other. This approach also encourages architectural design practice to remain agile and aligned with the disciplines it has historically synergised with. It nurtures an ongoing dialogue where architecture continually adapts to music's needs, and music acknowledges architecture's profound capacity to influence its performance and composition.

The stark contrast between the anechoic chamber and the BSF is pivotal in highlighting the potential richness of exploring and developing both practical and poetic relationships between sound and space. The extreme differences observed emphasise these relationships, much like examining something under a microscope to reveal previously unnoticed details. The insights from this project can also be applied to less extreme scenarios, providing valuable lessons in effectively exploiting acoustic conditions and refining methods to more directly activate architecture. Additionally, C002 demonstrates a method for developing cross-disciplinary tools and methodologies that benefit both composers and architects. These resources, nurtured through this project, can be used and refined in future Spatiosonic projects, where sound and space are engaged in active dialogue as equally valued components.

Sound files

These sounds are available on the Routledge E-Book page and a full-length recording is available at Emma-Kate's own website: https://www.ekm.works/construction002

MATTHEWS_1-ANECHOIC

Recording of opening few bars of C002, in the anechoic chamber

MATTHEWS_2-GENERIC-LARGE-CATHEDRAL

Anechoic recording of opening few bars of C002 processed with a 'generic large cathedral' effect in Ableton Live

MATTHEWS_3-CATT-POS1

Anechoic recording of opening few bars of C002 convolved with an IR simulated by CATT Acoustic, with musicians in position 1

MATTHEWS_4-EAR-POS1

Anechoic recording of opening few bars of C002 convolved with an IR simulated by EAR for Blender, with musicians in position 1

MATTHEWS_5-ACTUAL-POS1

Recording of opening few bars of C002 as physically captured in the Sagrada Familia, with musicians in position 1

Notes

1. This project was generously funded by the Australian Research Council.
2. Zhao et al., 'A Preliminary Investigation on the Sound Field Properties in the Sagrada Familia Basilica', p. 5.
3. The word 'Spatiosonic' is a self-coined portmanteau which (though not currently recognised by any dictionary) discusses a plethora of resonances between spatial and sonic practices, without explicitly disclosing the nature of that resonance. This enables an elastic usage, in relation to either conceptual or practical resonances.
4. Bagenal, 'Influence of Buildings on Musical Tone', p. 443.
5. Zhao et al., 'A Preliminary Investigation on the Sound Field Properties in the Sagrada Familia Basilica', p. 5.
6. Brant once stated that space is the 4th dimension of music, alongside these 3 listed elements: Brant, 'Spatial Music Progress Report', p. 22.
7. Vitruvius, 'Book V'.
8. RPBW Architects, 'Prometeo Musical Space'.
9. Sabine, 'Melody and the Origin of the Musical Scale'.
10. Zhao et al., 'A Preliminary Investigation on the Sound Field Properties in the Sagrada Familia Basilica'.
11. Dictionary, 'Extreme, Adj'.
12. Zhao et al., 'A Preliminary Investigation on the Sound Field Properties in the Sagrada Familia Basilica', p. 5.
13. Open Research in Acoustical Science and Education, *Pachyderm Acoustic*.
14. Krijnen, *EAR: Evaluation of Acoustics Using Ray-Tracing*.
15. Dalenbäck, *CATT Acoustic*.
16. Brant, *Textures and Timbres: An Orchestrator's Handbook*, p. xii
17. Refering to the process of creating musical content using artificial intelligence technologies.
18. Goodman, *Languages of Art: An Approach to a Theory of Symbols*, pp. 179–192.
19. Cook, *Beyond the Score: Music as Performance*, p. 240.
20. Cook, *Beyond the Score: Music as Performance*, p. 241.
21. C002 exists as part of a sequence of iterative works which learn from each other, and each pose a different question with regard to discovering and developing active relationships between the practices of making space and making sound.
22. The project is titled 'Architecture for Improved Auditory Performance in the Age of Digital Manufacturing'.
23. Dr. Jim Barbour gained his PhD from the Royal Melbourne Institute of Technology (RMIT) in 2017 with his dissertation titled: 'Spatial Audio Engineering: Exploring Height in Acoustic Space' – and has explored a range of recording techniques for capturing the sound of the uniquely tall interior of the BSF.
24. Avid, *Music Notation Software – Sibelius*.
25. Adobe Audition and SPEAR (by Michael Klingbeil) were used for visualising the existing IRs.
26. Spectrum analysis tools in computer programme 'Audacity' were also used to get a more fine-grained understanding of what was visibly apparent in the spectral frequency display.
27. For the anechoic chamber, there was no need for a simulation model, as it is completely absorbent to sound.
28. Sterne, 'Space within Space: Artificial Reverb and the Detachable Echo', p. 122.
29. Dalenbäck, 'Whitepaper: What Is Geometrical Acoustics (GA)?'
30. 'Acoustics – Max Fordham – SoundSpace'.
31. Yoshikawa and Narita, 'The Sagrada Familia Cathedral Where Gaudi Envisaged His Bell Music'.
32. Brant, 'Space as an Essential Aspect of Musical Composition', p. 238.
33. This possibility is discussed in detail in the following section.
34. Brant, 'Space as an Essential Aspect of Musical Composition', p. 224.
35. This device became particularly crucial in the BSF when the distances between the performers were large enough to hinder attempts to play 'together'.
36. Frequencies are based on the equal-tempered scale.
37. Sul tasto bowing is closer to the fingerboard and sul pont is closer to the bridge.
38. Ueno, Kato, and Kawai, 'Effect of Room Acoustics on Musicians' Performance. Part I: Experimental Investigation with a Conceptual Model'.
39. Gagne and Caras, *Soundpieces: Interviews with American Composers*, p. 60.

40 Brant and Crawford, 'An interview with Henry Brant: Spatial Music to Evoke the New Stresses, Layered Insanities, and Multidirectional Assaults of Contemporary Life on The Spirit'.
41 In music, a 'partial' is a component of the harmonic series which helps define the timbre or tone quality of an instrument. The fundamental frequency serves as the first partial, with subsequent partials being multiples of this fundamental frequency. The relationships between these partials can be either harmonic or inharmonic.
42 Meyer, *Acoustics and the Performance of Music: Manual for Acousticians, Audio Engineers, Musicians, Architects and Musical Instrument Makers: Manual for… (Modern Acoustics and Signal Processing)*, p. 88.

Bibliography

'Acoustics – Max Fordham – SoundSpace'. Accessed 26 February 2021. https://www.maxfordham.com/services/acoustics/details/#SOUNDSPACE.

Avid. *Music Notation Software – Sibelius*. Accessed 15 December 2018. https://www.avid.com/sibelius.

Bagenal, Hope. 'Influence of Buildings on Musical Tone'. *Music & Letters* 8, no. 4 (October 1927): 437–47.

Brant, Henry. 'Space as an Essential Aspect of Musical Composition'. In *Contemporary Composers on Contemporary Music*, edited by Elliott Schwartz, Barney Childs, and Jim Fox. New York: Da Capo Press, 1998.

Brant, Henry. 'Spatial Music Progress Report'. *Quadrille, A Magazine for Alumni & Friends of Bennington College* 12, no. 3 (1979).

Brant, Henry. *Textures and Timbres: An Orchestrator's Handbook*. New York: Carl Fischer Music, 2009.

Brant, Henry, and Caroline Crawford. 'An interview with Henry Brant: Spatial Music to Evoke the New Stresses, Layered Insanities, and Multidirectional Assaults of Contemporary Life on The Spirit'. Interview by Caroline Crawford. Oral History Center, 2006. https://digicoll.lib.berkeley.edu/record/218846.

Cook, Nicholas. *Beyond the Score: Music as Performance*. OUP, 2013.

Dalenbäck, Bengt-Inge. *CATT Acoustic*. Accessed 22 February 2024. https://www.catt.se/.

Dalenbäck, Bengt-Inge. 'Whitepaper: What Is Geometrical Acoustics (GA)?' *CATT Acoustic*, 7 May 2018.

Dictionary, Oxford English. 'Extreme, Adj', n.d. Accessed 20 February 2024.

Gagne, Cole, and Tracy Caras. *Soundpieces: Interviews with American Composers*. Scarecrow Press, 1982.

Goodman, Nelson. *Languages of Art: An Approach to a Theory of Symbols*. 2nd ed. Hackett Publishing Co, Inc, 1976.

Krijnen, Thomas. *EAR: Evaluation of Acoustics Using Ray-Tracing*. Github. Accessed 22 February 2024. https://github.com/aothms/ear.

Meyer, Jürgen. *Acoustics and the Performance of Music: Manual for Acousticians, Audio Engineers, Musicians, Architects and Musical Instrument Makers: Manual for… (Modern Acoustics and Signal Processing)*. Translated by Uwe Hansen. 5th ed. 2009 edition. Springer, 2009.

Open Research in Acoustical Science and Education. *Pachyderm Acoustic*. Github. Accessed 22 February 2024. https://github.com/PachydermAcoustic.

RPBW Architects. 'Prometeo Musical Space'. Accessed 7 October 2018. http://www.rpbw.com/project/prometeo-musical-space.

Sabine, W. C. 'Melody and the Origin of the Musical Scale'. *Science* 27, no. 700 (29 May 1908): 841–847.

Sterne, Jonathan. 'Space within Space: Artificial Reverb and the Detachable Echo'. *Grey Room* 60 (1 July 2015): 110–31.

Ueno, Kanako, Kosuke Kato, and Keiji Kawai. 'Effect of Room Acoustics on Musicians' Performance. Part I: Experimental Investigation with a Conceptual Model'. *Acta Acustica United with Acustica* 96, no. 3 (2010): 505–515.

Vitruvius. 'Book V'. In *Ten Books on Architecture*. Harvard University Press, 1914.

Yoshikawa, Shigeru, and Takafumi Narita. 'The Sagrada Familia Cathedral Where Gaudi Envisaged His Bell Music'. *The Journal of the Acoustical Society of America* 115, no. 5_Supplement (1 May 2004): 2529–2529.

Zhao, Sipei, Eva Cheng, Xiaojun Qiu, Pantea Alambeigi, Jane Burry, and Mark Burry. 'A Preliminary Investigation on the Sound Field Properties in the Sagrada Familia Basilica'. In *2nd Australasian Acoustical Societies Conference, ACOUSTICS 2016*, 2016. https://www.acoustics.asn.au/conference_proceedings/AASNZ2016/papers/p165.pdf.

26
IN PRAISE OF EMPTINESS
A future for performance venues

Fabricio Mattos

Introduction

In a recent visit to the Museu do Amanhã – Museum of Tomorrow – in Rio de Janeiro I came across the following sentence: 'what is normal is always transitory'. This sentence, allied to the spatial layout of that museum itself, made me think about our traditions in terms of how we use space in many settings, from museums to offices, from concert halls to airports. Taking museums, for instance, there has been considerable discussion recently about the functions of such institutions within our societies, forcing both specialists and society as a whole to reconsider their most basic assumptions about what museums actually are – their roles and responsibilities – to the extent that some 'museums' that we have today would barely have been recognised as such a few decades ago. In Brazil, one can find a few other representative examples, such as the Inhotim Institute, which offers pavilions and artworks spread over a vast outdoor park, and the extraordinary Museu Oscar Niemeyer, conceived by the famous architect himself. Although considerably different from one another, these have become institutions known for their alternative approach to the traditional conception of what a 'museum' should be, offering unexpected ways in which their layouts can be enjoyed, and a 'blank canvas' for the possibilities of new dynamics within their varied programme of highly conceptual exhibitions.

In this chapter, I will explore a series of ideas related to our understanding of space and place, and how a multicultural and deeper engagement with the idea of emptiness might illuminate both concepts and help in the design and use of alternative spaces for artistic performance. These ideas are based on a section of my PhD thesis, submitted to the Royal Academy of Music in 2023, and titled 'New Stages: revisiting the understanding of stage and performance in music events'; here, however, my intention is to trigger questions about our spatial biases when it comes to our performance venues, and how empty spaces could help redefine the relevance of performance events for today's performers and listeners.

Space and place

When referring to expressions such as 'performance space' or 'place of performance', we rarely think about the connotations inherent in the use of these words. Although in everyday life the terms space and place are usually exchangeable and generally used without much consideration, it is important to note some of the peculiar characteristics of each, and to consider some of the inconsistencies that emerge in their definitions. Even a simple exploration of three dictionaries, for example, reveals some useful observations.[1] In definitions of space, all three relate, to a greater or lesser extent, space to emptiness; the Cambridge Dictionary, for example, lists one of the definitions of space as 'an empty place', which might be a good example of the interchangeable qualities of the three terms discussed in this section. Macmillan describes space as 'an empty area between things' and 'the area in which everything exists' – which might lead one to think that the whole visible world is, in practice, 'space'.

In artistic performance, important attempts have also been made to understand these terms conceptually. One of the most important practitioners to put forward the idea that space might be more than a 'passive receptacle' was theatre director Tadeusz Kantor, who, in writings such as in the excerpt below from his 'Milano Lessons', imbued space with certain qualities related to what he identified as hyperspace and that guided and influenced his future works in the field of Performance art:

> Space is not a passive r e c e p t a c l e
> In which object and forms are posited....
> SPACE itself is an OBJECT [of creation].
> And the main one!
> SPACE is charged with E N E R G Y.
> Space shrinks and e x p a n d s.
> And these motions mould forms and objects.
> It is space that G I V E S B I R T H to forms!
> It is space that conditions the network of relations and tensions between objects.
> T E N S I O N is the principal actor of space.[2]

Kantor proposed the use of 'poor' spaces rather than the 'sacred' spaces of the theatre hall,[3] and his ideas of space as a 'network of relations and tensions between objects' resemble what the geographer Yi-Fu Tuan asserts about the 'meaning' of space in traditional societies, and which he contrasts with more modern understandings:

> In nonliterate and traditional communities the social, economic, and religious forms of life are often well integrated. Space and location that rank high socially are also likely to have religious significance. [...] Space in our contemporary world may be designed and ordered so as to draw one's attention to the social hierarchy, but the order has no religious significance and may not even correspond closely to wealth. One effect is the dilution of spatial meaning. In modern society spatial organization is not able, nor was it ever intended, to exemplify a total world view.[4]

Yi-Fu Tuan goes on to describe what he calls a 'mythical' space, describing it as an intellectual construct that can be very elaborate, and that 'differs from pragmatic and scientifically conceived spaces in that it ignores the logic of exclusion and contradiction'.[5] One of the kinds of mythical spaces listed by Tuan relates to cosmology, which helped societies in structuring and making sense of their spaces according to astronomical events. According to Tuan:

> Complex cosmologies are associated with large, stable, and sedentary societies. They are attempts to answer the question of man's place in nature. Practical activities seem arbitrary and may offend the gods or spirits of nature unless they are perceived to have their roles and place in a coherent world system.[6]

Such coherence can be observed in ancient ritual structures in many cultures geographically and culturally apart, such as the Greek theatres and the Inca ushnus, for example. In some cases, such organisation of space would also find its counterpart in the organisation of time, which could also be experienced as mythical. This is the case of the Kamayurá tribe in South America, for example, who can interpret 'time' as having characteristics of a logical, measurable sequence of events directly connected to everyday life — 'ãng' — but also in association to ancestral times, where events are continuous and immeasurable, characterised by their 'continuous potentiality' — 'mawe';[7] ritual events and their spaces would then function as a means of accessing 'mawe' from 'ãng', for which purpose an extremely complex musical system was developed and used ritualistically.

By contrast, in contemporary Western philosophy, Lefebvre explores the definition of space in relation to its social production. This was a key concept put forward by the French philosopher Henri Lefebvre, who connected spatial use to socially and politically defined modes of production. By connecting the very essence of space to the social practices that are carried out within it, Lefebvre not only established a strong basis for the social dynamics that inform our modern use and understanding of space, but also established the key idea that buildings also bring within their use the power relations subsumed by the relations of production in place in a society.[8] According to Lefebvre, spaces are inhabited not only by subjects but also by the power relations already in place, emulating and reproducing these relations within certain contexts. For him, spaces are the results of human actions while, at the same time, they offer the conditions for these same actions to take place.[9] And finally, another aspect emphasised by Lefebvre in his work is the fact that space is not a 'void', an 'all-accepting' passive receptacle, but connected to what he called 'social morphology':

> To picture space as a 'frame' or container into which nothing can be put unless it is smaller than the recipient, and to imagine that this container has no other purpose than to preserve what has been put in it – this is probably the initial error. But is it error, or is it ideology? The latter, more than likely. If so, who promotes it? Who exploits it? And why and how do they do so?[10]

It is interesting to note how Lefebvre's viewpoint resonates with Kantor's ideas, and that it is aligned to a wider understanding not just of space and place, but also, as we will see later, of emptiness. Another important philosopher who turned his attention to the intricacies of space and spatial practices was Michel Foucault. In a text written as a basis for a lecture in Paris in 1967, Foucault briefly analyses the historical understanding of space, and how distinct trends of knowledge such as structuralism and phenomenology have offered different, but somewhat complementary ideas. Foucault also demonstrates a particular interest in sites that have the 'curious property of being in relation with all the other sites, but in such a way as to suspect, neutralise, or invert the set of relations that they happen to designate, mirror, or reflect'.[11] Such sites, according to him, can be of two kinds, which he defined as utopias and heterotopias. The former is employed in a variety of situations and described by Foucault as 'sites with no real places', presenting society in a perfected form, and 'fundamentally unreal spaces'. The latter, however, are '[…] something like counter-sites, a kind of effectively enacted utopia in which [...] all the other real sites that can be found within the culture, are simultaneously represented, contested, and inverted', adding that '[...] places of this kind are outside of all places, even though it may be possible to indicate their location in reality'.[12] Such places offer a much larger ground for discussion, as these are probably present, according to the author, in every culture and civilisation. Foucault also sets six 'principles' for what could be considered heterotopias. The third principle is of particular interest to me, as it acknowledges the capability of such sites to juxtapose several different places that are themselves incompatible with one another, and it is interesting to note that Foucault found appropriate to consider heteropias examples from many civilisations and cultures that do not usually seem to be conceptually connected, such as theatres, cemeteries, gardens, and brothels, offering a fresh approach to the understanding of such everyday spaces.[13]

Another interesting aspect of these examples, and indeed of the discussions of many other authors concerned with the conceptual possibilities of space, is the interchangeability displayed between the ideas of space and place, in a similar way to some of the dictionary definitions that were explored earlier. For other authors, however, the distinction between these two terms offers significant conceptual possibilities. Yi-Fu Tuan, for example, although he sees the two terms as closely related, sees space as being more abstract than place, the latter which he likens to a sense of 'security' and 'stability'. He also argues that place can be defined in terms of the 'value' and 'experience' associated with the use of space:

> From the security and stability of place we are aware of the openness, freedom, and threat of space, and vice-versa. Furthermore, if we think of space as that which allows movement, then place is pause; each pause in movement makes it possible for location to be transformed into place.[14]

When associating place with attributes such as 'pause' and 'security', and space to 'movement' and 'direction', Tuan establishes a strong connection between these terms and the lived, physical experience of the world. The idea of 'stability' in connection to the definition of place is also present in Michel de Certeau's analysis of everyday practices. Place, for both authors, is understood in a way in which perception and familiarity become defining criteria of paramount importance. Space, for de Certeau, is expected to function as a 'polyvalent unity of conflictual programs or contractual proximities',[15] which we might argue resonates with our general understanding of performance spaces in terms of the functions and expected behaviour of performers and audience members, who we might consider to be in 'contractual proximity' for the duration of a performance event, in which they might have different — if not 'conflictual' — functions and expectations. In consonance with these ideas and the current use of the term 'performance space' to characterise our performance venues, it is easy to adopt the simplistic view that such venues are in fact spaces, and that the qualities attributed to place by Tuan and other authors are not applicable. We cannot forget, though, that such venues are special buildings 'stabilised' in space by architectural processes, which present the 'instantaneous configuration of positions' which Tuan saw as implied in the existence of places. In the context of theatre, Richard Southern connected the term place to its possible etymological origins in the English language — deriving from the Latin platea[16] — and understood it as a 'static concept', or, as postulated by Tuan, a type of object geometrically defined within space.[17]

Govan, Nicholson, and Normington, in turn, observed that such understanding has been inverted by contemporary theatre practitioners, who brought collections of material and objects from the 'external' world into performance venues, inviting audiences to reconsider the nature of the place where performances happen, and promoting the idea that '[…] the act of performance-making can shift the place to create a space […]', resulting in what they considered to be an 'act of liberation'.[18] Taking this viewpoint into account, we could argue that performance events act to change the qualitative parameters of the venues where they happen, a fact that would make such events intrinsically linked to our perception of the places and spaces that allowed these same events to occur. The 'act of liberation' offered by the authors is, in my view, a powerful idea connected to the understanding — conscious or not — of such a perception, which might be felt at some level by performers and audience members alike. The conversion from place to space would, in this way, become one of the most important hidden 'truths' of performance events, a shift that might further enrich expressions such as 'artistic freedom' for both performers and audiences. It is also interesting to consider that such a shift could happen in performance venues generally recognised as 'traditional', with a clear divide between performance area and audience area, but also in venues less concerned with such pre-conceived divisions, such as black boxes and 'found' sites. In all cases, it seems to me that a sort of 'activation' of the sites is necessary for this shift to take place, and such 'activation' is provided by the event itself, in conjunction with the performers' and audiences' acceptance of the performance 'conditions'. Performance sites would, in this way, go through shifting cycles in how we perceive them as places or spaces triggered by the presence or not of performance events and their constitutive elements.

However, I wonder if this dualistic understanding of place and space is enough to capture the kinds of performance environments that I would like to invoke in my own work. If we consider Tuan's idea that 'space is transformed into place as it acquires definition and meaning',[19] it could be suggested that the 'liberating' conversion of places back into spaces, as proposed by Govan, Nicholson, and Normington, might therefore render spaces 'undefined' and 'meaningless'; although this is not what the writers are proposing as such, it is a logical outcome of taking such a dualistic approach.

[PLACE]	→	[SPACE]	→	[PLACE]
Defined, meaningful		Undefined, meaningless		Defined, meaningful

At the same time, there are many forces — mainly social, political, and aesthetic — in place in performance venues, particularly related to the idea of stage in many cultures and periods in history, and it seems unlikely that such forces would simply 'recede' with the shift from place to space – the residues of 'definition' and 'meaning' would remain despite the impact of the shift. Therefore, I would like to suggest another possibility in which, instead of being converted back into spaces, such places for performance are converted into intermediary sites that cannot be defined as either space or place, but that would exist in a conceptual 'limbo' for the duration of a performance event. In such sites, whose existence would thus be temporary by nature, Foucault's 'places that are foreign to one another'[20] could be activated, acknowledged by all who take part, sustained throughout the performance events, and finally abandoned with the end of the event, which would bring the site back to its previous status of place. In this way, in the case of venues where a complete understanding of those sites depends on an event taking place — a category in which Foucault's examples of theatres and brothels could still be included, and to which we could add concert halls — Foucault's heterotopias would partially make sense. A site characterised by such a conceptual 'limbo' would, in my view, allow for the existence of as many artistic realities and event dynamics as can be conceived, thus being in deeper connection with the sense of freedom and liberation proposed by Govan et al.

[PLACE]　　→　　[?]　　→　　[PLACE]
Defined, meaningful　　Temporarily defined, meaningful　　Defined, meaningful

This subject needs, of course, further detailed analysis and historical contextualisation. However, I consider that, when contemplating the present and future of performance venues, it is important to have in mind such intricacies in the shifting perception of space and place, bearing in mind that, as Tuan himself acknowledges, 'in an area of study where so much is tentative, perhaps each statement should end with a question mark or be accompanied by qualifying clauses'.[21]

Emptiness

There is yet another shift in paradigm that I consider of utmost importance to the study of spaces and places for artistic performance, which relates to a concept with much deeper aesthetic, philosophical, historical, and even social implications. Such a shift is related to emptiness, a term that is currently applied, directly or indirectly, in the definitions of space and place in all three dictionary entries quoted at the beginning of this section. As we have briefly seen in the work of practitioners and theoreticians such as Lefebvre, the dominant European understanding of the concept of emptiness is an essentially negative one, a fact reflected in the dictionary definitions quoted above. Such definitions mainly associate emptiness with a lack of meaning, purpose, or substance, and even insincerity. Such negative connotation finds expression in terms such as horror vacui, which has been defined as 'a fear or dislike of leaving empty spaces, particularly in artistic composition'.[22] Although horror vacui has existed at least since the time of Aristotle, my current use of the expression relates to Mario Praz's critical description of the cluttered design of the Victorian Age, being also used pejoratively in reference to certain painting styles, interior design, and everything else that can be defined by overwhelming, content-saturated work, particularly with the need to fill up space[23] and 'an irrational aversion to the void'.[24]

Such fear or aversion towards emptiness might lead us to interpret it as a human phenomenon, rather than a cultural one. The understating of emptiness, however, finds in Japanese culture and aesthetics a very powerful counterpoint to the European horror. Kenya Hara, a Japanese graphic designer and one of the main figures responsible for the reconsideration and reconciliation of what might constitute emptiness in Japanese and European cultures, argues that this concept in Japanese aesthetics is associated with limitless potential, using the example of an empty vessel both as a metaphor and a practical application of such an attitude:

> An empty state possesses a chance of becoming by virtue of its receptive nature. The mechanism of communication is activated when we look at an empty vessel, not as a negative state, but in terms of its capability to be filled with something.[25]

Therefore, emptiness according to Hara is viewed in Japan not as the lacking of elements, but through its potential to hold new ones.[26] We could go even further in arguing that, if interpreted in a temporal frame, emptiness in the European context is associated not just with negative feelings, such as lacking or meaninglessness, but also to unrealised potential or the absence of essential elements associated with the past; in turn, the Japanese understanding might be linked to the future, open to the diverse possibilities generated by the innumerable variables in play at all times. This shift in the understating of such a complex concept is, according to Hara, due to the cultural and spiritual formation of the Japanese people throughout the centuries, particularly associated with Shinto and Buddhism, the two main pillars of Japanese spirituality. Jerrold McGrath also associates these two religions with the wider Japanese understanding of space. According to McGrath, '[…] from Shinto came the high value placed on harmony in relationships and a focus on the connections – spoken and unspoken – that tie people together. From Buddhism came the ideas of emptiness and selflessness'.[27] Shinto's architecture is very much connected to its spiritual practices, and this is reflected in the fact that Shinto shrines started with no structure at the beginning, with worship taking place first in an isolated, natural wood, and then in a spot thoroughly purified, planted with trees and fenced with stones around.[28] From this we can assume that the concept of emptiness in Japan as described by Hara might have been already closely linked to Shinto spatial practices from primitive ages, possibly articulating behaviours and relationships before the need for a physical structure to host them. For those participating in Shinto rituals in the early days of the practice, the invisible presence of the kami (spiritual beings) in a given empty space can be associated with the aspects of emptiness that Hara refers to as 'potentiality'. The spiritual practices taking place in such spaces might also provide a historical complement to McGrath's understanding of how, in Japan, spaces 'shape' relationships and have 'meanings prior to any activity that happens within them'.[29] In the case of Shinto shrines, the behaviour and relationship of the practitioners with the kami preceded the need for a physical structure for rituals to take place, a conceptual inversion that, although apparently paradoxical, can be closely linked to the definition of space as a 'facilitator' of relationships and social dynamics as referred to by McGrath. As Akiyama also notices, there are still a few Shinto shrines which have no main hall,[30] which is an example of how deeply and naturally such understanding of emptiness is still ingrained in Japanese spiritual and aesthetic sensibilities.

Another important root in the conceptual understanding of emptiness in Japanese aesthetics comes from Buddhist philosophy. The current Japanese word for space — 空間, ku-kan — literally means 'empty place', and it is of relatively recent origin, having been coined to express the concept of three-dimensional objective space which was imported from the West, for which the Japanese language had no word of its own.[31] The compound includes the character 空, ku, which today can be used for 'empty' in the simple physical sense, and for 'void' in Buddhist metaphysics;[32] and 間, a character that usually means mā when standing by itself. Mā is a complex concept, extremely difficult to translate, which integrates and embodies the Japanese aesthetic understanding of emptiness on many levels. Kiyoshi Matsumoto offers a useful conceptual explanation of mā as '[…] "space between," but rather than a static gap, it is the distance that exists between objects as well as between time', and also complementing that mā in architecture as '[…] the dimension of space between the structural posts of an interior', complementing with the fact that '[…] the layout is intentionally designed to encompass empty space – energy filled with possibilities'.[33]

When understanding mā as 'energy filled with possibilities', and attributing to it 'enabling' qualities, Matsumoto unlocks the real potential of the concept, which lies in its 'active' power when employed in the creation of a wide range of works of visual art, architecture, performance, design, language, and many other aspects of Japanese everyday life. This can be considered as directly linked to what Hara referred to as a 'creative receptacle' in relation to the understanding and application of the concept of emptiness in Japanese art and culture,[34] away from the negative context generally attributed to emptiness in the West. It might be tempting to consider mā as an isolated aesthetic concept linked specifically to Japanese culture; there are, however, very clear examples in Western culture of the application of the underlying concept of mā, with the gap between God's and Adam's fingers in Michelangelo's 'The Creation of Adam' in the Sistine Chapel being arguably the most famous and widely interpreted.[35]

A good example of such a shift in the understanding of emptiness is the often-quoted statement by theatre director Peter Brook who, at the beginning of his landmark book The Empty Space, creates what is, in my view, the best image of this turn towards 'active' emptiness and away from horror vacui in Western art, including performance, in the second half of the 20th century:

> I CAN take any empty space and call it a bare stage. A man walks across this empty space whilst someone else is watching him, and this is all that is needed for an act of theatre to be engaged. Yet when we talk about theatre this is not quite what we mean. Red curtains, spotlights, blank verse, laughter, darkness, these are all confusedly superimposed in a messy image covered by one all-purpose word.[36]

Still, in the field of theatre, the ideas exploring the 'active' power of emptiness resonate with Tadeusz Kantor's denial of space as a 'passive receptacle' as quoted earlier, in which some ideas can be related to some of the more complex aspects regarding space, place, and emptiness discussed in this chapter. In associating space with energy, for example, Kantor's interpretation is in line with the understanding of the concept of mā and emptiness associated with it; in recognising the 'network of relations and tensions between objects', we might be able to first identify social elements explored by Lefebvre and de Certeau in the social creation of spaces, and also the Japanese idea of the social qualities already present within everyday spaces and places even before these are used by McGrath; and, finally, the idea that space 'gives birth to forms' can be related to what I see as the 'conceptual pregnancy' — as related to the potentiality mentioned earlier — necessary for performance venues today. In my view, embracing such 'pregnancy' would offer a powerful means of representing our post-globalised world, or at least what is expected from our societies in the 21st century, following values connected to democracy, inclusiveness, and shifting ways of connecting individuals. I also consider that, if music performance is to be integrated into such a society, all involved, from the design to the final use of venues for music performance, would first need a broader understanding of the defining historical, social, aesthetic and cultural aspects linked to space, place, and emptiness, and how new paradigms could be integrated in order to converge with the values established by this same society; this could allow music events to elevate both the media and the message — assuming there is one — to a higher level of cohesion between conceptual ideas and their delivery.

Empty spaces

Having explored the possible definitions and conceptual interactions between space, place and emptiness, I identify a category of performance venues that I recognise as empty spaces, which in my understanding are places that do not rely on any pre-existing structural demarcation of specific areas for performers or audience and are instead activated for the duration of a given event. In relation to the need for a 'temporary' and 'defined' place as explored earlier, it is important to emphasise that a nuanced view of these terms is needed in order to understand not only the aesthetic and artistic characteristics of the events happening in empty spaces but also the creative possibilities involved in the design of such venues. This design should, in my view, take into consideration the many creative possibilities emerging from the combination of diverse cultural dynamics, technological advances, and multiple forms of interactions made possible by the rapid and constant evolution of our societal values, while also creating places where the creative potential of artistic creators can be fully unleashed. Many cutting-edge devices, methods and interactions included in this book can find in empty spaces a natural home, without needing to adapt to the relatively immutable structure and dynamics of, for example, concert halls.

Movements and projects related to my field of work, which involves experimentations with layouts and dynamics in music performance events, have been exploring the interactions between artistic performance and architecture. Two good examples are Performalism[37] and Permutations,[38] which can give a glimpse into the bright future that can be created by a closer interaction of both areas of expertise. Empty space venues such as the Angela Burgess Recital Hall and the Sands End

Arts and Community Centre, in London, and the National Sawdust, in New York, are very good examples of the need for a more open attitude in the design and use of spaces for performance in both educational and commercial settings (Figures 26.1 and 26.2).

Figure 26.1 On the left, Angela Burgess Recital Hall, at the Royal Academy of Music; on the right, the Main Hall at Sands End Arts and Community Centre, London, both in London.

Source: RIBA, "Royal Academy of Music – The Susie Sainsbury Theatre and The Angela Burgess Recital Hall," https://www.architecture.com/awards-and-competitions-landing-page/awards/riba-regional-awards/riba-london-award-winners/2018/royal-academy-of-music-susie-sainsbury-theatre-angela-burgess-recital-hall, accessed 24 January 2023. © Credit Adam Scott; Sands End Community and Arts Centre – Credit: SEACC.

Figure 26.2 National Sawdust, in New York.

Source: https://commons.wikimedia.org/wiki/File:NS_stage_up-1500x1018.jpg, accessed 9 May 2024.

It is, of course, important to consider the role of sound within such understanding of empty spaces for artistic performance. In the case of music, for example, we have grown used to the mighty presence of concert halls as the 'ideal' venue for the performance of music. The general layout of such halls was largely defined in the 19th century, with Wagner's Bayreuth Festival Theatre being particularly influential.[39] However, since the 17th century, the science of Acoustics increasingly assumed a leading role in the design of concert halls, having evolved, from the 19th century onwards, into a wide array of highly specialised fields.[40] In the field of music performance, however, I still observe a great predominance of old ideas and ideals of sound production and consumption amongst my peer performers, composers and promoters. Such predominance can be linked to the almost exclusive use of concert and recital halls as venues for music events, particularly in the field of classical music; in my view, there is a lingering contradiction in the fact that the science of Acoustics has evolved into a cutting-edge discipline while most of the music halls conceived with the help of acousticians are still connected to layouts and dynamics established centuries ago, for societies with ideas and values considerably different of our own. In fact, a growing number of my colleagues — performers and composers, particularly — express their discontentment with the limitations imposed by traditional halls and their frontal layout, which leads me to think that we need to think afresh about the design and use of spaces for music performance. Perhaps it is time for performance creators, particularly in the music field, to use more intensely the few available empty spaces in a way that is less connected to the past and more relevant to what can be created and expressed today. Such a change in attitude would perhaps trigger a reconsideration of the design of such spaces, integrating aesthetic and conceptual elements connected to the emptiness of the space that would slowly offer a change in the relationship between conception and final use. In a world where our societal values are reassessed on a daily basis, the products of our society are also, in one way or another, being reviewed in order to adapt to mindsets generated by new paradigms. Empty spaces might offer an alternative to traditional halls in being connected to values such as cultural inclusivity, equality and collaboration, while offering a platform to unleash the potential of creative ideas that would not find in concert halls a completely hospitable environment.

So where do my discussions of space and place leave us in terms of the sound that they house? It is fascinating to consider the roles that sound might assume within such a context, whereby it stops being the driver within an old paradigm to become an expressive tool in the collaborative environment nurtured by the concept of empty spaces. I hope we will soon witness more empty spaces being designed and built for artistic performance, and that, in the coming years, an increasing number of venues and creators will recognise and embrace their creative, social, and commercial potential.

Notes

1. The dictionaries consulted were: *Cambridge Dictionary*, accessed 17 November 2021, https://dictionary.cambridge.org/; *Merriam-Webster Dictionary*, accessed 17 November 2021, https://www.merriam-webster.com/; *Macmillan Dictionary*, accessed 17 November 2021, https://www.macmillandictionary.com/; *The Shorter Oxford English Dictionary*, vol. II (Oxford University Press, 1973).
2. Michal Kobialka, "'The Milano Lessons' by Tadeusz Kantor: Introduction," *TDR* 35, no. 4 (1991): 136–147, accessed 17 November 2021, https://doi.org/10.2307/1146169.
3. David Wiles, *A Short History of Western Performance Space* (Cambridge: Cambridge University Press, 2003), 14.
4. Yi-Fu Tuan, *Space and Place: The Perspective of Experience* (Minneapolis and London: University of Minnesota Press, 2018), 112–113.

5 Yi-Fu Tuan, *Space and Place: The Perspective of Experience* (Minneapolis and London: University of Minnesota Press, 2018), 99.
6 Yi-Fu Tuan, *Space and Place: The Perspective of Experience* (Minneapolis and London: University of Minnesota Press, 2018), 88.
7 Rafael José de Menezes Bastos, *A Musicológica Kamayurá: para uma antropologia da comunicação no Alto Xingu* (Florianópolis: UFSC, 1999), 109–110.
8 Henry Lefebvre, *The Production of Space*, translated by Donald Nicholson-Smith (Oxford and Cambridge: Anthropos, 1991), 33.
9 Henry Lefebvre, *The Production of Space*, translated by Donald Nicholson-Smith (Oxford and Cambridge: Anthropos, 1991), 57.
10 Henry Lefebvre, *The Production of Space*, translated by Donald Nicholson-Smith (Oxford and Cambridge: Anthropos, 1991), 93–94.
11 Michel Foucault, "Of Other Spaces (1967), Heterotopias," translated by Jay Miskowiec, *Architecture, Mouvement, Continuité* 5 (1984): 46–49, https://foucault.info/documents/heterotopia/foucault.heteroTopia.en/.
12 Michel Foucault, "Of Other Spaces (1967), Heterotopias," translated by Jay Miskowiec, *Architecture, Mouvement, Continuité* 5 (1984): 46–49, https://foucault.info/documents/heterotopia/foucault.heteroTopia.en/.
13 Michel Foucault, "Of Other Spaces (1967), Heterotopias," translated by Jay Miskowiec, *Architecture, Mouvement, Continuité* 5 (1984): 46–49, https://foucault.info/documents/heterotopia/foucault.heteroTopia.en/.
14 Tuan, *Space and Place*, 6.
15 Michel de Certeau, *The Practice of Everyday Life*, translated by Steven Rendall (Berkeley and Los Angeles: University of California Press), 117.
16 Richard Southern, *The Seven Ages of Theatre* (London: Faber and Faber, 1962), 99.
17 Yi-Fu Tuan, *Space and Place: The Perspective of Experience* (Minneapolis and London: University of Minnesota Press, 2018), 17.
18 Emma Govan, Helen Nicholson, and Katie Normington, *Making a Performance* (London: Routledge, 2008), 106.
19 Yi-Fu Tuan, *Space and Place: The Perspective of Experience* (Minneapolis and London: University of Minnesota Press, 2018), 136.
20 Michel Foucault, "Of Other Spaces (1967), Heterotopias," translated by Jay Miskowiec, *Architecture, Mouvement, Continuité* 5 (1984): 46–49, https://foucault.info/documents/heterotopia/foucault.heteroTopia.en/.
21 Yi-Fu Tuan, *Space and Place: The Perspective of Experience* (Minneapolis and London: University of Minnesota Press, 2018), 7.
22 Lexico powered by Oxford, s.v. "Horror Vacui," accessed 2 November 2021, https://www.lexico.com/definition/horror_vacui.
23 Mads Soegaard, "Horror Vacui: The Fear of Emptiness," Interaction Design Foundation, accessed 2 November 2021, https://www.interaction-design.org/literature/article/horror-vacui-the-fear-of-emptiness.
24 John Thorp, "Aristotle's Horror Vacui," *Canadian Journal of Philosophy* 20, no. 2 (1990): 149–166, https://www.jstor.org/stable/40231690.
25 Kenya Hara, *White* (Zurich: Lars Muller Publishers, 2018), 39.
26 A good example of how this understanding of [emptiness] is reflected in Japanese crafts, for example, is this sentence by Sakaida Kakiemon XIV, talking about the white areas in Kakiemon traditional pottery: '[…] I think the white areas are not merely empty space; they contain all the thoughts and emotions of humankind'. Sakaida Kakiemon XIV, *The Art of Emptiness*, translated by Gavin Frew (Tokyo: Japan Publishing Industry Foundation for Culture, 2019), 126.
27 Jerrold McGrath, "The Japanese Words for 'Space' Could Change Your View of the World", *Quartz*, 18 January 2018, https://qz.com/1181019/the-japanese-words-for-space-could-change-your-view-of-the-world/.
28 Aisaburo Akiyama, *Shintô and its architecture* (Tokyo: Tokyo News Service, 1955), 53.
29 Jerrold McGrath, "The Japanese Words for 'Space' Could Change Your View of the World", *Quartz*, 18 January 2018, https://qz.com/1181019/the-japanese-words-for-space-could-change-your-view-of-the-world/.
30 Aisaburo Akiyama, *Shintô and its architecture* (Tokyo: Tokyo News Service, 1955), 53.
31 Gunter Nitschke, "MA: Place, Space, Void," *Kyoto Journal*, 16 May 2018, https://kyotojournal.org/culture-arts/ma-place-space-void/.

32 Gunter Nitschke, "MA: Place, Space, Void," *Kyoto Journal*, 16 May 2018, https://kyotojournal.org/culture-arts/ma-place-space-void/.
33 Kiyoshi Matsumoto, "MA – The Japanese Concept of Space and Time," *Medium*, 24 April 2020. https://medium.com/@kiyoshimatsumoto/ma-the-japanese-concept-of-space-and-time-3330c83ded4c.
34 "Design Talk by HARA Kenya: SUBTLE – Delicate or Infinitesimal | Japan House London," YouTube, accessed 7 November 2021, https://youtu.be/iVaoDNEZwNg?t=855.
35 Among these interpretations there is an interesting analogy created by Paul Barolsky, in which the author compares the gap left by Michelangelo in 'The Creation of Adam' to Castiglione's ideas of musical 'imperfection', seeing dissonance not just as complementary, but also necessary to the perception of beauty. Paul Barolsky, "The Imperfection of Michelangelo's Adam." *Notes in the History of Art* 20, no. 4 (2001): 8, http://www.jstor.org/stable/23206730.
36 Peter Brook, *The Empty Space* (New York: Atheneum, 1968), 9.
37 Yasha Grobman and Eran Neuman, eds. *Performalism: Form and Performance in Digital Architecture* (Tel Aviv: Tel Aviv Museum of Art, cat. 8/2008).
38 "Projects: Permutations," Freya Waley-Cohen, accessed 24 January 2023, https://www.freyawaleycohen.com/projects/permutations.
39 Of course, with strong references to ancient Greek and Roman theatres.
40 The areas into which it has diversified are clearly exemplified by the great variety of ideas presented in this book. For a more detailed historical account, please check Dorothea Baumann, Music and Space: A systematic and historical investigation into the impact of architectural acoustics on performance practice followed by a study of Handel's Messiah (Peter Lang, 2011), 33–44.

Bibliography

Akiyama Aisaburo. *Shintô and its Architecture*. Tokyo: Tokyo News Service, 1955.
Barolsky, Paul. "The Imperfection of Michelangelo's Adam." *Notes in the History of Art* 20, no. 4 (2001): 6–8. http://www.jstor.org/stable/23206730.
Bastos, Rafael José de Menezes. *A Musicológica Kamayurá: para uma antropologia da comunicação no Alto Xingu*. Florianópolis: UFSC, 1999.
Baumann, Dorothea. *Music and Space: A Systematic and Historical Investigation into the Impact of Architectural Acoustics on Performance Practice Followed by a Study of Handel's Messiah*. Bern: Peter Lang, 2011.
Beranek, Leo. "Subjective Rank-Orderings and Acoustical Measurements for Fifty-Eight Concert Halls." *Acta Acustica united with Acustica* 89 (2003): 494–508.
Blesser, Barry, and Linda-Ruth Salter. *Spaces Speak, Are You Listening?* Cambridge, MA: MIT Press, 2007.
Britannica, accessed 14 May 2021, https://www.britannica.com/.
Brook, Peter. *The Empty Space*. New York: Atheneum, 1968.
Cambridge Dictionary, accessed 17 November 2021, https://dictionary.cambridge.org/.
De Certeau Michel. *The Practice of Everyday Life*. Translated by Steven Rendall. Berkeley and Los Angeles: University of California Press, 1980.
"Design Talk by HARA Kenya: SUBTLE – Delicate or Infinitesimal | Japan House London." Youtube, accessed 28 October 2022, https://youtu.be/iVaoDNEZwNg?t=855, Video, 14:15.
Foucault, Michel, translated by Jay Miskowiec. "Of Other Spaces (1967), Heterotopias." *Architecture, Mouvement, Continuité* 5 (1984): 46–49. https://foucault.info/documents/heterotopia/foucault.heteroTopia.en/.
Gann, Kyle. *No Such as Thing as Silence: John Cage's 4'33"*. New Haven, CT and London: Yale University Press, 2010.
Govan, Emma, Helen Nicholson, and Katie Normington. *Making A Performance*. London: Routledge, 2008.
Grobman, Yasha and Eran Neuman, eds. *Performalism: Form and Performance in Digital Architecture*. Tel Aviv: Tel Aviv Museum of Art, cat. 8/2008.
Hara, Kenya. *White*. Zurich: Lars Muller Publishers, 2018.
Hill, Leslie, and Helen Paris, eds. *Performance and Place*. New York: Palgrave MacMillan, 2006.
Kakiemon XIV, Sakaida. *The Art of Emptiness*. Translated by Gavin Frew. Tokyo: Japan Publishing Industry Foundation for Culture, 2019.
Kobialka, Michal. "'The Milano Lessons' by Tadeusz Kantor: Introduction." *TDR (1988-)* 35, no. 4 (1991): 136–147. https://doi.org/10.2307/1146169.

Kolarevic, Branko. "Back to the Future: Performative Architecture." *International Journal of Architectural Computing* 1, no. 2 (January 2004): 43–50. https://www.semanticscholar.org/paper/Back-to-the-Future%3A-Performative-Architecture-Kolarevic/f1d880ca54c91177fe1ec29aa3eb2e8d3d6f9d44.

Lefebvre, Henry. *The Production of Space*. Translated by Donald Nicholson-Smith. Oxford and Cambridge: Anthropos, 1991.

Macmillan Dictionary, accessed 17 November 2021, https://www.macmillandictionary.com/.

Marcantonio, Barbara. "Design Principles – Gestalt, White Space and Perception." *manifesto*, 6 February 2015. https://manifesto.co.uk/design-principles-gestalt-white-space-perception/.

Matsumoto, Kiyoshi. "MA—The Japanese Concept of Space and Time." *Medium*, 24 April 2020. https://medium.com/@kiyoshimatsumoto/ma-the-japanese-concept-of-space-and-time-3330c83ded4c.

Mattos, Fabricio. "New Stages: Revisiting the Understanding of *Stage* and *Performance* in Music Events." PhD diss., Royal Academy of Music, London, 2023.

McGrath, Jerrold. "The Japanese Words for 'Space' Could Change Your View of the World." *Quartz*, 18 January 2018. https://qz.com/1181019/the-japanese-words-for-space-could-change-your-view-of-the-world/.

Meddens, Frank, Katie Willies, Colin McEwan, and Nicholas Branch, eds. *Inca Sacred Space: Landscape, Site and Symbol in the Andes*. London: Archetype Publications, 2014.

Merriam-Webster Dictionary, accessed 17 November 2021, https://www.merriam-webster.com/.

"National Sawdust / Bureau V," *ArchDaily*, accessed 24 January 2023, https://www.archdaily.com/779989/national-sawdust-bureau-v?ad_medium=gallery. © Floto + Warner.

"National Sawdust in New York," accessed 9 May 2024, https://commons.wikimedia.org/wiki/File:NS_stage_up-1500x1018.jpg

Nitschke, Gunter. "MA: Place, Space, Void." *Kyoto Journal* 98 (2020): 8–23. https://kyotojournal.org/culture-arts/ma-place-space-void/.

Pätynen, Jukka and Tapio Lokki. "Concert Halls with Strong and Lateral Sound Increase the Emotional Impact of Orchestra Music." *The Journal of the Acoustical Society of America* 139 (2016): 1214–1224. https://doi.org/10.1121/1.4944038.

"Royal Academy of Music – The Susie Sainsbury Theatre and The Angela Burgess Recital Hall," RIBA, accessed 24 January 2023, https://www.architecture.com/awards-and-competitions-landing-page/awards/riba-regional-awards/riba-london-award-winners/2018/royal-academy-of-music-susie-sainsbury-theatre-angela-burgess-recital-hall. © Adam Scott.

"Sands End Community and Arts Centre," accessed 24 January 2023, https://www.seacc.uk/

Small, Christopher. *Musicking: The Meanings of Performing and Listening*. Middletown: Wesleyan University Press, 1998.

Soegaard, Mads. "Horror Vacui: The Fear of Emptiness." *Interaction Design Foundation*. https://www.interaction-design.org/literature/article/horror-vacui-the-fear-of-emptiness.

Southern, Richard. *The Seven Ages of Theatre*. London: Faber and Faber, 1962.

Stobart, Henry. *Music and the Poetics of Production in the Bolivian Andes*. Aldershot: Ashgate, 2006.

Swan, James, ed. *The Power of Place & Human Environments*. Bath: Gateway Books, 1993.

The Shorter Oxford English Dictionary, Vol. II. Oxford: Oxford University Press, 1973.

Thorp, John. "Aristotle's *Horror Vacui*." *Canadian Journal of Philosophy* 20, no. 2 (1990): 149–166. https://www.jstor.org/stable/40231690.

Tuan, Yi-Fu. *Space and Place: The Perspective of Experience*. Minneapolis and London: University of Minnesota Press, 2018.

Weber, William. "Did People Listen in the 18th Century?" *Early Music* 25, no. 4 (November 1997): 678–691.

Wiles, David. *A Short History of Western Performance Space*. Cambridge: Cambridge University Press, 2003.

27
OPERA IN THE BATHHOUSE
Exploring an acoustically led approach to dramaturgy and scenography

Rosalind Parker and Pedro Novo

Introduction

This chapter examines the understanding and creative application of acoustic intimacy from various disciplinary perspectives, focusing on the site of the (drained) Gala Pool at Moseley Road Baths in the Balsall Heath area of Birmingham. The observations of acoustic intimacy on this site are first explored intuitively and empirically and then analysed from an acoustic engineering perspective using a series of measurements relating to impulse responses.

This unique project began when opera director Rosalind Parker and theatre designer Leanne Vandenbussche embarked on a new commission for the Gala Pool space in Moseley Road Baths, Birmingham. They were struck by the pool's impressive and familiar aesthetic, which made a statement through both its visual and aural qualities. Recognising the reverberant nature of the space for an opera, they realised that the development of the operatic text's dramaturgical arc and the design's scenographic language needed to consider the behaviour of the sound. To gain a better understanding of 'the bounce', they assembled a team of specialists.

Exploring creative potential

Phase 1: Engaging with the local community

In the first phase of the project, underscore Collective ran a series of public and school workshops titled 'Discovering Acoustics through Intuitive Play'.

The public workshops enabled members of the local community to discover the acoustic signature of the space through intuitive vocal play. This enabled the team to explore how the space naturally inclined towards physical and acoustic inhabitation. The behaviour of the public revealed specific tendencies, as they were drawn to certain areas and occupied them in a particular manner with sound. However, they physically and vocally avoided other areas. Additionally, their testimonies provided further insights: 'it feels daunting at first [on the threshold] but as you're drawn in [to the basin] it becomes really welcoming'. This illustrates a synchronicity between how it felt to move around, and how the sound was behaving. The reverberation appeared to enhance their presence, making them feel 'welcome'. It soon became clear that sound played a significant role in shaping both the physical journey and emotional experience of the bathhouse occupants.

Phase 2: Creative development

In January 2022, this project was further developed by assembling a multidisciplinary research team. The team consisted of an architect-composer, an acoustic engineer, a professional sound recordist, a professional soprano singer, a trumpet player and a movement director. Each team member conducted their own practice-based research investigations on the acoustics of the space, while also exploring creative value in different aspects of the same space at Moseley Road baths.

The opera director and theatre designer took inspiration from the operatic text. They worked closely with the soprano, trumpeter and movement director to stage excerpts of the text in different areas of the pool, listening carefully to how the space responded to different musical stimulus. This allowed the team to experiment with different spatial relationships between the performers and explore how a choreography might influence their natural presence in the space, from an audience perspective. Concurrently, the acoustician, sound recordist, and architect-composer identified a number of 'acoustic hot spots' and captured a series of impulse responses in different places, as part of a quantitative analysis of the acoustics of the space.

The combination of intuitive explorations and quantitative analysis resulted in a comprehensive and detailed understanding of the acoustic potential of the space. This hermeneutic approach to praxis research tightens the feedback loop between practice and research, working laterally rather than consecutively. It sparks cross-disciplinary discoveries and leads to surprising conclusions and enriches the creative process of developing a site-specific opera for an acoustically unique space.

Welcoming acoustic intimacy

During phase 1, when we opened the space to community and schools, participants in our workshops noticed and appreciated the openness and invitation for engagement that the space offered. Pupils, teachers and members of the community felt free to experiment in the space, as both performers and listeners. The acoustic qualities of the pool basin and surrounding areas created a unique atmosphere where the performers' sonic presence could be directly felt, even from a distance. This allowed the listening audience to experience a close connection to even the most delicate and subtle sounds from performers occurring between the pool basin and the diagonally opposite balcony.

During phase 2, we made more nuanced observations, regarding specific sounds and their correlation to specific locations in and around the pool basin. As we listened to the soprano and the trumpet, performing in the gallery, we were surprised to discover that the sounds appeared to cascade from the gallery to the basin. Simultaneously, we were impressed by the close and intimate quality of the performers' sound, despite their placement on the opposite side of the pool.

Following these observations from two productive workshops, we became much more interested in exploring the possible causes and potential creative applications of this sense of acoustic intimacy. This observation was particularly noteworthy in contrast to the predominantly reverberant nature of the space, as it seemed to offer a wider range of choreographic possibilities for both performers and audiences.

Exploring architectural acoustics

Acoustic intimacy

Reference to intimacy in an acoustic context was notably first reported in Leo Beranek's seminal 1962 work, where he suggested that 'a hall has intimacy if music in it sounds as though it is being played in a small room'.[1] Michael Barron has also defined acoustic intimacy as 'one's degree of identification with the performance, whether one feels acoustically involved or detached from it'.[2] Beranek selected the term intimacy to 'characterize the listening attribute of the closeness of communication between the listener and the orchestra' or other sources of music.[3] He further proposed initial time delay gap (ITDG), i.e., the delay of the first strong reflections after the arrival of direct sound being below 20–30 ms as central to evoking in the listener a sense of Intimacy.

More recent work on acoustic intimacy indicates that this is likely to be a multi-modal phenomenon.[4] However, the sense of intimacy was also experienced in the dark at night in the Gala Pool with all lights switched off, which suggests that the aural component has on its own the capacity to evoke a strong sense of acoustic intimacy. Intimacy as experienced in the Gala Pool evokes both a perception of being close to/sharing the same space as the performer but also a sense that the whole space is activated by sound.

We observed that when a sound source (for example a vocalist) is located on the balcony there is a sense of sound cascading down from the balcony towards the pool tank. In this respect, Kuusinen et al. mentioned that 'In some halls the sound field seems as approaching, i.e., as looming, during big crescendos, while in other halls such looming does not occur'.[5] Kuusinen further adds that 'Whether such mechanism exists is not clear, but it seems reasonable that the dynamic variation of the spatial cues including auditory distance, induced by the hall acoustics, promotes an enhanced feeling of being in the same space with performers'.

Following such observations, we measured the acoustic properties of the Gala Pool and investigated how the information collected could help to explain the sense of intimacy that we all experienced in this space. This work falls within the field of Spatial Hearing studies, a field that has been effectively organised and established by the publication of Spatial Hearing by Jens Blauert, originally in German in 1974.[6]

Acoustic measurements

In the late 1960s, Harold Marshall discovered, through both listening experience and rigorous research, that early reflections arriving from lateral directions created a desirable sense of spaciousness.[7] This phenomenon, originally called 'spatial responsiveness' and later 'spatial impression', was extensively investigated by Barron and Marshall, resulting in the derivation of the 'early lateral energy fraction' (LF) as a linear measure of spatial impression.[8]

The investigation presented here follows a similar line, by aiming to relate objective spatial characteristics of the sound field with a subjective impression: acoustic intimacy.

The Gala Pool

The Gala Pool is approximately 31 m long, 17 m wide and 10 m high resulting in a volume of approximately 5,270 m^3 (Figure 27.1). The roof is made of wood panels and the walls and floor (including the empty pool tank) are covered with glazed ceramic tiles. The balcony seats and the changing room doors, located at walking level around the pool, are made of timber. A salient feature of the Gala Pool is its sloped ceiling, which helps directing sound towards the pool tank.

Impulse response (IR) measurement

An IR can be understood as the acoustic fingerprint of a specific source/receiver – pair in a room. The IR can be measured by generating an impulsive, frequency-rich sound (for example, with a balloon or a starter pistol) at a specific location and recording the resulting effect with a

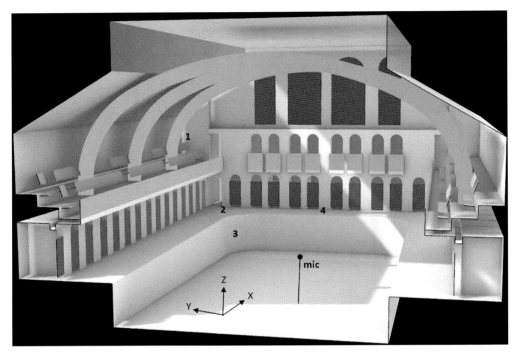

Figure 27.1 Gala Pool showing the location of the ambisonic microphone, and loudspeaker measurement locations (1–4) (Model Created by Emma-Kate Mathews).

microphone at a different location, in the same room. The recorded sound is composed of the 'direct sound' (sound that travels directly from the impulsive source to the microphone) followed by sound waves that have undergone one or more reflections off the room surfaces.

In practice, alternative methods are often employed instead of a balloon or a starter pistol, as they yield better results, particularly in relatively noisy environments, such as the Gala Pool where the ventilation system could not be completely turned off during measurements. For this project, the chosen method involved playing back a sine sweep through a loudspeaker. An overview of different approaches to measure IRs is given by G. B. Stan et al.[9]

In order to investigate the 3D spatial characteristics of a sound field, a microphone that can determine the direction from which the sound waves are coming is necessary. The Rode NT-SF1, has four microphone capsules oriented in different directions, and was used for this project. The format chosen to store the IR information was Ambisonics B-Format, which in 1st order encodes the 3D sound field in four channels: one channel per orthogonal (spatial) axis (X, Y, Z) and one omnidirectional channel (W).[10] Discriminating the direction of incidence of the sound waves is achieved by comparing the sound signals arriving at each microphone.

Measurement locations

It was decided that the microphone would be located at a fixed position, approximately at the centre of the empty pool tank, and that the loudspeaker would be moved between different positions (1–4) where measurements would be successively undertaken. Figure 27.1 shows the locations of these measurements.

Table 27.1 Acoustic parameters used to characterise performance spaces and their objective and subjective meanings

Parameter	Objective meaning	Subjective meaning
C50	Early to late arriving sound energy ratio – early: 0 ms to 50 ms; late: 50 ms to ∞	Speech Clarity
C80	Early to late arriving sound energy ratio – early: 0 ms to 80 ms; late: 80 ms to ∞	Musical Clarity
EDT	Early decay time – decay from 0 dB to −10 dB (multiplied by 6)	Related to reverberance/perceived reverberation
T20	Reverberation time – decay from −5 dB to −25 dB (multiplied by 3)	Late reverberation time

Table 27.2 Acoustic measurements results at four locations

Measurement number	C50	C80	EDT	T20
1	−4.1 dB	−2.4 dB	3.2 s	3.5 s
2	1.4 dB	2.3 dB	3.0 s	3.3 s
3	4.4 dB	5.6 dB	2.0 s	2.8 s
4	1.6 dB	2.2 dB	2.6 s	3.4 s

Measurements results analysis

Table 27.1 presents a brief description of four monaural (i.e., single channel) acoustic parameters used to characterise the acoustic properties[11] of the Gala Pool.

Table 27.2 presents the measured values for each of the parameters presented in Table 27.1 at the four selected locations (Figure 27.1).

It can be concluded that Speech Clarity (C50) is higher at the corner of the empty pool tank (Measurement 3) and that Musical Clarity (C80) follows a similar pattern.

Early decay time (EDT) is related to perceived reverberation during music play or speech and is at its highest on the balcony (Measurement 1). Measurement 3 exhibits the lowest EDT, which suggests that a singers' voice at Measurement 3 will be perceived as less reverberant than the same singers' voice located, for example, at the balcony (Measurement 1).

The above analysis focuses on monaural parameters, reflecting the behaviour of sound in time but not considering its spatial attributes. However, a spatial analysis is also possible thanks to the data stored in the measured 3D IRs.

The graphs shown below (Figures 27.2 and 27.3), make use of the information contained in the 3D IRs captured, respectively, during Measurement 1 and Measurement 3. The upper graph of each figure shows the IR, where only the omnidirectional W component of the 3D IR has been shown for clarity.

The lower graph illustrates the direction of sound incidence corresponding to the (W component) IR zones of higher sound pressure level (SPL). Dark gray indicates reflections with higher SPL and light gray indicates reflections with lower SPL. The axis orientation (see Figure 27.1) is such that the X-axis points forward, the Y-axis points to the left and the Z-axis points upward. In the analysis that follows the intersection of the 3 axis is nominally located at the measurement microphone (mic in Figure 27.1).

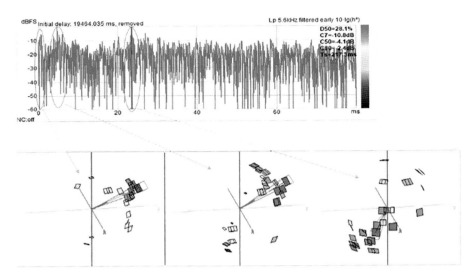

Figure 27.2 Measurement 1 – Upper Graph – Early impulse response, W component. Lower graph – direction of incidence of 3 sets of strong reflections identified on the upper graph (CATT Acoustic software).

Note: The horizontal axis represents time (ms) and the vertical axis represents sound pressure level (SPL). Zones of higher SPL have been circled in light gray.

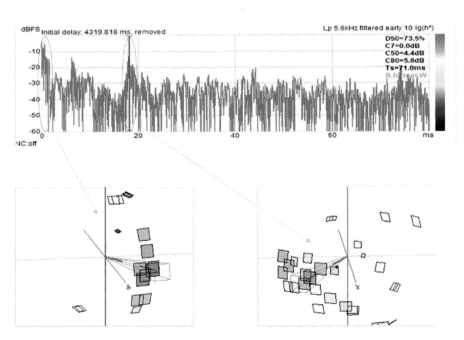

Figure 27.3 Measurement 3 – Upper Graph – Early impulse response, W component. Lower graph – direction of incidence of 2 sets of strong reflections identified on the upper graph (CATT Acoustic software).

The following aspects are of particular interest: (i) Beranek's considerations of the importance of ITDG – i.e., that a delay of the first strong reflections after the arrival of direct sound below 20–30 ms evokes a sense of intimacy[12] and (ii) Kuusinen's hypothesis on the relation between change in perceived distance/spatial cues, and intimacy.[13]

Measurement 1

It can be observed from Figure 27.2 that at around 5 ms (after the direct sound, which has been set at 0 ms) strong reflections arrive from the upper left direction (positive X, positive Y and positive Z), which corresponds to the walls and sloped ceiling adjacent to the corner where the loudspeaker has been located for Measurement 1 (see Figure 27.1). At approximately 25 ms there are strong lateral reflections from the lower-right direction (positive X, negative Y and negative Z), which correspond to the right-side wall of the pool tank.

Therefore, there are two sets of strong early lateral reflections that satisfy Beranek's criteria (ITDG below 20–30 ms). There is a change in direction of incidence from direct sound (0 ms) to the first set of early reflections (~5 ms) but the overall distance of the reflecting surfaces to the listener/microphone is approximately maintained. However, from the first set of early reflections (~5 ms) to the second set of early reflections (~25 ms) there is a significant change in both direction and distance of the surfaces providing the sound reflections, which according to Kuusinen's hypothesis, reinforces the sense of intimacy. The earlier observation, that the sound from a singer located on the balcony appears to cascade down towards the pool tank, seems to be consistent with these observations.

Measurement 3

In Figure 27.3 it can be observed that strong reflections occurring at around 17 ms originate from the lower-right direction (positive X, negative Y and negative Z), which corresponds to the right-side wall of the pool tank. The timing of these reflections meets Beranek's criteria for evoking acoustic intimacy, i.e., strong reflections with a ITDG below 20–30 ms.

Further artistic development

Research and development practices in theatre do not tightly predefine their outcomes. Instead, research in this area is more about exploration and creating spaces for discoveries and tangential curiosities to reveal fruitful results and open up new dramaturgical ideas. In a swimming pool in Birmingham in January 2022, we embarked on a multidisciplinary journey of discovery to understand what an acoustically led approach to dramaturgy meant.

> Intimacy begins with the most innocent of gestures
> A curious smile[14]

The notion of intimacy is inherently subjective. Intimacy is a word bound up with relationality. We cannot be intimate by ourselves, for it implies a connection. Intimacy is always an action that leans towards something or someone else.

Although not initially the main focus of our investigations at Moseley Road Baths, our project examines how large, reverberant spaces can create a sense of intimacy within the context of an operatic performance.

Initially, those who entered the space all commented explicitly on its vast visual appearance. The visitors were also stuck by a sense of hush and awe. These reactions all occurred at the entrance to the baths, beneath the balcony. This entrance space is an important threshold, as it still feels like an interior, which is both spatially and visually intimate. At this point, you have a clear visual link to the vastness of the bathhouse, but you don't yet hear it.

As the audience moved into the basin, something very different happened. This had the effect of unleashing the earlier tentative energies into a state of pleasant abandon. Shouts, splashes filled the air, and as the playful behaviour was set free, the acoustic presence of each person became much more noticeable. The visitors spoke of their experiences – 'There's a freedom here, an emotional freedom', 'it's almost like getting drunk on sound'. This newfound presence, on entering the space of the pool from beneath the balcony, was not only evident in their increased ease, but also in their extended duration of staying in the space. Those who spent more time in the space appeared to become more comfortable. Everyone there, whether listening or performing, perceive an 'amplified presence' in the space. They can produce a sound as vast as the space feels, but also feel a sense of intimacy and connection to their fellow participants. This dynamic is seemingly a direct result of the unique acoustic condition that the space provides.

Dramaturgy

The emerging dramaturgy, initially a response to creative exploration, naturally began to consider acoustics as the dominant feature of the space. This allowed us to move beyond our typical visual bias. Both the results from the empirical workshops and the acoustic analysis allowed us to make creative decisions about the placement of performers in the space. For example, the soprano could be on the balcony, facing away from the audience, directing her voice into the corner. This positioning conveys a deliberate sense of privacy, whilst the acoustics of the space enables the audience located in the pool basin to hear everything, despite being at a physical and visual distance from the performers.

This is where the unique capacity of aesthetics comes into play, particularly in operatic performances which offer an immersive experience that engages both our visual and auditory senses. Our perception of this experience is always influenced by multiple spatial factors. This project has raised several ongoing questions regarding intimacy in immersive and site-specific operatic works. Specifically, does a physical distance, coupled with acoustic intimacy, evoke uneasiness? What creative opportunities exist when decoupling acoustic intimacy from the spatial and visual proximity to a performer?

It would seem, then, that the idea of intimacy in operatic performances is more complex than we typically assume. Implicitly, as animals, we appear to believe that intimacy should be a function of physical closeness. When we feel the acoustic presence of the soprano in the corner, as if her breath is warming our neck, we might also become uneasy as she is physically turned away from us, as if deliberately defining her own private space in the corner. Despite this, we still experience the peculiarity of being spoken to so intimately, even though she utters her words towards the tiled walls and is not singing towards us or projecting her voice. We receive a 'sprechgesang', a German term for discussing a type of 'speech-singing' that has an interior quality of privacy.

From a directorial perspective, this uncanniness can be effectively explored. During our workshops and tests, we have observed a unique exploration of intimacy that blurs the line between public and private domains in a haunting and unsettling way. From a dramaturgical standpoint, it may be effective to allow these tensions to linger and interact, creating a captivating experience for both the soprano and the audience as they delve into the sonic content. This dynamism appears to encourage the audience to delve deeper into their emotions, particularly the sense of the uncanny.

While not typical in dramaturgical practice, an acoustically led approach offers a valuable departure from conventional assumptions about concepts like privacy and intimacy. Through this project, we learned that intimacy is not just an aesthetic quality, but an experiential and multi-model one.

This project presents a novel approach to directing opera by recognising that the performance space, including its acoustic properties, can play an active role in the direction's development. We learned that sound is not static, but rather an active behavioural element. The architectural form of the performance space is not just useful as a visual backdrop, but also in shaping the acoustic environment. Directorial ideas are then developed in collaboration and conversation with the space and its inhabitants, as opposed to imposing these ideas onto the space, as is typical. When guided by sound, the performance space is no longer a blank architectural canvas. The space comes to life through sound. As we work, we are also animating the Gala Pool.

Moving sound in space and time

In addition to capturing the impulse responses and exploring vocal and musical play with workshop participants, we also experimented with recordings of different musical instruments, to understand if there were any overlaps in the behaviour of both vocal and musical content. A recording of a solo classical guitar was particularly effective and appeared to cascade in the same way that the soprano's vocals did. We were simultaneously immersed in sound but could also hear the melodic detail of every note. As if the guitarist was standing right next to us.

The term 'cascade' implies a sense of movement in both space and time. Based on the insights gained from our workshops, the concept of content gradually cascading, or flowing into the space occupied by the audience emerged as a crucial directorial concept. This concept influenced decisions regarding the placement of sound, both in the physical space and throughout the duration of the performance. Furthermore, what was particularly intriguing about the concept of a 'cascade' was that it created a disconnect between the dynamic presence of sound appearing to move through the space (towards the audience) and the static stage direction for the performer. This duality creates an interesting juxtaposition between reverie and agitation, offering a fantastic opportunity for dramaturgical exploration in the uncanny. There is a similar disconnect between ideas of privacy and exposure in the space. The changing cubicles which line the edges of the pool, provide a visually privacy, yet sound made within them projects into the main space, allowing any performers within this space to be heard but not seen. These contrasting and sometimes confusing differences proved to be a valuable directorial tool, which aligns well with the content of the operatic text of Pierrot Lunaire.

Synergising the operatic text and acoustic insights

Our production focusses on Schoenberg's 'Pierrot Lunaire', a text that delves into the imagination, fantasy and the experience of non-reality within the mind of the central character Pierrot. It also explores the process of their 'landing' back down to earth.

If we were working with a more traditional narrative text, the acoustic character of the space at Moseley Road Baths could still be harnessed. For example, in Mozart's opera Don Giovanni, there is a solitary moment – the only one throughout the entire opera – where the Don sings alone. This moment has the potential to reveal his true, authentic self, as he is otherwise portrayed in relation to others. This particular moment is a simple serenade titled 'Deh vieni la finestra'. In the Gala Pool, that moment could be staged in a way that immerses the audience in sound while maintaining an intimate connection with the character. It would create the possibility of encountering the genuine man behind the Don, evoking intimacy without being eerie or unsettling.

The spatial configuration and acoustic properties of the Gala Pool at Moseley Road Baths offers a unique opportunity to explore the destabilising potential of the uncanny, which is particularly relevant to the nature of the text for Pierrot Lunaire. In this 21-movement song cycle, the main character Pierrot is intoxicated by the moon and transported into a fantasy space of symbolism and imagination. The disparity between the visual and acoustic features of the Gala Pool makes it a particularly relevant location for exploring the staging of such content, specifically by exploiting the acoustics. With this text, we are harnessing the building's natural acoustics, to not only inform the narrative but also enable a performance that discusses and evokes imagination and a sense of non-reality. The sensory disconnect that we identified creates an appropriate feeling of vertigo and disorientation that aligns with the intentions of the text. It occurs when the character experiences genuine sensory perceptions while being immersed in a fantastical or unreal environment or setting.

To recap, the vast visual appearance of the Gala Pool sets a number of audience expectations surrounding the issues of private and public occupation of the space. However, the acoustic properties of the space challenges and disrupts these expectations once we have passed beyond the threshold of the entrance and into the main volume of the Gala Pool.

Outlook

It should be noted that all of the insights shared in this chapter have arisen from a combination of first-person experiences and acoustic measurement and analysis. To further extend this approach the team intend to develop a number of auralisation tools which convolve the measured IR with a vocal signal in real time. The tool also allows for the flexible experimentation of placing different sounds in different locations. Typically, such tools are not yet used in the context of developing an operatic production.

Using the geometry and acoustic properties of the surfaces of the Gala Pool and the 3D IRs collected during phase 2, a 3D acoustic computer model[15] of the Gala Pool can be developed and calibrated. This would allow modelling any number of listener positions (and not just the location at the centre of the pool as originally used in the measurements). This digital model would also allow a simulation of the effect of the audience on reverberation time.

Another avenue to explore, which goes beyond the limitations posed by the physical/architectural characteristics of the performance space, is in the use of artificial reverberation systems, which provide virtual sound reflections. Artificial reverberation systems capture sounds with a microphone at one or more locations and play back these sounds (after amplification/delay/reverberation) through loudspeaker(s) located at the same or at different locations. The resulting effect consists of having the loudspeaker to act as if there was an actual surface providing sound reflections, therefore potentially replicating the conditions for enhanced intimacy.

In the legacy of this study, we have created an open-source web repository of 3D IRs, https://earlyreflections.weebly.com, starting with the 3D IRs measured at the Gala Pool. With this data, it is possible to convolve our IRs with any pre-recorded music or sound file, for playback binaurally (on headphones) or ambisonically (on an array of loudspeakers). We invite composers and acousticians to contribute to this database with their own samples. The measurement setup, procedures, software and methodologies employed in creating the 3D IRs are also included in this repository.

Conclusions

To conclude, we can recap on why the discussion of intimacy is so important. The Gala Pool is a large space, made of highly reflective materials, meaning that sound produced within the space is able to reflect for long periods of time. Typically, such spaces feel 'immersive' but sometimes also impersonal and indistinctive. The early reflections that we observed, both intuitively and in our measurements contribute to an acoustic fingerprint that enables the audience to feel a sense of connection with the space, and others who occupy it. This project creates a new creative approach to opera direction which embraces an active dialogue between architectural form and operatic content.

Endnotes

1 Beranek, Leo L. *Music, Acoustics and Architecture*. Wiley, 1962Beranek, *Music, Acoustics and Architecture*.
2 Barron, Michael. *Auditorium Acoustics and Architectural Design*. 1st ed. E & FN Spon, 1993.
3 Beranek, Leo L. *Concert and Opera Halls: How They Sound*. American Institute of Physics, 1996, p. 481.
4 Hyde, J. R. 'Acoustical Intimacy in Concert Halls: Does Visual Input Affect the Aural Experience', *Proceedings of the Institute of Acoustics*, Vol. 24, Pt. 4, July 2002.
5 Kuusinen, Antti, and Tapio Lokki. 'Auditory Distance Perception in Concert Halls and the Origins of Acoustic Intimacy'. In *Ninth International Conference on Auditorium Acoustics, Paris, France, October 29–31*. Institute of Acoustics, 2015Kuusinen and Lokki, 'Auditory Distance Perception in Concert Halls and the Origins of Acoustic Intimacy', pp. 151–158.
6 Blauert, Jens. *Spatial Hearing: The Psychophysics of Human Sound Localization*. MIT Press, 1996.
7 Marshall, A. H. 'A Note on the Importance of Room Cross-Section in Concert Halls'. *Journal of Sound and Vibration* 5, no. 1 (1 January 1967): 100–112.
8 Barron, M., and A. H. Marshall. 'Spatial Impression Due to Early Lateral Reflections in Concert Halls: The Derivation of a Physical Measure'. *Journal of Sound and Vibration* 77, no. 2 (22 July 1981): 211–232.
9 Stan, Guy-Bart, Jean-Jacques Embrechts, and Dominique Archambeau. 'Comparison of Different Impulse Response Measurement Techniques'. *Journal of the Audio Engineering Society* 50, no. 4 (15 April 2002): 249–262.
10 Zotter, Franz, and Matthias Frank. *Ambisonics: A Practical 3D Audio Theory for Recording, Studio Production, Sound Reinforcement, and Virtual Reality*. Springer, 2019.
11 *ISO 3382-1: Acoustics – Measurement of Room Acoustic Parameters. Part 1: Performance Rooms*. ISO, 2009.
12 Leo L. Beranek, *Concert and Opera Halls* (ASA Publishers, 1996), 481.
13 Kuusinen, Antti, and Tapio Lokki. 'Auditory Distance Perception in Concert Halls and the Origins of Acoustic Intimacy'. In *Ninth International Conference on Auditorium Acoustics, Paris, France, October 29–31*. Institute of Acoustics, 2015, pp. 151–158.
14 Davis, Clare. *Intimacy*. Poem, 2014. https://hellopoetry.com/poem/668330/intimacy/.
15 This computer model of a space simulates its acoustic characteristics by employing algorithms that account for sound wave propagation, reflection, absorption, and diffusion within the modelled environment. By inputting architectural and material details, the model calculates how sound behaves in that space. This process, known as acoustic modelling or room acoustics simulation, can then generate auralisations, which are audio representations that give listeners an impression of how sounds would be perceived within the modelled space.

Bibliography

Barron, Michael. *Auditorium Acoustics and Architectural Design*. 1st ed. E & FN Spon, 1993.
Barron, M., and A. H. Marshall. 'Spatial Impression Due to Early Lateral Reflections in Concert Halls: The Derivation of a Physical Measure'. *Journal of Sound and Vibration* 77, no. 2 (22 July 1981): 211–232.
Beranek, Leo L. *Concert and Opera Halls: How They Sound*. American Institute of Physics, 1996.
Beranek, Leo L. *Music, Acoustics and Architecture*. Wiley, 1962.

Blauert, Jens. *Spatial Hearing: The Psychophysics of Human Sound Localization*. MIT Press, 1996.

Davis, Clare. *Intimacy*. Poem, 2014. https://hellopoetry.com/poem/668330/intimacy/.

Hyde, J. R. 'Acoustical Intimacy in Concert Halls: Does Visual Input Affect the Aural Experience', *Proceedings of the Institute of Acoustics*, Vol. 24, Pt. 4, July 2002.

ISO 3382-1: Acoustics – Measurement of Room Acoustic Parameters. Part 1: Performance Rooms. ISO, 2009.

Kuusinen, Antti, and Tapio Lokki. 'Auditory Distance Perception in Concert Halls and the Origins of Acoustic Intimacy'. In *Ninth International Conference on Auditorium Acoustics, Paris, France, October 29–31*. Institute of Acoustics, 2015.

Marshall, A. H. 'A Note on the Importance of Room Cross-Section in Concert Halls'. *Journal of Sound and Vibration* 5, no. 1 (1 January 1967): 100–112.

Stan, Guy-Bart, Jean-Jacques Embrechts, and Dominique Archambeau. 'Comparison of Different Impulse Response Measurement Techniques'. *Journal of the Audio Engineering Society* 50, no. 4 (15 April 2002): 249–262.

Zotter, Franz, and Matthias Frank. *Ambisonics: A Practical 3D Audio Theory for Recording, Studio Production, Sound Reinforcement, and Virtual Reality*. Springer, 2019.

28
SOUND, SPACE AND THE IKO LOUDSPEAKER – THE APPARENT PARADOX OF DIVERSITY WITH UNITY

Angela McArthur and Emma Margetson

IKO: felt spatiality

The importance placed on diversity (in general terms) is evident in today's workplace cultures, and in a broader societal context. Technologies themselves can afford the realisation of this mandate. Particularly if they are novel, with no inheritance of conventions and the privileges which can accompany them. So too, can technologies enable diversity, if they reach a wide range of publics, and form part of an experience which is relevant and compelling to those publics.

The IKO loudspeaker (see Figure 28.1) achieves such things whilst also producing an affective awareness of place. It is an icosahedral (20-sided geometric) loudspeaker array. Each side houses a loudspeaker driver. The exceptional spatial experience it offers, is created through innovative beam-forming technologies, in a full 360 degrees. It is the most compact speaker system for higher-order ambisonics (HOA) available in the world. It has been designed so that the emission of direct beams combines with reflected sound from the walls, ceiling and floor of a site, to create and stabilise a distinct spatial impression, unlike any other. The spatial imaging of sounds becomes a felt presence. At times, this presence occurs in front of the listener but far from the IKO. At other times behind, it occurs above or beneath them. The IKO surrounds and immerses the listener in a space.

Place and space are terms used interchangeably in this text, to avoid a privileging of the 'scientism' of space, as outlined by Casey, where space is –

> …a neutral, pre-given medium, a tabula rasa onto which the particularities of culture and history come to be inscribed, with place as the presumed result.[1]

Bodies (taken non-literally to mean constituents which deserve consideration) interact in space (which, in Casey's critique of outdated thinking, gives rise to 'place'). These interactions form relations between bodies, which are themselves bodies. Whether human or other-than-human bodies, all are considered here as equally contributing actors. The bodily awareness which the IKO invokes serves to amplify one's sense of place.

What does a heightened sense of place mean, in such a context? This text asserts that it comprises of three aspects:

Figure 28.1 IKO loudspeaker.

1 sensitivity to one's own body;
2 sensitivity to bodies that are not one's own;
3 sensitivity to the way in which bodies are connected.

These aspects have ethical dimensions. Few personal or professional practices can situate themselves outside of such ethical concerns, including practices of sound-making, and of architecture. An awareness of the broad range of ways in which embodied approaches can be applied is increasingly documented. From the linguistic,[2] to the musical,[3] to the architectural[4] there are many scholars and practitioners who increasingly seek to (re)connect the mind and body in order to overcome the consequences of their disconnect. Such consequences are outlined shortly.

Out-of-body spaces

Sensitivity and a re-connection to one's own body in our information age (what can be called a re-balancing of the body's importance, relative to the mind) forms the foundation for the themes at work in this chapter. The individual space (self as space) is a core part of how the IKO works with space. The IKO creates a visceral listening environment. One becomes connected to oneself, to others, and to the environmental space. One is aware of these connections. The relational connectedness of these elements offers a sense of place. This relational unity can happen in diverse settings, with diverse publics, and diverse sound materials. More on this apparent paradox shortly.

Sensitivity to bodies that are not one's own, denotes awareness of 'other' (whether human or other-than-human. Whilst technologies have facilitated levels of customisation and personalisation previously unimagined and such specificity can be very helpful – we develop our interests and can pursue niche subjects in ways formerly impossible – issueswith such mechanisms arise. We are increasingly connected to the echoes of our previous choices (whether intentional or arbitrary). Our options are progressively curated, as we are funnelled into apparent individuation and extremity of position.

The filter bubbles which personalised recommendations create, carry with them tendencies to increasingly isolate and inure us from perspectives which diverge from our own. This includes reactions to existing intellectual hierarchies. Mistrust of the apparatuses behind knowledge production are, in a sense, responses to a lack of diversity in the historical ways we generate and value information. How can we hold the impetus towards diversity alongside the drive towards consent, on matters which affect communities? How can we greet the increased exposure to risk, difference, and uncertainty, which diversity ushers in? Our sensitivity to 'other' (whether human or other-than-human) may be a vital component for seeking ways beyond the ecological crises of the early 21st century, one of many complex issues for which multiple perspectives are *crucial*. Technologies need not invoke zero-sum equations. From loudspeakers to material surfaces in buildings, we must reframe our dominion 'over' them, and listen to their affordances without conceit. They too are 'other'.

> When our ability to decode spatial attributes is sufficiently developed using a wide range of acoustic cues, we can visualize objects and spatial geometry: we can 'see' with our ears. [...] The composite of numerous surfaces, objects, and geometries in a complicated environment creates an aural architecture.[5]

The aural architecture[6] Blesser and Salter are proponents of, situate architects within the ethical imperatives of sonic spatial awareness. Listening becomes an ethical act. It requires, and engenders, a sensitivity to the way in which bodies are connected. Listening as an offering, which connects. Collective listening cultivates this quality by creating a space in which all bodies are connected. Our awareness of the listening of others, of those bodies, of our connection to them and theirs to each other, is heightened. The connections between bodies have their own place and being, they materialise as a network of relations in and of, themselves. This is the paradox. Individuation of bodies, and at once a coherent, unified network, to bond them.

Awareness over control

Listening environments for the IKO take this even further. Being as they are – uncontrolled spaces – environmental sounds necessarily form a part of the collective, interconnected listening body. A dialogue with the architecture ensues. This doesn't occur in cases where unwanted sounds are controlled out. The IKO helps achieve the aims of aural architecture, accounting for both active and passive 'aural embellishments'.[7] Active embellishments, say Blesser and Salter, include –

> ...water sprouting from a fountain, birds singing in a cage, or wind chimes ringing in a summer breeze – active sound sources functioning as active aural embellishments for that space.[8]

Passive embellishments by distinction include –

> …interleaved reflecting and absorbing panels that produce spatial aural texture, curved surfaces that focus sounds, or resonant alcoves that emphasize some frequencies over others, create distinct and unusual acoustics by passively influencing incident sounds.[9]

Many of these embellishments may reproduce unwanted acoustic effects. An unwanted effect in a controlled space presents as unwanted. The difference with IKO environments is not only that the space is uncontrollable (a fact registered in the awareness of listeners). It is also the way in which sound, wanted or not, is activated as a connecting material alongside 'sanctioned' acoustic effects. This is due to the IKO's unique sound reproduction. The affective result of its engineering is a felt connection between listener's environment and sound, which overcomes the cognitive dissonance arising from hearing unwanted sounds.

The IKO holds both cognitive and sensual experience for the listener, simultaneously, without privileging either. The 'bottom-up' processing of the body combines with 'top-down' mental activity: both present, both leading the auditive experience. Tacit knowledge fused with intellect. The IKO somehow transcends apparent polarities. We can consider two contending perspectives – one which asserts that sounds can be used to illuminate and augment space, and one which asserts that sounds produce space.[10] These positions are seemingly in opposition. However, both perspectives are upheld by the IKO's relationship to sound and space. The IKO illuminates space as well as producing it. The unique sonic characteristics of a space – its materials, dimensions, volume, and so on, are elicited. The IKO 'sounds' the space. It does this by forging an embodied relationship to a space and its various components, on the part of the listener, through the visceral reproduction of sound. It extends our spatial awareness to an order which exceeds prior experience, an order which does not uphold previous binaries.

Overcoming binaries in this way, a listener's sensitivity to space and sound is not only enhanced but liberated. They can curiously explore the nuance of space, in a manner not so readily available when evaluating sound in terms of valence or utility. Evaluative listening – more common in codified settings – causes us to set a frame of reference which necessarily limits our perspective. Allowing sounds to work through us without such evaluation, we re-position space as dynamic, as somewhere that shapes outcomes beyond our creative intent/extent, beyond our involvement in the design of the sounds or spaces. This, combined with the sonic illumination and production of space via the IKO, makes it distinct. Visionary composers have considered these themes. Stone writes about Morton Feldman's 'inhabitable relationship between sound and space a characteristic of musical modernity'[11] made possible by –

> …placing acoustic materials in such a way that they became architecture themselves, while at the same time transforming the existing architecture of a place and, in the process, facilitating experimental, speculative modes of association […] these acoustic formulations, by turns utopian, and quotidian, came to take on the properties of architecture itself.[12]

A firm commitment to experimentation and spatial discourse is required for this. Feldman lived in another era to the one we find ourselves in now, with all its challenges and opportunities, eroding the time and attention needed for such commitment. Stone goes on to discuss the historical context and recognition of sounds' 'capacity to produce concrete spaces'[13] as well as how modern subjectivities arise from such spaces. What does (or could) this mean for our age? In which ways are we still tethered to previous ideologies in our practice and listening?

IKO: relationality

In conventional listening, reflections are 'controlled out' of spaces. Concert halls, performances venues and electroacoustic spaces which house large loudspeaker arrays, or recording studios which are precision-designed to acoustically optimise for control. These aim at an 'ideal' sound which forms the pretext for the intrinsic architectural properties of a space, and the place of listeners. Such exactitude directs us towards an abstract neutrality in the relationship between space and sound, as well as uniformity in listening. Space, and the mechanisms of reproduction (the loudspeakers) are ignored/ are not present. The vacuum created by their absence separates listener, technologies and space. Only the work is valued, as if it exists autonomously. With the IKO, space cannot be ignored. The IKO answers previous calls to '…encourage and open up spaces for novel, diverse and as yet unforeseen articulations between subject positions and technological assemblages in digital music and sound art'.[14] It allows for the audience to be a valued as integral part of the cycle – they are a part of the work.

Such an approach is fundamentally ecological, in the sense that an environment's information is structured in such a way that disagrees with the idea that a listener needs to construct a unified perceptual experience via an internal cognitive schema.[15] In an ecological approach, our sensoria are attuned to the information's structuring, to the environment. This indicates the relational nature of perception, aesthetics, environment and audience.

> …sound represents an ever-expanding nexus of relations, practices, memories and other shared and transmitted forms of knowledge which cluster around it and form a complex environment. The relationality of sound perhaps represents its most salient attribute as it is in the creation (or co-creation) of sound, its presentation and its subsequent reception that mutuality between animate and non-animate objects is established and relational ties proliferate.[16]

The IKO offers an 'inside-out' system (see Figure 28.2), in contrast to the customary 'outside-in' multichannel speaker systems (such as BEAST,[17] ZKM,[18] Sonic Lab at SARC[19] or the Acousmonium[20]) (see Figure 28.3). This reversal fundamentally alters the relationship between sound, space, and listener. Evocatively, the IKO's beams activate the spatial properties of the environment, and bring to life the materiality of space, in relation to the materiality of sound and listening bodies –

> The first time I listened to the IKO I perceived forms, actual moving sound bodies, a very specific tangibility, distinct from other 3D spatialisation systems. I immediately related it to the sense of touch, which had a profound impact on the choice of sonic materials I began using to compose for both the IKO and other spatialisation systems.[21]

Considerations centred around these elements therefore also profoundly shift. Composition, architectural awareness and listening experience are deepened, and unsettled. The bodies and processes for the reproduction of sound within a space are intrinsically linked (see Figure 28.4). These processes and methods include (1) the technology (IKO loudspeaker), (2) the place, (3) the sound and (4) the listener (see Figure 28.4).

IKO – The loudspeaker/instrument

- Design – A compact loudspeaker array, which projects focused sound beams in a freely adjustable direction.

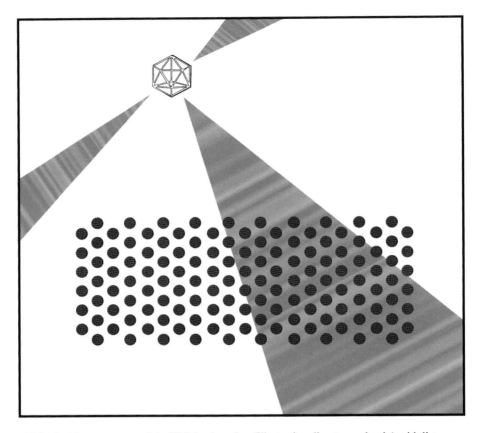

Figure 28.2 Inside-out system of the IKO loudspeaker (illustrating direct sound only) with listeners.

- Reflectors – Inside a space, beams can be directed to excite selected wall reflections, or combinations of reflections. Large reflector panels can be placed within a space to affect the sound-propagation paths and stabilise the imaging of the sound.
- Placement – The positioning of the IKO in a space. Creative decisions can be made within an environment to alter the paths of the sound beams and imaging, in order to alter the perception of their direction and localisation, with support of the reflector panels.

Place – Taken here as the other-than-human environmental elements of a space which hold a unique acoustic signature

- Dimensions – The physical shapes and geometry of the acoustic space e.g. volume, form, scale, etc.
- Surfaces/Materials – The materials which alter sound propagation, through, for example, reflection or absorption. These may cause sound waves to change direction, become diffused or focused, etc.
- Objects – Bodies which interact with the behaviour and reception of sound waves. Objects can be placed creatively within the space to impact the sound, e.g. chairs, curtains, etc.

Sound, space and the IKO loudspeaker

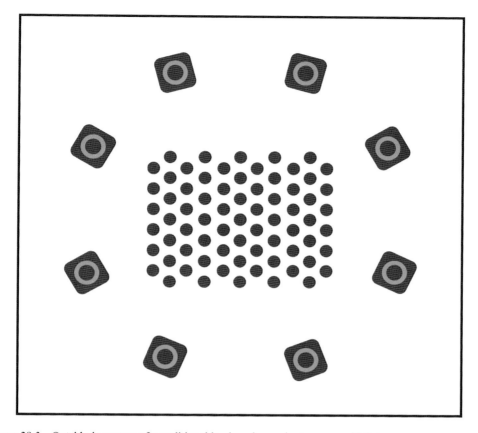

Figure 28.3 Outside-in system of a traditional loudspeaker orchestra array with listeners.

Sound – Described primitively, it is a mechanical disturbance (vibration) that propagates through a medium. In describes both singular and complex combinations of sounds. Example properties of sounds relevant in this discussion are:

- Spectra – Frequency (pitch) and amplitude to create a perceived timbre.
- Duration – The length of a time a pitch and/or tone is sounded.
- Trajectory – The spatial movements of sounds which produce perceived location of the sound beams, over time. Where sound locations are static, no trajectory is present.

Listener – One or more human/other-than-human listener(s) – though only the human is dealt with here

- Semantic References – A sound may have multiple meanings dependent on the context in which it is presented, and the frames of reference in the mind of a given listener. For example, meanings may alter depending on extra-musical associations.
- State (physical, mental, emotional) – Listeners enter environments in various states of arousal, mood, and mindset. These in turn impact their attention and experience.

Angela McArthur and Emma Margetson

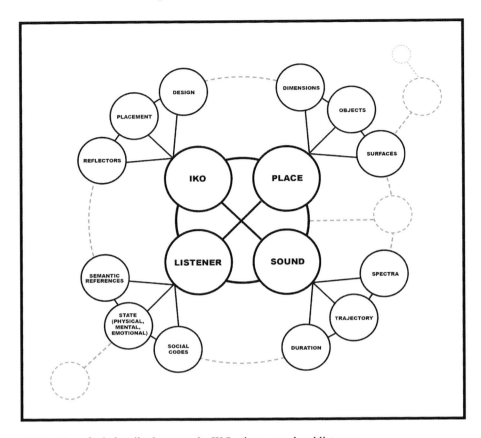

Figure 28.4 Map of relationality between the IKO, place, sound and listener.

- Social Codes – Environments for listening are often social, and highly codified. Such conventions impact the listener's behaviour and listening. Due to the IKO's novelty, we are freed from such conventions (at least to some extent, though they may persist through the built environment, the social context, the format of presentation, etc.).

As we observe in Figure 28.5, the listener inadvertently becomes part of both sound and place. Their physical presence affects the sound (its propagation paths, spectra, and so on). This interaction cultivates heightened bodily awareness, amplifying their sense of themselves, and of themselves within a place. The sound, being directed by the IKO and redirected by the space and bodies within it, meanwhile enhances the acoustical qualities and 'unheard' characteristics of a space.

The combination and recombination of elements highlight the dynamic interconnectedness that the IKO affords, not only for artists working with the IKO loudspeaker but also for listeners. At once, it encourages us to consider (whether in real-time or in reflection after the experience) an individual and collective space. It promotes a sharpened response to place and the listener, as elements of place. Such contemplation encourages artists to challenge more traditional ways of considering these elements. More on this shortly.

Sound, space and the IKO loudspeaker

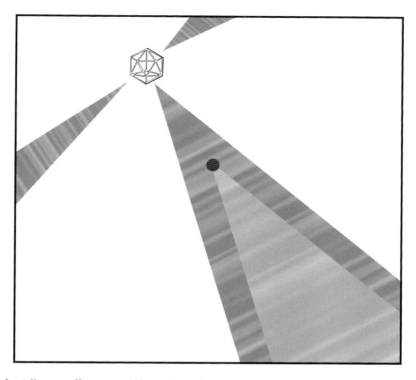

Figure 28.5 A listener affects a sound beam by casting an acoustic shadow.

Beyond the self: re-imagining the creative process

There's something incredibly precious about the journey you take, personally, with the IKO, when you spend time with it. I've been working with the IKO for more than five years, and I'm only scratching the surface of what it can do. It's the one technology that has significantly altered my practice, in terms of artistic concepts, and also the way that I think about process.[22]

The lack of established conventions and communities of practice with the IKO has resulted in it attracting a diverse range of artists.[23] As creators of sound art and compositions, we (the authors) usually work in isolation. The IKO has brought us together. During residencies, workshops, concerts, listening sessions, and practical demonstrations, we have felt inspired to discuss the impact that listening via the IKO has had on us, and on others. As we start to incorporate it into our practice, our discussions frequently extend to how we are working with it, as an instrument. How it has significantly altered our compositional process. What does it mean to have a composition extend beyond a singular creative intent, to take on previously unimagined qualities, particularly at the stage where it is placed into a reproduction environment for a performance? A multitude of potentialities in the composition and performance of a work open up and we find ourselves asking – how much do we actually control? What materials, approaches and techniques are best suited to a particular environment? Such creative freedom has its own challenges: an increased complexity of interplay between production and reproduction of a work; the introduction of aleatory factors in performances; and the commitment to re-organising one's creative thinking and techniques.

Our creative processes are altered because the IKO requires space to be treated as a compositional parameter. This means that the physical environment (place) needs to become part of the creator's materials. Arrangements and materials may need to be altered, and a listening journey which unfolds in space (as part of the work) needs to be accommodated. This is unfamiliar territory, even for those who have worked with multichannel sound and Ambisonics technologies for years. Such a lack of convention is exciting – there are no fixed ways of doing things. This is enhanced by the way in which the IKO can travel outside traditional reproduction environments, into a diversity of settings.

Common changes to one's process include the alteration of sound materials (for example, spectral content) to achieve one's desired spatial impression. Spatial imaging – placement and movement – may require more time to be fully realised. The arrangement of the work may need to be extended. Sounds that mask spatial imagining may need adjustment. The material elements in an environment meanwhile, being key to how the sound propagates or is absorbed, may require their own attention.

The portability of the IKO (one speaker, amplifier and playback system, plus reflectors if required), enables artists to transport their work to any suitable place (indoor or outdoor). The logistics of moving just one compact loudspeaker array are relatively trivial – the costs in time and person power are far lower than they would be with the transportation of other loudspeaker arrays. This possibility, while exciting, can discompose one's sense of efficacy.

Depending on the environments (both that in which work is composed and the one in which it is performed, often different spaces) aspects of the composition can change. Coming to a different reproduction environment, one may need to spend a significant amount of time re-working a composition, or creating a site-specific mix for work. The creative and practical process of these adjustments is not linear. As materials change so too do environmental concerns, feeding back on one another in a way that one's workflow becomes cyclical.

Some composers choose to create site-specific works to make fuller use of the architectural characteristics (and therefore spatial sound affordances) of environments. For artists who like new ways of working and the introduction of unpredictability to their work (and with the time to invest), the IKO is an instrument to work with.

> When having worked with the IKO in a fairly small room with a low ceiling height, and later moving the loudspeaker to a Church for a performance, the spatial height of the composition came to life in unimaginable ways. The composition changed, it felt elevated and more 'alive' by the large gestural swooshes in height which weren't as apparent in the rehearsal location. This could have been a spatial parameter that I may have exaggerated whilst in the Church, but it was already there, and I wouldn't ever have realised it until I was in the space – it's those unexpected encounters with the IKO which really excite me![24]

The introduction of unpredictability can keep one's creative practice alive. It also demands a leap of imagination in the compositional process – one can conceptually stretch into the concert event but cannot know in advance how things will sound. Even when one has experienced the sound in that space the acoustics of the space can change – with temperature or humidity. When numbers of bodies enter the space, when they position or re-position themselves within it, again the composition is altered. As a consequence of these and other 'disturbances' (if framed as such) the relinquishment of control is called for. An ethical dimension is again at play.

The fine grain of spatial detail planned by the composer (intentional 'spatialisation' of sounds via electronic or material means), as opposed to the (unplanned) spatial detail created by environmental acoustics, becomes a processual artistic concern. The foregrounding and demanded consideration of such details not only meet a composer's attention but may in fact elicit it. This then extends beyond sound, to an orientation to details in the built environment (as one relational element at work in the composition). Work may be specifically designed for a space. Such a space may be more peculiar than the kind one would select if aiming for an idealised space. Work is 'installed' into a space, constituting a process in itself. This circles back to the unified assemblage which the IKO orchestrates. Claire Bishop articulates the important difference between the installation of art and installation art, how 'a work of installation art, the space, and the ensemble of elements within it, are regarded in their entirety as a singular entity'.[25] The work, as a combination of elements, becomes part of the same perceptual unit for the audience.

Renewed focus on the listener, and the IKO's creative unpredictability, encourages us to think 'beyond the self' as artists, challenging preconceived ideas of the 'genius' individual composer. Instead, we become more aware of the elements in Figure 28.4, and how these contribute to a network, which creates a 'combined' and 're-combined' work. Rather than diminish any individual actor, it situates us, sustains us, and lessens the burdens of individualism.

The ethical and creative advantages of moving beyond the self can be seen in the evolution of one's journey with the IKO and is often part of a wider movement in any creative practitioner's biography. The narratives propagated in a field about its professional heroes, those that form 'the canon', is always, in part, a fiction. The lauding of the singular virtuoso in contemporary culture should induce scepticism. Outdated, partial, reductive: most of our historical protagonists were supported by networks, whether economic, cultural, interpersonal or otherwise. Widening the canon is a cause for relief. In this age of isolation and anxiety, we can reduce the weight on our creative shoulders, and feel connected in the process, by acknowledging the extension given to work as a result of supportive networks, and other bodies. By acknowledging the way in which we cannot control the work as it leaves our hands and is birthed into the wider world (see Barthes 1968). If we embrace such reality, it can serve our continual engagement with, and curiosity about, the wider environment. Rather than see a dilution of our vision, instead we can see a novel recombination of the work's elements with elements from the broader ecosystem. In this way, the creative process extends into a horizon unbounded.

Unsettling space

The way in which the IKO connects bodies through space and sound can be applied to listeners and creators. Listeners experience a sense of freedom, no longer tied to the sweet spot of outside-in systems, where a limited number of positions reproduce the work as imagined in the mind of the creator. Beyond these positions, one can suffer distortions in the work being performed. The IKO overcomes this positional privileging of space –

> During the creative process, it allows for an expansion and redefinition of the concept of space, beyond its conventional role as a compositional parameter. It creates an immediately multidimensional space, continually renewed through a negotiation between the resonances of the hosting structure, the flexibility in choosing a listening position, not predetermined by a sweet spot, and the spatiality within the piece.[26]

Our valuing of space is unsettled, re-configured, and enriched through the IKO. This of course can disorient an artist, for example when listening back to an IKO work in a space different to the one it was created in, or in a performance when many more listeners occupy a space. As we have heard, the IKO orchestrates and combines such elements to create the work. Varying the elements in the orchestra varies the results, in sometimes striking ways. This can elicit anxiety. However, artists auditioning their work in a new space can be reassured about the contribution that the space is making to the work, through sharing the experience with the others (artists, technicians, and so on) who are also present in the space. In cases where the space is unfamiliar, this works best if those others know the work (their own or another's).

This heralds a significant shift from the singular approach artists usually traverse during site-specific performance. Generally, sound-checks are undertaken in relative isolation. This novel collective orientation of IKO works during audition – as a part of the composition – provides shared sensibility towards the built environment.

The listener too experiences an embodied re-orientation in their relationship and to the space, sound, and other elements. They sense a more active contribution to the performance. This creates a lasting impression. Without prompt – creators are moved to navigate the space as actors. The diversity of spaces and other elements in play with the IKO is compound, and resists fixity. Yet it also unified and coherently connected, through bodies. It enables us to rethink the ritual of the concert, and encourages affect for all listeners, including the artist. It produces more engagement with space and a curiosity which extends beyond the scope of a 'singular' work, involving and including the 'other' – an ethical act. This involvement, though immersive, does not subsume (as outside-in arrays can). One can see the IKO, it is a visual presence. It provides a finite reference point and piques curiosity about its workings. It forms an assemblage, giving rise to agency, for its elements.

The IKO encourages us to have more advanced listening practices, not just as composers, but as listeners of any kind. These practices re-embody and situate us, connecting us at once to the interior and exterior. Such dipole connection is an ethical act. Particularly in an age of polarisation.

If 'planners are lacking an adequate design vocabulary such as aural evocative concepts and tools to integrate an acoustic consciousness in the design process of urban spaces' as Leus[27] tells us, then built environment designers can enhance their awareness of sound, and its place and impact in those environments, through a practice of listening. Artists, already with a practice, can extend their practice by working with the IKO.

The IKO generates new creative possibilities, by an order of difference. The complexity of space and sound, combined, resists control. The IKO follows this trajectory to rebalance the place of the individual. No longer isolated or autonomous, but part of a network. It connects diverse elements in an embodied, coherent manner, while sensitising listeners to their bodies, the bodies of others, and the connections between bodies. These ethical directions are prescient and can be applied to other fields.

Notes

1 Casey 1996, 14.
2 Lakoff and Johnson 2003.
3 Leech-Wilkinson 2009.
4 Robinson and Pallasmaa 2015; Mallgrave 2013; Emmons 2019.
5 Blesser and Salter 2006, 2.
6 Blesser and Salter see aural architecture as the way in which 'properties of a space that can be experienced by listening' Blesser and Salter. "Spaces Speak, Are You Listening? Experiencing Aural Architecture," 5.
7 Blesser and Salter 2006, 5.
8 Blesser and Salter 2006, 51.

9 Blesser and Salter 2006, 51.
10 See Fowler 2015.
11 Stone "Auditions: Architecture and Aurality," 4.
12 Stone 2015, 4–6.
13 Stone "Auditions: Architecture and Aurality," 6.
14 Born and Devine "Gender, Creativity and Education in Digital Musics and Sound Art," 14.
15 See Gibson 1966 via Clarke 2005.
16 Lauer 2019.
17 birmingham.ac.uk/facilities/BEAST
18 zkm.de/zirkonium
19 qub.ac.uk/sites/sarc/AboutUs/TheSARCBuildingandFacilities/TheSonicLab
20 inagrm.com/accueil/concerts/lacousmonium
21 Giulia Vismara, email message to authors, 27 November 2023.
22 Angela McArthur, email message to authors, 3 June 2023.
23 Experimentations with the IKO have brought a range of different artists together at the SOUND/IMAGE Research Centre at the University of Greenwich. Primarily artists have undertaken residencies (led by Dr Angela McArthur) in exploring the use of spatial sound for fixed and interactive creative methods, which has included collaborative group compositions, to interactive works with noise objects, to installation-based works in unusual places. It has provided a laboratory for exploration into the potential of sound reproduction and has begun to challenge traditional surround sound methods.
24 Emma Margetson email message to authors, 1 September 2022.
25 Bishop "Installation Art: A Critical History," 6.
26 Giulia Vismara, email message to authors, 27 November 2023.
27 Leus "The Soundscape of Cities: A New Layer in City Renewal," 356.

Bibliography

Barthes, Roland. "The Death of the Author." In *Roland Barthes: Image Music Text*, translated by Stephen Heath, pp. 142–149. London: Fontana Press, 1977 [1968].

Bishop, Claire. *Installation Art: A Critical History*. 2005.

Blesser, Barry, and Linda-Ruth Salter. *Spaces Speak, Are You Listening? Experiencing Aural Architecture*. Cambridge: The MIT Press, 2006.

Born, Georgina, and Kyle Devine. "Gender, Creativity and Education in Digital Musics and Sound Art." *Contemporary Music Review* 35, no. 1 (2016): 1–20. https://doi.org/10.1080/07494467.2016.1177255.

Casey, Edward. "How to Get from Space to Place in a Fairly Short Stretch of Time: Phenomenological Prolegomena." In *Senses of Place*, edited by Steven Feld and Keith H. Basso, 13–52. Seattle: University of Washington Press, 1996.

Clarke, Eric F. *Ways of Listening: An Ecological Approach to the Perception of Musical Meaning*. Oxford: Oxford University Press, 2005.

Emmons, Paul. *Drawing Imagining Building: Embodiment in Architectural Design Practices*. Abingdon: Routledge, 2019.

Fowler, Michael. "Sounds in Space or Space in Sounds? Architecture as an Auditory Construct." *Architectural Research Quarterly* 19, no. 1 (2015): 61–72. https://doi.org/10.1017/s1359135515000226.

Lakoff, George, and Mark Johnson. *Metaphors We Live By*. Chicago; London: University of Chicago Press, 2003.

Gibson, Eleanor. *Principles of Perceptual Learning and Development*. New York: Appleton–Century–Crofts, 1966.

Lauer, Martin. *Acoustemology – Sound as an Inexhaustible Source of Knowledge Report from a Master Class by Steven Feld Johannes Gutenberg University, Mainz*. Agosto Foundation: Acoustemology. Last updated October 3, 2019. Accessed September 10, 2023. https://agosto-foundation.org/steven-feld-acoustemology.

Leech-Wilkinson, Daniel. *The Changing Sound of Music: Approaches to Studying Recorded Musical Performance*. London: Centre for the History and Analysis of Recorded Music, 2009.

Leech-Wilkinson, Daniel. "Compositions, Scores, Performances, Meanings." *Music Theory Online* 18, no. 1 (2012). https://doi.org/10.30535/mto.18.1.4.

Leus, M. "The Soundscape of Cities: A New Layer in City Renewal." *WIT Transactions on Ecology and the Environment*, 2011. https://doi.org/10.2495/sdp110301.
Mallgrave, Harry Francis. *Architecture and embodiment: The implications of the New Sciences and Humanities for Design*. London: Routledge Taylor & Francis Group, 2013.
Robinson, Sarah, and Juhani Pallasmaa. *Mind in Architecture: Neuroscience, Embodiment, and the Future of Design*. Cambridge, MA: The MIT Press, 2015.
Smith, John. Interview by author. Amsterdam. September 19, 2019.
Stone, Rob. *Auditions: Architecture and Aurality*. Cambridge, MA: The MIT Press, 2015.

29
INTIMATE SOUND
Making known, curating and composing for small spaces

Lawrence Harvey

Encountering design, influence on modes of teaching

When I first entered a design school as a research assistant in the mid-1990s, I was struck by the breadth of designers' obsessions with visual representation, which might connect building facades to the texture of a landscape, to the finish of an object to detailing in urban furniture, or a physical textile to a texture map in a digital environment. Design education continually linked shape, colour, texture, finish, detail, envelope, and material at different scales and a plethora of contexts. A general music education in the 1980s didn't have a sonic equivalent merging lived experience, acoustic environments, music, materials, and sound. Practice and teaching in electroacoustic music studios, especially through musique concrete, then computer music and soundscape studies and composition came close. The work of the World Soundscape Project (WSP) was not as widely known as now and had only recently been re-invigorated through The Tuning of the World Conference in Banff in 1993[1] then further by the World Forum for Acoustic Ecology[2] and affiliates. In 2005, the work of CRESSON (Le Centre de Recherche sur l'Espace Sonore et l'environnement) urbain became more widely accessible with the English publication of Sonic Experience (Augoyard and Torgue, 2006).

This broad and rich landscape of influences, along with practice-based learning and research was deeply influential to development of a curriculum to make known sound and sonic experience. While working closely with design practitioners from landscape, interior, industrial, interaction design and architecture, it was clear they view the world and approach creative work in distinct ways. This suggested different approaches to sound and the acoustic environment would also be required to teach sound across these key design programs.

Introduction to teaching and spaces

The central topics of our design studios and electives evolved to: acoustic design, soundscape research and design, sonification and auditory display, and spatial sound composition and performance. The teaching activities strongly emphasise working with sound as a material for design. Students engage in creative production, listening, situated learning, alongside traditional modes of lectures, demonstrations and reading. Sound-based studies in design and architecture departments at the Royal Melbourne Institute of Technology (RMIT) date back to the 1990s (van Schaik, 1994; Lines and McLachlan, 1997), and accelerated with the establishment in 2004 of Spatial Information Architecture Laboratory (SIAL) Sound Studios. Teaching practice has since evolved through experimentation, and approaches borrowed from instrumental and electroacoustic composition, music performance, acoustics, soundscape studies, sonic and design practices (Harvey, 2017).

In an international survey of teaching acoustic design, our approach certainly fits Milo's analysis that, 'The two main methods in acoustic design education could be identified as the analysis of existing settings and the design of new settings' (Milo, 2020, 14). Like sound courses in other design schools, we draw heavily on R. Murray Schafer, the WSP, and in parallel Kitacpi's four phases based on principles from Schafer.[3] These are the technical lecture, preliminary research and sound walk phase, initial design phase, and the holistic soundscape phase (Kitapci, 2019, 4).

Each semester we set a unique brief for the sound-based design studios, although some have shared titles or subtle differences in brief such as Museum of Sound (2004) & Museum for Spatial Sound (2014), Sound Craft (2019) and Acoustic Craft (2022). The briefs and details of all design studios named in this chapter are available on the Studio's website.[4] The electives were established in the 2000s to build technical skills and introduce sound-based language and technologies to support project work in design studios. Electives, also on the website, can be taken as part of a design degree, or as a single, standalone subject.

As we work with all design disciplines, developing a singular approach or workflow for teaching wasn't feasible. Instead, we have evolved a suite of activities that can be composed or assembled into a repertoire for teaching a particular course or discipline. Activities might elucidate the brief or respond to the site, or a problem set by an industry partner, or explore a sound approach suited to a specific discipline e.g. landscape, architecture, interior, or investigate a particular form of sound-based representation using a 3D game environment, spatial soundscape, auralisation, or museum exhibition. The activities are listed here organised under four groups against location (Table 29.1).

Table 29.1 The 22 key teaching activities described in the next section. In practice, most are not stand-alone but combined with two or more others

Group	Description location	Activity
Class-based	Equipped with 16-channel sound system for lectures, demonstrations, listening, extending to home-based reading, videos and online content.	The Brief; TexturalResources (incl theory, domain); Repertoire; Video Tutorials; Sharepoint
Studio-based	In a sound studio or sound software on home computer.	Spatial Composition; Nebula; 3D Modelling; Acoustic Modelling; Sonification; Resonant Objects
Fieldwork	In offsite locations including urban, home, and environmental locations	Recording; Soundwalking; Sound Mapping; Sound Diary; Tempo Exercise; Enacting Change; Impulse responses; Site visits
Practice-based	Public-facing exhibitions and performances.	Sound Gallery; Sound Exhibitions; Speaker Orchestra

Class-based activities

The brief – textural resources – repertoire – Sharepoint™ site and technical videos[5]

Of the four groups, Class-based activities are more traditional modes of teaching through lectures, reading, listening and watching. The content includes technical, theoretical or contextual information to support studio and fieldwork, and delivered in the classroom, or in students' self-directed learning time.

The Brief is the focus for a 13-week design studio where sound design and auditory experience are central. The sound brief might include a description of transformed acoustic or electroacoustic mediated conditions, a type of resonant object, a sonic activity and intended auditory experience, a particular type of virtual acoustic space, or parameters for the intervention or technology that creates those conditions. Where the brief uses a local site students might document existing conditions through recordings, soundmaps, images, videos and Sound Pressure Level (SPL) measurements, and if an interior space, impulse responses (IRs) or spectrograms.

Textural resources and repertoire listening might be the first time a design student encounters sonic thinking and language. Or for a music student, a new approach to sound, listening, culture and space. The main textural resources in use include CRESSON's, Sonic Experience (Augoyard and Torgue, 2006); R. Murray Schafer's Tuning of the World (Schafer, 1977), and Barry Truax's Acoustic Communication (Truax, 2001) and accompanying websites. The resources are updated each semester with recent journal and conference papers, contemporary media reports and websites on sound. Annotated bibliography of personal reflections and connections between texts is used to assess progress through the reading. With some texts, reading and listening are combined, for example with Barry Truax's Genres and techniques of soundscape composition as developed at Simon Fraser University (Truax, 2002). A lecture is dedicated to group reading of this paper and listening to excerpts from the cited compositions, some in their original multichannel versions.

The idea of a Repertoire is borrowed from music and includes electroacoustic compositions, soundscape compositions, and ambisonic or spatial environmental recordings made on staff research projects. The in-house repertoire collection has been incrementally built from significant historical works and contemporary compositions previously presented in Speaker Orchestra concerts (see below). These range from classic works by for example Karlheinz Stockhausen or Iannis Xenakis to the Groupe de Recherches Musicales (now INA/GRM), late 20th century and contemporary works, and commissions from the RMIT Sonic Arts Collection (Harvey et al., 2020) and other SIAL Sound Studios concerts.

The Technical Video collection was established when student numbers increased to mitigate teacher fatigue from repeated technical demonstrations. International students have reported the videos are easier to follow than a class as translation captions and pausing can be used while attempting technical tasks in their own time. Existing online videos of, for example, Reaper™[6] techniques were unnecessarily long and focussed on commercial music production, not multichannel soundscape composition or sound to accompany a walkthrough of a virtual model. The 28 titles in the collection cover topics such as audio production for reverberation, simple and advanced recording techniques, taking SPL measurements, visualisation, spatial sound production techniques in Reaper, Nebular spatial sound modelling, and other software introductions and tips. During the pandemic the collection was moved to a new Sharepoint Site where we expanded to support teaching students in extended lockdowns across four time zones with no access to a sound studio, and with limited or no opportunity to undertake fieldwork activities. The Sharepoint Site includes studio guides, additional reading, announcements, special lecture videos, links to other resources, recommended software, technical and logistic support resources.

Studio-based

Spatial composition-design – soundscape composition – 3D environments (games, Blender) – nebula – spatial sound production – acoustic modelling – sonification – resonant objects

In these activities, students work directly with sound as a design material. They learn to record, produce, model and compose with sound, or to make sound-producing objects in a dedicated sound studio that would be familiar to an electroacoustic music practitioner. The facility includes three production spaces – The Pod (a 12-channel sound studio), nSpace (with seating for 22 students with 18 channel sound system for teaching, workshops, rehearsals and production work), and The Gallery (a large format projection for 3D environments with 4 channel sound system). Audio production tools include Reaper and Visual Studio Team Services (VSTs),[7] the IRCAM Forum software,[8] Max/MSP,[9] GRMTools[10] and CATT acoustic.[11] Studio and location recording use a suite of Zoom and SoundDevice recorders, Soundfield ambisonic microphones, binaural, dummy head and more standard microphones including heavy-duty windshields.

In Soundscape Composition and Sound Design,[12] students produce a linear spatial sound work. For sound-based or soundscape compositions students use spatial sound techniques for personal exploration and to directly experience the communicative influences of sound as a material. Sound-design projects are required to address a brief more closely. Sound production and spatialisation are also used to investigate over-used terms like envelopment and immersion. Assignments include technical exercises, perceptual explorations, a composition or installation or virtual walkthrough of real or speculative space.

Since the early 2000s 3D Environments have been used in several design studios and electives, where the virtual environments are positioned as a soundscape laboratory (More et al., 2002; Moloney and Harvey, 2004). After completing exercises in notation-documentation, students produce a 3D model in a games engine and produce all sounds and their performance using the sound engine of the 3D environment. Different organising principles for both time and sound in linear and non-linear platforms arise. Although a virtual environment is populated with sound, the final unfolding is determined by the user's journey through the environment, similar to a real-world scenario. Whereas the temporal organisation of linear soundscape composition is determined by the composer.

The Nebula platform developed by Jeffrey Hannam, sought to improve teaching in 3D environments. As modelling in CATT Acoustic and AutoCAD can be time-consuming and software-expensive, Nebula was conceived with real-time interaction in mind enabling immediate aural feedback during the process of design. The application is used to manage sound spatialisation design, sound materials, playback parameters of individual and groups of sounds and timing values for all sounds. Acoustic Modelling[13] (auralisation) is established in sound-design-architecture pedagogy and discussed in literature as a teaching tool, as are Grasshopper plugins for acoustic simulation. Using his research into Sonification for sight-impaired astronomers, Jeffrey Hannam also led the design Studio Astrosonics 1 and 2 (2018) based on a citizen science model. The studios explored StarSound (Foran et al., 2022) a Max/MSP environment for sonification of large astronomical data sets collected from optical telescope arrays across the globe.

Design studios for interior and industrial students have explored the creation of Resonant Objects. In 2004, Ross McLeod and Nicholas Murray led a studio titled Resonant Objects, that saw '…students embark on a series of experiments into the nature of sound and its relationship with materiality, form and spatial volume' (McLeod, 2023). Working with Anton Hassel, co-designer of the Federation Bells and director of Australian Bell (Hasell, 2023) the Spatial Sonorities Design Studio (2016) produced tuned resonant plates as an acoustic 'solo', recorded, transformed and embedded in a multichannel soundscape for a museum exhibition.

Fieldwork

Site visits – notation-documentation (soundmaps, sound diaries) – sound recording – soundwalking; IRs; enacting change – tempo exercise

Fieldwork can test new skills and thinking patterns in a range of acoustic environments separate from the classroom and studio. Auditory spatial awareness (Blesser and Salter, 2007, 11–66) can be exercised in diverse places of work, recreation, study or rest. So combinations of Sound Maps, IRs or Soundscape Recordings might be undertaken in a park, train station, shopping precinct, culture centre, street or laneway, corporate foyer, public baths or home.

Techniques such as the Tempo Exercise or Soundmapping are intended to 'slow down' time and encourage students to pay closer attention to the qualities of an acoustic environment. Contextual knowledge of a project site can be assembled with these techniques, and complement acoustic analysis, spatial sound recording, or for interior spaces, capturing IRs.

Sound Recording is positioned within the design workflow and encourages students to explore sound as having a critical role in design communication of the project. The format of Sound Recording is in part determined by the final output being a linear or non-linear project. This might involve recordings of ambient layers, interiors or sound objects for a 3D walkthrough. For a linear soundscape composition or video with spatial sound, recording of events with distinct temporal envelopes is required e.g. a dawn chorus, or city soundscape to construct a narrative.

Soundwalks, Soundmaps and Sound diaries are borrowed with some variants from the original WSP techniques. We've extended the Soundmap technique into three parts; onsite map capturing event, time and location; sound list and line-score timeline.

A Soundmap is a perceptual exercise. By way of broad comparison – a sound recording is like a photograph whereas a sound map is similar to hand made sketch. Soundmap analysis can lead to insight into perhaps previously un-noticed an acoustic environment, highlight the diversity of sound sources along with their temporal and intensity qualities and spatial locations surrounding a listener. For example, Soundmaps and associated analysis can demonstrate that a single object in the world such as a tram, isn't a single sound, but comprised of an ensemble of sounds.

Time is rarely considered as a spatial design parameter. The Tempo Exercise is a way to explore duration, tempo and distance of a listener on a site. First explored in an early Master of Design Innovation and Technology (MDIT) studio Polyquarters (2013), and Museum of Sound (2016) and used extensively in Promenade Sonique (2023) with the Melbourne Recital Centre this exercise is highly suited to project work in urban sites, a museum, gallery or circulation system of a cultural centre. On a selected site, students use a metronome to synchronise steps in beats per minute (BPMs) and use a stopwatch for timing overall duration, and a tape-measure to record distance of both very slow yet comfortable, and a fast-walking pace. The data can be aligned to produce two connected representations: a section diagram, a line score with grid in BPMs, and a temporal grid in Reaper for a soundscape.

Site visits include semi-permanent or enduring sites with particular acoustic or sounding features. Melbourne-based examples include Fairfield Amphitheatre,[14] the Federation Bell installation[15] or Proximities urban sound installation.[16] Useful semi-permanent sites include exhibitions in galleries and museums, spatial sound concerts or films with a notable sound design. Visits have also been to listening-auralisation rooms in local acoustics companies Marshall Day Acoustics[17] and ARUP Acoustics, high-end performance spaces or sound studios, or dedicated facilities such as the RMIT Reverberation Chambers. Site visits might include location recording or IR capture.

Practice-based

The gallery – sound exhibitions – speaker orchestra concerts

The Gallery has a large format projection inside a four-channel sound system, and was established for the design-studio ELEMENTAL (2022) and since used in three others. In this first studio, students collated eye and earwitness accounts of climate change-induced extreme weather, then used those accounts for abstract audio and visual designs for a large-scale exhibition. The final design stage was for a museum room or exhibition system of the abstract work, displayed as a 3D walk-through with four-channel soundscapes.

Since 2004, SIAL Sound Studio has presented over 30 concerts on the RMIT Speaker Orchestra and collaborated on spatial sound exhibitions with RMIT Gallery (2013) and Mclelland Sculpture Park and Gallery (2021–2022). The work of over 200 composers and sound artists has been presented in these events and exhibitions. Concerts and exhibitions have been used for design studios and elective projects. In many instances, exhibitions and concerts are the basis for practice-based research and produce reportable non-traditional research outputs (NTROs). Assessable tasks can be set on an event or exhibition, such as documenting curatorial or spatial design of the speaker orchestra configuration, or a personal reflection on the spatial sound experience of an event.

Within an educational setting spatial Sound Exhibitions and Speaker Orchestra Concerts can play several roles. Firstly, they move the educational experience beyond the classroom and complement sound studio and site activities. A concert or exhibition requires extended durations of listening, whether to a series of spatial sound works or time spent in an exhibition. Concerts and exhibitions can form connections between the academy-based practitioners, research ideas, industry partners including other arts organisations, and audiences. Through these connections can flow creative works ranging from explorations of everyday listening to highly propositional or speculative ideas about sound, or new spatial electroacoustic works. And finally, there is the experience of shared listening for both students and audiences beyond the academy.

Practice – an intimate speaker orchestra

Introduction: curating and composing on small systems for intimate listening

Speaker orchestras were first developed in French electroacoustic studios in the 1970s (Roads, 2015, 249–250) and since spread to other university studios or cultural organisations (Deruty, 2012). They are like instruments, spatially designed, tuned, and performed in a space, making them an ideal approach for public-facing engagement of an educational or research-experimental organisation. For introduction and discussion on the practice see (Harrison, 1999; Austin and Smalley, 2000). Prior to starting at RMIT, I had curated two large electroacoustic music series in Melbourne on speaker orchestras (The Reflective Space, 1996 and Next Wave Festival, 1998). A few years earlier, I had attended nearly all the winter series of Acousmonium concerts at the GRM concerts in Paris, listening and observing immediately behind the central mixer location for most concerts.

Although developed for use in large concert halls for audiences of 200 or more, the cost of larger venues can be prohibitive, and hire durations must be long enough for bump-in, setup, rehearsal and at least two or more concerts to be presented. One solution is to use smaller venues that are affordable and accessible for three or more days. While Harrison (Harrison, 1999, 123) notes even in a small venue the approach remains the same as the goal is to maintain a group spatial listening experience, even for audiences of 50–80 people which might use 10 to 20 loudspeakers. The larger number of loudspeakers relative to room size is not for maximal acoustic energy as in dance music, but to provide point sources for a sophisticated sound spatialisation. In a concert, this speaker field geometry is the sonic scaffold, landscape, or stage onto which a composition is spatialised.

Principles of speaker orchestra design – creating an intimate listening space

We have developed five key research and development phases for preparing a speaker orchestra performance: curatorial, loudspeaker spatial design, software control, mixer desk assignments and the performed spatialisation. Some events require more or less work per phase. A multi-work concert exploring a specific theme (Oceans, 2017; Site and Sound, 2021) requires more research in the curatorial phase than a concert with 1–2 longer works (The Planting, 2023). If spatialisation is partially or fully automated, then software control requires more development than a real-time diffusion performance with stereo channels assigned to specific loudspeakers. Each spatial configuration of the speaker orchestra responds to the repertoire for the event, venue architecture, audience seating locations, and the sonic characteristics of loudspeakers. Pre-planning draws on prior experience and evolves in collaborative sketching, or a combination of digital plans, 3D models used for spatialisation data for Max/MSP, or test setups with subgroups of loudspeakers. Also in this phase are several weeks of detailed analytical listening in the studios to individual works to explore strategies for sound diffusion. Ideas would be captured on a paper-based sound diffusion score. Since the Oceans concert (2017) paper scores were replaced with sound diffusion directions annotated into the sound-file display using the Reaper 'Insert Item' function and markers.

The goal is to craft a concert design with a novel palette of sonic-spatial perspectives for the unique gestural and textual qualities of each composition. This is explored through combinations of loudspeakers for various front images, envelopment from the side speakers, or immersion from the sides and above, loudspeakers to deliver sound trajectories across the space or using equalisation to emphasise sound spectra in spatially separated subgroups of loudspeakers. Although preparing a sound performance, the process seems to have more in common with a lighting design and plot, but to illuminate a choreography of sound. In an intimate space, these decisions must be refined and precise.

Intimate speaker orchestras

Since 2004 we evolved an 'in-the-round' presentation format for events in smaller intimate venues. Two curving audience sectors or hemispheres face a central area. In some versions, the mixer is placed centrally, while in others there are just loudspeakers. For each sector, the signal feeds are duplicated to the front, sides and rear speakers for each audience-half, which can also form a large ring of speakers for the whole space. This configuration has several useful aspects. On entering the space, there is a subtle shift to the usual audience location and focus as the audience members are facing each other, and not looking in the same direction, perhaps at a stage empty except for loudspeakers. Splitting the audience in two hemispheres means there are two front rows and more of the audience can be centrally located between front, side and rear speakers divided across the two hemispheres.

Figure 29.1 3D model and image of Speaker Orchestra installed in the Primrose Potter Salon, Melbourne Recital Centre, 2009. The performance shown is for La lontananza nostalgica utopica futura by Luigi Nono, where the violinist journeys between music stand 'stations' positioned through the space.

This type of configuration with variations for venue architecture and repertoire has proved highly versatile. In 2009, SIAL Sound Studios was invited to present five concerts of electroacoustic music in Opus 1, the inaugural concert series of the Melbourne Recital Centre. Four of the concerts used the Primrose Potter Salon, a 15 m × 10 m room with 8 m high ceiling. We used the height to maximise the number of loudspeakers and maintain distance from the audience, by placing five Meyer UMP 1 loudspeakers in the ceiling grid. This gave a full spectrum sound forming a canopy over the whole space. Other loudspeakers on stands and at different heights created a dome-like arrangement and layers, although not a smooth curving arrangement. With some variation for each performance, this approach was used for La lontananza nostalgica utopica futura by Luigi Nono (see Figure 29.1), a concert of Xenakis' electroacoustic tape works, and two survey concerts of 20th-century electroacoustic music.

Where height was the spatial opportunity in The Salon of the Melbourne Recital Centre in 2009, room width and a long window with views over a sculpture park were the conditions for a concert in 2021 at Mclelland Sculpture Park and Gallery. This concert was part of the SITE and SOUND: Sonic Art as ecology practice and presented compositions inspired by solar winds, glaciers, tectonic plates, rain and the spectra and timing of animal colonies. This setup used only a single hemisphere, but the idea of audience in smaller subgroups was maintained. Three audience sectors were used, where each sector had a front stereo and side speakers. To ensure direct and more localised stereo fields for each sector we used CODA Tube loudspeakers placed in close as front stereo and close side locations. Further outside these were Genelec 1030A loudspeakers for both fuller spectrum sound, and JBL Curves raised on stands and road cases to provide height. Although minimal in acoustic power the system was elegant and not overpowering although subs provided more low-end energy (Figure 29.2).

The planting – 2023 – embedding and configuring layers

This next example merges spatial composition and sound diffusion performance on an intimate speaker orchestra. It is an example where a composition was spatially re-interpreted to create an intimate listening experience for long-form work. The Planting is a 45-minute speculative audio artwork set in 2029, exploring the future impacts of climate change, social movements and caring for Country across the continent of so-called Australia (The Things We Did Next, 2022). Interview

Intimate sound

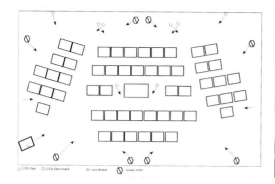

Figure 29.2 SITE and SOUND: Sonic Art as ecology practice. Diagram produced for early planning and image of final installed Speaker Orchestra. Note the elevated CODA Tube speakers and audience in three sectors with side speakers.

participants projected their activism into 2029, describing successful social, environmental, and political events drawn from a future sound archive. The current version is for 14 loudspeakers and was developed with Alex Kelly (Lead Artist), and David Pledger (Dramaturg) and the author as composer/sound designer. The Planting was created during the pandemic years of 2021/2022, a time when access to a spatial sound studio was either non-existent or very limited.

The work comprises extremely diverse and detailed sonic environments including single voices, a Greek chorus, electroacoustic transformations of voices, environmental recordings, spatialised sound synthesis, and layered combinations of all these. The synthesis and spatialisation of the electronic textures were made in the OM-sharp version (OM-Sharp, 2023) of IRCAM's OpenMusic platform, mainly with OMChroma and OMSpat libraries. The voices, granulations in Roads' Emission Control (Roads, 2021), and sound transformations were spatialised with the GRM Spaces VST and ReaSurround. One way to articulate the narrative journey of the work was through spatialisation to gently, and at times dramatically shift listeners' attention to different locations or speaker groupings. The final spatialisation is based on nine interlocking subgroups of speakers. The six different sonic layers of the composition revolve between speaker subgroups to maximise spatial separation of the layers and fluctuating points of focus.

The 2022 premiere in Storey Hall at RMIT University in Melbourne used a circular in-the-round setting and central mixing desk. The 2023 Castlemaine State Festival regional Victorian venue was much smaller, so the circle arrangement was flattened to gently curving seating banks traversing the venue. A line of speakers divided the space and the hemispheres. Each side had Coda Tubes as front stereo image mainly for the untreated voices, bookended by JBL Curves raised slightly to provide both side envelopment and subtle height sources and to widen the front stereo field. This central axis is then intersected at 90 degrees with another pair of JBL Curves as centre-front/centre rear – but also a raised diamond shape. The final main group are four Genelec 1032A loudspeakers placed on the diagonal axis forming an X across the room (Figure 29.3).

Exhibitions: intimacy and listening while walking

Although the acoustic conditions of most art galleries are not suited to sound or detailed listening, installations have great potential for non-linear forms of experience and intimacy. The latter is achieved through connections between visitors, the work and its creator (Batchelor, 2019, 307).

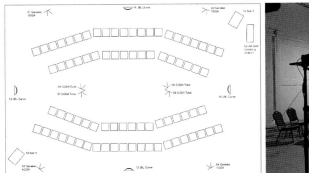

Figure 29.3   The Planting, Castlemaine State Festival, 2023. The diagram on the left shows two wide seating banks. In the final version pictured, the seating banks were narrowed and made into three rows. However, the gently curving layout and hemispheres can be seen in the image.

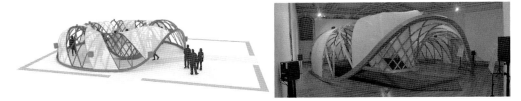

Figure 29.4   Sound Bites City. Early render of the space and image of final installation. Due to deadlines for exhibition opening, some loudspeakers were positioned outside and not on the structure: Image Nick Williams.

Simply placing loudspeakers in front of a seat is a missed opportunity. The gallery invites strolling, pausing, stopping and backtracking. An exhibition design can foster a unique journey while listening, but only if the designer and curator are willing to create conditions where audiences have options. Providing multiple vantage points to an exhibition suggests openness and trust, both necessary for intimate relationships, are valued. There is no insistence on a singular relationship between listener and sound. This emotional perspective on intimacy requires a curator to think through what is being afforded to the audience, and what is there for them to explore, to discover. So, conditions for extended listening time must allow an audience to tune in and entrain to a different type of experience. Listening while walking is one way we can shift our physical and aural perspective in a space, and allow individual visitors to create proximity or distance to the work.

Listening while walking formed key parts of the brief to architect Nick Williams for Sound Bites City (2013) at RMIT University Gallery (see Figure 29.4), and designer Ross McLeod for the exhibition design of Site and Sound at Mclelland (2021–2022). The first part of the Sound Bites City brief called for an exhibition solution in a gallery as a prototype sound gallery, an iconic easily identifiable site to present the recently established RMIT Sonic Arts Collection (2012). The brief also asked for a design that would encourage the audience to not only walk but also sit or lie down and listen inside a spatial sound field. Williams solved this with a Torus-like structure that could hold loudspeakers and create a circulation system and central area for listening, including underfloor speakers for the subs. (Williams, 2017, 92–120).

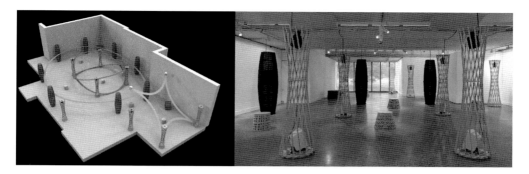

Figure 29.5   SITE and SOUND 3D model and exhibition image. The diagram indicates the speakers sharing different spatialisations, intended to operate as zones or pools of sound: Image Ross Mcleod.

The brief for Site and Sound drew on the surrounding sculpture park, calling for a design inviting strolling between and stopping at key points, so the interior exhibition would be a journey albeit over a much smaller landscape, and one articulated by sound, not sculpture. The brief also called for a solution to an issue of electroacoustic sound in galleries – rectilinear loudspeaker boxes rarely look attractive in a gallery. And they are often pushed up against walls or mounted on a ceiling – neither of which is optimal acoustic position for creating a more complex spatial listening environment.

McLeod's elegant solution was acoustically transparent sculptural forms, into which the loudspeakers could be placed and so positioned anywhere in the room, while acknowledging the visual aesthetics of gallery practice (SIAL Sound Studios, 2021) (Figure 29.5). Each sound work occupied the whole gallery, as,

> The room is designed … like a swimming pool: with a shallow end of listening at one end, and the deepest section in the middle of the room where the largest concentration of sound converges from 24 speakers. Listening is thus shifted away from the atomised experience so common of gallery headsets into a shared, visceral experience.
>
> (Beudel, 2021)

Concluding statements

Although university management compartmentalises academic work into teaching, research and engagement, practice and imagination roam more freely. To counter the rigidity of the academy, we have pursued a pedagogical model to make sound known in design, through the four broad groups described here, and applied in design studios and electives. Making, listening, observing, and reflecting are interwoven to build knowledge through students' auditory spatial experience. The core activities are extended by the cultural program of concerts and exhibitions that bring detailed spatial sound events directly into the annual academic program. The creative sound-based works in this program might be proximate to our daily acoustic environments using soundscape compositional techniques or situated way beyond those to highly speculative and complex electroacoustic compositions. Wherever a work is situated across this spectrum, it is the first-hand intimate experience of sound that is shaping a student's knowledge.

Notes

1. https://albertaonrecord.ca/tuning-of-world-conference or https://www.youtube.com/watch?v=YkfF4HkqSyw
2. https://www.wfae.net
3. We had hosted and recorded a round table discussion with Schafer and Hildegard Westerkamp as part of the AFAE Conference held in Melbourne in 2003.
4. See https://sialsound.studio/study/
5. https://www.microsoft.com/en-us/microsoft-365/sharepoint/collaboration
6. https://www.reaper.fm/
7. https://azure.microsoft.com/en-us/products/devops
8. https://forum.ircam.fr/collections/detail/technologies-ircam/
9. https://cycling74.com/products/max
10. https://inagrm.com/en/store
11. https://www.catt.se/
12. For discussion on differences between sound design and sound art, see Harvey (2013, 122).
13. In the late 1990s, Neil Mclachlan established an eight-channel studio using early versions of CATT Acoustic and the Huron Lake Audio system for auralisation.
14. https://www.yarracity.vic.gov.au/facilities/fairfield-amphitheatre
15. http://federationbells.com.au
16. https://leberandchesworth.com/public-spaces/proximities/
17. https://marshallday.com.au/innovation/the-listening-room

Bibliography

Augoyard, Jean-François, and Henri Torgue, eds. *Sonic Experience: A Guide to Everyday Sounds*. Montreal, Quebec: McGill-Queen's Press-MQUP, 2006.

Austin, Larry, and Smalley, Denis. "Sound Diffusion in Composition and Performance: An Interview with Denis Smalley." *Computer Music Journal* 24, no. 2 (2000): 10–21. http://www.jstor.org/stable/3681923.

Batchelor, Peter. "Grasping the Intimate Immensity: Acousmatic Compositional Techniques in Sound Art as Something to Hold on To." *Organised Sound* 24, no. 3 (2019): 307–318.https://doi.org/10.1017/s1355771819000372.

Beudel, Saskia. "Site & Sound: Sonic Art as Ecological Practice." Review in *Artlink* (12 March 2021). https://www.artlink.com.au/articles/4892/site-and-sound-sonic-art-as-ecological-practice

Blesser, Barry, and Salter, Linda-Ruth. *Spaces Speak, Are You Listening? Experiencing Aural Architecture*. Cambridge, MA: MIT Press, 2007.

Deruty, Emmanuel. "Loudspeaker Orchestras: Non-Standard Multi-Loudspeaker Diffusion Systems." *Sound on Sound* (2012). https://www.soundonsound.com/techniques/loudspeaker-orchestras.

Foran, Gary, Cooke, Jeff, and Hannam, Jeffrey. "The Power of Listening to Your Data: Opening Doors and Enhancing Discovery Using Sonification." *Revista Mexicana de Astronomıa y Astrofısica Serie de Conferencias RMxAC* 54 (2022): 1–8.

Harrison, Jonty. "Sound, Space, Sculpture: Some Thoughts on the 'What', 'How' and 'Why' of Sound Diffusion." *Organised Sound* 3, no. 2 (1999): 117–127. https://doi.org/10.1017/S1355771898002040.

Harvey, Lawrence. "Improving Models for Urban Soundscape Systems." *SoundEffects-An Interdisciplinary Journal of Sound and Sound Experience*, no. 3 (2013): 113–137. https://www.soundeffects.dk/article/view/18444

Harvey, Lawrence. "Teaching Spatial Sound." Uploaded 31 October 2017. YouTube video 34:51 min. https://www.youtube.com/watch?v=BB73DXMlOuA.

Harvey, Lawrence, Jon Buckingham, Lisa Bartolomei, Gillian Lever, and Josh Peters. "Voicing the Sonic: a case study of the RMIT Sonic Arts Collection and the Speaker Orchestra." In *2019 ACUADS Conference: Engagement*, pp. 1–18. Australian Council of University Art and Design Schools (2020). https://acuads.com.au/wp-content/uploads/2020/05/Harvey-Lawrence-et-al_2019.pdf

Hasell, Anton. "Australian Bell." Accessed 4 December 2023. https://ausbell.com.au

Kitapci, Kivanc. "Room Acoustics Education in Interior Architecture Programs: A Course Structure Proposal." *INTER-NOISE and NOISE-CON Congress and Conference Proceedings* 259, no. 2 (2019): 7713–7721. Institute of Noise Control Engineering.

Lines, Robyn, and McLachlan, Neil. "Acoustic Design at RMIT University." *Acoustics Australia* 25, no. 1 (1997): 29–30.

McLeod, Ross. "Resonant Objects." Accessed 4 December 2023. http://www.rossmcleod.com/resonantobjects.htm

Milo, Alessia. "The Acoustic Designer: Joining Soundscape and Architectural Acoustics in Architectural Design Education." *Building Acoustics* 27, no. 2 (2020): 83–112. https://doi.org/10.1177/1351010X19893593

Moloney, Jules and Harvey, Lawrence. "Visualization and 'Auralization' of Architectural Design in a Game Engine Based Collaborative Virtual Environment." In *Proceedings. Eighth International Conference on Information Visualisation, IV 2004*, pp. 827–832. IEEE, 2004.

More, Gregory, Harvey, Lawrence, and Burry, Mark. "Understanding Spatial Information with Integrated 3D Visual and Aural Design Applications." Palmona, CA: ACADIA, 2002. https://papers.cumincad.org/data/works/att/7032.content.pdf

OM-Sharp. 2023. Accessed 5 December 2024. https://forum.ircam.fr/projects/detail/om-sharp/

RMIT Sonic Arts Collection. 2012. Accessed 5 December 2024. https://artcollection.rmit.edu.au/?p=rmit-gallery-explore#browse=enarratives.14

Roads, Curtis. *Composing Electronic Music: A New Aesthetic*. New York: Oxford University Press, 2015.

Roads, Curtis, Jack Kilgore, and DuPlessis, Rodney. "Architecture for Real-Time Granular Synthesis with Per-Grain Processing: EmissionControl2." *Computer Music Journal* 45, no. 3 (2021): 20–38.

Schafer, R. Murray. *The Tuning of the World*. New York: Knopf, 1977.

SIAL Sound Studios. "SITE & SOUND: Sonic Art as Ecological Practice" (2021). Accessed 5 December 2024. https://sialsound.studio/exhibition/archive/site-sound/

The Things We Did Next. (2022) Accessed 5 December 2024. https://www.thethingswedidnext.org/the-planting/

Truax, Barry. *Acoustic Communication*. Westport, Conneticut: Greenwood Publishing Group, 2001.

Truax, Barry. "Genres and Techniques of Soundscape Composition as Developed at Simon Fraser University." *Organised Sound* 7, no. 1 (2002): 5–14. https://doi.org/10.1017/S1355771802001024

van Schaik, Leon. "The Sound of Space." *Art and Australia* (Summer 1994).

Williams, Nicholas. "Plugin Practice: Recasting Modularity for Architects." PhD diss., RMIT University (2017). https://researchrepository.rmit.edu.au/esploro/outputs/doctoral/Plugin-practice-recasting-modularity-for-architects/9921863816301341

URLs

https://azure.microsoft.com/en-us/products/devops
https://www.catt.se/
https://cycling74.com/products/max
http://federationbells.com.au
https://forum.ircam.fr/collections/detail/technologies-ircam/
https://inagrm.com/en/store
https://leberandchesworth.com/public-spaces/proximities/
https://marshallday.com.au/innovation/the-listening-room
https://www.yarracity.vic.gov.au/facilities/fairfield-amphitheatre

30

LISTENING WITH, LISTENING TOWARD

Proposing graphic transcription as a means of (re)hearing space

Ben McDonnell

Introduction

The space of this chapter will be used to propose listening and graphic transcription as a methodology for understanding the 'event' of the crit (shorthand for a dialogic, peer-led critique) in studio-based higher education practice. More broadly, this chapter will ask what value transcription can have as a means of learning through listening. The text will form an intentionally messy relationship between space, image, sound, and pedagogy focusing on the studio as a site in which spaces of collaborative pedagogy can be made. This work will specifically look at the transcription of critique sessions at a university in London known for art and education and a central London design school and seek to make connections between the transcriptions, the questions they ask, and the connection this has to contemporary writers and artists working in similar fields.

Georges Perec's series of short explorations of text and a series of ever-expanding spaces A Species of Spaces and Other Pieces intertwine text and writing as both a generation of and a reflection on a variety of spaces. He begins by delineating the space of his page with horizontal and oblique text as a means of creating a few signs/a top and a bottom, a beginning and an end, a right and a left, a recto and a verso.[1] It would seem appropriate to start this chapter by using a few signs to delineate this space as a place primarily for listening. We shall start just as minimally as Perec did, with an invitation to listen, guided by a time frame and a suggestion of what might be heard (Figure 30.1).

The above score is an invitation to begin to aurally identify features of the space in which this book is being read. The act of engaging with this score, listening to space, is the most appropriate introduction to this chapter as it demonstrates, through practice, the main themes and questions that will be explored here. Initial questions that arise from the exercise of inhabiting through listening to the space that we currently occupy could include:

- What did the score enable me to do?
- Why did it not use traditional music notation?
- What did I hear?
- How could I transcribe what was heard?
- What would be the value of transcribing it?

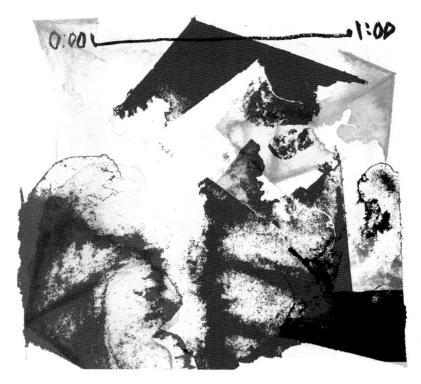

Figure 30.1 'Listen' (2023) graphic score.

The above are indicative of the types of questions that initiated the author's practice/research, which seeks to establish a dialogue around the interconnections between sound, image, and pedagogy.

This text will present graphic transcription as a means of analysing events in space that have already existed. This differentiates it from the more established graphic score that uses elements of traditional Western notation as well as more interpretive visual elements as a means of provoking a performance of a piece of music. The transcriptions looked at here are not intended to be an objective recording of an event that could then be re-made or performed (although this could also be an outcome). Instead, the transcriptions form a subjective '(re)hearing' (listening and listening again) of the event that suggests and provokes new ways of understanding and interpreting the sessions or events transcribed.

This text is predominantly drawn from practice/research based in the spaces of arts and higher education in the UK, but the intention for the methodology presented here, is that it be applicable in other areas or spaces of research and practice. The listening methodologies introduced here are activated as a way of occupying or creating spaces and questioning what we hear when we inhabit them. The interconnected spaces of this book and the ideas within it, the physical space inhabited by the reader, and the spaces of our own practices can be heard as voices in a chord; each has its individual character, its timbre. When read together or heard through each other, re-interpretations and re-hearing of each are possible. The voices of the chord interact with each other, creating overtones that are unique to the particular time and space occupied by you, here and now.

This text starts by positioning these ideas within a critical and creative constellation of artists, theorists and musicians; in doing so, it will give some context to terms such as 'event', 'space', 'listening', etc. that the introduction has given rise to. The transcriptions of crit sessions (Figures 30.2–30.5) will be used as case studies of practices that generated the ideas that led to the work presented here. Finally, conclusions will be drawn by way of asking further questions and suggestions for further avenues of practice/research.

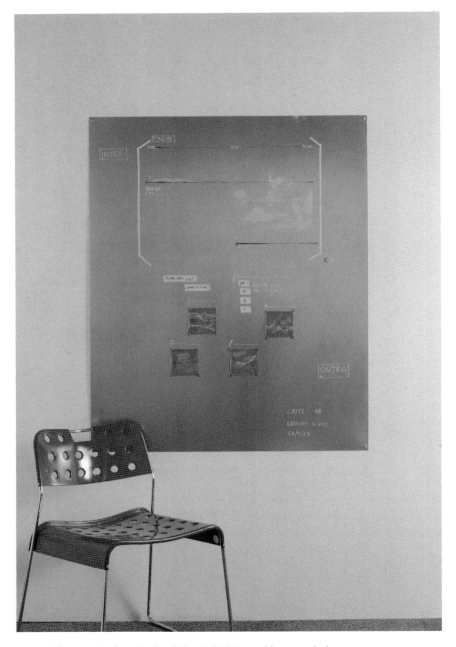

Figure 30.2 'Crit Transcription, Design School' (2023) graphic transcription.

Listening with, listening toward

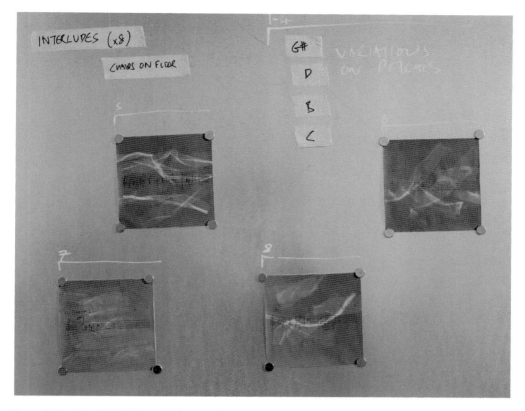

Figure 30.3 Detail 'Crit Transcription, Design School' (2023) graphic transcription.

Constellation

This practice/research sits within a constellation of contemporary and canonical theorists and artists whose approaches have helped steer this work. This brief introduction will start to identify and give context to terms used throughout the chapter as well as help clarify the reason for their use and provide a structure that helps give a broader positioning of this work.

This chapter takes Henri Lefebvre's Rhythmanalysis as a theoretical framework and point of departure. Lefebvre proposes Rhythmanalysis as a method for analysing the rhythms of social space, going as far as to describe the aim of writing Rhythmanalysis as nothing less than to found a science, a new field of knowledge [savoir]: the analysis of rhythms; with practical consequences.[2] By dissecting how social, economic, and political forces shape and are shaped by rhythms, Rhythmanalysis became a tool for understanding and critiquing social space.

> **Rhythm** reunites **quantitative** aspects and elements, which mark time and distinguish moments in it – and **qualitative** aspects and elements, which link them together, found the unities and result from them. Rhythm appears as regulated time, governed by rational laws, but in contact with what is least rational in human being: the lived, the carnal, the body.
>
> (p. 18)[3]

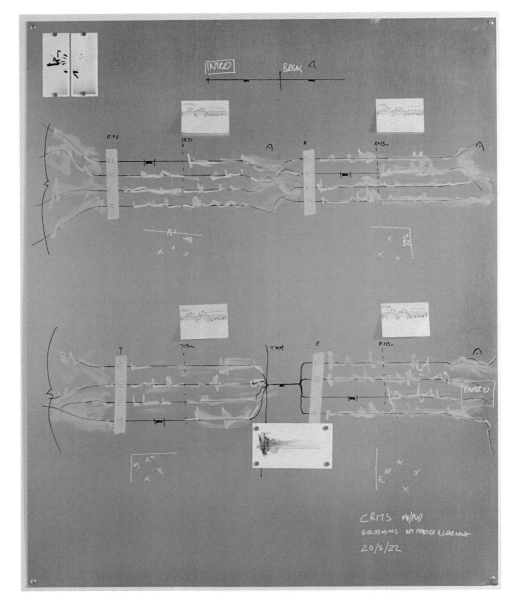

Figure 30.4 Crit Transcription, Art/Education School' (2023) graphic transcription.

Lefebvre understands rhythm not just as a metronomic unfolding of measured time but as a concept that encompasses a complex interrelation of both the marking of time (the quantitative) but also messy, lived (qualitative) rhythms. Analysis through listening to a polyrhythm that defines social space gives a precedent for using listening as a methodology for understanding the spaces we occupy, but also using specific devices for doing so (rhythm). If the analysis of rhythm can

Listening with, listening toward

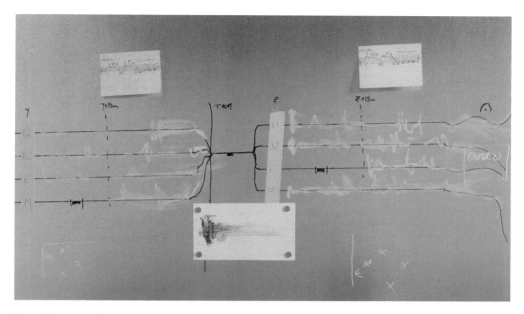

Figure 30.5 Detail 'Crit Transcription, Art/Education School' (2023) graphic transcription.

provide the basis to found a 'science' then what about the other elements of music? Could there be a precedent for a harmonicanalysis, timbralanalysis or melodicanalysis for example?

Lefebvre discusses music, making a case for rhythm as the dominant element of music and stating that harmony and melody have been written about a lot whereas rhythm has been less thoroughly covered. He says Music and musical rhythms will not, following this, take on **immeasurable** importance.[4] There is a sense in which Lefebvre is quick to distance himself from music. It could be that this is intentionally done to give space to a broader interpretation of rhythm as a social construct rather than solely a musical one. This chapter builds on the concept of understanding non-musical social spaces through a musical framework as a basis for analysis – thinking not just about rhythm, but other elements of music that Lefebvre was maybe too quick to dismiss, such as musical structure, melody, timbre, and resonance.

Lefebvre is present also in the analysis of space. The Production of Space offers a complex and interdisciplinary analysis of how space is socially constructed, experienced, and contested and this chapter makes use of his triadic model of space that includes perceived, conceived, and lived spaces (which he names social space, representation of space and representational space) as a means of understanding what spaces are at play in the transcriptions that form the basis of this chapter.

In discussing what is being transcribed, this chapter makes reference to the 'event' to refer to the activity of the studio-based crit. This refers specifically to the definition of event by Slavoj Žižek, who writes.

We should begin by reminding ourselves that an Event is a radical turning point, which is, in its true dimension, invisible.[5]

To describe a crit as a radical turning point could sound hyperbolic; however, the pedagogic space (or event) is, or can be, a space for radical transformation. Elizabeth Ellsworth describes this transformation as a space of pivoting or hinging – a turning point. She states:

The designs of particular objects, environments and social gestures do just that: They actualize, in objects, spaces, and event times, the abstract pedagogic pivot point that sets inner realities in relation to outer realities.[6]

In both the crits transcribed here, an environment and series of social gestures aim to encourage learning through putting 'inner realities in relation to outer realities' that could be described as enabling a radical turning point for those involved. The invisibility of the event that Žižek refers to is of relevance, as if it cannot be seen then rather than looking at it or reflecting on it (or using other visual metaphors) perhaps we would be better off listening to it. Brandon LaBelle pulls at this further, referring to the acousmatic (hearing without seeing the source of the sound)

The acousmatic, from my perspective, situates us within a complex space by which recognition is shaped less through visual identification and face-to-face relation, but rather through a concentrated appeal to the listening sense.[7]

Taking Lefebvre as a departure point for listening as an analytic tool and a guide to understanding space we can start to see (or rather hear), the pedagogic event or pivot point as an invisible or acousmatic space that asks to be engaged with through listening.

As this work is based on practice/research, as well as gives a theoretic positioning it is important to situate the work among practitioners who make use of graphic transcription that starts to suggest transcribing beyond music.

Samson Young is an artist and composer who has a broad practice that includes making 'sound drawings' that represent contested space and/or war sounds. He specifically draws attention to the value of transcription in the text On Drawing Sound.

Notation does not need to result in an action, which is to say, function as a score. The reserve is possible. Other operative logics are equally probable. Notation as transcription is particularly under-explored.[8]

Christine Sun Kim also uses a graphic form of notation as a means of transcribing events that are beyond the sonic. As a deaf artist, her use of notation is drawn from traditional music notation, graphic notation, and the gestures of American sign language, giving voice to a sonic world that cannot be heard. In the catalogue for the 2013 exhibition Soundings at MOMA, Sun Kim describes transcribing a feeling in her work.

I often employ my own voice in my work. I can feel it inside of my body, and in this way, it is accessible to me.[9]

Both Young and Sun Kim's work notation is used as a means of making accessible – using listening and transcription as a methodology for uncovering and making known the unseeable.

Transcription – it started with a screech and a crack

The 'crit' (short for critique) is an art school staple of studio-based dialogic feedback[10] that can take a number of forms but is commonly based around a presentation of work with subsequent discussion feedback from peers. This is often carried out in a studio setting, with the work being discussed and a small (4–12) group of artists/students and a smaller number of tutors. The crit is so ubiquitous yet hard to define as it changes depending on who is leading it, where it takes place, the course and institution in which it occurs, and so on, and as such requires further thought. What is the sound of the spaces that the crit occupies, and what can be learned through listening to them?

The images in Figures 30.2 and 30.3 depict transcriptions that form part of a series of ongoing practice/research that proposes listening as a tool for better understanding the often tacit, or unspoken, material of teaching and learning in the studio-based art school learning environment. Musicians are trained to transcribe, that is to understand a sonic event through listening, repeating, and often writing it down. A well-known example of this could be the composer Messiaen, who was famous for transcribing birdsong and using it later as a basis for melodies in his work in the seven-volume Catalogue D'Oiseax Pour Piano.

In the examples given here, each transcription was instigated by a single sonic event that in some way defined the session and prompted the work. This 'punctum' for the crit (detail in Figure 30.3) was when one of the artists stepped on a piece of charcoal as a way of initiating her presentation. There was a loud crack of the snapping charcoal, which was followed by the silence that this made in the space. At the design school (detail in Figure 30.5) the sound of chairs being dragged across the floor as the participants moved around the studio between presentations, from one student's work to the next, lead to the transcription discussed and shown here. In each case, this initial sound was transcribed using traditional notation, much as Messiaen did with the birdsong. This then led to a broader interpretation of the whole event of each crit, made using graphic notation.

In the first transcription (Figures 30.2 and 30.3) there were four artists taking part and the transcription reflects this by giving each artist a stave, a little like the SATB score of a choral piece of music, with each voice having a line of notation. There was no tutor or overall guide through the session, other than the timekeeping which was done by each artist in rotation. The group chose to discuss the work while the artist stayed quiet, denoted by the rest symbol in each of the parts at the corresponding time. After 15 minutes, an alarm set on one of the participant's phones would sound, as shown in a direct transcription using the traditional Western notation of the 'crystals' alarm tone on the yellow post-it notes midway through each session. After this marker, the artist would respond and join in the dialogue around the work.

Other notable elements include the transitions between artists' presentations. As there were only four artists and no one leading the sessions the transitions were not orderly and would take some time as the discussion broke down and reformed. This is indicated by the dissolution of the staves with a 'pause' mark between presentations. The exception to this is the transcription made of the charcoal being broken by the fourth artist; this has been made on a rhythm stave in traditional notation with charcoal marks indicating time and timbre – it is followed by a tacit rest for all staves (participants), which indicates the silence this mini-event initiated before the final presentation.

Through the transcription, four main repeated forms are evidenced. If analysed as a piece of music, the session is similar to a variation on a theme or improvisation based on an AB song form, the A section being part in which the artist was quiet and the B section being where all four artists were responding to each other. This repetition of a basic form is similar in approach to a jazz standard, where the form is rigid and repeated but the content shifts as the musicians take it in turns to lead or guide an improvisation based on the fixed form. This structure allows for the four-part harmony of the conversations to take shape around the constraints of the 15-minute time limit and enforced silence or rest for the artist whose work is being presented.

The second transcription (Figures 30.4 and 30.5) follows a standard musical structure akin to a Sonata form (a common three-part musical form often used in the classical period of Western music) as it starts with an introduction or exposition. This was the procedural, hello and welcomes and described the structure for the day. The session concludes with a short outro or recapitulation which consists again of the more formal thanks and goodbyes that summarise and conclude the session. The main section of the day (or 'development' section if we continue to follow the sonata form) consists of a 20-minute structure repeated eight times (once for each student/designer) with a short interlude of chair scraping as the participants moved around, the sonic event that initiated the work between each.

The transcription lists the intro and outro simply with text delineated by a box. The bulk of the work is predominantly given to the main repeated structure and notation of the interludes. The main structure is surrounded by the parenthesis that indicates a repeated section in traditional notation. Within these are three time-based lines – one indicating the timings for each section, the other two indicating the 'voices' of the presenter (the student/designer) and the group response. The intensity of activity for each of these voices is indicated by the white marks on each line. The interludes themselves are transcribed in more detail. The pitches of the chairs on the floor are indicated. In four of the interludes, a more granular transcription of the noise (pitches, dynamics, and rhythm) is made using traditional notation; this is drawn onto a cyanotype (a photographic process developed to make blue prints) image of translucent tape, the texture of which indicates the timbre of the interlude.

Listening back

After a take in a recording session, it is common for the performers to gather in the control room and listen back to what has been recorded. Both the transcriptions themselves and any subsequent analysis function similarly. In listening back, or (re)listening, we hear things we did not notice during the performance. By transcribing the crit sessions and analysing them as if they were a piece of music, we notice elements that compositionally do not make much sense and could be changed, as well as structures emerging that would possibly not have been done if the session were not transcribed in this manner.

Form (verse one, art/education university crit)

The crit session was peer-led and had a small number of participants. As a result, it was fluid in its development – the only fixed metric was the time of each presentation. This approach to building a session provided a light framework that was still rigid enough to allow rest of the session to grow around it. The transcription highlights the in-between sections of the event: these are the moving around, re-setting and 'off-topic' discussions that bookended each timed, formal presentation. Using a more traditional form of notation or research method (an interview, photograph, or recording), these sections could easily be missed or discounted, but in the transcription, these become one of the most visually striking elements.

Analogous to the spaces between movements in a concert setting; these un-scored, and to some degree unintended, transitional sections give preference to the importance of the in-between, the need for space to transition from one event to the next. The tradition of not clapping between movements in Western art music gives space for each section to breathe, the silence of the concert hall (often defined by audience coughing) gives weight to the movements proceeding and preceding it. The transcription shows the 'silences' as transitions that allow a decompression between the discussions in each artist's presentation. The concert hall is a hierarchical space in which the conductor initiates the silences and decides how much space is given for each movement. The audience too is controlled by the conductor's baton, the coughing stops when the music starts. In the crit, there was no conductor, and no articulated higher authority, so it is of interest that the gaps are undefined and not given a time in the transcription. But they are notated and given an indication of texture and a breakdown and emergence of the individual voices. How could a less hierarchic ensemble/studio/classroom/event occur? How would it start?

Under each of the main sections of the transcription there are references to a more directly spatial aspect of the session, the locations of the participants in relation to one another. In most musical settings there are conventions that dictate the location of the performers – the drums in the middle (or to one side in some jazz settings) the layout of the orchestra with the percussion at the back and so on. These locations are not often scored or transcribed but could have real significance to the outcome of the performance. In a fully scored hierarchically organised work like a concerto, the soloist and the orchestra need to see the conductor, but the soloist does not need to see the third violin (for example) whereas a less scored and more improvised piece such as a free jazz performance may necessitate line of sight for the cues to be given during the performance. The artist in a crit stands in front of their work or at the back of the group does this dictate aspects of the discussion. Are there positions that could be scored that would go some way to counteracting the various privileges (height, physical ability, volume of voice, eyesight) of the participants?

Form (verse two, design school crit)

The session could loosely be described as a Sonata form – an exposition followed by the development and a recapitulation. The main repeated section, the development, took the form of call and response – a musical form that has its origins in the church (the priest says a line and the congregation responds). In this case, the person presenting their work was the caller, and the rest of the students and tutors were the respondents. The caller was often silent during the responses, and the responses largely came from a panel of tutors and guest artists, the experts, and in this case, the rest of the participating students (or congregation) were largely silent observers. Through listening to and transcribing this form, the rigidity and didactic quality of the dialogue became apparent. Could the congregation be involved all the way through? Could the call and response be broken up differently? What other structures could work? A fugue perhaps, with its interwoven melodies?

The other compositional 'error', highlighted through this transcription, was the repetition of the main theme eight times. Even with short interludes of chair scraping, this would be a repetitive piece of music to listen to – there was no chorus, shout chorus, breakdown, middle eight, key change, dance break to break it up. Imagine the drudgery of listening to a pop song with eight verses and no chorus. Similarly, if this were a sonata the development of the main theme does not really develop or evolve – each student's presentation was largely unrelated to previous student's work and often would not be referred to again by the students or tutors. What would a 'chorus' be in a crit setting? Is there an art school equivalent of a breakdown or a bass drop? Would there be a way of restructuring this session that more consciously draws on compositional tools or techniques that encourage some meaningful development? Explicitly stating a theme at the start of the day could be a way of giving a hook to the participants allowing them to refer and develop past ideas – something as simple as asking participants to keep in mind 'texture' (for example) throughout the session could provide enough of a theme to develop. What would retrograde, inversion, or reharmonising do to this theme in this context?

Beyond the observations that arise from listening back to the two transcriptions and the events they depict, there are some broader observations and questions that emerge from the way the work is made.

Why graphic?

The process of listening and transcribing these events was initially done using traditional Western notation. As the work developed, the transcriptions expanded to encompass the whole of the event, not just the initial sound that prompted the transcription. This in turn necessitated a more openly interpretive form of graphic notation. This idea is not a new one; there is a long history of graphic notation (notably Cage, Cardew et al.) as well as contemporary composers and artists (such as Young and Sun Kim mentioned earlier in the text) that give precedence for expanded notation representing expanded ideas around a musical event.

> As soon as the temporal course of music is congealed in a picture so that the temporal connections become spatial, so that the quality of sequential events in the structural relations are transferred to an optical impression, the communication of music gains simultaneous, extra-musical attraction. Temporal experience allows itself to be transposed into spatial experience.[11]

Here, the composer Stockhausen develops this by making links between the temporal, the spatial, and the picture. The extra-musical attraction refers to a contemporary trend of exhibiting scores as visual art in galleries, which suggests that graphic notation can be more easily accessed by both an audience of trained musicians and a non-musical layperson (the probable audience of the transcriptions detailed here). It also suggests that there is something extra that the notation is capable of conveying, the extra, in this case, is the temporo-spatial rhizome of the event being transcribed. Using non-traditional notation allows for a transcription space that can encompass traditional notation of sonic events such as charcoal cracking and chairs scraping, as well as spatial and temporal events of dialogic conversation between people, artworks, and the studio space that comprises the events transcribed.

Structure

One of the most prominent features of the transcriptions is a focus on structure. Timings and delineations of sections are often more clearly indicated than what happened within the timeframe or section. Listening, in this instance, is a way of revealing tacit structures at play in the spaces being transcribed. Thinking about the structures or places of pedagogy in this way, that is making them explicit, is reminiscent of Jo Freeman's essay The Tyranny of Structurelessness. She uses the essay as a space to critique informal structures – arguing that tacit and unspoken structures produce elitism and a system which is harder to critique and change. In the text, she examines the formation and evolution of feminist working groups in the 1970s but this work is directly translatable to pedagogy, politics, or any other arena in which there are systems of power that are at play that are unstructured, unarticulated, and unheard.

For everyone to have the opportunity to be involved in a given group and to participate in its activities the structure must be explicit, not implicit. The rules of decision-making must be open and available to everyone, and this can happen only if they are formalised.[12]

Through the examination of listening, what we listen for, and the structures we listen to, this work proposes listening initially as a means of knowing what is there, that is otherwise unseen or unarticulated. This chapter doesn't seek to suggest better structures or improvements, but the prominence of structure in the work highlights the importance of knowing what is not seen but is nevertheless present in a social space.

Space

In The Production of Space, Henri Lefebvre categorises social space into Spatial Practice, social interactions – Representation of Space, that is the design and planning of space and Representational Spaces, those of art or images and systems of symbols.[13] The transcriptions directly deal with the representation of space and the social, spatial practice within them. The physical studio spaces (the representation of space) were quite large with temporary/movable walls that allowed students to install work and have their own sense of space. In the design school, as the students/designers moved around, the space itself challenged the tutors and other students/designers (the audience) to position themselves in relation to them. There was no stage or seating and no single focus to the event, which challenged those participating in the work to ask – where should we be? In a sense, this relational dialogue was happening with the crits at art/education university too, as there was no named hierarchy that decided where the participants were to be positioned. As well as the size, other qualities such as the resonance of the space exerted an influence over the events. The resonance that was key in helping the chairs make such a remarkable sound at the design school could be detrimental to those speaking. In larger spaces, softer voices become more diffuse and indistinct. This highlighted the dynamic range – that is the difference between loud and soft sounds. Those more nervous about 'performing' would tend to get quieter, reinforcing their anxiety around the introduction of their work and louder voices seemed amplified. It could be that thinking about where we position ourselves and how we make the space around us, initiated by listening, for example, could start to help address some of these issues. These negotiations start to allude to the relationship between the representation of space (the design of the space) and the social space, the interactions that define the event. The transcriptions themselves are not representations of space but representational spaces:

> space as directly lived through its associated images and symbols, and hence the space of 'inhabitants' and 'users', but also of some artists and perhaps of those, such as a few writers and philosophers, who describe and aspire to do no more than describe.[14]

The 'living through' in this case could be the pedagogy that is inherent in the act of translating the signs and images. Both translating the event into a series of images and signs and inhabiting this space through looking at/listening to the transcriptions open a transitional space as defined by Ellsworth, where inner and outer meet.

Material

There are two predominant materialities in this practice/research: the material of the transcriptions themselves and the material of teaching and learning in the spaces transcribed. The materiality of the transcriptions reflects the subject matter. Tape, magnets, and chalk pen are quick, temporary, and re-workable, qualities that reflect the fluidity and temporality of the event but also qualities that allow for. An intuitive working process allows for revisions and repositioning of different elements as the work is developed. The transcriptions are forms of assemblage, loose agglutinations, coming together of constituent parts. In the introduction to A Thousand Plateaus, Capitalism and Schizophrenia Deleuze and Guattari show a score under which they describe the lines that form the assemblage of a book.

> In a book, as in all things, there are lines of articulation or segmentarity, strata and territories; but also lines of flight, movements of deterritorialization and destratification. Comparative rates of flow on these lines produce phenomena of relative slowness and viscosity, or, on the contrary, of acceleration and rupture. All this, lines and measurable speeds, constitutes an assemblage[15]

This helps summarise the materiality of the work. If a book can be described in these terms, why then can't the materiality of teaching and learning practices? Maybe transcribing them, listening, in the way suggested here, is a way of articulating the assemblage of social, spatial and material practice of art education; allowing the tacit structures that operate in these environments to be heard.

In her 2018 paper Teaching and Learning with Matter Dr Tara Page, an academic based at Goldsmiths University, proposes we better understand

The relationalities of learning and teaching so that we can become conscious of our ways, materials and spaces of pedagogy. From this, we can then ensure that we explicitly acknowledge and support the creation of these places to enable a sustained pedagogic engagement for all learning environments that can be transformative and emancipatory.[16]

The proposition here that the materiality of pedagogy is relational and that this needs to be made explicit to create and support learning spaces describes the transcription process. Through transcription, the material of teaching and learning is made explicit through being listened to and this act of listening then creates a structure, the material of the transcription itself that is a pedagogical space, a space that puts inner and outer in relation and has the potential to be transformative and emancipatory.

Conclusions (Outro)

Through establishing a constellation of theory, contemporary practice, and the subsequent analysis of two transcriptions, this chapter positions transcription as a viable method and listening as a framework for further understanding the event of the crit. Defining pedagogy as an event in Žižek's definition is to encompass Ellsworth's hinge and transitional space and to allow the crit to be a space of radical transformation that it has the potential to be. In looking at (listening to) the transcriptions of these events, we start to recognise and define the social space, representation of space, and representational spaces that are occupied in this work. Through listening to the acousmatic sound of these spaces as an event (rather than purely a sonic experience), invisible structures and relationships become apparent.

The transcriptions are an initial exploration in starting to understand what is being listened to and what spaces we are listening in. They define a point from which to start to focus on the qualities of listening that could be used in future explorations. If the tacit structures that Jo Freeman warns against are of primary concern then listening to the silences and asking who or what occupies these silences could be a way to develop this methodology. Listening to the timbre or the rhythms that define the transitions of the first crit transcribed would lead to further questioning of how an event begins and what the conditions are that allow it to occur. The transcription of the resonance of the space in the design school crit specifically highlighted the importance of the representation of space on the event. Further work could entail a more resonance-focused transcription of studio spaces that are used for teaching and learning.

The introduction began with an invitation to listen and established the themes in the chapter as notes in a chord that resonated together. As a recapitulation, the 'outro' proposes a (re)listening that focuses on specific qualities of this chord, the harmonic content, timbre, resonance, and rhythms inherent within it. Through this focused (re)listening, the complex and often invisible rhizome of space(s), structures, qualities, and dynamics that comprise the spaces in which we work, can be heard. If they can be heard they are explicit and they can be challenged, championed, reproduced, followed, and fought against – engaged with.

Notes

1. Perec, Georges. *Species of Spaces and Other Pieces*. Translated by John Sturrock. London: Penguin Press, 1999, 10.
2. Lefebvre, Henri. *Rhythmanalysis, Space, Time and Everyday Life*. London and New York: Bloomsbury, 2016, 13.
3. Lefebvre, Henri. *Rhythmanalysis, Space, Time and Everyday Life*. London and New York: Bloomsbury, 2016, 18.
4. Lefebvre, Henri. *Rhythmanalysis, Space, Time and Everyday Life*. London and New York: Bloomsbury, 2016, 24.
5. Žižek, Slavoj. *Event*. London: Penguin Press, 2014, 179.
6. Ellsworth, Elizabeth. *Places of Learning Media, Architecture, Pedagogy*. London & New York: Routledge, 2005, 48.
7. LaBelle, Brandon. *Sonic Agency, Sound and Emergent Forms of Resistance*. London: Goldsmiths Press, 2018, 33.
8. Samson Young. "On Drawing Sound." Accessed September 6, 2023. https://www.thismusicisfalse.com/text.
9. Christine Sun Kim. "Soundings Exhibition Artists" Moma. Accessed September 6, 2023. https://www.moma.org/interactives/exhibitions/2013/soundings/artists/5/works/.
10. Orr, Susan. "We Kind of Try to Merge our Own Experience with the Objectivity of the Criteria: The Role of Connoisseurship and Tacit Practice in Undergraduate Fine Art Assessment." *Art, Design and Communication in Higher Education*, 2010, 4.
11. Stockhausen, Texte zur Musik, 178.
12. Jo Freeman. "Tyranny of Structurelessness." Accessed September 6, 2023. https://www.jofreeman.com/joreen/tyranny.htm
13. Lefebvre, Henri. *Production of Space*. Translated by Donald Nicolson-Smith. Oxford: Blackwell, 1991, 39.
14. Lefebvre, Henri. *Production of Space*. Translated by Donald Nicolson-Smith. Oxford: Blackwell, 1991, 39.
15. Deleuze, Gilles. Guattari, Felix. *A Thousand Plateaus, Capitalism and Schizophrenia*. Translated by Brian Massumi. Minneapolis and London: Minneapolis University Press, 2005, 3–4.
16. Page, Tara. Teaching and Learning with Matter. *Arts*, 2018, 1.

Bibliography

Christine Sun Kim. "Soundings Exhibition Artists" Moma. Accessed September 6, 2023. https://www.moma.org/interactives/exhibitions/2013/soundings/artists/5/works/

Deleuze, Gilles. Guattari, Felix. *A Thousand Plateaus, Capitalism and Schizophrenia*. Translated by Brian Massumi. Minneapolis and London: Minneapolis University Press, 2005.

Ellsworth, Elizabeth. *Places of Learning Media, Architecture, Pedagogy*. London & New York: Routledge, 2005.

Jo Freeman. "Tyranny of Structurelessness." Accessed September 6, 2023. https://www.jofreeman.com/joreen/tyranny.htm

LaBelle, Brandon. *Sonic Agency, Sound and Emergent Forms of Resistance*. London: Goldsmiths Press, 2018.

Lefebvre, Henri. *Production of Space*. Translated by Donald Nicolson-Smith. Oxford: Blackwell, 1991.

Lefebvre, Henri. *Rhythmanalysis, Space, Time and Everyday Life*. London and New York: Bloomsbury, 2016.

Messiaen, Olivier. *Catalogue D'Oiseax Pour Piano*. Paris: Alphonse Leduc, 1958.

Orr, Susan. "We Kind of Try to Merge our Own Experience with the Objectivity of the Criteria: The Role of Connoisseurship and Tacit Practice in Undergraduate Fine Art Assessment." *Art, Design and Communication in Higher Education*, 2010, 9(1), 5–19.

Page, Tara. Teaching and Learning with Matter. *Arts*, 2018, 7(4). https://doi.org/10.3390/arts7040082

Perec, Georges. *Species of Spaces and Other Pieces*. Translated by John Sturrock. London: Penguin Press, 1999.

Samson Young. "On Drawing Sound." Accessed September 6, 2023. https://www.thismusicisfalse.com/text

Stockhausen, Texte zur Musik, Köln: M. DuMont Schauberg 1963, 178.

Žižek, Slavoj. *Event*. London: Penguin Press, 2014.

31
SITE-ORIENTED MUSIC CURATION. CONTOURING THE LISTENING SPACES

Sasha Elina

Introduction

Music curation is a relatively new phrase, or term, applied in performance arts and sound-based practices. In order for us to understand what it means to curate music, or complex music situations,[1] we should think of this one essential aspect of a listening experience: sound penetrates space. Space then becomes a force, material, and both a physical and social 'container', or context, presenting a curator with different ways to creating and facilitating events. Music curation, therefore, should be viewed as a site-oriented practice: with reflection on the characteristics and conditions of the site and milieu in which the event takes place. I see this text as a rather personal collection of notions, examples, crossovers and reflections that provide some paths to exploring the role of the music curator.

The role of a facilitator of listening experiences has been steadily increasing in its importance. Similar roles could be seen in early concert promoters and impresarios, composers and musicians self-organising events, curators working with concepts and spaces, across different media and within various sound-based practices. The term 'curator' in music was borrowed from the sphere of visual arts. For example, exhibitions are 'curated', and such formulation comes as no surprise for the audience. However, the 'music curator' is something that has been introduced and used increasingly more often only in the last decade or so. The term itself indicates an originally formulated role of a curator as someone who, since the late 19th century and for some period of time, used to be taking care of and managing collections in museums and galleries, often just 'behind the scenes', and with no control over the display decisions. Gradually, the curator began to step into that area of working – with various spatial arrangements that helped in achieving the desired effect on the viewer's experience.

Space is seen and heard, operating in resonance with all our senses. Both built and natural spaces suggest and contain multiple events to be noticed, responded to, experienced. Individually and collectively, in private and in public, music is heard, located, listened to, and thought of. The act of curating sound and music in space can be a powerful way for building or navigating social relations. For as long as music was performed for the audience, different musical formats remained 'milieu specific',[2] moving from the physical confines of one site to another, whereas the audience would reflect upon the conditions of a particular type of music presentation, to form certain demands for it. In this way, musical space can be described as an 'assembly of sound producers, listeners and sound producing objects',[3] ultimately acting as an instrument for aural experiences.

It is often assumed that the work of the music curator is focused exclusively on the musical content of an event, but ultimately, and quite naturally, the curator steps into an intricate relationship between the listener and the listening site, to produce an experience that ties many elements together. The bond between performed music and the physical space of its performance presents the opportunity for a curator to navigate the complex of spatial and temporal narratives that arise within the listening experience, built into the fabric of an event. However ambiguous and fluid the music curator's role may still be in a contemporary context, it is worth reflecting upon their sphere of influence on the culture of musical performance and continue developing what could in time become a discipline in its own right.

Early listening spaces. Sounds contained and propagated.

Shaping the concert venue

The long history of the development of musical genres, and along with them musical formats and architectural structures, had broadened and diversified the music industry existing today to a substantial degree. Over the last few centuries, the Western musical landscape, with its various modes of listening offered in various kinds of settings, has been consistently aiming at both satisfying and provoking expectations of concertgoers. Different types of performance settings formed in the 17th, 18th or 19th centuries reflect in most contemporary formats. Some types welcome the audience to pay attention to a performance whilst also flexibly regulating the social side of an event, making music a pleasurable accompaniment to a situation. Other types demand a social discipline, regulated by the architectural arrangement, strict rules of behaviour and an attentive, intellectual engagement with the programme showcased.

London, from the end of the 17th century and throughout the 18th, was one of the epicentres of early development and further establishment of the public concert. Early concert venues introduced and explored a range of formats, diverse in their spatial, acoustic, and social arrangements, as well as different types of music programming. Many of those formats are now familiar to us, using familiar strategies of engaging the audience.

The first formats of the public concert (accessible to everyone by purchasing what was considered 'affordable' tickets) appeared in places such as taverns or private buildings with separate rooms repurposed for running regular music events. In Britain the latter were usually called 'music rooms', and if a room was an extension to a tavern, it was often regarded as a 'music house'. The demand and popularity of this type of entertainment across Europe grew with the growth of the middle-class concert audience along with professionalisation of musical performance practice.

The architecture of the first purpose-built concert spaces 'emphasized the social nature of concert going'.[4] Early music rooms were often built with an oval shape, and with the stage to one end, with the audience seated along the perimeter facing each other, sometimes with space left in the middle for promenading between the acts. Such oval-shaped halls included Oxford's Holywell Room (opened in 1748), and Scotland's oldest purpose-built music room St Cecilia's Hall in Edinburgh (opened in 1763). Throughout the 18th century, the seating area of a concert hall was often designed with the audience's gaze directed towards other listeners rather than the stage, whether the shape of the hall was oval or rectangular. The Altes Gewandhaus in Leipzig (built in 1781), one of the earliest public concert halls in Germany, serves as a distinct example. London's very popular concert hall Hanover Square Rooms, established in 1774 and accommodating between

500 and 700 people, also demonstrates a type of spatial arrangement welcoming a more relaxed and socially interactive kind of musical experience: one part of the space offers seats directed towards the stage, whilst the other is turned towards other visitors, leaving a physical gap between them, this layout perhaps also metaphorically points to the increasing diversity of concert formats that developed during this period.

The concert experience for audiences slowly became more and more disciplined by the mid-19th century, and, although secular, one could describe the experience as spiritually charged. This, in some way, can be evidenced by the audience seating being gradually redesigned to focus the listeners' full attention towards what was happening on stage. The noises and conversations that were once common amongst the audience were now prohibited, and the programme displayed the performed works as objects of high cultural value, matching that kind of self-identification of bourgeois concertgoers. As the cultural importance of public musical performances grew, so did the seating capacity. The Great Hall in the Vienna Musikverein opened in 1870 with a capacity of over 1,500 people (doubling or even tripling the norms of a previous century), establishing this shift.

Outdoor entertainment. The case of London's pleasure gardens

Music, with its intimate ties to architecture, is curated both in and out of dedicated concert spaces. Music is now presented to us in different kinds of venues; transmitted straight into the headphones or speakers of our private rooms, played outdoors on the streets and in parks to people casually strolling by, or to large gatherings of thousands of concertgoers at a festival stage, and so on. Spaces for music performance do not necessarily mean spaces enclosed by walls and set under a roof, nor does a musical experience necessarily dictate a seated audience experience. Consequently, the role of the music curator has been evolving, in parallel to music culture and the many types of spaces it occupies.

The most popular early commercial outdoor leisure venues – pleasure gardens – started taking shape in the second half of the 17th century in the musically thriving city of London. Pleasure gardens existed roughly for two centuries until their popularity, for a number of reasons, gradually faded. Throughout the 18th century, some pleasure gardens were amongst the most attended places for music listening and general recreation by people of diverse social backgrounds and milieus. Opposed to a hectic and overcrowded London life, disciplined and secure pleasure gardens promised to reflect an English urban society's wish to become immersed in nature and peaceful leisure activities, establishing 'quasi-utopian spaces in which to envisage a whole society at play, yet under control'.[5]

Vauxhall Gardens, established in 1661, was one of the most prominent pleasure gardens, with a wide-ranging influence on the European sphere of outdoor entertainment. Whilst popular from its opening, the gardens are best known during their grand reinvention by the patron and impresario Jonathan Tyers (1702–1767), who became the landlord of the gardens in 1728. This started a journey of much experimentation in what was to quickly become one of the most significant public entertainment venues of its kind.

A socially unifying exercise, crucial for the ideals of contemporary society, was reflected in various outdoor activities gathering thousands of people throughout the summer seasons. Events at Vauxhall Gardens – from concerts, fireworks, masquerades, and so on – were notoriously elegant and tasteful, and set a high standard for similar venues. The gardens not only facilitated new kinds of social interaction such as promenading (or strolling in a dedicated space like a park or similar), but also did 'an immense amount to propagate the idea of public concerts, both in England and abroad'.[6] The concert programming at Vauxhall, and the other similar notable venues in Ranelagh, and Marylebone, amongst others, helped to promote the names of musicians and composers who performed there, thus reinforcing an overall interest in musical performance amongst the public, and steadily becoming a more present and respectable part of public society.

Vauxhall Gardens were accessible by admission fee for the public of all classes[7] – which was a crucial characteristic of the gardens' promotional strategy and image. During the golden age of the Vauxhall pleasure gardens and similar venues, there was significant attention given to the diversity of their audiences, as originally intended, and enforced by Tyers. However, the mixing of different social classes and backgrounds in these spaces was carefully regulated. The promotion of these events emphasised the presence of nobility, creating a sense of commercial success by fostering a feeling of inclusivity and reducing social barriers. Historian Hannah Greig highlights this point:

> On the one hand, it might be presumed that mingling suggests a common experience, a coming together of different groups to create a new body and/or the circumstances of meaningful exchange. On the other hand, it might more simply imply a momentary co-presence of different types of people, a short-term encounter during which the original separateness of those involved was either retained or easily restored once the moment of mingling has passed.[8]

The grand opening of the Vauxhall Gardens in June 1732 presented a glorious event titled the Ridotto al Fresco (which loosely translates from Italian as 'a place for entertainment located outdoors'). Following a period of inclusivity, this time it was not meant for social mixing, and was designed to be a more exclusive event, accessible by a much higher price and in all ways tailored for wealthier members of society – which would, in the eyes of the management, work towards attracting more patronage. The opening event is reported to have gathered around 400 people, including royalty such as Frederick, Prince of Wales, who later became a patron of the gardens. One of the first pavilions built in the gardens was the Prince's Pavilion, frequently used by him when attending following events.

Tyers assigned the task of developing a sophisticated music programme to the leading impresarios in England at that time. Vauxhall Gardens was a truly music-centred venue, programming a variety of contemporary composers' works, in some contrast to many of London's other venues and entertainment facilities that focused on either more traditional or at least popular music. David Coke and Alan Borg write:

> Tyers exposed a substantially larger audience to serious music than had ever been possible or even conceivable before. The fact that he did so in a setting where the audience could choose to listen or not, and could choose where to listen from, fundamentally transformed the public's experience of musical performance, and led to a much wider and easier acceptance of the concert as a public entertainment.[9]

In addition to creating an event programme, Tyers also commissioned many unique architectural structures, which demonstrates a space-oriented curatorial strategy in devising an overall experience for the visitors. Those structures were permanent and temporary and were established over years across the grounds of the Vauxhall Gardens: from pavilions, ballrooms and supper-boxes, to bandstands and several purpose-built music rooms. The gardens were designed with careful consideration for every aspect, taking into account the various activities of strolling, listening, dining, and socialising. Importantly, all these elements work together to create a sophisticated and immersive experience that engages all the senses, beyond merely the visual. One crucial element that ties everything together, socially, spatially, and experientially, is the curation of the musical programme. The curator, or event director, therefore, plays a vital role in shaping the overall 'architecture' of the space, through the means of music and other activities that engage people's imagination and instigate emotional response to the site.

In terms of structures dedicated to supporting the performance of music, in 1735, the first open-air bandstand, called the Orchestra, was built, and became one of the major architectural symbols of the gardens. The Orchestra was an octagon-shaped building accommodating around 30 musicians, serving as the stage for regular concerts of composers' newest creations. Another building dedicated to music in the gardens was the Organ. Opened two years later than the Orchestra, in 1737, the Organ's design presented a more complex three-level square structure, connected to the Orchestra by a bridge, with four openings on each side. Such outdoor music events were a spectacle rarely seen before by the public.

These concerts met with considerable success, however, some complaints from the audience were present in regard to the lack of clarity with which the sound was propagated in the open-air environment of the gardens. In an attempt to improve the acoustics, the same year as the Orchestra was erected, a number of trees surrounding the building were felled, with the intention of improving the passage of sound between the performers and the audience, and to better distribute the sound across the gardens.

The Orchestra and the Organ were elevated above the ground, which was intended to improve the sound for the audience, but also to regulate the listening format and social interaction around the performance. In other performance venues of the time, it was common for audience members to directly address musicians during their performance, to request specific pieces to be played – a manner and behaviour that could cause an unwelcome interruption to a performance in more modern concert settings. The elevation of the buildings in the pleasure gardens articulated a clear separation between the performance space and the listening space, forging a more respectful attitude on part of the listeners.

With London's climate, event organisers were also incentivised to provide indoor alternatives. One of the most famous alternatives to the outdoor pleasure garden was the Ranelagh Gardens' Rotunda – a large, richly decorated concert space that proved memorable to its visitors. In the centre of the hall was a structure that supported the roof, at first used as a stage for the orchestra, but later on repurposed for a large fireplace, surrounded by benches people could sit on during those particularly cold days. It was for reasons related to acoustics, perhaps a deficient sound propagation, that the orchestra was moved to the side, now positioned together with a pipe organ erected here in 1746. With this architectural composition, the Rotunda also famously acted as an 'eternal', circling promenade pavilion.[10] Upon entering the space filled with music performed by an orchestra or sang by an opera singer with accompaniment, visitors would join in the meticulously arranged, controlled act of walking. Promenading arranged in such a strict manner eventually became a matter of joking for the boredom it brought onto participants.

> The Rotunda undeniably kept Ranelagh's visitors' feet dry, and eliminated any element of risk, but once everybody had seen and been seen by everybody else, the circular promenade became profoundly dull and repetitive. The walks at Vauxhall, although all straight and at right angles to one another, were at least all different, and allowed visitors to choose which way they walked, and to escape the bright lights, while still hearing the music.[11]

Both Ranelagh and Marylebone, whilst successful for several decades, did not have the enduring popularity of Vauxhall Gardens and eventually, both closed, leaving Vauxhall Gardens to remain the most influential and long-standing until its eventual closure in 1859. Before its closure, from the beginning of the 19th century, every few decades and under different management, Vauxhall Gardens exercised different programming strategies to stay competitive, providing opportunities for new artistic formats by constructing new purpose-built concert halls, theatres, galleries, and exhibition pavilions. Some fashionable new acts were included, reflecting London's appeal to theatrical performances, circus, variety shows, and ballet. This led to rising expenditures, for 'such acts to be properly appreciated', it is noted by Coke and Borg, 'they needed to be seen in enclosed spaces before a seated audience'.[12] This resulted in building temporary theatres, as well as an illuminated fountain, sculptures, and other attractions. Eventually, the gardens closed with little trace, and the music scene was taken over by concert halls, opera houses, galleries – enclosed spaces delivering great comfort to the audience, whilst also regulating it to an even greater extent.

Inside, though, and out of the frame

The space traced by the ear in the darkness becomes a cavity sculpted directly in the interior of the mind.[13]

The organism is a force, not a transparency.[14]

Following the examination of the dynamic development of London's pleasure gardens, we may ask ourselves, what factors are active in shaping our experiences of musical events, and what is the role of curation in establishing the framework for those experiences? Furthermore, how might we direct these observations into identifying the curatorial practice in music, in the interest of calibrating spatial, social, and musical content?

The context, or the framework which the curator attempts to outline, has the power to provoke or contradict the audience's expectations, or set different levels of engagement with content communicated through it, ranging from passive reception to a more active participation with the work or a music situation. This is achieved through certain staging parameters, or contextual cues. With the standard concert format, and (what has become) the typical concert hall, we arrive at a space fundamentally designed and constructed to produce an escape for the audience from an external context, whilst facilitating the ceremonial aspects of music performance, listening and socialisation governed by accepted and prescribed behavioural norms. Here, the framework is outlined by architectural structures, with the classical separation of the listening and performance zones articulated by architectural elements such as the stage, the proscenium arch, or anything indicating a 'fourth wall'. Observed from a curatorial perspective, organising musical events within such well-defined and inflexible spatial organisations becomes an act of 'meeting in the middle'. Events are tailored by a balance of audience expectations and the curator's expectations of the audience's response.

The space in the middle, however, is very fluid and requires constant negotiation, depending on the nature of a given music situation (and on how far can negotiating go in response to circumstances), and on the kind of goal, or effect the curator wants to achieve. In the previous part of this text, we looked at a case study of music events organised beyond the confines of the concert hall, which still demonstrated some clear spatial organisation. This was just one of various ways for music to leave a concert space, finding itself on the streets of a city, or in a quiet park, or in a dim, crowded bar, or a vast and brightly lit gallery space, or any other space suddenly transformed into a listening site. For a music curator, this means finding the right touch between the music and the site and creating mechanisms of producing framing foundations for the audience's experience.

Moving through a proposed constellation of ideas that, I believe, can inspire, or resonate with contemporary curatorial strategies in music, I would like to divert to the realm of visual arts. Rosalind Krauss recognises one very specific approach to negotiating a work of art with the spectator (or, in the musical context, the listener) in modernist sculpture. She regards to its character as 'nomadic', or at least functioning through the loss of a dedicated place, or site, 'producing the monument as abstraction, the monument as pure marker or base, functionally placeless and largely self-referential'.[15] In this approach, the base, or the pedestal, elevates a sculpture from the ground, and forces it to behave autonomously from its context. Henry Moore, one of the most prominent British modernist artists, talks about the advantages of setting a sculpture outdoors 'so that it relates to the sky rather than to trees, a house, people, or other aspects of its surroundings'.[16] One could think of an analogy of such a disposition in a concert space, provided there is a stage that serves as that 'pedestal' for a performance. In similar ways, we can think of both a concert hall (with its walls and rules) and an open-air stadium. The sky, as one of the backdrops to these spatial and sonic arrangements, presents few limits here, if any at all.

Often, we are met with music situations that aim to neutralise the context of everyday life, by detaching a formal listening experience of a concert in an acoustically idealised space from a more hectic listening experience of sound environment which acts with no warning or pre-composition. Some music is not meant to be listened to with attention. Some is built into the background for other events to happen and be experienced: music heard on the street, or in a shop, a calibration of noises into a musically meaningful material. Neutral spaces were built in the 19th century and proceeded through the 20th, and into the modern days, to provide the type of a 'clean', perfected space, both for the ears and for the eyes: 'black box' theatres, 'white cube' galleries,[17] and numerous forms of technologically advanced concert spaces that acoustically have been getting more and more close to the character of a recording studio. Contemporary music curation deals more and more with something opposite to that: with taking music out of a concert hall.

When music curation is approached as a site-oriented practice, the notions of ambience and atmosphere (as they are regarded in certain readings through this differing terminology and definitions), can help direct this thought, as well as suggest some important perspectives into possible strategies to working with staging listening experiences. Sociologist Jean-Paul Thibaud describes ambience as 'a space-time qualified from a sensory point of view',[18] or a mood appearing as 'a complex mixture of percepts and affects, a close relationship between sensations and expressions'.[19] In the words of the philosopher Tonino Griffero, atmosphere appears as a sort of precondition, or a 'sensible encounter with the world'.[20] Atmosphere is evoked subjectively, although the influence within a group gathering, when we speak about a musical event, cannot be ignored. This opens a discursive space for music curation that ultimately seeks out some alternative, or loosened approaches to staging music situations, especially those that take place outdoors or are embedded in space in the form of a sound installation, and so on.

American artist Robert Irwin poses the question of seeing and not seeing as, perhaps, one of the central notions to understanding the so-called site-conditioned art, or phenomenal art, of which he was one of the pioneers in the 1970s. In his writings, Irwin discusses how the 'meaning' of an artwork is tied to the cultural traditions and the system of values that ground and shape the observer's reasoning of the work, and its experience. He proposes that change, in conditional art, acts as a central parameter to phenomenal perception, prioritising the sensory experience of an artwork as opposed to a purely intellectual one (stating that intellectualisation is traditionally more conducive to the creation and perception of modernist art). This attitude is not only relevant to the creation of new kinds of artistic objects, Irwin explains. It is a matter of questioning the subjects that art negotiates, especially when examined under the lens of the relationship between the observer and the hosting space (which we can refer to as a frame or a framework); between our senses, learnt experience and imagination: 'all perceptual knowing is knowing in action (change) and the equivalent of the phenomenal'.[21] In his 1985 text, he addresses the two states set in the title of the text – being and circumstance. Being, in a way, acts as the source of reference (sociocultural, personal), to our presence, and immediate response to the physical context we are living in, which is the circumstance. The state between being and circumstance, appears in a constant dynamic flux, through the process of ever-discovered change, and in this state, in this awareness, lies the experience of an artwork. 'What applied to the artist now applies to the observer'.[22] That is something Irwin calls a 'social implication of a phenomenal art'.[23]

Transitioning the question of seeing and not seeing into music, we could, on the one hand, speak of hearing and not hearing. But there is a different take on the terminological intricacies in the studies of acoustics and aural perception, such as the one proposed by composer and researcher Barry Truax. He suggests that the processes in human aural experience are based on the juxtaposition of hearing and listening. Hearing, he explains, is 'a sensitivity to […] the detail of physical vibration within an environment', whereas listening enacts our ability to 'interpret information about the environment and one's interaction with it, based on the detail contained within those physical vibrations'.[24] In this way, listening is addressed as a process of more determination, as an interface between a person and their surroundings, in which some sounds are brought into the foreground of our aural experience when they have some significance to us. Truax also distinguishes the acts of 'listening-in-search' and 'listening-in-readiness',[25] both following specific cues in the surrounding environment. In studying acoustic spatiality, Brandon LaBelle proposes the idea of zones of sonic intensity that contrast with conventional view of the inside and the outside having a distinct border between them:

> Given sound's vitality, its propagating verve, it readily puts into play a less clear distinction between rooms, and between buildings, between the distinctness of separate spaces. Instead, we can understand acoustic spatiality as 'zones of intensity', that is, as timbral identity by which differences are brought into play. What is inside then, as an architectural space, is less defined by sightlines or by the appearance of walls. Rather, sound ripples through space to easily occupy multiple areas, immediately bridging one space with another, and often leaking over lines between in and out, back to front, below from above.[26]

This demonstrates another crucial point to understanding how a music situation might work for a perceiving body under different circumstances – on a range of modes of interpretation and active exploration of the sounds in space that happen to come into attention. This individual space, that from which attention grows, is a mental space, addressed by the artist Robert Morris:

> The experience of mental space figures in memory, reflection, imagining, fantasy – in any state of consciousness other than immediate experience. And it often accompanies direct experience: one imagines oneself behaving otherwise, being somewhere else, thinking of another person, place, time, in the midst of present activity.[27]

A curator has no direct influence on mental space but can and should take its certain powers into consideration, aiming at building conditions that would spread through and beyond that inaccessible and unpredictable space – conditions that would forge a dialogue with a music situation presented before the listener. Listening, seeing, being positioned in a space in a specially suggested way is of that concern for a curator. To talk about senses in this context, therefore, is to talk about what we encounter through all our senses in the surrounding environment, and what is communicated through those encounters.

Some sound-based works merge with their surroundings and allow audiences to gradually discover them, when the space is formed, and understood with all its fragility. It is a process that erases, or at least weakens, the boundaries of a possible event framework, making it more flexible for the audience to explore intuitively and individually. In this kind of attuning to the environment, a constant flux of events and actions within it is met by our senses and our mind.

> […] making one's mark consists in leaving an imprint on the ambiance of a place, depositing a perceptible trace or behaving in a singular or unexpected manner. From this point of view an ambiance cannot be reduced to the sum of individual output or self-expression. Rather it engages a continuous, back-and-forth movement between what is ordinary and what becomes remarkable. In this respect the micro-events of daily life serve as a reminder that at any moment an atmosphere may be requalified.[28]

LaBelle regards the spatial framework as one of the things that holds together sonic events, propagating and merging them. Time, pace, along with the notions of the beginning and ending, are the other parameters that play a crucial role in both the perception and curation of a music situation. Time as a part of its framing mechanism. But when does the music situation start for its attendant? For example, when considered a social event centred around music listening, does it start with the audience entering the space and getting themselves prepared for the anticipated listening experience whilst having a conversation with someone, observing the architectural features of the venue, having refreshments, or reading the booklet? Or does it start only with the first musical sounds, or those musical silences of anticipation, the tension of a quiet murmuring accumulating in the atmosphere? The notion of anticipation in an event should be emphasised in this discussion as something that can both be a part of temporal and spatial experiences, can shape the ways they influence one another, and ultimately dictate the strategies employed by the curator in engaging with a music situation.

***

Acknowledging the relationship between sound and space gives the curator the opportunity to create the conditions for the listeners to experience a music event in the most engaged way. Sometimes, those conditions suggest a specific framework that works as a guide for the listeners even in the ways very distant from a standard concert experience: lying on the floor, being in complete darkness, moving through the space in a particular way, or provoking the audience's senses and perception through other staging methods. Sometimes, instead of adding explicit cues for the audience to recognise the framework of a music situation, those cues can be inferred, by the gentle choreography of some immediate aspects unfolding over time. This instigates a process of creation through adaptation for curators, and a process of discovery through self-awareness for the audience. Either by accident or by attuning to the subtleties of that elusive situation, or bringing an active imagination into play, with a more flexible scenario formed on both sides. For example, the increasingly popular act of soundwalking, or site-specific sound installations, can invite this kind of engagement on the part of the audience. Here, the curator not only closely engages with the site, but also with the individual's presence, perception, with their active contribution to their own listening experience as a more participatory practice, as well as the collective.

The music curator outlines specifics of the site and musical content, creating avenues to experiencing both. This practice lies in applying existing formats, creating new forms, and communicating otherwise invisible frameworks to help establish and shape the relationship between the music situation and the listener. The music curator may set conditions that would welcome the listener into the space of an event, whilst exploring different possibilities of its relationship within a bigger context. The music curator may intentionally position the music situation, with certain conditions applied, as either contrasting to or resonating with the context, articulate and unambiguous on both intellectual and sensory levels. The music curator may intentionally leave as much untouched, uncontrolled, unattended, and open, as it takes to build the music situation, as well as its framework, layer by layer, actively by the listeners themselves. The role of the curator here is not eliminated, but its affordances enter the realm of chance and coincidence, playing with individual imagination and perspective.

Notes

1. Here and later in the text I refer to 'music situations' alongside 'events' and other terms to highlight a wide range of possibilities in contemporary musical formats. The term 'situation' offers a perspective directed more towards the context that frames many autonomous elements, or multiple 'events', and into a complex of parameters in which a state of active perception is enacted and the audience's experience of music is being formed. By using the term 'situation', I draw different types of musical formats into a vast, and yet distinct, system instigating curatorial inquiry.
2. Kirchberg, 'A Sociological Reflection on the Concert Venue', p. 191.
3. Roden and LaBelle, *Site of Sound: Of Architecture and the Ear*, p. 8.
4. Forsyth, *Buildings for Music*, p. 53.
5. Conlin, 'Vauxhall on the Boulevard: Pleasure Gardens in London and Paris, 1764–1784', p. 44.
6. Forsyth, *Buildings for Music*, p. 43.
7. This was subject to the affordability of the ticket of one shilling per person, rather than a social rank.
8. Greig, 'All Together and All Distinct: Public Sociability and Social Exclusivity in London's Pleasure Gardens, ca. 1740–1800', p. 55.
9. Coke, *Vauxhall Gardens: A History (Paul Mellon Centre for Studies in British Art) (The Association of Human Rights Institutes Series)*, p. 141.
10. At Vauxhall Gardens, Jonathan Tyers commissioned his own Rotunda in 1748, to become the space used when the weather conditions did not allow the musicians and the visitors to be outside.
11. Coke, *Vauxhall Gardens: A History (Paul Mellon Centre for Studies in British Art) (The Association of Human Rights Institutes Series)*, p. 82.
12. Coke, *Vauxhall Gardens: A History (Paul Mellon Centre for Studies in British Art) (The Association of Human Rights Institutes Series)*, p. 292.
13. Pallasmaa, *The Eyes of the Skin*, p. 49.
14. Dewey, *Art as Experience*, p. 246.
15. Krauss, 'Sculpture in the Expanded Field', pp. 30–44.
16. Kwon, *One Place after Another: Site-Specific Art and Locational Identity*, p. 63.
17. See the 1976 series of articles 'Inside the White Cube' by Brian O'Doherty published in Artforum magazine.
18. Thibaud, 'A Sonic Paradigm of Urban Ambiances'.
19. Thibaud, 'A Sonic Paradigm of Urban Ambiances'.
20. Griffero, *Atmospheres: Aesthetics of Emotional Spaces*, p. 5.
21. Irwin and Simms, *Notes Towards a Conditional Art*, p. 204.
22. Irwin and Simms, *Notes Towards a Conditional Art*, p. 219.
23. Irwin and Simms, *Notes Towards a Conditional Art*, p. 219.
24. Truax, *Acoustic Communication*, p. 18.
25. Truax, *Acoustic Communication*, p. 22.
26. LaBelle, 'Acoustic Spatiality'.
27. Morris, 'The Present Tense of Space', p. 176.
28. Thibaud, 'The Three Dynamics of Urban Ambiances', p. 48.

Bibliography

Coke, David E. *Vauxhall Gardens: A History (Paul Mellon Centre for Studies in British Art) (The Association of Human Rights Institutes Series)*. Illustrated edition. Yale University Press, 2011.

Conlin, Jonathan. 'Vauxhall on the Boulevard: Pleasure Gardens in London and Paris, 1764–1784'. *Urban History* 35, no. 1 (2008): 24–47.

Dewey, John. *Art as Experience*. New York: Putnam, 1980.

Forsyth, Michael. *Buildings for Music: The Architect, the Musician, the Listener from the Seventeenth Century to the Present Day*. CUP Archive, 1985.

Greig, Hannah. 'All Together and All Distinct: Public Sociability and Social Exclusivity in London's Pleasure Gardens, ca. 1740–1800'. *The Journal of British Studies* 51, no. 1 (January 2012): 50–75.

Griffero, Tonino. *Atmospheres: Aesthetics of Emotional Spaces*. 1st edition. Routledge, 2016.

Irwin, Robert, and Matthew Simms. *Notes towards a Conditional Art*. 2nd edition. Getty Publications, 2017.

Kirchberg, Volker. 'A Sociological Reflection on the Concert Venue'. In *Classical Concert Studies: A Companion to Contemporary Research and Performance*, edited by Martin Tröndle. Routledge, 2020.

Krauss, Rosalind. 'Sculpture in the Expanded Field'. *October* 8 (1979): 31–44.

Kwon, Miwon. *One Place after Another: Site-Specific Art and Locational Identity*. MIT Press, 2004. https://doi.org/10.7551/mitpress/5138.001.0001.

LaBelle, Brandon. 'Acoustic Spatiality'. *The Zone and Zones – Radical Spatiality in Our Times*, June 2012. https://www.sic-journal.org/Article/Index/123.

Morris, Robert. 'The Present Tense of Space'. In *Continuous Project Altered Daily – The Writings of Robert Morris (October Books)*, edited by Robert Morris. MIT Press, 1994.

Pallasmaa, Juhani. *The Eyes of the Skin: Architecture and the Senses*. 2nd edition. Chichester; Hoboken, NJ: Wiley, 2005.

Roden, Steve, and Brandon LaBelle. *Site of Sound: Of Architecture and the Ear*. 2nd edition. Smart Art Press, 1999.

Thibaud, Jean-Paul. 'A Sonic Paradigm of Urban Ambiances'. *Journal of Sonic Studies* 1, no. 1 (1 October 2011): 1–14.

Thibaud, Jean-Paul. 'The Three Dynamics of Urban Ambiances'. In *Sites of Sound: Of Architecture and the Ear Vol. 2*, edited by Brandon Labelle &. Martinho, 2:43–53. Errant Bodies Press, Berlin, 2011.

Truax, Barry. *Acoustic Communication*. Bloomsbury Academic, 2001.

32
NOTES FROM THE FAR FIELD

Philip Samartzis

Introduction

The contemporary approach to fieldwork as a method of observing and registering environmental sound is in many ways an elaboration of practices established by musicologists, ethnographers and bioacousticians, who since the late 19th century have experimented with new applications of technology. Bela Bartók,[1] and Alan Lomax[2] are renowned for their early deployment of portable sound technology, often in remote rural contexts, to study and preserve people and traditions including folk music and stories, customs, and rituals. In these instances, the recordings are as interesting for the places in which they occur, as for the principal subject matter, for their integration of social, material, and spatial information. American herpetologist Charles Bogert[3] applied a similar approach to recording frog vocalisations to understand the way different species communicate within densely sounding biospheres. The opportunities made available by technology to document different places afforded new forms of encounter and expression affected by location attributes and dynamics. Advances in portability and fidelity furthered engagement with the world that was increasingly attractive to audiences interested in cultural heritage, politics, technology, and natural history.

Radio and film producers also used the immediacy of fieldwork to create playful and poetic outcomes such as the Hörspiel (radio play) work Weekend (1930)[4] by Walter Ruttmann which drew on Berlin Street life, and Etudes aux Chemins de Fer (1948)[5] by Pierre Schaeffer based on the choreography of steam trains at the Gare des Batignolles in Paris. Both composers drew on the agility of being in situ to establish new methods for composition using editing, juxtaposition, and haptic manipulation to advance experimental narrative forms. While real-world sound became the raw material for abstraction as evidenced by the evolution of musique concrète and its transformation of sound matter through expressive electroacoustic process, a turn towards more transparent representations of place gradually emerged. Led by artists interested in understanding the anthropogenic impact on urban and natural ecologies, new compositional methods evolved articulating expansive and nuanced evocations of space. Straddling sound art and acoustic ecology, A Sound Map of the Hudson River (1989)[6] by Annea Lockwood, offers an aural journey from Lake Tear of the Clouds in the high peak area of the Adirondacks, all the way downstream to the Lower Bay and eventually the Atlantic. The work traces the specific contours and textures contained along the river formed by the terrain and weather, and, downstream, the human environment whose sounds are woven into the river's porous soundscape. Peter Cusack on the other hand concentrates on the environmental catastrophe caused by invasive human practices by documenting nuclear, military and greenhouse gas sites for his Sounds from Dangerous Places project (2003 – current).[7] Whereas these examples are concerned with clearly delineated subject matter, more recent approaches to sound recording have considered the interaction between human and non-human actors, liminal states, and amorphous behaviours as important. Intérieurs (2020)[8] by Eric La Casa explores the architectural spaces of Paris to express his encounter with the everyday by positing fullness against emptiness, control against indeterminacy, to map the spatial dimensions of these enclosed worlds in which a confluence of acoustic interactions complicates the way place is experienced. By applying different methods and philosophies of recording and sharpening the focus on specific networks of sound agents, and their causes and affect, field recording offers new access points and wayfinding to navigate increasing social and environmental complexity.

The ability of sound recordists to work effectively in variable conditions and challenging topography is due to significant advances in electronics that have underpinned the production of increasingly portable and durable equipment. These developments have enabled new opportunities to work in a wider range of locations drawing on discrete and adaptive recording processes. While the direct-to-disk recording systems used by Alan Lomax in the 1930s offered the means to undertake location recording, the technology itself was heavy, cumbersome, and generally insensitive, constraining the parameters of what could be recorded. In addition, the poor fidelity and inherent noise of the recording system, and the limited capacity of the disk media established a methodology suited for short form, intensely performed material. The introduction of light-weight portable tape-recording systems in the 1950s facilitated new opportunities to record in increasingly difficult terrain as demonstrated by Charles Bogert's work in deserts, mountains, and wetlands. The tape recorder also had the capacity to produce longer recordings which accommodated temporally complex source material. As tape recorders evolved, they became smaller and cheaper, and better sounding, enabling a wider range of users to apply them for different purposes that reflected divergent formal and aesthetic interests. Advanced Environmental Control (1995)[9] by Marcus Behrens provides an enigmatic encounter of an urban environment using a tape recorder hidden inside a pocket. In order to question systems used for traffic control and pedestrian behaviour, Behrens deploys the hidden recorder while walking to discreetly pick up traces of external sound, rustling clothes and body movement that culminates in an oppressive study of urban space. With the introduction of new digital audio tape and hard disk recording systems in the 1990s, and more efficient battery technology, sound recordists gradually edged towards the margins of the planet encountering increasingly volatile conditions. Douglas Quin travelled to Antarctica three times to make hydrophone recordings of Weddell seals and emperor penguins around McMurdo Station culminating in the seminal CD Antarctica (1998).[10] On the opposite side of the world, Max Eastly travelled three times to the high Arctic in the 2000s on a sailing schooner to make recordings of bearded seal and walrus, ice, and wind as part of the Cape Farewell Climate is Culture project.[11]

In parallel with the evolution of portable recording systems are the advances made in microphone technology that has enabled complex environmental behaviours and characteristics to be registered in increasingly sensitive ways. For Compositions Ornithologiques (1996)[12] Bernard Fort recorded woodpeckers, thrush and wren sounding in a Canadian forest, demonstrating new levels of detail through the focused lens of a parabolic microphone system. Brames, Et Autres Mouvements D'Automne (2012)[13] by Marc and Olivier Namblard is a study of space in which several synchronised recording devices are used to capture the movement and vocalisations of moose. The final composite mix establishes complex spatial behaviour operating within the animal's woodland habitat achieved through a play of depth and panorama. Led by Ludwig Berger, the Institute of Landscape and Urban Studies at Zurich University undertook a multiyear investigation called Bodies of Water (2023)[14] documenting glaciers, snowfields, lakes, rivers, and a hydroelectric scheme, to understand the spaces, materials, and technologies comprising the Swiss soundscape. Sensing technologies including geophones, hydrophones, electromagnetic receivers, stereo, and ambisonic microphones were deployed to render immersive experiences of the powerful forces operating in these locations. Expanding on the material investigation of things and spaces Toshiya Tsunoda uses a ceramic piezoelectric contact microphone to record solid vibration occurring in glass, plastic, and metal objects, and structures such as warehouses and cyclone fences to reveal the way these materials absorb and disperse sound. Extract From Field Recording Archive (2019)[15] is an expansive collection of recordings made in and around the Port of Yokohama since 1995.

The convenience and sophistication of contemporary audio technology offers artists the opportunity to work exclusively in the field eschewing the affordances of the traditional sound studio. While it offers advantages for some types of recording and monitoring situations, the recording studio generally constrains how these tasks are performed due to its highly engineered environment designed to promote sound isolation, dispersion, and absorption. The field on the other hand demands highly adaptive responses to negotiate indeterminate conditions and hidden tensions. Rather than consider these as undesirable, fieldwork offers an opportunity to be in the world, to observe and contest dominant narratives, and through improvised responses and participatory practices ameliorate aleatoric forces. These destabilising factors become more acute when extreme climate and weather, human incursion, and social and cultural differences converge at the margins of our planet. A zone of rupture and divergence marked by tension, disquiet, and exclusion. Here on the edge of elsewhere, I share my investigations of three places comprising remarkable landscapes and resilient communities unsettled by environmental dissonance and invasive human action.

Polar patterns

Casey Station, Eastern Antarctica
66° 16' 55" S, 110° 31' 36" E
28 January to 19 February 2016

Every sound seemed frozen and the silence stagnant.

(Frank Hurley, 1912)[16]

Located 3,880 km south of Perth and overlooking the low, rocky Windmill Islands is Casey Station, the biggest of three Australian Antarctic research stations situated in Eastern Antarctica. Established in 1969, the station is located on a craggy outcrop at the base of the Antarctic ice sheet. Casey is the site of large lichen and moss beds which are recognised as among the most significant vegetated areas on the continent. East of the station is Law Dome gently rising to a peak of 1,400 m, and in the west is Shirley Island colonised by a boisterous population of Adélie penguins. Directly across is Newcomb Bay where the abandoned and mostly buried US station Wilkes is located, and just beyond it a horizon filled with icebergs of assorted shapes and sizes. Wilkins Runway is approximately 70 km southeast and serves as a desolate terminal for the Intercontinental Air Service. It takes a four-and-a-half-hour flight to reach Wilkins from Hobart and a further four hours to reach Casey by Hägglund. The icy terrain between Wilkins and Casey is flat with only rutted caterpillar tracks and sparse waypoint markers disrupting the pristine vista of white striae set against a deep blue sky (Figures 32.1 and 32.2).

Figure 32.1 Wilkes Station.

Figure 32.2 Casey Station.

At the height of summer, daytime temperatures oftentimes hover above freezing while in winter they fall into the minus 20s and 30s °C. The hours of daylight similarly vary throughout the year. In mid-summer the sun stays above the horizon almost continuously and in mid-winter, the sun appears for less than an hour each day. Katabatic wind is prominent at Casey due to its proximity to Law Dome whereby its velocity fluctuates between temperate and intense. Katabatic wind is a low gravity wind commencing at the south pole that radiates outwards towards the coast, gaining force as it travels down elevated slopes. When the cooler temperature of a katabatic mixes with the warmer temperature of the onshore wind, a very unstable weather system emerges making Casey a fascinating place for meteorological observation and weather forecasting.

The research station which can support up to 120 expeditioners, comprises a series of colour-coded prefabricated modular buildings used for accommodation, recreation, operations, science, meteorology, utilities and storage. Inside the braced steel framed and insulated panel buildings, a silence pervades that imposes a profound sense of isolation from the external environment. Outside the volatility caused by extreme weather and climate including freezing temperatures and high velocity wind is expressed through a variety of resonances emitted by assorted structures and materials undergoing tremendous stress. Located at the edge of the main complex is an incinerator used to burn station waste and a geodesic dome that conceals a dish antenna for terrestrial communication. The dome made of finely perforated mesh protects the satellite from the harsh conditions and produces low-frequency beating in response to powerful wind gusts. A quiet zone marks the end of station limits where a multi-frequency spaced array radar system is used to measure upper atmospheric conditions through the transmission and reception of coded sine tone pulsations. The time difference between these signals as well as the returned pulse shape characteristics provide information on the velocity and direction of dynamic middle atmospheric forces such as turbulent winds caused by the effects of heating and cooling. Adjacent to the interconnected series of towers and cables of the radar system is a fuel dump and a corroded concrete dock used for water operations around Newcomb Bay. Bordering the dock are a series of ice cliffs that extrude large chunks of ice into the bay, accumulating, and dissipating with the wind, current and tide (Figure 32.3).

My fieldwork in Antarctica draws on the photography of Herbert Ponting and Frank Hurley for inspiration. Their capacity to work in extreme conditions to capture otherworldly landscapes demonstrates the capacity of the medium to render profoundly poetic representations of place. I am particularly intrigued by Hurley's depictions of life on the ice through two iconic photographs, The Blizzard, and Leaning on the Wind both taken in 1912. The photographs convey the ferocity and atmospheric effects of the polar environment using a mix of techniques including staged scenes and composite printing to viscerally express conditions that are mostly impossible to articulate through conventional documentary photography. The construction of narrative through darkroom manipulation allowed Hurley to compose dramatic scenes by merging different details and vantage points. Inspired by these evocative depictions of abstract landscapes and volatile conditions I attempt an equivalent account using sound recording to render an embodied and affective experience of the ice continent.

The time I spend reading the polar environment acutely sharpens my sense of audition which I am constantly drawing on for self-preservation. The dynamic wind and biting cold create variable modes of encounter operating at different states of amplitude and intensity. The intermittent stillness that shrouds Casey provides a baseline for attentive listening to discern local signifiers sounding within the perimeters of the station. The quiet achieved by the accumulation of ice and snow however is often disrupted by the inclement weather which affects sound propagation leading to complex aural and spatial cues. The diesel engines of the main powerhouse that are so vital

Figure 32.3 Newcomb Bay.

for sustaining life generate a sound continuum that operates as another baseline to measure the environment. Their deep omnipresent thrum radiating all the way out to station limits offering an audible means of wayfinding during poor visibility. Within its radius assorted mechanical and industrial sound is produced by construction and remediation works, infrastructure distributing water and power, and transport operations. In addition, activity occurring in workshops and labs and around radar and scientific instrumentation, provide additional sources of anthropogenic sound. The morphology of these human and technological sounds filtering through the generally muted polar soundscape provides an abrasive but comforting experience to ward off the deep isolation enveloping the station.

With the advent of snowfall, the industrial exuberance is firmly suppressed. As the snow intensifies it becomes more audible, triggering various surfaces with a gentle but persistent patter. As the wind exerts its influence the patter becomes increasingly strident with fuel drums, restraining cables, and steel crates filtering the sound into a syncopated series of resonant patterns. Depending on temperature and wind speed snow can rapidly transform into hardened particles to generate showers of oscillating noise as ice granules forcefully collide into various objects. The effect is particularly notable on large sheets of heavy plastic used to protect various building materials and tools. The sound of people moving around Casey is also quite distinct as they negotiate snow flurries, dirty ice, and rock-strewn paths wearing heavily reinforced Baffin boots, and stiff, oversized winter jackets.

The variable intensity of katabatic wind determines the way sound is experienced through different gradients of definition and coherence. At its most intense it simply obliterates everything within its path. A high-intensity collision with the research station transforms the environment into an intense series of ascending and descending pitches that sound like a supercharged Aeolian harp. Inside time and space are distilled into a series of discrete events: a howling air vent, convulsing wall, bursts of radio static. Each sound appears to occur in complete solitude. While sheltering in an ice-encrusted cold porch, I am told that wind gusts are exceeding 185 KPH. The piercing shrieks of a stricken anemometer emerging from the white abyss are testimony to its ferocity.

Over three weeks I recorded an assortment of material activated by wind and shaped by the cold. Wind gusting across desolate ice fields, billowing plastic sheets, murmuring cables, ice granules striking objects, and the transformative effect of warming and cooling on the polar environment. Over a century has passed since Hurley documented the dramatic events at Cape Denison where he pushed the limits of photography to express the chaos he incurred on the ice. As Hurley has demonstrated, katabatic wind is as intriguing for its evocation of atmosphere, as it is for its capacity to express nature in extremis. Wind and cold are deeply visceral forces not easily rendered then or now, but I am hopeful my recordings offer access to the frozen sounds and stagnant silences identified by Hurley as being integral to the experience and understanding of the ice continent.

Topology of dreams

Newman, Western Australia
23.3575° S, 119.7303° E
September 1 to 16 2022

> A deathly silence prevailed. There was not a breath, not a birdsong to be heard, not a rustle, nothing. And although it now grew lighter once more, the sun, which was at its zenith, remained hidden behind the banners of pollen-fine dust that hung for a long time in the air. This, I thought, will be what is left after the earth has ground itself down.
>
> (Sebald, 2002)[17]

Located 1,200 km north-east of Perth and bordering the Great Sandy Desert is the remote mining town of Newman servicing The Pilbara region of Western Australia. The Pilbara is a vast stretch of land comprising one of Earth's oldest blocks of continental crust dating back more than 3 billion years. The morphology of the ancient crust leads to a distinctive landscape, with light-coloured granite domes surrounded by dark belts of volcanic and sedimentary rocks.[18] The region is well known for its rich, ancient Aboriginal history extending over at least 40,000 years. The Nyiyaparli people are the traditional owners of the land and waters of the East Pilbara region, which includes the township. Bordering Newman is Mount Whaleback Mine, owned and operated by Broken Hill Propriety Company Ltd (BHP). It is the largest single open-cut iron ore mine in the world at 5½ km long, 2½ km wide, and 500 m deep. The township services this and other significant iron ore producers including the Roy Hill and Tom Price mines operating in the vicinity (Figure 32.4).

Newman was established in 1966 as a company town to support the development of iron ore deposits at Mount Whaleback. Its architecture reflects the modernist styles of the period, practical and free of embellishment. The houses usually comprise two prefabricated halves inserted together into a steel I-section frame, a design that resists cyclone winds which can affect the region. Most houses are elevated by a few steps to help with air circulation and have large air-conditioning

Figure 32.4 Mount Whaleback Mine.

units to provide cooling against the very hot summer temperatures. Newman has a desert climate with temperatures regularly exceeding 38°C in summer. The annual average rainfall indicates a semi-arid climate except that its high evapotranspiration makes it a desert climate. Precipitation is sparse, but the influx of monsoonal moisture in the summer raises humidity levels and can cause occasional heavy storms.

The flightpath from Perth to Newman offers an expansive view of the interior topology of The Pilbara comprising abstract geometry devoured by time and climate. A patina of red dust greets me upon landing, lightly covering the regional airport, and eventually settling in the back of my throat. A feeling that would persist for the duration of my trip. I have travelled to Newman to participate in a collaborative art project called Kulininpalaju which includes Martumili artists who when in town are based at the East Pilbara Art Centre, an impressive multipurpose steel building in which various cultural activities occur including exhibitions, workshops, painting, gathering and public events. It is my first time in the region, and I am impressed by the industrious nature of the mining town. Mount Whaleback Mine is only a couple of kilometres from town centre which is easily identified by the amount of activity operating within its vicinity. From Radio Hill Lookout I view plumes of red dust emerging from the crest of the mine similar to that of a smouldering volcano. Surrounding it are vehicles and infrastructure of enormous proportions, purposefully designed and implemented to maximise the yield of iron ore. I feel very small in this supersized part of the world. The unique process of gravitational upturn in which assorted rocks and minerals fused in peculiar ways while the planet was still forming provides an otherworldly landscape resembling Mars more than it does Earth. The network of public and private roads in and around town are populated by an abundance of white utility vehicles, road trains, oversized trucks laden with mining equipment and spare parts, campervans and caravans, and buses ferrying workers to and from workplaces. A dense arrangement of rail tracks and sidings is used to service trains transporting iron ore to Port Hedland located 450 km north of Newman. The average train measures 3 km long and consists of four locomotives, 268 iron ore cars and one driver. A fully loaded train pulls a payload of around 42,000 tonnes of ore. Six trains fill a bulk carrier.

My time in The Pilbara offers a unique opportunity to get to know Martu artists and observe their painting practice in which different features of the surrounding landscape are expressed through the deft use of colour, form, and repetition. I use their lexicon of dreams to guide me in the field where I focus on the sounds of technology, nature, and atmospheric events. I soon get in the habit of recording between dusk and dawn in a place that never seems to sleep to register phenomena unique to this part of the world. The widespread nature of mining however means I can never quite escape its obtrusive presence regardless of how far I travel. Rather than ignore these sounds, I end up tracing them back to their source – the mines, the vehicles, the trains – to explore ways in which they play across the landscape over vast distances. In this way, I seek to create a conversation between the anthropogenic and the natural to reveal the complexity of the industrialised soundscape that restlessly occupies this ancient place (Figure 32.5).

The operations at Mount Whaleback Mine are central to the prosperity of the town and therefore become a focus of my recordings. While it is off limits to the public, a tour is provided daily that offers a close encounter with the mine and its attendant infrastructure and technology. The mine which originally stood at 805 m is now 482 m above sea level and is one of several mountain ranges in the vicinity being excavated. A steep red road leads to the crest of the mine winding

Figure 32.5 Fortescue River.

past piles of rocks, discarded machinery, and stationary vehicles. Water carts discharging 110,000 L of water per hour are used for dust suppression caused by unsealed roads. A viewing platform provides a panoramic sweep of mine operations in which a fleet of excavators, loaders and trucks extract and transport ore to the top of the mountain. Drill rigs used for exploration and blasting are set across the bottom of the recursively contoured site. The thrum of machinery is modulated by strong gusts of wind causing clouds of red dust. Adjacent to the viewing platform an anemometer groans and cyclone fencing and metal railing rattle. The road eventually winds back down towards the base of the site where a series of crushers process the ore to the size of grapefruit. A network of conveyor belts transports the crushed ore to a beneficiation plant for further processing and sorting before finally being loaded onto an ore cart. The site is saturated by machine noise produced by different material and spatial interactions that express the enormous power and scale of the operation.

Despite widespread human activity, the landscape contains many natural attributes and abundant wildlife. On my last night, I travel to Ophthalmia Dam to make recordings of water holes and caves located between the Eastern Ridge, and Jimblebar Mines. The night is filled with the call of green tree frogs sounding from various locations across the dark. Occasionally, a leaf-nosed bat appears to emit a series of high-frequency pulses. As dawn nears various birdcall penetrates the bucolic atmosphere including the wedge tail eagle, whistling kite, Australian reed warbler, spinifex pigeon and little grassbird. The gloom that signals the transition between night and day is punctured by a shock wave of infrasound produced by a slow-moving locomotive. Concealed by the woodland it radiates a piercing metal-on-metal screech as it slowly negotiates a curve leading to the main trunk line to Port Hedland. It takes seven minutes for the train to shudder and grind past, absorbing the delicate arrangement of birdsong composing the morning chorus. I eventually relocated to a nearby rock shelter used by 1,600 generations of Nyiyaparli people. It offers a place of refuge from the increasing heat, and a means of establishing distance from the mines by embedding myself into the rounded folds of liquified rock. While the vista of red sand, rocky domes, spinifex, and ghost gums suggest little has changed, the amorphous sound of heavy industry, and the succession of vehicles and trains moving across the horizon are indicators of a transformed landscape. My recordings evidence its rapid degradation by registering the attenuation of bioacoustic sound caused by ever-widening bands of noise disrupting the space–time continuum.

Energy fields

Bogong High Plains, Victoria
36.8647° S, 147.2792° E
July 11, 2021, to May 21, 2022

> We came to coax ghosts of the unsaid, the unsayable or the quietly said from our surrounds, tease out the traces of what is no longer here.
>
> (Taylor, 2016)[19]

Located 350 km north-east of Melbourne are the Bogong High Plains which form a part of the Great Dividing Range, a cordillera system consisting of an expansive collection of mountain ranges, plateaus and rolling hills, that runs parallel to the east coast of Australia. In winter it is one of the largest snow-covered areas in the country. The Bogong High Plains comprise heathlands, which cover some of the steeper and more wind-protected areas, and grasslands, which grow in exposed areas, and are typically more resistant to wind and frost. Some of the richest soils are

Figure 32.6 Bogong High Plains.

found in wetlands, defined as places where water is stationary for at least one month per year. The presence of water typically leads to highly organic soils, from the decay of vegetable matter, and very little mineral matter.[20] Snow patch herb fields occur in places where snow remains for a large portion of the summer but are very rare. Clustered across the plains are snow gums which are the hardiest and most cold-tolerant species of eucalyptus. The Bogong High Plains were extensively damaged by the Eastern Victorian alpine bushfires in 2003. The bushfire was started by 87 fires caused by lightning that were unable to be contained. They eventually joined to form one of the largest and longest fires in Victoria burning over 1 million hectares over a period of two months (Figure 32.6).

The traditional custodians of the land surrounding the Bogong High Plains are the indigenous Bidawal, Dhudhuroa, Gunai–Kurnai and Nindi–Ngudjam Ngarigu Monero peoples. Before the arrival of settler colonialists, these groups would travel to the high plains during summer to feast on the Bogong moth – an important seasonal food source and to participate in intertribal gatherings. Aside from the practice of agriculture and pastoralism, the biggest change to the landscape is due to the development of the Kiewa Hydroelectric Scheme, which commenced construction in the 1930s. It is the first scheme of its kind, and the second largest overall in mainland Australia. A series of aqueducts are used to capture mountain streams, which then channel their flows into catchments including Pretty Valley Pondage and Rocky Valley Dam. Since its inception, the scheme has evolved into four interconnected hydropower stations serviced by substations, transmission towers, rail sidings, dams, pipes, and networks of roads and tunnels (Figures 32.7 and 32.8).

Philip Samartzis

Figure 32.7 Pretty Valley.

Figure 32.8 Kiewa Valley.

Contained within the hydro scheme and located at an elevation of 1,800 m is Falls Creek Alpine Resort which serves as a gateway to the Bogong High Plains. The resort comprises chalets, lodges, and apartment complexes most of which were built from the 1980s onwards from corrugated iron, profiled metal, timber cladding, stone, plastered masonry and painted concrete block. Major wall colours are recessive in tone and sympathetic with vegetation and rock colours, with an emphasis on darker tones to blend in with the natural environment during the non-winter period and to provide a contrast against the winter backdrop during the snow season. Falls Creek is situated within sub-alpine woodland, dominated by snow gums and a dense shrubby understorey. Other vegetation types include alpine bog and vegetation associated with creeks and ephemeral streams. Commonly seen and heard around the resort is the Crimson Rosella, Flame Robin, Australian Magpie, Pied Currawong, and Australian Raven. A dense silence envelopes the resort during the off season broken only by the occasional sound of snow-melt pooling in drains, assorted materials and structures straining against the wind and cold, and birdcall and insect song sporadically sounding from different elevations within the landscape.

A notable feature of the region is the number of tree trunks still standing after the fires. Their grey, sepulchral forms dominate the sub-alpine woodland, and provide stark monuments to the ferocity of the fires that burnt here. When the north-east wind gusts across the ghost canopy, a melancholic sigh circulates through the withered landscape. New growth emerging from the root system of hollowed-out snow gums adds textural complexity caused by leaves and branches scraping against one another. The altitude and undulating topography further contours air flow and direction so it is experienced initially as an omnipresent hum that accelerates in intensity as it traverses the exposed landscape before receding into a distant, restless murmur. The sound of water is ever-present in the landscape – flowing, melting, dripping, splashing. Rain and snowmelt pools across lumpy heathlands slowly releasing water into streams, creeks, and billabongs. It is collected by aqueducts that gradually wind their way into large water catchments. The landscape, which mostly appears natural has been shaped by major earthworks to enable different structures and technologies to harness the collected water. The combination of exposed and concealed infrastructure adds to the complexity of the alpine soundscape through the activation of introduced materials. Water sounds across concrete aqueducts and retaining walls, and within underground pipes and drainage systems before being channelled into power-generating turbines drawing on a process of gravitational force. Wind plays restlessly through infrastructure including communication towers, gates and safety barriers, and various types of signage, causing them to groan, shudder and bang depending on wind velocity and direction. Moisture, wind and cold are unrelenting in this exposed part of the Alps. The harshness produced by altitude and climate culminates in a highly agitated soundscape in which the material sound of anthropogenic manipulation and environmental degradation is palpable.

Recreational tourism is popular in the region, particularly in winter sports. The Bogong High Plains is transformed by the quantity of visitors during the ski season which can number 10,000 people per day. The increased visitation is initially expressed by the flow of vehicles travelling on snow chains which are necessary to traverse the steep icy road leading into Falls Creek. Another signifier is the clamour generated by crowds of skiers wearing plastic boots, and carrying assorted skies, poles, and snowboards. The infrastructure used to transport people to the ski fields and to maintain the various runs provide an endless source of mechanical sound produced by chair lifts, snow-making machines, and snow grooming and clearing vehicles. Regular medical evacuations by a fleet of over-snow ambulances and helicopters are a solemn feature of the active winter soundscape. Many of the resort staff operating infrastructure use portable sound systems to play music to enhance the convivial atmosphere, although the din is generally ignored by most people.

I spend a year recording this site of energy production, adventure tourism, and ecological disaster and regeneration, in the company of the spirits of traditional owners displaced from this culturally significant place. Their absence made tangible by the Bogong moth I encounter during summer. Within its extremes is a solitude that is pervasive. One conveyed through the gentle ripples of naturally formed billabongs, and snow on columnar basalt formations formed 40–60 million years ago. By latticed icicles clinging to posts marking the highest publicly accessible road in Australia, and by wind ceaselessly roaring through gullies and over mountain peaks. Sound is ever present and available through the materials and spaces that make up this fragile place in which climate and weather, industrialisation, and tourism converge into a complex chorus of human and non-human exchanges. Tracing the many sounds that comprise the Bogong High Plains provides a way to consider these complex histories and interactions to construct new narratives and access points. Ones that counter the tenuous argument circulated by energy suppliers and resort management regarding the low impact of activities portrayed as sustainable. The environmental dissonance caused by these tensions offers a different form of encounter shaped by concrete and spectral realities and contrasts in atmosphere, tone, and texture.

Conclusion

Contemporary audio technology provides the capacity to record in places seldom seen or heard. By being in the world sound artists bear witness to change that is usually incremental but increasingly exponential evidenced by accelerated social and environmental transformation. The practice of fieldwork offers a mechanism to engage with and register change, and to preserve and advocate for these mutable ecologies and their attendant stakeholders. The recordings also afford audiences a chance to experience often highly regulated, hard-to-access places through different aesthetic and narrative forms, and immersive and affective encounters.

The ways people live and work in remote places progressively resemble the broader contemporary experience, in which strict protocols and hyper-vigilance are used to mitigate risk. The unpredictable nature of life in extremis that necessitates constant adaptation is in many ways how we live on the rest of the planet where our assumptions are regularly tested. The resilient communities who occupy these distant and fragile environments provide models of resistance that can help deepen understanding of loss and decay. Sound artists play an increasingly vital role in observing and recording the tension between climate, landscape, technology, and human action, to demonstrate the interconnectedness of things.

Sound files

These sounds are available on the Routledge E-Book page:

1. Casey Research Station was recorded during the most intense blizzard ever registered during the Austral summer at the station at that time (2016) with winds gusting around 100 knots.
2. Jimblebar Iron Ore Mine was recorded around 4 a.m. directly outside the gatehouse used for traffic control. The train featured in the recording is around 3 km long and comprises 230 cars each containing 100 tonnes of iron ore. The train is bordered by bat and frog call which progressively gets louder in an attempt to communicate above the sound of the passing train.
3. Bogong Powerstation Turbine Hall comprises a series of recordings made inside the power station while it was generating hydroelectricity. I managed to locate a microphone adjacent to the penstock used to spin the blades in a turbine, which, in turn, spins a generator that produces electricity.

Notes

1. Bartók, B. *Dance Suite/Hungarian Pictures/Two Pictures/Romanian Folk Dances/Romanian Dance* (2000).
2. Lomax, A. *Lomax Digital Archive* (2023).
3. Bogert, C.M. *Sounds of North American Frogs* (1998).
4. Ruttmann, W. *Weekend* (1994).
5. Schaeffer, P. Étude aux chemins de fer (1990).
6. Lockwood, A. *A Sound Map of the Hudson River* (1989).
7. Cusack, P. *Sounds from Dangerous Places* (2012).
8. La Casa, E. *Intérieurs* (2020).
9. Behrens, M. *Advanced Environmental Control* (1995).
10. Quin, D. *Antarctica* (1998).
11. Eastley, M. *Arctic* (2007).
12. Fort, B. *Compositions Ornithologiques* (1996).
13. Namblard, M & Namblard, O. Brames, *Et Autres Mouvements D'Automne* (2012).
14. Berger, L. *Bodies of Water* (2022).
15. Tsundoa, T. *Extract from Field Recording* (2019).
16. McGregor, A. *Frank Hurley: A Photographer's Life* (2019), p. 66.
17. Sebald, W.G. *The Rings of Saturn* (2002), p. 229.
18. Murphy, D., Allen, C., Schrank, C., & Wiemer, D. *How the Pilbara Was Formed More than 3 Billion Years Ago* (2018)
19. Taylor, E.K. *Journal #3* (2016)
20. Slattery, D. *The Australian Alps: Kosciuszko, Alpine and Namadgi National Parks* (2015), p. 68–70.

Bibliography

Bartók, B. *Dance Suite/Hungarian Pictures/Two Pictures/Romanian Folk Dances/Romanian Dance* [CD]. Ganarew: Nimbus Records, 2000.

Behrens, M. *Advanced Environmental Control* [CD]. Koblenz: trente oiseaux, 1995.

Berger, L. *Bodies of Water* [LP]. Zurich: Institute of Landscape Architecture, 2022.

Bogert, C.M. *Sounds of North American Frogs* [CD]. Washington DC: Smithsonian Folkways, 1998.

Cusack, P. *Sounds from Dangerous Places* [CD]. Thornton Heath: ReR Megacorp, 2012.

Eastley, M. *Arctic* [CDr]. Dorchester: Cape Farewell, 2007

Fort, B. *Compositions Ornithologiques* [CD]. Grenoble: 38e Rugissants Productions, 1996.

La Casa, E. *Intérieurs* [CD]. Paris: Swarming, 2020.

Lockwood, A. *A Sound Map of the Hudson River* [CD]. New York: Lovely Music, 1989.

Lomax, A. Lomax Digital Archive. Available at: https://archive.culturalequity.org/collections (accessed 10 February 2023).

McGregor, A. *Frank Hurley: A Photographer's Life*. Canberra: National Library of Australia, 2019.

Murphy, D., Allen, C., Schrank, C., & Wiemer, D. How the Pilbara Was Formed More Than 3 Billion Years Ago. Available at: https://theconversation.com/how-the-pilbara-was-formed-more-than-3-billion-years-ago-94977 (accessed 21 December 2022).

Namblard, M., & Namblard, O. Brames. *Et Autres Mouvements D'Automne* [CD]. Périgueux: Ouïe-Dire Production, 2012.

Quin, D. *Antarctica* [CD]. Seattle: Miramar, 1998.

Ruttmann, W. *Weekend* [CD]. Lormont: Metamkine, 1994.

Schaeffer, P. 'Étude aux chemins de fer', *L'Oeuvre Musicale* [CD]. Paris: INA-GRM, 1990.

Sebald, W.G. *The Rings of Saturn*. London: Vintage, 2002.

Slattery, D. *The Australian Alps: Kosciuszko, Alpine and Namadgi National Parks*. Clayton: CSIRO Publishing, 2015.

Taylor, E.K. *Journal #3*. Available at: https://bogongsound.com.au/artists/erin-k-taylor/entry-3 (accessed 20 January 2023).

Tsundoa, T. *Extract from Field Recording* [CD]. New Jersey: Erstwhile Records, 2019.

33
FLUID ARCHITECTURES AND AURAL SCULPTURALITY – TOWARDS AN AESTHETIC OF SONIC SPATIO-TEMPORAL ENVIRONMENTS

Gerriet Krishna Sharma

Sonic spatiology – traces and indications

Thinking about sound reminds the architect that human beings do not live in silence.
(Pallasmaa, *The Eyes of the Skin*, 1996)

We stand at the dawn of an era that will see the emancipation of architecture from matter. Music, especially computer music, will have much to teach the new 'liquid architecture' of cyberspace.

Together, architecture and music will stand as the arts closest to the functioning of the human cognitive and affective apparatus
(Novak, *Liquid Architecture in Cyberspace*, 1991)

Listen! Interiors are like large instruments, collecting sound, amplifying it, transmitting it elsewhere…
(Peter Zumthor, *Atmospheres: Architectural Environments*, 2006)

These three preceding quotes outline the field of tension in which the title of this chapter positions itself: architectures, whether virtual or real, embody ideas of how we understand, design, and sensually perceive living spaces of all kinds of use. In this context, the acoustic aspect, with its own traditions, utopias, and instruments, plays an increasingly significant and personal role, leaving traces in the design and experience of these different and new spaces. The following will pursue these acoustic traces.

Listening to the status quo of mediatised space

Daily life experiences are already augmented and will be increasingly informed by 'artificial' spaces such as the inside of cars, working spaces, laboratories, concert halls and even 'intelligent homes'. The central conceit of 'immersive audio'[1] is predicated on the use of auditory cues and their plausible spatial reproduction from an 'objective' or 'real' world, that is highly idealised, and questionably real. The development of advanced sonic (loudspeaker) environments – whether virtual or real – is most frequently directed by technological concerns, a focus upon physical modelling of signals and realistically simulated reproduction – less often towards aesthetic or perceptual characteristics, and seldomly, the creative potentials inherent to these systems. Virtual reality and spatial audio technologies have ushered in a new paradigm in the fields of games, music and sound art. Media developed with these technologies produce experiences beyond what is perceivable in the physical world, thereby extending our capacities to design and compose as well as our sensibilities for spatial and temporal perception. By operating in the spatio-temporal domain, these new media cause us to question our disciplinary understandings of space and time, as well as their aesthetics, which requires an altogether new post-disciplinary conception of design, composition, and experience. To fully understand the outlines of this new field one should concentrate on both: the reproduction of highly idealised 'natural' everyday life sound fields and at the same time developing strategies for arts, design and architecture to create alternative spaces with and in the environments yet to come.

What might it mean to design everyday XR environments with sound as a design parameter, using loudspeakers and virtual sound sources? How could sound practices aid us here?

Composing zones of intensities – reality in need of artistic Utopia

The technical evolution has been enormous within the past 60 years when it comes to sound field reproduction and also the creation of fictive spatial constellations. For example, how different reverb-qualities and times within one and the same pop song or ambient track, are forming a utopian co-existence of spatialities, became a part of the artistic sonic vocabulary. Accepted by a broad audience as common and normal, these constellations became part of the everyday experience. When dealing with sound that has been designed, we necessarily do so aesthetically (aesthetic appreciation as perceptual experience). Yet our symbiotic absorption of mediated 'realities' means we soon incorporate them into an updated range of realism (McArthur et al., *Sounds too true to be good*, p. 31).

Furthermore, by the use of loudspeaker arrays using Ambisonics, Wave Field Synthesis, Spatial or Dolby Atmos[2] artists can produce spaces and sculpture-like phenomena never been perceived outside the technical setup before, as such. We can determine that with today's technologies engineers' and artists' shared utopias from the past came close and we are only experiencing the beginning of possibilities.

Thus, historically we are now inside Edgard Varèse's utopia from 1936 (!)[3]:

> When new instruments will allow me to write music as I conceive it, the movement of sound-masses, of shining planes, will be clearly perceived in my work, [...]. Today, with the technical means that exist and are easily adaptable, the differentiation of the various masses and different planes as well as these beams of sound could be made discernible to the listener by means of certain acoustical arrangements... [permitting] the delimitation of what I call 'zones of intensities'.

With so-called 'immersive sound systems' we now have entrance to these zones of intensities. The materiality and the spatiality of sound have been changed dramatically. However, the substances of these spaces are multi-layered and need a different knowledge compared to former ideas of navigation, composition and research. Experience spaces, homes, meeting places and n-dimensional worlds – in the future, more than ever, we will decide what we surround ourselves with. But by that, we will not alter a virtual reality, we will change the everyday and how we perceive and understand it.

As composers of these spaces, we have to find a way of sharing experiences in immersive environments and ask ourselves what we really know about perception and perceptibility of spatial phenomena in n-D environments.

The following, I would like to illustrate, using two acoustic spatialisation techniques as examples, the new possibilities as well as challenges for composers and architects in the composition of space. The icosahedral loudspeaker (IKO) as an instrument for generating sculptural sound forms in resonance with the existing acoustic properties of architectures that I have been working with since 2009, and the artistic utilisation of the awarded Odio-App for creating reactive binaural sound architectures since 2022 as alternative everyday environments.

This constitutes a proposition aimed at advancing an alternative epistemological framework for the examination of spatial practices within the context of forthcoming (sonic) architectural developments.

Instrumentalising IKO – sounding out the space

Buildings do not react to our gaze, but they do return our sound back to our ears.
(Pallasmaa, *The Eyes of the Skin*, 1996)

The IKO consists of a housing carrying 20 individually driven loudspeakers. It was built in 2006 by Franz Zotter at IEM Graz, originally with the idea to holographically model musical instruments (Figure 33.1).[4] Gerriet K. Sharma and Franz Zotter developed the IKO into an instrument

Figure 33.1 IKO speaker.

and composition tool from 2009–2012 targeting new manifestations of spatial sound in contemporary computer music. As a result of this fruitful composer-acoustician dialogue, the compositions grrawe and firniss were developed. These two pieces were stepping stones towards thinking about the device and its acoustical principles as a medium of artistic expression and the orchestration of sound as space. From 2014 to 2018, Orchestrating Space by Icosahedral Loudspeaker (OSIL)[5] was a research project funded by the Austrian Research Fund within the Framework of the Programme for Arts-based Research (FWF/PEEK) at the Institute of Electronic Music and Acoustics Graz (IEM). Members of the research team were: Franz Zotter, Gerriet K. Sharma, Matthias Frank, Florian Wendt, and Markus Zaunschirm supported by Frank Schultz and supervised by Robert Höldrich. During this time, IKO was tested and performed within over 40 international concerts and was the subject of over 30 scientific and artistic research publications. In 2016 the Graz-based start-up sonible[6] developed a new amplification and loudspeaker system taking the IEM concept to a state-of-the-art production, performance, and marketing level. The IEM-sonible IKO is now available as musical instrument with unique and characteristic features for sound spatialisation and sonic sculpting. Common spatialisation systems for computer music employ loudspeaker arrays that surround the listening area, such as the BEAST (University of Birmingham), the Espro (IRCAM), the Klangdom (ZKM), the CUBE and the MUMUTH (KUG), to name only a few. These either use the psychoacoustic phenomenon of a phantom source to create auditory objects between the loudspeakers, e.g. VBAP (Vector Base Amplitude Panning) and Ambisonics, or aim at recreating a physically accurate sound field, e.g. wave field synthesis. The novelty in IKO in contrast to common surround loudspeaker systems lies in controlling the strengths of the wall reflections that can be excited from a single performer's location. That is, an IKO is employed as an instrument of adjustable directivity at this location. In electroacoustic music, the notion of adjustable-directivity loudspeakers was introduced in Paris in the late 1980s by researchers at IRCAM. For the renowned concept study 'la timée',[7] a cube housing six separately controlled loudspeakers was built to achieve freely controllable directivity. Despite the ingenious idea and theory, loudness and focusing strength weren't convincing enough to be employed in a lot of concerts. In 2006, researchers at IEM reconsidered the theory aiming at an acoustically correct and powerful reproduction of musical instruments in their lower registers, including the entire 3D directivity pattern. The resulting IKO is more powerful, of larger size, and deploys a larger number of loudspeakers than the IRCAM system. Moreover, success in quality was achieved by reconsidering algorithms and acoustic calibration to control sound beams that are three times narrower than those beams of earlier systems (Figure 33.2).

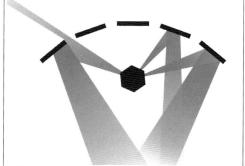

Figure 33.2 IKO beamforming graph.

Aural sculpturality

The IKO permits the formation of three-dimensional auditory objects. By applying the beamforming algorithm developed by Franz Zotter in Ambisonics[8] (3. Order), it is possible to project tightly bundled sounds onto floors, ceilings and walls, so that one not only hears sounds at or from the source itself (such as a violin or a conventional loudspeaker), but at projected and reflective point (mirror source). Here, individual loudspeakers are weighted by superposition so that sound projection in a desired (main)-direction is possible. Constructive interferences in the desired direction are formed during sound propagation, while the individual signals superimpose themselves where little or no sound projection is desired. The beams are freely adjustable in terms of direction. Therefore, different beams can be blended, or their beam width manipulated. Resulting plastic sound objects[9] can be moved around towards the reflecting surfaces or collapsed onto the IKO. These sculpture-like phenomena can be also composed to include remarkable gradation of depth, which was discovered during the composition process and proved in research with audience-listener responses.[10] Thus, with the IKO we can form flexible spatial objects not only as the mere output from a chain of membranes, but we are able to compose spatial sound entities that are almost impossible to create with any other instrument. Experiencing the IKO, first-time listeners describe sculptural sound phenomena that they have not been able to imagine before and even specialists in the field experience spatial events that are hardly describable by common terminology.[11] The sculptural possibilities have been investigated in several listening experiments[12] and disseminated in scientific papers as well as my publications: Composing Sculptural Sound Phenomena in Computer Music, also summarised in Aural Sculpturality – Spatio-temporal Phenomena within Auditive Media Techniques.[13]

These phenomena intricately hinge upon the acoustic characteristics of the space. As a result, the sound artist employing the IKO must adeptly interpret the architectural properties of the venue prior to each performance or installation. This involves utilising the soundcheck process to fine-tune the composition, ensuring that it reproduces the sculptural essence intended by the artist. The foundation underlying the articulation of these inherent 3D audio phenomena is accordingly encapsulated in the following thesis: The establishment of traditional body-space relationships[14] negotiated over centuries by sculptors and audiences can currently also be found, traced and even empirically proven in the auditory arts. Beyond materiality, it is about the specific formations that can create different spaces and spatial constellations. We call this potential 'Sculpturality'.

This process does not only include the understanding of the acoustic nuances but also the placement of IKO and the audience, as well as lighting and the co-presence or absence of a performer. During the cause of the OSIL investigations people involved in the research process gave workshops and lectures as well as engineering and compositional classes. One particular observation that was repeatedly made was that participants reported their changed awareness of architectural environments by listening to IKO. Thus, in the case of an IKO performance, the loudspeaker does not transmit the formerly composed and experienced, rather it creates experiences by activating and revealing the acoustic properties of the building. Herewith, we can become aware of architecture's materiality, shape and atmospheric potentials, acoustically.

Exploring shared perceptual spaces – intersubjectivity as the *spectrum of perception* in mediatised space

The flexibility that the IKO affords, in terms of creating bespoke environments, allows sound artists to explore a wide range of aesthetic possibilities, in acknowledgement of the differences in socialisations and expectations between members of the listening audience. It is therefore a matter of finding parameters for an intersubjective space for the perception of three-dimensional sound phenomena. The composer grapples with the extent to which a communicable and self-explanatory composition of dynamic sound objects is conceptually, theoretically, and practically achievable in the context of evolving (augmented) architectural scenarios, varying spatial descriptions, and diverse perceptions. Is there a realm within the domain of space-sound composition where my perception during the compositional process aligns with both the engineers' and the audience's perspectives? Can we articulate an approximate convergence in this space? And, furthermore, from which facets – be they linguistic, technical, artistic, or otherwise – can we approach and explore this intricate field?

Anyone who has spent a while working in a music laboratory has experienced the specialisation of their own perception that has very little in common with third parties' listening experiences and habits. This subjective experience can sometimes also take the form of acoustic illusions. My experience of teaching composers has often revealed to me that such distortions are frequent (Smalley, 'Spectromorphology: Explaining Sound-Shapes', p. 111). To convey this impression effectively, it is essential to explore strategies that foster a more consistent perception among third parties. In doing so, my emphasis is not on enforcing a singular description or a universally fixed form that is expected to resonate with everyone in an identical manner. Such an approach would be deemed unacceptable and regressive within the realms of art and music. Consequently, the focus is not on aligning perceptions or rigidly fixing modes of how perception should unfold. In this context, the artistic research frequently engages in a purposeful and productive conflict with engineers who advocate for the fixation of 'auditory objects' within a Euclidean space for their models. The essence lies in the nuanced layering of diverse perspectives and the incorporation of their respective descriptions of spatial sound objects throughout the compositional process. Therefore, we need an approach that tries to understand what is triggered, i.e., what perception spectrum we create, and which spatial categories exist for and in the listening experience of audiences, scientists and composers. The aim is to better understand variability and, through research (constructing models, verbalisation, new compositions), to get to know one's own sound material, forming objects with certain instruments and their staging in a different way through the assumption of a shared perceptual space (SPS).[15] The concept of the SPS, defined as the intersubjective space wherein perceptions of different listening groups intersect was utilised in the OSIL project incorporating artistic experience and psychoacoustic research.[16] OSIL conducted listening experiments that provide evidence for a common intersubjective perception of spatio-sonic phenomena created by IKO[17] and excited room reflections. The experiments were designed based on a hierarchical model of spatio-sonic phenomena that exhibit increasing complexity, ranging from single static sonic objects to combinations of multiple partly moving objects. The SPS is further proposed to research different aspects in the field of spatialised sound in so-called auditory virtual environments (AVEs). However, this is only one of a few exceptions[18] of how an investigation starting from the domain of music and sound art fostered a series of investigations in the scientific domain. The knowledge gained by these experiments was available and applicable in both 'worlds' but would not have been brought to the surface without the research questions intrinsic to the artistic practice. This continuous feedback then yields a plethora of further procedural questions that represent the very textures of spatial practices with AVEs. How do we describe these phenomena? How can we reproduce them on different AVEs with different spatialisation techniques (and if not, why?), and how do we communicate the knowledge and archive the results? To be able to describe this SPS, the search for references has to be extended to the adjacent fields of music and engineering, sciences like sociology, philosophy, cognitive linguistics, and architecture. In an artistic practice this can happen through provocation of experiences in the areas bordering terminology. Therefore, we shall look out for traces and artefacts resulting from ongoing processes in these fields.

Mobile headphonespaces – binaural worldmaking

In the present day, there is a growing availability of mobile media devices that enable individuals to have control and customisation over their sonic surroundings. They have the capability to filter, alter, and enhance the sounds encountered in our everyday lives, offering a more personalised and immersive auditory experience. As the newest development with the advent of on-board head-tracking in consumer headphones since 2021, it is now possible to adjust virtual sound source positions to the rotation of the head to the effect that we can compose and produce even more realistic or plausible soundscapes and composed scenes as the water stream, the reading voice, the fireplace, the coffee machine, or the piano stays in its position, in relation to the head. While traditional stereo displays cause sounds to rotate along with the head movement, a new category of audio experiences called 'binaural'[19] and 'head-tracked 3D-audio scenes' maintains the position of sound sources in virtual coordinates, creating a vivid sense of being in a lived space. Currently, these technologies are relatively costly, making them inaccessible to many. However, it is anticipated that they will become more affordable and widely available in the near future. As a result, most existing devices are likely to incorporate these advancements, as they become standardised and commonplace. In other words: The spatiality of sound has climbed another level of impact and importance in our mediatised[20] societies (Figures 33.3 and 33.4).

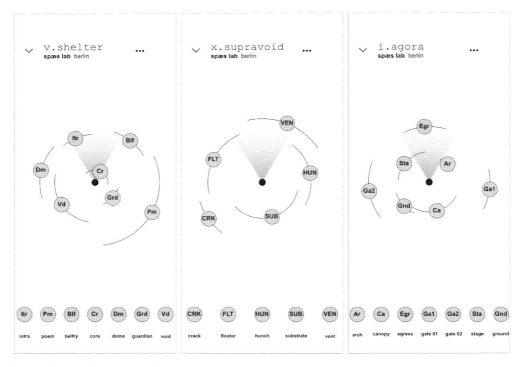

Figure 33.3 Odio-App (graphical schemes).

Fluid architectures and aural sculpturality

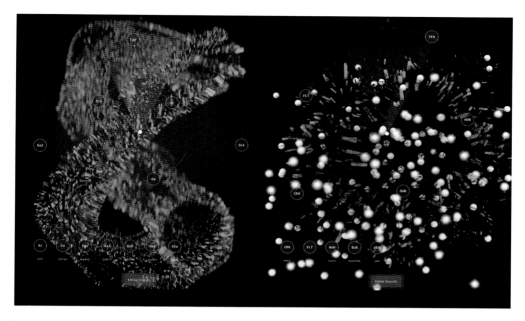

Figure 33.4 Odio-App (screenshots).

Since 2021, spæs Lab Berlin[21] and I have been collaborating with Volst, a Dutch software company that has developed Odio, a surprising 3D-Audio App utilising head-tracking built into the Apple AirPods, AirPods Max, Beats Studio Pro and virtual interactive sound sources with a visual interface for mobile phones and tablets. Binaural soundscapes are treated and distributed much like traditional tracks or songs, offering individually adjustable sound sources in a headphone-centric world. Through a simple touch and smooth finger-swipe on the circular representation of each sound source on the screen, listeners can seamlessly remix the aural architecture.[22] This interactive process engages with their auditory perception, granting them the power to reshape the soundscape according to their preferences. Seeing this technology as an artistic challenge and a tool for investigating alternative world-building, I worked with spæs lab Berlin on the creation of liquid architectures I. – X., a series of soundscapes for Odio-App. These exploratory compositions delve into the intersections of technologically induced realities that intertwine with our daily lives. Through the creation of liquid architectures, I constructed ten distinct utopian environments, each reactive and inherently unique. These synthetic and artificial spaces possess elements of familiarity but exhibit ever-changing qualities. They serve as commentary on the compelling shift away from traditional spatial concepts that have been emphasised and augmented through audio-visual mediations over centuries.

Shifting terminology within liquid architectures

In liquid architectures, every soundscape comprises five to eight distinct sound sources that collaboratively shape the aural architecture of the virtual environment. Each of these sound sources is identified by a specific indication, such as guardian (grd), core (cor), poem (poe), belfry (bel), void (voi), dome (dom), and intra (int), corresponding to the title e.g. shelter, of the particular soundscape. Each indication, linked to specific sounds within the scene, serves a dual purpose, assuming both concrete architectural functions and figurative connections with other sounds and their indications. As a result, every soundscape becomes a complex helix of terms, forming a spatial entity through the intricate layering and positioning of sounds. It seamlessly weaves itself into the built architecture, cocooning the listener within another shell, a realm of both plausible and utopian spatial descriptions. The morphing between these auditory shells is not technically induced; instead, it is triggered by the listener's shifting attention and focus. Each sound is developed to fit its soundscape, thoughtfully composed as a part of the larger spatial ensemble. Additionally, each sound varies in length and is seamlessly looped. Consequently, the soundscapes possess distinctive characteristics, atmospheres, and recognisable elements, continuously altering their textural consistency over time. The positions of the virtual sound sources are visually represented on the screen by objects, each depicted as circles bearing abbreviations of their respective indications. This visual mapping provides an intuitive means for the listener to engage with and navigate the multi-layered auditory environments. When a listener opens a soundscape, the arrangement of these objects appears as a pre-composed structure, forming an aural architecture that radiates concentrically from the idealised head position, represented at the centre of the screen (see graphic). These objects are positioned at distinct directions and distances from this central point, creating a unique spatial texture. Despite this initial arrangement, the listener enjoys complete freedom to interact with the soundscape. They can swipe and reposition these sound sources, multiplying or erasing them based on their personal situations, preferences, and needs. The volume of each object diminishes as they are placed farther away from the head, contributing to a sense of spatial depth and distance. This dynamic interaction allows the listener to actively shape and mould the auditory environment, creating a highly personalised and enveloping experience, tailored to their desires and sensibilities.

Shifting awareness within augmented spatial practices

While working on the series we asked people visiting spæs lab to try out these environments and comment on them. Participants from various backgrounds and with different preconceptions (e.g., disliking the brand, headphones, synthetic sounds, or having negative experiences with head-tracking) all exhibited a strikingly similar reaction when trying out this seemingly simple setup. Even before expressing their thoughts verbally, their bodies responded to the medium's impact. They effortlessly assimilated into the experience, their actions aligning with the altered soundscape, moments before their intellect began analysing the situation and comprehending the profoundness of the auditory shift. The visceral response surpassed any need for persuasion or convincing. Regardless of initial preferences or opinions about soundscapes and audio technology, participants intuitively embraced the transformative power of this binaural sound propagation. The effect was immediate and profound, making way for a deep and direct encounter with the 3D audio medium. Now, we can purchase consumer devices that we deliberately place into or onto our ears and forget about the act and the fact only moments after this. In the realm of binaural audio experiences with headphones, the certainty of sound presence transcends the notion of physical sources. Instantly after activating applications like 'Odio'[23] or 'Endel',[24] the mind constructs an augmented environment, seamlessly interweaving haptic and visual elements with everyday sounds that define the surroundings. From the gentle hum of a radiator to the faint rustle of a half-open window, from the subtle creaking of a door to the static appearance of a wall, and even the spaciousness of a lofty ceiling or the chirping of birds in a tree entangled with an Ambient drone, all these elements come together like a magical cape that we put on and disappear into a world morphed into the world. Consequently, there will be numerous instances where it becomes difficult to distinguish between reality and the virtual realms of the 'Metaverse'.

Seamless spatial drift

With the projected decrease in the size of average homes in urban areas,[25] the need for hyper-realistic spatial audio solutions becomes crucial for creating the illusion of larger spaces and for effectively blocking out disruptive noises from neighbours, streets, and skies. Such solutions will play a vital role in fostering a healthy and peaceful living environment. As a result, the trend towards living in personalised 3D audio bubbles will continue to grow. We realise that technologies and labour practices reshaped perception, absorbing and immobilising subjects through attentive practices aimed at production or consumption.[26] From this perspective, the act of isolating our ears using spatial audio technology appears less like a rebellious or entertainment-driven gesture and more like an essential and radical necessity for coping with the demands of modern living. As we adapt to smaller living spaces and increasingly busy urban environments, the ability to retreat into our own private auditory realms becomes increasingly valuable for maintaining peace of mind and well-being. Once more, the query arises concerning the expansion of our epistemological boundaries to encompass the continual processes, transformative shifts, fluid transitions, and drifting phenomena, with the aim of seamlessly incorporating them into the domain of spatial practices.

Toward aesthetic research in spatial practices

Reality is as much about aesthetic creation as it is about any other effect when we are talking about media.

(Sterne, *The Audible Past*, 241)

We can observe that a vivid research field deals with the perceptual aspects of spatial audio from an engineering viewpoint. It looks at their quantitative and qualitative analysis, and how it expands spatial audio vocabularies. Considering the broad field of computer performance, 3D spatial stability, loudspeaker layouts and projection paradigms, this seemed to be more than exigent and appropriate within the past. For an aesthetic debate – and no art can be spared by this – we need additional and another kind of knowledge, a certain typological consolidation of the intrinsic perceptive phenomena of musical works and these at a preferably high degree of generalisability.[27] Although listener-based research is not totally new within the field of Electronic Music[28] it is the exception rather than the rule. I claim that today with the powerful virtual environments working strongly on perceptual cues, investigations into aesthetic response must surely be at least as significant as the discovery of a composer's and designer's strategies, working methods, and tools.[29] We as composers and sound artists have never confronted a media machine of the collective, networked, and externally defined – (spatial) perception design as we experience it today. Concepts, definitions, and interpretations of space and of spatiality are essentially how we construct the world and by extension, how we create means to intervene into that world, with our daily practices. What do we share with our audience, the engineers and scientists working on perceptions in these media spaces, and how can we still detect potential for aesthetic experiences and make them useful for the sonic arts? Space multiplies its meanings and can become poetic artefacts that we collectively produce and reproduce through time, within our cultures, our sciences, and our arts. Through the dimension of time, these are the spaces that serve as vehicles for adaptation and transformation, from the scale of the individual to that of collective societies and of the environment at large. To facilitate the exploration of the conditions underlying these potentials, it is imperative to acknowledge the fundamental evolution that the concept of space has undergone in the past century.[30]

Experiencing and articulating hybrid spaces – verbalising sound as space

Space is existential,
existence is spatial

[Merleau-Ponty, 1966 (1945)]

The perception of sound emerges primarily within the auditory cortex, a consequence of intricate spatio-temporal processes within an embodied system. Notably, the term 'space' assumes a pivotal role in the compositional realm involving loudspeakers. Along with the introduction of the multi-channel technique and the first performances with complex loudspeaker arrangements, e.g. at the 1958 World's Fair, the subject was increasingly discussed.

> [...] In the context of analogue tape composition, the composer tries [...] to shape the direction and the movement of the sounds in space and to develop it as a new dimension for the musical experience. [...] We notice more and more that all musical ideas are becoming increasingly spatial.
>
> (Stockhausen, *Musik im Raum*, p. 153)

InTowards Understanding and Verbalizing Spatial Sound Phenomena in Electronic Music[31] I addressed the problem of terminology by summarising, comparing and evaluating different attempts to verbalise electro-acoustic phenomena in music. It is stated that authors with very different backgrounds over the last 100 years felt the need to describe and categorise sound phenomena but using very different methods and only very few of these address the subject of space directly.[32] Moreover, where the subject space is indeed addressed, the use is not contingent in most cases.[33]

Chion distinguishes between internal space, which is created in the composition, and external space that arises during the performance [Chion, *Les deux espaces de la musique concrete*, 1988]. Risset describes the fragility of an illusory space produced by the composer and the real space of the performance in which the illusory space is presented [Risset, *Quelques observations sur lèspace et la musique ajourd'hui*, 1998]. Smalley counts over 20 different spaces in electroacoustic music, for example, the composed space, the listening space, and the superimposed space [Smalley, *Space-Form and the Acousmatic Image*, 2007, 35ff]. Emmerson speaks of nested spaces [Emmerson, *Aural Landscape: Musical Space*, 1998] and space frames [Emmerson, *Local/Field and Beyond, The Scale of Spaces*, 2015], and Roads distinguishes between virtual and real spaces [Roads, *Composing Electronic Music – A New Aesthetic*, 2015]. Due to the fact that the concert space could be designed and played in differently by the positioning of loudspeakers, other spatial concepts were considered: Pierre Boulez speaks in favour of exploring more flexible spatial concepts that can change over the course of a piece:

> It seems to me that one of the most urgent objectives of present-day musical thought is the conception and realization of a relativity of the various musical spaces in use. [...] [T]he time has obviously come to explore variable spaces, spaces of mobile definition capable of evolving (by mutation or progressive transformation) during the course of a work.
>
> [Boulez, Boulez on music today, 1971 (1963)]

The versatile use of the term is not surprising, considering the fact that, parallel to the development of music, the concept of space has been given interdisciplinarily new and historically noteworthy consideration in the so-called Spatial Turn.[34]

Fluid architectures and aural sculpturality

Within contemporary discourse, a notable upswing is discernible in the examination of the concept of space, spanning various domains including philosophy, art, geography, cognitive linguistics, and diverse academic disciplines. A more general theory[35] of space distinguishes between absolutist and relativistic concepts of space. In the absolutistic concept, space exists independently of matter. Movable bodies and things are in a space that remains unmoving itself. The space exists continuously, for itself, and forms an equal, homogeneous basis for action for all. This idea of a container space has been replaced in science with the development of the relativity theory of relativistic spatial concepts. However, it still characterises the everyday understanding of space and is usually an indispensable condition for psycho-acoustical studies and ingenious scientific research in the area of spatial audio. In the relativistic concept of space, the space does not exist independently of the bodies. Instead, space is understood as a relation, as a relational structure between bodies. The bodies, whose arrangements give rise to one another, are in constant motion. Thus, the space itself is no longer static, but becomes processual and constantly changes over the course of time. Since the arrangement of bodies cannot be thought of independent from the observer's reference system, space is not absolute, rather always exists relative to the consciousness of the observer.[36] These seemingly disparate views on space come together harmoniously in every musical composition. Without the inclusion of physical space or its simulated representation, a thorough experiential engagement with a given work becomes unattainable. Simultaneously, the arrangement of musicians, orchestration of instruments, and the positioning of the audience create intertwined socio-cultural, historical, and conceptual configurations closely linked with individual or communal spatial perceptions.

In order to conceptualise and simultaneously approach prospective interpretations and compositions of sonic spaces within the domains of music, sound art, and architecture, I posit a hybrid spatial model for the creation of spatial sound compositions. This model is predicated on three interrelated stages (Figure 33.5):

1 Composition in Space
2 Composition with Space
3 Composition of Space

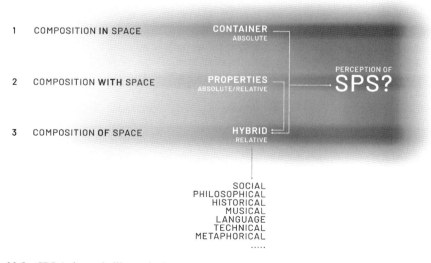

Figure 33.5 SPS (schematic illustration).

These three stages collectively constitute the formative elements of a synthesis of spatial concepts, contributing to the delineation of an SPS. This framework is not only intended to interpret already existing hybrid spaces. It is also designed to facilitate the creation of such spaces in the compositional process across various contexts and with different approaches. As space assumes an increasingly pivotal role in mediating divergent media and lived positions, the framework seeks to elucidate the procedural essence of this phenomenon and foster transparency in its understanding.

In the context of working with the IKO, the construction of space hinges on its acoustic properties (stage 2: Composition with Space), coupled with the characteristics of the employed sound materials. These sounds reference diverse origins and histories, such as environmental sounds, musical genres, and synthetic sounds, reintegrating traditional spatial models (e.g., social, historical, linguistic) in the process (stage 3: Composition of Space).

In contrast, the liquid architectures I. – X. created with the Odio-App diverge from incorporating the sonic attributes of the physical environment. Instead, they exclusively generate aural architecture in a virtual realm (stage 1: Composition in Space) through the projection of binaural sound sources. Nevertheless, these sounds also evoke distinct origins and histories, and they can be technically spatialised (e.g., through artificial reverb) in the process of space composition (stage 3). The proposed schema here aims to enumerate the various factors that contribute to the manifestations of (mediatised) hybrid space within the respective composition. Simultaneously, this endeavour aims to enable a more precise verbal characterisation of the utilised instrument or tool concerning its space-forming properties and facilitate a discourse on the subject.

FACIT – the phenomenological shift

If we wish to follow knowledge to the tips of its tentacles, then we must insist on examining this new nature on its own terms, postponing the assertion of what we know for the sake of finding out what is, and more, what can be.

(Novak, *The Music of Architecture*, 1992–2007)

According to Goodman worlds are made through processes of eversion that include composition and decomposition; weighting; ordering; deletion and supplementation; and deformation.[37] All these processes could be considered a confrontation with the world and a means whereby to produce alternatives to it. The domain of worldmaking is that of possibility. When we make worlds, we conject the Other. In doing so, we also help shape the world and its trajectory.[38]

In recent years, spatial audio technologies have gained prominence in various fields, including game design, online platforms like YouTube, and major companies such as Apple, Google, Sony, and Facebook. These technologies have also made their way into consumer electronics, event venues, and architectural spaces. It is highly probable that in the near future, AVEs will become a common part of everyday life, seamlessly blending into the reality of countless individuals.

A paradigm shift is underway; as technological constraints are rolled back, so must conceptual constraints be re-evaluated. Some of these are concerned with what spatiality actually is. Although we think of three-dimensional Euclidean space, it is by no means clear that this is anything other than a conceptual latecomer.

(however useful)[39]

The pervasive integration of spatial audio technology does not necessarily foretell a complete severance from physical reality or an abandonment of three-dimensional mathematical models in software development. However, it undeniably marks a noteworthy paradigm shift in phenomenology. As these tools become more affordable and accessible, they will undoubtedly impact our bodily experiences and mindsets beyond our common terminology. The emerging question is, who creates these environments and with which intentions, and how can music, sound art and architecture contribute with their own strategies to such a reality?

> However, to make a (different) world is to know (differently). Shifting perception then is the means by which to critique and question the world. We change the frame, change the perspective, and thereby change our understanding of it.[40]

We now find ourselves at a pivotal juncture, where we have access to a mobile, interactive, and deeply personal world-making experience that draws inspiration from past concepts of Euclidean and constructed spaces while seamlessly delving into the realms of procedural and imaginary environments. This convergence of possibilities simultaneously offers plausibility and a sense of lived reality within these novel liquid spaces. While some may view this as a playful and innovative journey, the consequences of these advancements are serious and unpredictable. Aural perception, sonic worldmaking, hearing and listening modes, and the relationship between the perceiving body and these new hybrid spaces will all face profound challenges and transformations. This is not a mere game; we are confronting the future of social engagement and interaction, the production and sharing of cultural experiences, and the growing sense of isolation and loneliness experienced by individuals.

The technical and aesthetical efficacy of IKO has been exemplified through more than 60 performances across diverse venues, featuring over 30 compositions within the last 15 years.[41] A succession of auditory investigations has accompanied these endeavours, revealing that both users and audiences undergo a discernible shift in their interaction with acoustic environments following a concert or an encounter with an IKO installation. This transformative experience not only fosters an enhanced awareness of a location's acoustic nuances but also serves as a testament to the concept of reinterpreting the built architecture into an alternative auditory structure. The seamless integration of IKO into the composition and mixing processes extends beyond the immediate sonic experience, leaving an enduring perceptual impact. Notably, this impact persists even in instances where the IKO is not actively employed. Individuals who have worked with the IKO continue to possess an acute awareness of architectural acoustics. The fact that only one speaker is used for this purpose, instead of an array of speakers distributed around the audience, emphasises the direction of observation towards personal perception and the characteristics and acoustic potential of the architecture, rather than the possibilities of technology.

The soundscapes of liquid architectures I. – X. seamlessly integrate with all architectural environments in which we find ourselves. Users can freely arrange the spatial sound elements of these soundscapes much like furniture, while also augmenting windows, doors, and the spatial environment itself with sound. This process emancipates the user, on the one hand, from the architectural constraints and acoustic exigencies of their surroundings and, at the same time, prompts a transformation in their personal spatial strategies concerning appropriation and active spatial configuration.

These two examples are intended to elucidate how continuous and interconnected artistic practices can open up alternative perspectives and experiential environments. The current objective is to share these experiences, engage in interdisciplinary discussions, and foster the creation of new works.

To be able to understand single aspects of instrumentality,[42] verbalisation, and sculpturality and how they are working together on spatial perception as well as to formulate their impact on the hybridisation of the composed space we need a shared knowledge base combining diverse backgrounds. We need shared teaching and research programs in which compositional, anthropological, and technological knowledge can be experientially conveyed and dissected. We need programs that aim to break down an experientially based vocabulary (glossary) for describing the phenomena for all participants. Verbalisation of spatial sound phenomena is crucial for developing an awareness of sound as space and can foster future conceptualisations and collective developments in the field. And we urgently require a shared conceptualisation of the preceding types of spaces we mean when we talk about 'space' in mediated spatial experiences. Therefore, I proposed a Hybrid Space Model for Spatial Sound Compositions integrating different backgrounds and approaches in the field of (sonic) spatial practices.

Composing sound as space

The thoughts and questions explored in this text are mere glimpses into a vast and intricate landscape, obscured by the jargon of marketing and multi-media outlets, attempting to tout this as the next 'grand innovation'. In essence, 3D sound serves as the vital ingredient to bring XR spaces to life, infusing them with presence and personality. We have to find a way of sharing experiences in immersive environments. What do we know about perception and perceptibility of spatial phenomena in n-D environments? Surely there is more than basic directional descriptions indexed to outdated reminiscences like 'Kick' or 'Snare' or 'Voice'. So how can we communicate spatial qualities intrinsic to these systems? To benefit from varying technical and artistic viewpoints, individuals involved in artistic practice and those involved in theoretical or applied research would need to engage in regular dialogue. This would have to happen on a level from which we could derive an original use of the immersive tools of our time, producing fundamental different and challenging artefacts. What we truly require are designers, architects, and composers who can shape this uncharted territory, cultivating a terra incognita that abounds with surprises and challenges, offering not just smooth and entertaining experiences in AVEs, but also thought-provoking and unconventional concepts for alternative social engagement and future cultural pathways. These approaches may serve as a counterpoint to the polished appearances of mainstream media products, instigating a fundamental aesthetic discourse on the immersive effects and phenomena we encounter. In doing so, we can extend an alternative and inclusive invitation to listeners, audiences, users, amateurs and specialists, beckoning them to share in an expanded ontology of sonic spatial arts, where sound is wielded as a medium to compose and shape space itself.

Notes

1. A critical essay on the notion of *immersion* can be found in: "Surrounded by Immersion – Means of Post-Democratic Warfare", *Ultra Black of Music*, Mille Plateaux, Frankfurt am Main 2020.
2. Spors, S., Wierstorf, H., Raake, A., Melchior, F., Frank, M., Zotter, F., "Spatial Sound With Loudspeakers and Its Perception: A Review of the Current State", *Proceedings of the IEEE*, 101(9), 1920–1938, 2013; Peters, N., Marentakis, G., McAdams, S., "Current Technologies and Compositional Practices for Spatialization: A Qualitative and Quantitative Analysis", *Computer Music Journal*, 35(1), 10–27, 2011.
3. Varèse, E., *The Liberation of Sound, Contemporary Composers on Contemporary Music*, Perseus Publishing, p. 197, 1998(1936).
4. Zotter, Franz, *Analysis and Synthesis of Sound-Radiation with Spherical Arrays*, PhD thesis. Institute of Electronic Music and Acoustics, University of Music and Performing Arts Graz, 2009.
5. PEEK (FWF/AR 328), online documentation:https://www.researchcatalogue.net/view/385081/958807

6 www.sonible.com.
7 http://www.entretemps.asso.fr/Timee/ (last visited Dec. 2023); Misdariis, N. et al., "Radiation Control on Multi-Loudspeaker Device: La Timée", in: *Proceedings ICMC*, 2001.
8 Zotter, F., *Analysis and Synthesis of Sound-Radiation with Spherical Arrays*, 2009; Zotter, F., Frank, M., "Investigations of Auditory Objects Caused by Directional Sound Sources in Room", *Acta Physica Polonica A*, 2015; Keller, B. D., Zotter, F., "A New Prototype for Sound Projection", *Fortschritte der Akustik*, DAGA, Nürnberg, 2015.
9 González-Arroyo, R., *Towards a Plastic Sound Object*, Nomos Verlag, 2012.
10 Frank, M., Sharma, G. K., Zotter, F., "What We Already Know about Spatialization with Compact Spherical Arrays as Variable-directivity Loudspeakers", in: *Proceedings inSonic*, 2015; Wendt, F. et al., "Perception of Spatial Sound Phenomena Created by the Icosahedral Loudspeaker", *Computer Music Journal*, 41(1), 76–88, 2017; Sharma, G. K., Frank, M., Zotter, F., "Evaluation of Three Auditory-Sculptural Qualities Created by an Icosahedral Loudspeaker", *Applied Sciences*, 9(13), 2698, 2019.
11 Landy, L., *Understanding the art of Sound Organization*, The MIT Press, 2007; Nyström, E., *Topology of Spatial Texture in the Acoustic Medium*. PhD thesis, City University London, 2013; Sharma, G. K., Frank, M., Zotter, F., "Towards Understanding and Verbalizing Spatial Sound Phenomena in Electronic Music", in: *Proceedings inSonic*, 2015.
12 Wendt, F., et al., "Perception of Spatial Sound Phenomena Created by the Icosahedral Loudspeaker", 2017; Sharma, G. K., Frank, M., Zotter, F., "Evaluation of Three Auditory-Sculptural Qualities Created by an Icosahedral Loudspeaker", 2019.
13 Sharma, G. K., *Aural Sculpturality. Spatio-temporal Phenomena within Auditive Media Techniques*. ZKM/Hertz-Labor, 2019; Sharma, G. K., *Composing with Sculptural Sound Phenomena in Computer Music*. PhD thesis, University for Music and Performing Arts Graz, 2016.
14 These relationships contrast the physical mass with the space. Both exist in a relationship of reciprocal relations. If one observes the development of sculpture in the history of the visual arts in this context, one can see how the volume of the body gradually opens up towards the space, tries to conquer it, and at last almost dissolves in it. Therefore, the space is not only space and a shell, but also an active co-designer of the sculpture since the modernity.
15 Sharma, G. K., *Composing with Sculptural Sound Phenomena in Computer Music*. PhD thesis, University for Music and Performing Arts Graz, 2016.
16 Wendt, F. et al., "Perception of Spatial Sound Phenomena Created by the Icosahedral Loudspeaker", 2017.
17 Zotter, F. et al., "A Beamformer to Play with Wall Reflections: The Icosahedral Loudspeaker", *Computer Music Journal*, 41(3), 50–68, 2017.
18 E.g.: Barrett, N., Crispino, M., "The Impact of 3-D Sound Spatialisation on Listeners' Understanding of Human Agency in Acousmatic Music in Acousmatic Music", *Journal of New Music Research*, 2018.
19 Blauert, Jens (ed.), *The Technology of Binaural Listening*, Berlin and Heidelberg: Springer, 2013.
20 I use the term Mediatization in accordance with: Hepp, A., Hjarvard, S., Lundby, K., "Mediatization: Theorizing the Interplay between Media, Culture, and Society", *Media, Culture & Society*, 2015.
21 Lab for Spatial Aesthetics in Sound Berlin: www.spaes.org // For 20 years, the members of spæs work through cross-disciplinary collaboration and experimentation with practitioners, artists, scholars, engineers, and researchers from all disciplines including arts, architecture, XR, games, music, and design, to explore the relationships between sound, space, and human experience from the perspective of aesthetics. This exchange of ideas, skills, and knowledge on the interface of science, practice, and academia allows us to cultivate a critical interdisciplinary reflection and to push the development and dissemination of practices within the field of aesthetics, space, and sound further. spæs thinks of spatial aesthetics as a day-to-day practice of world-making inherent to every compositional act with sound as one of its raw materials, and with loudspeakers in particular.
22 For the concept of *Aural Architecture*, I refer to Blesser, B., Salter, L., *Spaces Speak, Are You Listening?* MIT Press, 2006.
23 https://odio.app/
24 https://endel.io/
25 https://www.rentcafe.com/blog/rental-market/market-snapshots/national-average-apartment-size/ (website entered, Dec. 2023)
26 Hagood, M., *Hush: Media and Sonic Self-control*, p. 14, 2019.
27 Lynch, H., Sazdov, R., "A Perceptual Investigation into Spatialization Techniques Used in Multichannel Electroacoustic Music for Envelopment and Engulfment", *Computer Music Journal*, 41(1), 13–33, 2017;

Sharma, G. K., Zotter, F., Frank, M., "Towards Understanding and Verbalizing Spatial Sound Phenomena in Electronic Music", in: *Proceedings inSonic*, 2015; Normandeau, R., "Timbre Spatialisation: The Medium is the Space", *Organised Sound*, 14(3), 277–285, 2009.

28 Thies, W., *Grundlagen der Typologie der Klänge*, Hamburg: Verlag der Musikalienhandlung K.D. Wagner, 1982.
Merlier, B., "Vocabulary of Space Perception in Electroacoustic Musics Composed or Spatialised in Pentaphony", in: *Proceedings SMC*, 2008; Grill, T., *Perceptually Informed Organization of Textural Sounds*. PhD thesis, University for Music and Performing Arts Graz; Ratti, F. S., Bravo, C. F., "Space–Emotion in Acousmatic Music", *Organised Sound*, 22(3), 394–405, 2017.

29 Weale, R., "Discovering How Accessible Electroacoustic Music Can Be: the Intention/Reception Project", *Organised Sound*, 11(2), 189–200, 2006; Hill, A., *Interpreting Electroacoustic Interpreting Audio-visual Music*. PhD thesis, De Montfort University, 2013; Sharma, G. K., Zotter, F., Frank, M., "Towards Understanding and Verbalizing Spatial Sound Phenomena in Electronic Music", in: *Proceedings inSonic*, 2015; Landy, L., *Understanding the Art of Sound Organization*, MIT Press, 2007.

30 Guenzel, S., *Raum, Eine kulturwissenschaftliche Einführung*, transcript, 2018.
Guenzel, S., Liebe, M., Mersch, D., "The Space-Image – Interactivity and Spatiality of Computer Games", in: *Conference Proceedings of the Philosophy of Computer Games*, pp. 170–189, 2008.

31 Sharma, G. K., Zotter, F., Frank, M., "Towards Understanding and Verbalizing Spatial Sound Phenomena in Electronic Music", in: *Proceedings inSonic*, 2015.

32 Smalley, D., "Space-form and the Acousmatic Image", *Organised Sound*, 12(1), 35–38, 2007; Normandeau, R., "Timbre Spatialisation: The Medium is the Space", *Organised Sound*, 14(3), 277–285, 2009; Nyström, E., *Topology of Spatial Texture in the Acoustic Medium*. PhD thesis, City University London, 2013; Emmerson, S., "Aural Landscape: Musical Space", *Organised Sound*, 3(2), 135–140, 1998; Merlier, B., "Vocabulary of Space Perception in Electroacoustic Musics Composed or Spatialised in Pentaphony", in: *Proceedings SMC*, 2008.

33 Born, G., *Music, Sound and Space: Transformations of Public and Private Experience*, Cambridge University Press, 2013; Sharma, G. K., *Composing with Sculptural Sound Phenomena in Computer Music*. PhD thesis, University for Music and Performing Arts Graz, 2016.

34 Soja, E., *Spatial turn, Postmodern Geographies: The Reassertion of Space in Critical Social Theory*, London: Verso Press, 1989.

35 See Löw, M., *The Sociology of Space, Materiality, Social Structures, and Action*, p. 15, Palgrave, 2016(2001).

36 Löw, *The Sociology of Space*, p. 188, 2016.

37 Goodman, Nelson, *Ways of Worldmaking*. Vol. 51. Hackett Publishing, 1978.

38 Hosale, M-D., Murrani, S., de Campo, A., *Worldmaking as Techné, Participatory Art, Music, and Architecture, Library and Archives Canada Cataloguing in Publication*, Riverside Architectural Press, 2018.

39 Lennox, P., "Spatialization and Computermusic", in: *The Oxford Handbook of Computer Music*, ed. by Dean, R., p. 259, New York, 2009.

40 *Worldmaking as Techné*, Introduction, p. vii, 2018.

41 www.ikoweave.com

42 Hardjowirogo, S. I., "Musical Instruments in the 21st Century – Identities, Configurations, Practices", in: *Instrumentality, On the Construction of Instrumental Identity*, Singapore: Springer Nature, 2017.

Bibliography

Aarseth, E. J., "Allegories of Space. The Question of Spatiality in Computer Games", in: *Cybertext Yearbook 2000*, ed. by Koskimaa, R., Eskelinen, M., pp. 44–47, Jyväskylä: University of Jyväskylä, 2001.

Barrett, N., "Spatio-Musical Composition Strategies", *Organised Sound*, 7(3), 313–323, 2002.

Barrett, N., "Spatial Music Composition", in: *3D Audio*, 1st ed., ed. by Paterson, J., Lee, H., New York: Routledge, 2021.

Blauert, Jens (ed.), *The Technology of Binaural Listening*, Berlin, Heidelberg: Springer, 2013.

Blesser, B., Salter, L., *Spaces Speak, Are You Listening? Experiencing Aural Architecture*, Cambridge, MA: MIT Press, 2006.

Born, G., *Music, Sound and Space: Transformations of Public and Private Experience*, Cambridge: Cambridge University Press, 2013.

Boulez, P., *Boulez on Music Today*, [Musikdenken / Penser la musique aujourd'hui.] translated by Bradshaw, S., Bennett, R., 83–84, Cambridge, MA: Harvard University Press, 1971(1963).
Chion, M., *Les deux espaces de la musique concrète*, in: *Lien: L'espace du son,* LIEN I, ed. by Dhomont, F., Bruxelles: Musiques et recherches, 31–33, 1988.
Emmerson, S., "Aural Landscape: Musical Space", *Organised Sound*, 3(2), 135–140, 1998.
Emmerson, S., "Local/Field and Beyond, The Scale of Spaces", in: *Kompositionen für hörbaren Raum, Die frühe elektroakustische Musik und ihre Kontexte*, ed. by Brech, M., Paland, R., Bielefeld: transcript, 13–26, 2015.
González-Arroyo, R., "Towards a Plastic Sound Object", in: *Raum: Konzepte in den Künsten, Kultur- und Naturwissenschaften*, Baden-Baden: Nomos Verlag, 2012.
Goodman, N., *Ways of Worldmaking*. Vol. 51., Indianapolis, IN: Hackett Publishing, 1978.
Grill, T., *Perceptually Informed Organization of Textural Sounds*. PhD thesis, University for Music and Performing Arts Graz, 2012
Guenzel, S., Liebe, M., Mersch, D., "The Space-Image – Interactivity and Spatiality of Computer Games", in: *Conference Proceedings of the Philosophy of Computer Games*, pp. 170–189, Potsdam: University Press, 2008.
Guenzel, S., *Raum, Eine kulturwissenschaftliche Einführung*, Bielefeld: transcript, 2018.
Hagood, Mack, *Hush: Media and Sonic Self-Control*, Durham. NC: Duke University Press, 2019.
Hardjowirogo, S. I., "Musical Instruments in the 21st Century – Identities, Configurations, Practices", in: *Instrumentality, On the Construction of Instrumental Identity*, Singapore: Springer Nature, 2017.
Hepp, A., Hjarvard, S., Lundby, K., *Mediatization: Theorizing the Interplay between Media, Culture, and Society*, in: *Media, Culture & Society*, 37(2), 314–324, 2015.
Hill, A., *Interpreting Electroacoustic Interpreting Audio-visual Music*. PhD thesis, De Montfort University, 2013.
Hosale, M-D, Murrani, S., de Campo, A., *Worldmaking as Techné, Participatory Art, Music, And Architecture*, Toronto, Ontario: Riverside Architectural Press, 2018.
Keller, B. D., Zotter, F., "A New Prototype for Sound Projection", *Fortschritte der Akustik*, Nürnberg: DAGA, 2015.
Kendall, G. S., "Meaning in Electroacoustic Music and the Everyday Mind", *Organised Sound*, 15(1), 63–74, 2010.
Kendall, G. S., "Spatial Perception and Cognition in Multichannel Audio for Electroacoustic Music", *Organised Sound*, 15(3), 228–238, 2010.
Landy, L., *Understanding the Art of Sound Organization*, Cambridge: The MIT Press, 2007.
Lennox, P., "Spatialization and Computermusic", in: *The Oxford Handbook of Computer Music*, ed. by Dean, R., p. 259, New York: Oxford University Press, 2009.
Lynch, H., Sazdov, R., "A Perceptual Investigation into Spatialization Techniques Used in Multichannel Electroacoustic Music for Envelopment and Engulfment", *Computer Music Journal*, 41(1), 13–33, 2017.
McArthur, A., Stewart, R., Sandler, M., "Sounds Too True to be Good: Diegetic Infidelity – The Case for Sound in Virtual Reality", *Journal of Media Practice*, 18(1), 26–40, 2017.
Merleau-Ponty, M., *Phenomenology of Perception*. Berlin: de Gruyter 1966 [Paris 1945], 339.
Merlier, B., "Vocabulary of Space Perception in Electroacoustic Musics Composed or Spatialised in Pentaphony", in: *Proceedings smc08,* Berlin, 1–11, 2008.
Misdariis, N., Nicolas, F., Warusfel, O., Caussé, R., "Radiation Control on Multi-Loudspeaker Device: La Timée", in: *Proceedings ICMC 2001* (https://www.researchgate.net/publication/228987796_Radiation_Control_on_Multi-Loudspeaker_Device_La_Timee).
Normandeau, R., "Timbre Spatialisation: The Medium is the Space", *Organised Sound*, 14(3), 277–285, 2009.
Novak, M., "Liquid Architectures in Cyberspace", in: *Cyberspace: First Steps*, ed. by Benedikt, M., 225–254, Cambridge, MA: MIT Press, 1992.
Novak, M., *The Music of Architecture: Computation and Composition*, Santa Barbara: University of California, Media Arts and Technology Graduate Program, 1992–2007.
Nyström, E., *Topology of Spatial Texture in the Acoustic Medium*. PhD thesis, City University London, 2013.
Otondo, F., "Contemporary Trends in the Use of Space in Electroacoustic Music", *Organised Sound*, 13(1), 77–81, 2008.
Pallasmaa, J., *The Eyes of the Skin: Architecture and the Senses*, 3rd ed., Hoboken, NJ: John Wiley & Sons Inc, 2012 (1996).

Peters, N., Marentakis, G., McAdams, S., "Current Technologies and Compositional Practices for Spatialization: A Qualitative and Quantitative Analysis", *Computer Music Journal*, 35(1), 10–27, 2011.

Ratti, F. S., Bravo, C. F., "Space–Emotion in Acousmatic Music", *Organised Sound*, 22(3), 394–405, 2017.

Risset, J. C., *Quelques observations sur lèspace et la musique ajourd'hui*, in: *Lien: L'Espace du Son*, LIEN I, ed. by Dhomont, F., Bruxelles: Musiques et recherches, 21–22, 1998.

Roads, C., *Composing Electronic Music – A New Aesthetic*, p. 261, Oxford: Oxford University Press, 2015.

Sharma, G. K., *Composing with Sculptural Sound Phenomena in Computer Music*. Ph.D. dissertation, Univ. Music and Performing Arts, Graz, 2016, https://phaidra.kug.ac.at/o:66171

Sharma, G. K., "Surrounded by Immersion – Means of Post-Democratic Warfare", in: *Ultra Black of Music*, ed. by Mille Plateaux, Frankfurt am Main, 2020.

Sharma, G. K., Frank, M., Zotter, F., "Towards Understanding and Verbalizing Spatial Sound Phenomena in Electronic Music", *Proceedings inSonic 2015, Aesthetics of Spatial Audio in Sound, Music and Sound Art*, 2015.

Sharma, G. K., Frank, M., Zotter, F., "Evaluation of Three Auditory-Sculptural Qualities Created by an Icosahedral Loudspeaker", *Applied Sciences*, 9(13), 2698, 2019.

Sharma, G. K., Zotter, F., Frank, M., "Orchestrating Wall Reflections in Space by Icosahedral Loudspeaker: Findings from First Artistic Research Exploration", in: *Proceedings ICMC-SMC,* Athens, 2014.

Smalley, D., "Spectromorphology: Explaining Sound-Shapes", *Organised Sound*, 2(2), 107–126, 1997.

Smalley, D., "Space-Form and the Acousmatic Image", *Organised Sound*, 12(1), 35–58, 2007.

Spors, S., Wierstorf, H., Raake, A., Melchior, F., Frank, M., Zotter, F., "Spatial Sound with Loudspeakers and Its Perception: A Review of the Current State", *Proceedings of the IEEE*, 101(9), 1920–1938, 2013.

Sterne, J., *The Audible Past*, Durham, NC & London: Duke University Press, 2003.

Stockhausen, K. H., "Musik im Raum", in: *Texte zur elektronischen und instrumentalen Musik*, Band I, Köln, p. 158, 1958.

Thies, W., *Grundlagen der Typologie der Klänge*, Hamburg: Verlag der Musikalienhandlung K.D. Wagner, 1982.

Varèse, E., *The Liberation of Sound, Contemporary Composers on Contemporary Music*, New York: Perseus Publishing, 1998(1936).

Wendt, F., Sharma, G. K., Frank, M., Zotter, F., Höldrich, R., "Perception of Spatial Sound Phenomena Created by the Icosahedral Loudspeaker", *Computer Music Journal*, 41(1), 76–88, 2017.

Zotter, F., *Analysis and Synthesis of Sound-Radiation with Spherical Arrays*. Doctoral Thesis, Institute of Electronic Music and Acoustics, University of Music and Performing Arts Graz, 2009.

Zotter, F., Frank, M., "Investigations of Auditory Objects Caused by Directional Sound Sources in Rooms", *Acta Physica Polonica* A, 128(1-A), A-5–A-10, 2015.

Zumthor, P., *Atmospheres: Architectural Environments, Surrounding Objects*. Basel: Birkhäuser, 2006.

34

ACOUSTIC ATLAS – AN ORCHESTRA OF ECHOES

Cobi van Tonder

Mythical beginnings

The Greek myth of Echo, with its classical depiction in the Metamorphoses in 8 AD by the Roman poet Ovid, the myth of Echo begins with Juno (who is Hera in Greek mythology) cursing Echo, causing her to lose sovereignty over her voice so that she can no longer do anything but 'repeat the last of what is spoken and return the words she hears'.[1]

Echo's love for Narcissus is unrequited:

> Scorned, she wanders in the woods and hides her face in shame among the leaves, and from that time on lives in lonely caves. Her love endures, her sleepless thoughts waste her sad form, till only her voice remains, her bones, they say, were changed to shapes of stone. No longer to be seen, but to be heard by everyone. It is sound that lives in her.[2]

By becoming a phenomenon (or the metaphor for phenomena),[3] Echo becomes omnipresent, ubiquitous and a multiplicity by 'losing' herself to mirror 'the other'. Could there be a lesson in this? To listen, we must momentarily relinquish our sovereignty and become silent and empty. Perhaps even hold our breath to listen in moments of absolute stillness? In the same way that the interiors of caves and spaces all over the world hold the capacity to echo whichever sound enters it, by listening, we hold the capacity to echo (and colour with our individual set of experiences, memories, and abilities) in the interior spaces of our consciousness.

In his Musurgia Universalis, Athanasius Kircher focused on the 'capability to provoke the marvellous by means of sounds'. In Phonurgia Nova (1673), Kircher examined acoustic phenomena. He defined two meanings for echo – in the first case, echo is – reflected (or repeated) voice. The second meaning, as expressed in Latin, is resonance. This second meaning of echo refers to the behaviour of air in the cavities of a body, as within the Vitruvian vases or in sound chests.[4]

Kircher's definition of echo is helpful in combining **resonance** and reverberation under one term, 'echo', and simplifies an otherwise clumsy collection of phrases. In this chapter, I refer to echo in this way whilst also giving homage to the nymph as a metaphor for the invisible but felt 'genius loci' or feeling of a place.

Describing the world as echoes

To start this chapter, which is a loop, we must start in the middle and consider that there is no beginning, no introduction, nothing linear to this speculative exploration. Observations are presented based on my practice as a sound artist/composer/listener as a walk through a labyrinth with Echo. We walk simultaneously through ancient cities, current cities, hybrid old and new cities, deserts, mossy forest trails, and mountainous open sky views. Wherever we go, Echo is there. The intention is to describe the world as an orchestra of echoes, to listen, to hear with physical and imaginary ears, the potential of each pathway, street, corridor, doorway, entrance, transition, interior, and exterior, how sound moves like water, its interactions with materials causing resonance, spectral glow, quiet absorption, electric shocks of sonic light touching our senses. How sound waves move, collide, reflect, interfere, resonate, and get absorbed into the world is not so different from how sound moves inside a musical instrument. There is just a difference of scale, magnitude, and relativity. The fact that we, a singular listener, cannot hear it all and cannot be omnipresent is no longer an obstacle. We are a species with recording devices, and many sound artists and enthusiasts record every possible place on Earth. Can we hear every single keyboard tapping away right this moment sounding like a fish-eating-coral-reef rhythmic buzz? Most devices have built-in mics; technically, most devices are connected to the internet; technically, the multiplicity ear is growing every moment.

Whereas, in the past, I focused on field recording and sampling (of environmental sounds) as the source of sound worlds and compositional ideas, when viewing the world as an orchestra of echoes, the natural progression is to record impulse responses (IRs),[5] also termed acoustic fingerprints and use these as (meta) compositional process and/or material. An impulse such as a starter pistol, balloon pop or frequency sweep is used to stimulate the acoustics of a space. The reply contains an acoustic fingerprint of that source–listener spatial relationship with its values of frequency and resonance responses. It is also possible to completely simulate such IRs mathematically from visual information. The IR information is the equivalent of a 3D digital visual model of space but for sound. Through convolution, incoming audio signal can be convolved with the response of a space, thus placing us in a simulated acoustic virtual reality copy of that space.[6] As composers, we can create recombinant sonic architectures based on places that morph into each other or are juxtaposed next to each other. We can sing into a hybrid skyscraper-duomo-temple-tunnel that morphs into a mineshaft-silo-subway-staircase. We can place the percussion in an empty warehouse. We can sonically immerse our mezzo-soprano on the highest top of the Himalayas whilst our brass section echoes through the Grand Canyon. Attention can suddenly shift to a small Khoisan vocal click inside a seashell. We can go sonically where the cinematic eye has taken us. Via forensic acoustics, we can reverse engineer sound recordings from the past to use clips with footsteps to find the IRs of places from film scenes. We can try discovering what acoustic landscapes are revealed by lightning thunderclaps. We can go underwater to include underwater echoes. We can bury Echo underground in a tiny box, consider miniature and micro spaces, transpose a sound file, record the echo, and transpose it back to the original (Figure 34.1).

Recombinant acoustics – shapeshifting through space – morphology through acoustics

Changes are the most exciting moments in composition – how something changes, the duration, the rate of change, what is revealed and what stays the same, the surprise, the transformation. Similarly, changes in reverberation are most exciting when transitioning from one space to the

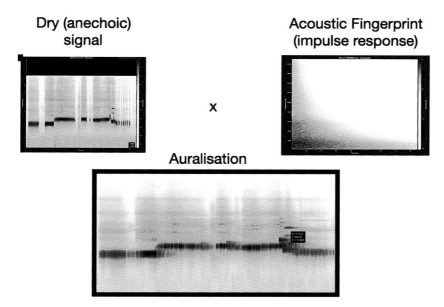

Figure 34.1 When convolving the incoming dry signal with the impulse response (IR) the end product is an 'Auralisation' – which sounds as if the dry signal is now in the virtual acoustic space.

next – think, for example, about being outdoors in a hilly landscape and entering a narrow opening into a cave. The sonic transition is as stark as going from daylight into darkness. Sonically, it might mean going from a vast soundscape of far-off insects, birds, wind in treetops or grass, perhaps the drone of a distant highway or aeroplane, with multiple subtle far-off echoes and shorter nearby reflections into a narrow sounding muffled tunnel; next into a very reverberant, large chamber of darkness and echo. Had we lived in a permanent state of that inner chamber of being acoustically enveloped, our ears would fatigue, and we would adjust the sounds we make. We might no longer notice it. Because cathedrals or caves are so starkly different from our everyday acoustic reality, they create awe and amazement when transitioning from a cityscape into such a liquid acoustic fabric.

When describing an interior space, the IR is, in essence, the embodiment of everything that is not the boundary – if one could remove the walls, ceiling, and floor of, say, a cathedral and only keep everything that is not these boundaries, you have an inverted space or 'emptiness' in the shape of the building. When collecting an IR of a site, one is, in a way, taking a picture of Echo. It is a deliberate word choice to emphasise an awareness that sounds, and images are not as different as we are inclined to think upon impulse: sound is also an 'image' processed by the brain based on sensory input. If the surfaces and borders were a mould and instead of pouring wax or melted metal into this mould to find a sculpture of the interior shape, we now pour sound. Without this action, the shape remains 'invisible' (inaudible). Some of these nuances are defined and recognisable, some hardly perceptible.

My practice and spatial sound awareness started roughly 20 years ago when a friend lent me in-ear binaural microphones and a mini-disc recorder. I was in and around Kyoto, Japan, and recorded hours and hours of sound, streets, people, and trains – to listen back in fascination with the incredible spatial detail and hyper-realism. It was genuinely exciting material. Especially since they were my ears, meaning that to me, the HRTF (head-related-transfer function) filtering fitted like a glove, and playback was my ears listening to ambiences. Today, technology and knowledge have evolved tremendously. Simultaneously, it feels like a mammoth task ahead to understand space as material due to the complexity of listener and space relationships and the ephemeral nature of sound.

Leaping forward in time, I developed an increasing desire to shapeshift (and travel) through places on earth, to position a musical note in one place and then the next in another and another, as if playing a keyboard of spaces. Next, I wanted to play them all simultaneously or in a chord (in a triad of spaces). Consider the possible dynamic range of acoustic size: massive recombinant architectural scale. Say we have a tiny box, a small room, a hall, a cathedral, a stadion, following ten cathedrals combined, following 1,000 cathedrals. When I use 'scale' in this sense, I refer to proximities, the size or intensity which is relative. For example, it could be the 'distance' between the softest and loudest sound in amplitude dynamics or the 'distance' between a crisp, close sound and an extremely long reverberation tail. These elements can help create the idea of scale, but I am also talking about a conceptual sense of scale and considering imaginary sounds. 'Imagine musical instruments the size of military weapons – flying gongs in the sky and percussive bullets flying into it'. (Imagine spending a military budget on the arts)! The motivation for scale comes from an essential need to explore bigger musical spaces. Returning to the idea of the IR as a container for space, I think my need for more vastness in musical space is connected to it being a 'container' for consciousness. To explore further, I need to make the boundaries of the space bigger, in order to sense the boundaries of consciousness – because for sure, I have not reached them yet. Perhaps consciousness is boundaryless, but the only way we can feel or express it is when its boundaries (and material qualities) are revealed. (Which is often limited by the boundaries of our human bodies with its limitations). Just like the IR mirrors the space, music is a mirror to consciousness.

My desire for scale in sound is undoubtedly driven by the post-internet world we live in and my life experiences. From growing up in a small town in South Africa, then moving to big Johannesburg, then suddenly to many countries, continents, and islands. I want to express how big something is: the interdependence of emotions. Displaced from an 'origin', there is an intensity of longing. Distances have emotional qualities. Such expression is seemingly effortless in words: 'From here to the moon' – we just covered a vast conceptual distance. But how can we express this in sound?

I have attempted to express scale through accumulative reverberation by working with the physics of acoustics and actual sound. Via the development of recombinant reverberation and feedback in pieces titled The Audio Tunnel (2009), and The Persistence of Sound (2018) where a live audio signal looped through seven daisy-chained internet-connected physical spaces, or Goodbye Anthill (2018), and other drone pieces where voices were sent through loops of multitudes of daisy-chained virtual acoustic spaces, and new works developed for Acoustic Atlas (2019–present).[7]

Conceptual scale happens by situating the listening mind in these real-world places as part of the musical material. When using a mathematically abstract reverb, it differs from when convolving the signal with an actual space. The awareness and memory, history, social, political, and sonic complexities that immediacy come with this awareness add more depth and reality to the sound product. My work overlaps with sonic archaeology research fields, also termed cultural acoustics or archaeo-acoustics. However, whilst I am curious about listening to the 'past', I aim to consider the data produced in such research as material with immense potential for contemporary composition (and sonic arts). What does hearing a sound in a reconstructed virtual acoustic version of the Taj Mahal mean? Or of a place that no longer exists, such as a simulation of Pompei. Now, we are composing with cultural memes or icons and what they embody, simultaneously with the physical properties of the sound. Similarly, fields such as building acoustics and materials acoustics and the physics of sound and vibration feed into sound-space-human material blurring composition with architecture.

Besides acoustics, there are various connected approaches to scale: duration, range, complexity, rates of change, size of wavelengths themselves, the intensity of vibrations. Starting with wavelengths: through creating different tones or third tones that result from when two frequencies are played together, and a third tone appears with a frequency of the difference between the two tones.[8] The lower the wave frequency, the larger the wavelength: for example, a 1 Hz wavelength in air at 20°C is roughly 340 m long![9] Whilst we cannot hear that low, we can experience it by playing two tones together that are 1 Hz apart, and the interference pattern that will form between these two waves will be that low. Scale can be created via massively large sound waves and by infrasound transduced through surfaces/objects/spaces. For example, one can create earthquake-like vibrations by attaching specific transducers to the structure of a building and shaking the structure. We can go even larger: 0.1 Hz will be 3,400 m long, and so on.

Musically, scale can be expressed through duration too, with extreme duration, such as the thousand-year-long piece Longplayer[10] and various drone pieces. Feldman called his compositions' time canvases' and described the subject of his music as 'surface'. He said that he 'more or less primes' the music with an overall hue of the music' and tried to let 'time undisturbed' control his music.[11] Eliane Radique used minimal change over time, with subtle and slow changes that there is no way to understand how one got from A to B.[12] Gradually, even slower than the sky at dawn, change creeps over you. Then there is the option of loudness, such as death metal drone music, such as Sun(((O)))).[13] Or overwhelming complexity or information density – density as scale.

Colour, proportion, abstraction

Sound without space is a bizarre, anechoic reality. Like a raisin instead of the fresh juicy grape, a skeleton without the flesh, a fish without water – unable to move and breathe … or an image without colour. Depending on the space, the acoustics animate the sound. Waves collide, bounce back and forth, and interference happens. Resonant frequencies get energised to sustain much longer, sometimes significantly longer (such as inside the Inchindown oil tanks in Scotland, with reverb decay times of 75–80 seconds).[14] Entire genres of music have developed around the reverb of large cathedrals, such as Gregorian chants.[15] Even when the reverberation of a space is quite subtle, for example, outdoors, in the woods, in an open field, or a valley, upon closer listening and comparison, the variations and characteristics are unambiguous and recognisable.

A room is a filter. It adds or subtracts spectral colour to the incoming sound, which happens four-dimensionally (where time is the fourth dimension). Thus, surrounding or immersing the listener. Let us consider colour in sound. As interesting as rhythm, colour is, by default, microtonal vertical information. It can float in space separately from time. It can exist as a glow or an impression. Working with a composition's spectral (or harmonic) colour allows for abstraction working with a sound field.

The highly articulated and psychological use of colour in colour field painting (a subcategory of Abstract Expressionism) has a lot in common with how the internal characteristics of sounds are used in the context of spectral music. For example, a single note, played at a specific volume, with a particular attack, sustain and decay, in a specific space (which includes the body of the instrument plus the body of the space): these are all elements that influence the timbre which can be scientifically observed via spectral analysis and creatively used in various applications of spectral music and synthesis.

Rothko compared the painter to a philosopher and said that, like philosophy, the work of art is the creation of a particular notion of reality. He continued to call painting 'plastic speech' by using colours and forms. At a mature point in his life and work, he considered that plastic languages only serve art when generalising beyond the particular. That the artist needs to reduce phenomena to shed light on human sensuality. He concluded that sensuality is neither objective experience nor subjective experience but something which exists outside of both and, therefore, contains both.[16] There is a similarity to what I have noticed about my process in composition: there is the physical material that is sound and then also this almost telepathic essence when thinking of sound and working with sound – as already described previously – the journey that happens in the mind, the inspiration, the idea to travel or shapeshift through spaces. (Perhaps I should add that I experience a form of synaesthesia when listening to music).

Also, listening to composers who work with spatial sound speak about their work, there are often words used about spatial attributes that are connected to the desires of the composer to express a feeling or a spatial trajectory of a sound that makes complete sense visually, but that is impossible to achieve auditorily due to limitations in how humans hear. Yet, this dichotomy is fascinating. Similarly, when a piece is composed about a place or a historical event, this double-layered outcome also exists – one is the conceptual or even narrative layer, and the other is the material. To return to Rothko, I think his definition (that sensuality is neither objective experience nor subjective experience but something which exists outside of both and, therefore, contains both) rings very accurate for music and listening.

Continuing my speculative thought journey, I would substitute 'subjectivity' with 'our telepathic essence', where the external world, which, when listening in this way, is no longer exterior/external, but rather a synaesthesia synthesis of sound orgasm (to emphasise audio-physicality although third ear could perhaps also work as image). Our ability to imagine abstractly is a remarkable characteristic and an essence of consciousness. In the context of how we interact with various objects such as machines, we tend to anthropomorphise, to see human-like characteristics, as is, for example, a lot of the case with machines and machine learning. But in sound, specifically it is as if we can synthesise a dimension that is not possible through visual input, it is physical, and this is a place where our relationship with time especially, can lead to altered states, utter pleasure, mood changing experiences, transcendence, dream state, experiences of shapeshifting, synaesthesia, morphing, going beyond the limits of our bodies, of looping, of becoming vast mathematical patterns, of entering parallel (be it make-believe) realities that are convincing enough to form platforms for connections (for example in music sub-cultures) and mostly utterly positive experiences.

Rothko said one more thing I would like to highlight: plasticity is how an artist creates 'the effects of movement in space, a sensation or experience of reality as something which moves through time and space'. That 'without colour, form, and space', we cannot perceive the 'sensation of movement' or the 'artist's reality'. If we perceive this, we become aware of the painting as something that has a life. The painting is the premise for a journey we follow and take, or, as he says, the picture is a vehicle for an experience that lives outside the picture. That experience, he says, is the 'experience of plastic continuity'.[17]

Three short text scores

Piece 1: Think of four spaces. Think of four notes. Sing them to yourself. Imagine each note in a matching space.
Piece 2: Close your eyes. Imagine clapping your hands (don't clap, imagine yourself clapping) – each time you clap you teleport to a new place. Your eyes remain closed. You must

recognise the place of arrival purely through the echo in the new location. Try seven times. Where did your imagination take you?

Piece 3: Imagine a fast succession of very short, sharp, bright, notes. They start from one note a second and accelerate within 7 seconds to 100 notes. From very soft to very loud, from very close to very far. Perceptually, even though the notes get much louder, they proportionally sound the same, as the louder they get the further they are.

Acoustic Atlas

Intro

Acoustic Atlas – Cultivating our Capacity to Listen, a Marie Skłodowska-Curie funded research project,[18] has allowed me to develop a unique web-based software where visitors can listen to virtual acoustic natural and cultural heritage sites (Figure 34.2). The app enables the Acoustic Atlas to share IRs and, in some cases, soundscapes, field recordings or music connected to a specific space via a globe-based interface. Locations can be found geo-spatially, and the acoustics can be experienced via a convolution reverb in the browser (termed auralisations). The focus is to have an audible experience of the IRs in the browser.

A particular field of sonic archaeology and architectural acoustic studies focuses on high-precision quality acoustic measurements of heritage sites. A branch of these studies extends into forensic acoustic simulations of ruins that no longer exist in the prior form. This work is fascinating, and I have been following researchers in this area for the past decade. However, only recently

Figure 34.2 Globe UI for Acoustic Atlas.

has such research been published with audible content. In the past, papers that described in detail the findings of these researchers' measurements of a small engineering society existed only as texts. A wider audience cannot appreciate their work without specialist knowledge of what this data means. I find this a great pity, and that's where Acoustic Atlas contributes helpfully – to access this research directly as sound. The importance of this work is that it aims to 'cultivate our capacity to respond'[19] by cultivating our capacity to listen. As our lives extend and expand into the digital, preserving listening as a heritage is crucial so that our future digital ears expand and remain sharp. Environmental field recordings or site-specific music can be embedded into the listening experience of locations. Next, these environments and their echoes can be used as musical material and settings to create musical works. To use the archive as an orchestra of echoes. Working with spaces in this way can be helpful in many ways. For example, I have had feedback from a theatre production that used the app to aid in their sound design for a play.

Structure and tech

The project offers user accounts to researchers in architectural acoustics, heritage acoustics, and sound art, enabling them to upload and manage their research and share it easily in audible format. Each location (space) has a unique URL, allowing it to be embedded in other sites and allowing researchers to share their work in various ways. I also manage the site to add IRs that people send to me. This means the acoustic heritage of Acoustic Atlas has a growing database.

Upon selecting a globe location, an interior view of the site is loaded, either in the form of a still image, a 360 panoramic image, a 3D image or an artistic drawing/sketch. The visual image orientation, or point of view (where this information is known), corresponds to the room impulse response (RIR) listener position (or microphone position). Selecting a location also turns on the audio signal chain, unmutes the microphone and activates the convolution reverb with the corresponding RIR. The user can now listen to their voice as a real-time auralisation from the mic input. The app has a built-in record function so visitors can record and save their auralisations to their local devices.

The user can switch to other interior or exterior listening points via a search function or by clicking directly on the globe. Additionally, the user can play and control levels for the embedded field recordings or music compositions. Now, the experience focuses on having the listener/participant in a single sonic sweet spot, able to move their head in 360, but not virtually 'walk through' a space. The latter may come later when the archive has more sites with sufficient source/listener positions and fast enough processing to handle such a task in the browser. (Listen: https://acousticatlas.de. It is best to use wired headphones on your mobile device or computer, select the dots on the globe, and sing/talk into spaces).

An in-depth technical overview of the back-end code design and libraries used is available.[20]

Beyond the actual sonic effects of the space as a filter and effect, as already mentioned, there is often a multitude of other exciting dimensions: the history of a place, its politics, perhaps various musical traditions connected to that space, mythology, memories, architectural form, or entirely imaginary associations driven by the space. The option also exists to fabricate imaginary IRs with imagined spaces as a result.

Contributors

Acoustic Atlas currently contains virtual acoustic of 60 locations contributed by 13 users. Locations are organised via their longitude and latitude values embedded on a globe user interface. The criteria for inclusion are heritage value and unique acoustic properties for both cultural and

Acoustic Atlas

Figure 34.3 Some of the sites in the current archive.

natural sites. The collection has cathedrals, duomos, churches, amphitheatres, theatres, caves, mines, tunnels, land art sculptures, and beaches. A location consists of the place name or building name, city, country, a general description, one or many IRs, descriptions about the technology and formats chosen, source/listener positions, photos, sketches and information with weblinks about the contributor. The highest current resolution for RIRs is first-order Ambisonic, which works with matching panoramic images. (Higher order has been tested and will become a feature soon). Besides the real-time auralisation in each location, there also exists a track player which contains, when available, field recordings, music, or any other relevant sound files. There is also a section for artistic works, with a first album released in December 2022, where nine artists were commissioned to select locations and produce music for them or even create wholly imaginary/poetic places.[21] The archive includes locations in Europe, the UK, India, Africa, Iceland, Bahrain, and Japan. The current and future list of RIR contributors can be viewed under the community page on Acoustic Atlas (Figures 34.3 and 34.4).[22]

Examples

Beyond general browsing, I suggest five marvellous sites to visit (and sing into):

1 **San Giorgio Maggiore (San Zorzi Mazór in Venetian)** is a basilica on the island of San Giorgio Maggiore in Venice, part of the homonymous monastery. The church, designed by Andrea Palladio, who also built the refectory, overlooks the San Marco Basin. Recorded by Cobi van Tonder & Giulia Vismara. https://www.acousticatlas.de/locations/venice_chiesa
2 **Santa Maria della Salute (or Chiesa della Salute or simply La Salute)** is a basilica in Venice erected in the Punta della Dogana area, from where it stands out against the panorama of the San Marco Basin and the Grand Canal. Recorded by Cobi van Tonder & Giulia Vismara. https://www.acousticatlas.de/locations/venice_santamariadellasalute

Figure 34.4 Artwork for the album Echoes and Reflections, created by nine artists together with the author.

3 **The Taj Mahal** is on the river Yamuna's right bank in the Indian city of Agra. It was commissioned in 1632 by the Mughal emperor Shah Jahan (r. 1628–1658) to house the tomb of his favourite wife, Mumtaz Mahal; it also houses the tomb of Shah Jahan himself. The Taj Mahal was designated as a UNESCO World Heritage Site in 1983 for being 'the jewel of Muslim art in India and one of the universally admired masterpieces of the world's heritage'. It has an exquisite reverb. Recorded by Ableton. https://www.acousticatlas.de/locations/taj_mahal

You can also listen to an artist performance inside the virtual Taj Mahal here: https://www.acousticatlas.de/locations/echoes-album-Iti

4 **Ingleborough Cave, UK** A large part of the Ingleborough Cave system that we worked in was an underwater river, and the whole 500 m stretch was visually exhilarating. Walking down the curving tunnel, the last 100 m of the area accessible to the public sounds the most enticing. This area is named 'Long Gallery, '2nd Bells', 'Skittle Alley' and 'Pool of Reflections'. The play of light, darkness, echo, reflections, shapes, and sound is mesmerising. The connected underground otherworldly network of tunnels, halls and passageways has a tranquillity and magical atmosphere, far away from our everyday reality and immerses the senses completely.

Recorded by Mariana Lopez and Cobi van Tonder. https://www.acousticatlas.de/locations/ingleborough_cave/rirs/deepend

5 **Inchindown Oil Tank Nº 1** The Inchindown Oil Tanks are an underground oil depot in Invergordon, Ross-shire, Scotland, built for war purposes. They hold the record for the longest recorded reverberation in any human-made structure. Recorded by Sofía Balbontín & Mathias Klenner.
https://www.acousticatlas.de/locations/inchindown

Notes

1 Ovid, *The Metamorphoses*, book III, v, translated by A.S. Kline, https://ovid.lib.virginia.edu/trans/Ovhome.htm.
2 Ovid, *The Metamorphoses*, book III, v.
3 François J. Bonnet, *The Order of Sounds: A Sonorous Archipelago*, translated by Robin Mackay (London: Urbanomic Media Ltd, 2016), 21.
4 Lamberto Tronchin, "The 'Phonurgia Nova' of Athanasius Kircher: The Marvellous Sound World of 17th Century," *Proceedings of Meetings on Acoustics* 4, no. 1 (2008).
5 For a technical overview of IRs, convolution reverb and auralisation (terms frequently used in this chapter) see Jens Holger Rindel, Claus Lynge Christensen, and George Koutsouris, "Simulations, Measurements and Auralisations in Architectural Acoustics," *Proceedings of the Acoustics* (2013); or F. Rumsey, *Spatial Audio* (Focal Press, 2012); or Huseyin Hacihabiboglu, Enzo De Sena, Zoran Cvetkovic, James Johnston, and Julius O. Smith III, "Perceptual Spatial Audio Recording, Simulation, and Rendering: An Overview of Spatial-Audio Techniques Based on Psychoacoustics," *IEEE Signal Processing Magazine* 34, no. 3 (2017): 36–54.
6 Jens Holger Rindel, Claus Lynge Christensen, and George Koutsouris, "Simulations, Measurements and Auralisations in Architectural Acoustics," *Proceedings of the Acoustics* (2013); or F. Rumsey, *Spatial Audio* (Focal Press, 2012).
7 Van Tonder Artist, https://otopllasma.com. Accessed on 30 July 2023.
8 Hermann L.F. Helmholtz, *On the Sensations of Tone as a Physiological Basis for the Theory of Music* (Cambridge University Press, 2009).
9 Alfred Bedard and Thomas Georges, "Atmospheric Infrasound," *Acoustics Australia* 28, no. 2 (2000): 47–52.
10 https://longplayer.org/about/overview/. Accessed on 30 July 2023.
11 Morton Feldman, "Give My Regards to Eighth Street," *The New York School* (2000): 181.
12 Richard Glover, "Minimalism, Technology and Electronic Music," in *The Ashgate Research Companion to Minimalist and Postminimalist Music* (New York: Routledge, 2016).
13 Owen Coggins, "Mountains of Silence: Drone Metal Recordings as Mystical Texts," *The International Journal of Religion and Spirituality in Society* 2, no. 4 (2013): 21.
14 https://www.acousticatlas.de/locations/inchindown. Accessed on 30 July 2023.
15 Vitale, Renzo, Raffaele Pisani, Paolo Onali, and Arianna Astolfi, "Why Does the Acoustic Space of Churches Exalt Gregorian Chant?" in *Proceedings of the 31st International Computer Music Conference (ICMC)*, Barcelona, Spain, September, 2005, pp. 4–10.
16 Mark Rothko, *The Artist's Reality: Philosophies of Art* (Yale University Press, 2023). Kindle edition, 'Particularization and Generalization'.
17 Rothko, *The Artist's Reality: Philosophies of Art*. Kindle edition, 'Particularization and Generalization'.
18 Visit https://acousticatlas.de. Accessed on 30 July 2023.
19 M. Kenney, "Anthropocene, Capitalocene, Chthulhucene – Donna Haraway in Conversation with Martha Kenney," in *Art in the Anthropocene: Encounters Among Aesthetics, Politics, Environments, and Epistemologies*, edited by H. M. Davis and E. Turpin, 1st ed. (2015).
20 C. van Tonder and M. Lopez, "Acoustic Atlas – Auralisation in the Browser," in *2021 Immersive and 3D Audio: From Architecture to Automotive (I3DA)*, Bologna, Italy, 2021, pp. 1–5. doi: 10.1109/I3DA48870.2021.9610909.
21 Listen to Echoes and Reflections either via Bancamp: https://telepathicbeing.bandcamp.com/album/echoes-and-reflections or via Acoustic Atlas: https://www.acousticatlas.de/echoesalbum. Accessed on 1 May 2023.
22 https://www.acousticatlas.de/community. Accessed on 1 May 2023.

Bibliography

Acoustic Atlas. https://www.acousticatlas.de/. Accessed on 30 July 2023.

Bedard, Alfred, and Thomas Georges. "Atmospheric Infrasound." *Acoustics Australia* 28, no. 2 (2000): 47–52.

Bonnet, François J. *The Order of Sounds: A Sonorous Archipelago*. Translated by Robin Mackay. London: Urbanomic Media Ltd, 2016.

Hacihabiboglu, Huseyin, Enzo De Sena, Zoran Cvetkovic, James Johnston, and Julius O. Smith III. "Perceptual Spatial Audio Recording, Simulation, and Rendering: An Overview of Spatial-audio Techniques Based on Psychoacoustics." *IEEE Signal Processing Magazine* 34, no. 3 (2017): 36–54.

Helmholtz, Hermann L.F. *On the Sensations of Tone as a Physiological Basis for the Theory of Music*. Cambridge: Cambridge University Press, 2009.

Kenney, M. "Anthropocene, Capitalocene, Chthulhucene – Donna Haraway in Conversation with Martha Kenney." In *Art in the Anthropocene: Encounters Among Aesthetics, Politics, Environments, and Epistemologies*, edited by H. M. Davis and E. Turpin, 1st ed.. Open Humanities Press, 2015.

Long Player. https://longplayer.org/about/overview/. Accessed on 30 July 2023.

Ovid. *The Metamorphoses*, book III, v, translated by A.S. Kline. https://ovid.lib.virginia.edu/trans/Ovhome.htm.

Jens Holger Rindel, Claus Lynge Christensen And George Koutsouris. "Simulations, Measurements and Auralisations in Architectural Acoustics." In *Acoustics 2013 New Delhi, India*, November 10–15, 2013.

Rothko, Mark. *The Artist's Reality: Philosophies of Art*. New Haven, CT: Yale University Press, 2023.

Rumsey, Francis. *Spatial Audio*. New York: Taylor & Francis, 2012.

Tronchin, Lamberto. "The'Phonurgia Nova'of Athanasius Kircher: The Marvellous sound world of 17th century." *Proceedings of Meetings on Acoustics* 4, no. 1 (2008). AIP Publishing, 4185–4190.

Van Tonder. https://www.otoplasma.com/. Accessed on 30 July 2023.

Van Tonder, C., and M. Lopez. "Acoustic Atlas – Auralisation in the Browser." In *2021, Immersive and 3D Audio: From Architecture to Automotive (I3DA)*, Bologna, Italy, 2021, pp. 1–5. https://doi.org/ 10.1109/I3DA48870.2021.9610909.

Vitale, Renzo, Raffaele Pisani, Paolo Onali, and Arianna Astolfi. "Why Does the Acoustic Space of Churches Exalt Gregorian Chant?" In *Proceedings of the 31st International Computer Music Conference (ICMC)*, Barcelona, Spain, September, pp. 4–10, 2005.

INDEX

Note: **Bold** page numbers refer to tables; *italic* page numbers refer to figures and page numbers followed by "n" denote endnotes.

Ablinger, P. 143, *143*, 144
absorption 18, 33, 39, 40, 68, 71, 74, 98, 99, 101, 102, 104–105, 255
acousmatics 202–203, 418; intimacy 87–88, 93–94 *see also* intimacy; listening 205; space 18
acousmatique 203
acoustic atlas, echoes: colour, proportion, abstraction 475–476; contributors 478–479; examples 479–481; globe UI *477*; myths 471; observations 472, 477–478; reflections *480*; shapeshifting 472–475; sites *479*; structure and technique 478; text scores 476–477
Acoustic Craft (2022) 400
acoustics 2; architectural *see* architectural acoustics; consultant 57; echoes *see* acoustic atlas, echoes; ecology 209–210; environments 326, 330; extrapolation 45; extremity 18, 19; gallery 21; hot spots 374; insights 381–382; intimacy 373, 374; modelling 383n16; and noise control 230; politics of space 8; privacy and intelligibility 5–6; products 104; propagation 5; reconfiguration 238; resonances 15; response 348; reverberation 7; scale 475; shadows *see* acoustic shadowing; simulation tools 5, 6; sounds 301; spaces 165, 301; spatial design 9; testing 53
acoustic shadowing: confidentiality 36; FabPod I 32, *32*; FabPod II 34–35; health well-being, privacy and open-plan work areas 30–31; human speech privacy and intelligibility 28; interior architecture and urban design 29–30; inverse of sound 35; listening environments 36; mapping, fisheries and health diagnostics 29; phenomenon 28–29; plan and section *34*; projects 31–35; shells 31; sound perception and privacy experiment 33–34; theory 28
active reverberation systems 7
act of liberation 364
aesthetics 55, 144, 154, 156; Japanese 365, 366; material sound transmission 241–244; motivations 71; musical 246, 347, 356; perceptual characteristics 453; political vectors 314; qualities 139, 381; research 461
affectivity 324–326
affordances: acoustics and noise control 230; architecture 229; artistic practice 229; landscape architecture 229; open framework 225–227; peripheral perspectives 231–232; plural approach 232–234; policy and legislation 230; resources 234; sound, space and public 225; urban development 230–231; urban sound designer and acoustic planner 224; *see also* value chains
Afternoon Delight (Prest) 156
Akiyama, M. 203, 366
Alberti, L.B.: De Pictura 130
alienation mechanism 147
Allen, J. 266
Amacher, M. 240, 245; Living Sound, Patent Pending 241–242
ambisonics: audio technologies 345; binaural synthesis 111, 114; microphone array 353, *353*; recordings 4, 353; software encoding 54; sound installation *63*; spatial audio 58
The Ancient Mariner (Coleridge) 214
Anderson, P.T.: Phantom Thread 271

Index

Anderson, S. 210
anechoic signals 79, 107, 148, 169, 305, 346, 348–350, 352–353, 355–357
Angela Burgess Recital Hall *368*
anti-monumental architectures 245
anti-noise sentiment 278
Antrobus, R. 160
Aquinas, T. 291
architectural acoustics 9, 10, 15–18, 240; applications 68; case study 80–84; Gala Pool 375, *376*; goal of 68; intimacy 374–375; IR measurement 375–376; measurement locations 376, **377**; measurements 74–80, 375; measurements results analysis 377–379; parameters **377**; RIR measurement 77–80; room acoustics 72–74; room impulse responses *76, 76*–77; scene setting 74–76; sound generation 68–69; sound propagation 69–71; studies 477
architectural design 7, 52, 318
architectural energies 245–250
architectural shells 31
architecture 229; concert spaces 427; and engineering 10; evolutionary algorithms 183; framing 203; interior and urban design 29–30, 255; landscape 229; multidisciplinary 7; and music *see* music; organism 287; performer 348–350; poetics 133; renaissance 108; silence 260; togethering 3; vibrational *see* vibrational architectures
Arendt, H. 199, 200, 206
Aristoxenus of Tarentum 289
Ars Cantus Mensurabilis 128
Art Basel Miami (2009) 57
artificial intelligence (AI) 10, 281
artificial reverberation systems 93, 382
artistic development 379–380
artistic freedom 364
Arup SoundLab 24, 27, 53–55, *54,* 66
astronauts 110, *111,* 116, 122
asymmetrical hearing 172–173
atmosphere 31, 61, 111, 115, 168, 259–261, 268, 325, 432, 443
Attali, J. 147–149
attention circuits 302–303
audial perception 9
audience 155, 156, 405, 408; acceptance 364; contemporary technology 55; coughing 420; crumples 265; electrifying effect 214; expectations 382, 431; experience 66, 428; intimacy 22; listening 109, 153, 174, 374; mind 62; object-based spatial audio 48; perception 331; performer relationships 5, 64, 344; sculpturers 456; trained musicians 422
Audio Definition Model (ADM) 38
Audio Smut 156

audition 141–144; awareness 142, 143; map *147*; masking effect 79; sense 142; space 272; streams 144, 148
auditory virtual environments (AVEs) 457
Auger, J. 331
augmented reality technologies 120
augmented spatial practices 460
Augustidis, T. 161
auraldiveristy 170, 171
Aural Diverse Spatial Perception: From Paracusis to Panacusis Loci (Drever) 109
auraltypical perception 170
autonomous sensory meridian response (ASMR) 211, 265, 266, 270, 271

Babbitt, M. 172
Bachelard, G. 17, 88; The Poetics of Space 219
Bagenall, H. 346
Bain, M. 240, 241, 242
Balbontín, S. 481
Baldwin, J. 244
Barbour, J. 353, *353*
Barrett, N. 265
Barron, M. 172, 374, 375
Barry, K. 331
Bartók, B. 437
Bassuet, A. *56, 192*
Batra, L. 172
Bauby, J.-D.: The Diving-Bell and the Butterfly 175
Bauman, H.-D.L. 199, 200
Bavister, P. 109
Beach, D. 140
Beckett, S. 154, 269
Behrens, M. 438
Belanger, Z. 18
Bell, A.G. 199
Bell, J. *56*
Beranek, L. 172, 374–375, 379
Berger, L. 439
binaural listening 24, 171–172
binaural recordings 17, 77, 112, 114
binaural RIRs (BRIRs) 41, *42,* 43–44, 47, 77
binaural room impulse response (BRIR) 114
binaural synthesis technologies 111, 116
biometric evolutionary computation: background 181–182; design 182–183; digital model 180, *181, 193*; experiments 192; future research 193; IEA 184–191; influence of test site 191; practical response 193–194; research design 192; sample size 191; search processes 183–184; sound caption 194; trends 191
biometric sensing 186–187, *189*
biomineralisation 312, *312*
biotech developments 281
Bish Bosch: Ambisymphonic (2013) 62, *63, 65*

Bishop, C. 395
Biziorek, R. *59, 61*
Björk 16, 19, 22–25
B&K OmniPower™ Sound Source (Type 4292-L) 81
B&K Sound Calibrator Type 4231 81
Blauert, J. 111, 166, 375
Blesser, B. 387
Bodies of Water (2023) 439
Bogert, C. 437, 438
Bogong High Plains 447, *447,* 450
Böhme, G. 258, 259
Bolton, A. *56*
bone conduction headphones 118, 121
Boning, W. 192
Borg, A. 429, 431
Bosco, H. 219
Boubaki, N. 289
Boulez, P. 462
Boullée, E.L. 133
Bowen, E. 264
Bracciolini, P. 128
Braidotti, R. 277
Brant, H. 347, 354, 356
brass instruments 69
breaking plane 101–103
Bresson, R. 157
broadcasting radio 153, 154, 156, 204, 260
Brook, P. 366
Brown, E. 134
Brown, N. *61*
Bruhn, M.J. 215
BSF performance layout diagrams *352*
Buck, D. 108
Bühlmann, V. 220
'Building, Dwelling, Thinking' 88
Burrows, D. 332
Burry, J. 16
Burry, M. 17
Byrne, D. 18, 91–93; 'My Love is You' 91–93

Cage, J. 143, 144, 148, 245
Camuni 110
Camunian Rose 110
Cantor, G. 288, 289
Cardew, C. 136
Cardinal, L. 282
Carpenter, E. 165, 166
Carter, P. 210
Cartesian theory 144
cascade 21, 308, 374, 379, 381
Casey, E. 385
Casey Station *440*
Caulkins, T. *59, 63,* 62
Cave, N. *59,* 59–60

Central Processing Unit (CPU) 56
Certeau, M. de 89, 363, 367
Challenger Deep programme 117
Chanan, M. 90
Chion, M. 203, 462
Chipperfield, D. 238
Chladni, E. 242
Choisy, A. 134
Chopin, F. 288
chromosome 184, 186, 193
Chronically Sick and Disabled Persons Act (1970) 200
Cimini, A. 241
cinematic montage theory 134
clarity index at 50 ms (C50) 45
Clary, J. 132
class-based activities 401
classical metaphysics 287
coccoliths *312*
cochlear implants (CI) 197
cocktail party effect 168, 174
cognition theory 216
cognitive dissonance 388
cognitive space 57
Coke, D. 429, 431
Cole, H. 134
Coleman, P. 41, 46
Coleridge, S.T.: The Ancient Mariner 214
Colman, P. 16, 17
Colònia Güell Chapel (1898–1912) 81
communal rituals 62
communal space 153, 265
communicational model 145, 147
composing sound 149, 466
composition 10, 55, 81, 115, 268, 280, 393, 472; architectural 130, 430; artistic 365; challenges 347; decisions 354; dynamic sound objects 457; electroacoustic 401, 409; electronic 270, 291; error 421; failures 109; imagination 394; by montage 134; music 8, 348; musical passages 92; parameter 348, 395; performance 349; poetic 216, 218; simulations 351; sonic layers 407, 463; soundscape 402–403; space and sound 1, 156; time canvases 475; universal 146
Computer-Aided Theatre Technique (CATT) 348, 350
conceptual pregnancy 367
condenser microphones 88
conflictual programs/contractual proximities 363
confluent love 91–92
Connor, S. 198
consciousness 17, 157, 158, 215, 218, 219, 256, 259, 306, 463, 474, 476
constellation 413, 415–418, 432, 453
Construction 002: Tracing/Occupying' (C002) 346–351, 354, 356, 357

Consulting the Oracle (Waterhouse) 205, *205*
contemporary architectural design 323
contemporary audio technology 439, 450
contemporary immersive technologies 324
Continuous Drift 225, *226*
continuous uniform function 290
CoolEdit 54
Corbusier, L. 133, 239, 287
Corner, J. 131
Cornucopia 16, 24, 25
COVID-19 pandemic 47, 120
Cox, C. 245, 246
Cox, T. 6
creative potential: acoustic intimacy 374; development 374; local community 373
creative receptacle 366
creative reciprocity 3
Cubasch, U. 333
cultural memory 260
Cusack, P. 438

Damisch, H.: 'The Origin of Perspective' 130
'Dancing to the Radio' 159
darkness 64, 108, 164, 261, 293, 431, 473
Darwin, C. 182
Davenport, I. 55
Davenport, W. 55
Davies, C. 167
Davies, S. 148, 149
Davis, L.J. 199
Davis, S. *61*
deaf gain 199
deaf space 200
Death of Bunny Munro (2009) *59,* 59–60
The Debt project 224, 235n3
deconvolution method 77
Deconvolved Electrodermal Activity *190*
deep learning algorithms 29
deep listening practices 279–281, 330
Deleuze, G. 288
Delgado, D. 25
Delikaris-Manias, S. 115
De Pictura (Alberti) 130
deregulatory neoliberalism 149
descriptive experience sampling (DES) 157
dialogues: civilised 84; composer-acoustician 455; developing 5–8, 12; film 144; relational 423; resonant 238; rigidity and didactic quality 421; spanning 17; verbal chiaroscuro 168; *see also* spatiosonic dialogues
Dibben, N. 91
Dicken, C. 255
Dietz, B. 241
diffraction 30, 71, 165
diffuse field 98, 169
digital audio workstations (DAWs) 93

digital fabrication techniques 2
digital simulations 5, 33, 355
digital technologies 77
dimensionless space: breathing night 268–269; intimate space 269–272; sensory response 266–268; tissue 264–265
disciplinary resonance 2–4
dissonant intervals 354
distributed subjectivity 307
The Diving-Bell and the Butterfly (Bauby) 175
domestic space 9, 88–91, 156, 211
Doppler effect 167–168
Doyle, P. 90
dramaturgy 380–381
dreams topology 443–446
Drever, J.L. 271, 277; Aural Diverse Spatial Perception: From Paracusis to Panacusis Loci 109
Duffy, M. 331
Dufour, F. 28
dummy head microphones 112
dummy head recording 78
Dunn, D. 142, 144
Dunne, A. 331
Dupetit, G. 331
dynamic range compression 88
dynamics system theory 76

early decay time (EDT) 45, 80, *83,* 377
early lateral energy fraction (LF) 375
ear wiggling 167
Eastly, M. 438
echolocation 170, 174, 318
Eisenstein, S. 134, 135
electroacoustic reverberation 4
electroacoustic systems 84, 437
electrodermal response (EDA) *190*
electromagnetic induction 197, 198, 201, 206
electromechanical recording 90
electronic dance music (EDM) 94
Elina, S. 344
Ellis, W. *59,* 60
Ellsworth, E. 417
Emmerson, S. 462
emotional engagement 53, 87, 88
emotions 19, 53, 57, 66, 109, 158, 210, 474
emptiness 361, 365–367, 473
empty spaces 344, 367–369
energy fields 446–450
energy transfer model 148
engineering 10; acoustic 17, 171, 347, 373; compositional classes 456; consulting 225; electrical 313; listening 175; physics 27; problem 145; sound 29
Engineering Acoustics and Vibration 52
environmental frequency 309

Environmental Noise Directive 223
environmental sounds: attunement relationships 329–330; bodies and landscapes sensing 330–331; imaginary dimensions 329; and listening 331–332; shaping 332–336; sonic states 336–337; speculative sonic process 329
Euler, L. 289
European Broadcasting Union (EBU) 38
Evaluation of Acoustics using Ray-tracing (E.A.R.) 348
Evans, R. 130
evolutionary algorithm (EA) 183
evolutionary programming 184–185, *185*
evolutionary search processes 183–184
evolutionary soundscape experiment *188*
excitation signals 78–79
exhibitions 407–409, *409*
experimental setup 33, 81–82, 154, 190, 264, 289, 313, 348, 437
exponential sweep sine (ESS) signal 81
exquisite corpse process 60
exterior spaces 9, 320, 324, 326
extra-musical attraction 422
extrapersonal space 112
extreme environments 111
extreme futures 121–122
extreme spaces 9; acoustics 19; hangar 22, *23*; hazardous 118; Headlands tunnel 20; Orbit Pavilion 25–26, *26*; precarious 119–121; remote and unfamiliar 116–118; reverberation chamber 22–25; scaling problem 27; SFMOMA stair 21, *21*; sound column *20*, 20–21; speech and formant patterns *119*; 2D virtual speaker mappings *116*; unbreathable 115–116; wave physics 19

FabPod I 32, *32*
FabPod II 34–35
fabrication 18, 35, 98, 101, 103, 104
FACIT, phenomenological shift 464–466
falsification principle 285
Faraday, M. 198
Farina, A. 330
Favilla, S. 108
Feldman, M. 388, 475
Ferrando, F. 279
Fessenden, R. 154
fictioning 331, 332, 334
finger tapping 271
finite element method 73
Fordham, M. 351
formal listening tests 41, 44–46
Forsyth, I. 17, 52, 53, 55, 57–59, 62–66
Forsyth, M. 2
Fort, B. 438
Fortescue River *445*

fossilised whale inner ear (tympanic bulla) 307, *308*
Foucault, M. 362–363
Fowler, M. 108
Foxwell, A. 6
Frank, M. 455
Fredericksen, A. 130
free-field model 70, 169
Freeman, J. 422
Friedner, M. 199
fundamental musical structure *see* Ursatz
FX track in film 59

Gala Pool 375, *376, 377*, 381, 382
galvanic skin response (GSR) 187, *187*
Ganchrow, R. 211
Gandy, J.M. 131
gardens' promotional strategy 429
Garrefls, I. 174
Garthwaite, N. 108
Gaudí, A. 17, 32, 71, 74–76, 81, 84, 133, 351
Gehry, F. 180
genes 184
genetic algorithm (GA) 186
genetic algorithmic search 184
geobatteries 303
Gershon, R.R.M. 145
Gerzon, M. 114, 121
Giard, L. 89
Gibson, J.J. 29, 219
Giddens, A. 91
Giedeion, S. 286
Giedion, S. 287
Gilmurray, J. 330
Gilpin, W. 134
Gisborne, J. 216
glass 18
Goffman, E. 218
Goodman, N. 136, 349, 464
Goodman, S. 240
Goods, D. 25
Govan, E. 363, 364
Granö, J.G. 165
granular synthesis 333
gravitational waves 326n1
Greig, H. 429
Griesinger, D. 89
Griffero, T. 432
Griffiths, P. 136
Guiterrez, A. 276–277
Guthrie, A. *59*

Hall, A. 154
Hall, E.: 'The Hidden Dimension' 9
Hall, S. 199
Halmrast, T. 87, 90
Hamilton, J.M. 330

Han, B.-C.: Psycho-Politics 268
hangar 22, *23*
Hannam, J. 402
Hara, K. 365, 366
Harrison, D. 287, 405
Harris, Y. 330
Harvey, L. 344
Hassel, A. 402
Haydn 182
Hayton, J. 160
head and torso simulator (HATS) 78
headphone training rituals 120
head-related impulse responses (HRIRs) 43–44; *see also* binaural RIRs (BRIRs)
head-related transfer function (HRTF) 111, 114, 166, 474
head-tracked 3D-audio scenes 458
health well-being 30–31
Heaney, S. 214
hearing 141; ambisonics and binaural synthesis 108; frequency ranges 107; identity 276–277; long-distance 304–305; modes of 302–303; (re)prioritisation 164; shared 313–314; space 111–114; spatial experience 107; urban modes 303; *see also* spatial hearing
hearing aids 197
Heidegger, M. 88, 285, 294
heightened/diastematic neumes 127
Helmholtz, H. von. 149–150
Helmholtz resonance 102–103
Helmreich, S. 199, 241
Henry, J. 198
heterotopias 363
'The Hidden Dimension' (Hall) 9
high angular resolution planewave expansion (HARPEX) 117
higher-order ambisonics (HOA) 114, 385
high-quality speech intelligibility 57
Hogarth, W. 278
Höldrich, R. 455
Hornak, L. 155
Houd, R. 154
Hovestadt, L. 220
Howard-Birt, D. *56*
Hubbard, T. 142
Huddy, H. 173
Hull, J.M. 175
human cognition 214, 218
human perceptual system 321
human speech 68–69
Hurlburt, R.T. 157
Hurley, F. 441
Husserl, E. 287
hybrid spaces 462–464
hyper-asymmetry 174
hyperobject 3

icosahedral loudspeaker (IKO) 344, *386, 454, 464*; awareness over control 387–388; beamforming graph *455*; creative process 393–395; inside-out system *390*; instrumentalising 454–455; listener *393*; out-of-body spaces 386–387; outside-in system *391*; portability 394; relationality 389–392, *392*; spatiality 385–386; technical and aesthetical efficacy 465; unsettling space 395–396
Ihde, D. 258, 322
Ilk, Ç. 245
imaginative audio 64
imagining: cinema for sound 153–154; the dark event 154–155; inner experience of sound 157–158; listening body 158–162; spaces 155–156
immeasurable framework 417
immersive: outdoor sound system 25; quality of audial experiences 4
immersive sound systems 453
impairment theory 170
imperceptible sounds 268
impression of spaces *see* room impression
impulse responses (IRs) *21*, 350, 472–473, *473*: measurement 375–376, *378*
Inchindown Oil Tanks 481
indigenous listening practices 280, 282
induction loop (IL): deafness infrastructure 197; effectiveness 198; Euston station case study 204–206; help points 202; marketing display *204*; PA listening points 202–203; sound as spatial solid 198–200; ticketing windows 201–202; types of *201*; typologies 201
informal minimalist strategies 265
informal structures 422
Ingleborough Cave, UK 480–481
Ingold, T. 3, 274–275, 330
initial time delay gap (ITDG) 375
innovation 2, 7, 93, 127, 181, 201, 403, 466
inorganic sensation 304
instrumentality 466
intensity 260–261
interactive evolutionary algorithms (IEA): acoustic reflection data *190*; biometric sensing 186–187, *189*; contextual limitations 190–191; drawbacks 185–186; genetic algorithm 186; musical limitations 188–190; phenotypes 185; practical limitations 191
interaural cross-correlation coefficient (IACC) 45, 77, 80
interaural level differences (ILDs) 114
interaural time difference (ITDs) 114
interdisciplinarity 3, 11
Intergovernmental Panel on Climate Change (IPCC) 333
interior spaces 9, 16, 319, 320, 324

intimacy 9, 265, 379; acoustics 373, 374–375; defined 87, 89; domestic space 88–89; home studios 93–94; music recordings 87–88; portrayed subjective 91–93; recorded domestic space 90–91; types, distance relations 91; video sharing/social media 270; virtual recorded space 87
intimate sounds 87
intimate space 269–272
Irwin, R. 432
Iversen, M. 130

Jackson, P. 16, 17
James, K.: 'Prepared to Love' 156
James, R. 149
Janus, A. 141

Kadow, C. 333
Kahn, D. 247, 250
Kanngieser, A.M. 330, 331
Kantor, T. 361, 367
katabatic wind 441, 443
Keats, J. 215
Kelly, A. 407
Kent, W. 132
Khait, I. 333
Khannanov, I. 211
Kiewa Hydroelectric Scheme 447
Kiewa Valley *448*
Kim, C.S. 160
Kim, S. 418
Kircher, A. 471
Kish, D. 174
Kleiner, M. 90
Klenner, M. 481
Klimoski, J. 26
Klorman, E. 92
Knowles Electronics Manikin for Acoustic Research (KEMAR) 112, 114
Kommune, S.: *Lydens By* 227
Krakowiak, K. 238, 239, 240
Kramer, E.M. 265
Kraugerud, E. 17
Krause, B. 330
Krauss, R. 133, 432
Krebs, S. 167
Kubisch, C. 245
Kulininpalaju project 444
Kurth, E. 289
Kuusinen, A. 375, 379

LaBelle, B. 8, 330, 418, 433, 434
laboratory excitement 98–101
laborious/time-consuming participatory tests 44
La Casa, E. 438
Lacey, J. 211
LaFave, K. 142

Lai, T.-Y.S. 135
Lake, K. *58*
Lanois, D. 90
Laocoon 133
Larsson, P. 90
Laurel, S. 167
Le carré (Square) 291
Ledoux, C.N. 133
Lee, H. 117
Lefebvre, H. 362, 365, 367, 415–418, 422
Lessing, G.E. 133
Levin, D.M. 7, 217, 218
Lewis, A. 145, 175
Libera, M. 238, 239
Ligeti, G. 267
light, precision 103
linear chromatic progression 294
linear prediction coding (LPC) 41
Linehan, F. *63*
liquid architectures 459–460
listening 141–142; acousmatic 205; audience 109; binaural 24, 171–172; body, imagining 158–161; case study 161–162; constellation 415–418; devices 4; environments 36; evaluation 388; fictional elements 332; form 420–421; and graphic transcription 412, *413–417*; material 423–424; methodologies 413; PA 202–203; participatory tests 47; pedagogy 424; perspectives 335; posthuman *see* posthuman listening; practices 344; practice strategies 396; reasons for graphic notation 421–422; repertoire 401; resonance 424; small systems, curating and composing 404–405; sound 331–332; space 422–423; speaker orchestra design principles 405; status quo, mediatised space 453; structure 422; subject recording 78; transcription 418–420
listening-in-readiness 433
listening-in-search 433
Living Sound, Patent Pending (Amacher) 241–242
Lobachevsky, N. 289
Lobkovsky, A.E. 265
Lobo, M. 331
Lockwood, A. 437
Lokki, T. 193
Lomax, A. 437, 438
Lombard reflex 168
Long Range 98, *99,* 100–105, *102*
Lopez, M. 481
loudspeakers 43, 46
Louro, I. 331
Lucian, P. 108, 139, 141, 142, 144–149
Lucier, A. 245
Lydens By (Kommune) 227

machine learning 10, 29, 45, 476
Macmillan 361

MacNeice, L. 154
macro-auditory environments 31
Maes, N. 268
Maier, C.J. 332
Main Hall at Sands End Arts and Community Centre *368*
Malpas, J. 211
Manifesto of Transdisciplinarity (Nicolescu) 2
Manolopoulou, Y. 3
Margetson, E. 344
Marshall, B. *63*
Marshall, H. 375
Martin, G. 265
Master of Design Innovation and Technology (MDIT) 403
materiality 7, 15, 31, 107, 126, 132, 148, 156, 180, 231, 238, 243, 245–247, 267, 279, 299, 389, 423, 456
material sound transmission: aesthetics 241–244; politics 243–244
Matsumoto, K. 366
Matthews, E.K. 343
Mattos, F. 344
maximum length sequences (MLS) 78–79
Mayol, P. 89
McArthur, A. 6, 344
McDonnell, B. 6, 344
McDowall, E. 158, 160
McGrath, J. 365–366
McLeod, R. 402, 409
McLuhan, M. 165, 166
McWhinnie, D. 154, 158
mechanical/acoustical recording 90
media content 49
Medwin, T. 216
megaphones 278
Meyrowitz, J. 88
Michelangelo 366
microaudial listening 266
micro-auditory environments 31, 33
microelectromechanical (MEMs) microphones 121
microphones 68
microphone technology 438
micro-sound practices 211
microsounds 265–267, 269
Mills, M. 199, 201
Milton, J. 215
Mirschel, W. 333
mobile headphonespaces 458–459
Mocquereau, A. 127
modal density 73
modal superposition method 73
modal theory 72–73
mode shape 72
Monocular Abyss (1982) 174
Moore, H. 432

Morris, R. 433
Morton, T. 3, 274
Mount Whaleback Mine *444*
Moving Picture Experts Group (MPEG) 38
Mozart 141, *187*, 381
Mrs Dalloway (Woolf) 256
Mulhallen, J. 216
multi-channel technique 462
multidisciplinary 3, 7, 11
multimodal resource 11
Murch, W. 168
Murray, J.J. 199
Murray, N. 402
Museum for Spatial Sound (2014) 400
Museum of Sound (2004) 400
museums 280, 360, 403, 426
music 10, 343–345; acoustic/compositional actions 194; atonal and tonal 290–295; composition 8, 348; limitations 188–190; mathematical topology 288–290; meta-scientific language 295; orchestral 3; performance of 369; performances 27, 64; practice and theory 52; production technology 93; recordings 87–88; space problem 285–288; stand, togethering 3
musical documentary drama 67n10
Musical Instrument Digital Interface (MIDI) 350
musical instruments 69
musical modernity 388
music curation: concert venue shaping 427–428; defined 426–427; frame 431–435; listening experience 426; outdoor entertainment 428–431
music notation: auditory symbols *133*; autumn afternoon transcription 132; beyond rhythm 133–136; distorted perspective *131*; origin of 126–128; space representation 128–132
music theory 211, 287
Myers, N. 330
'My Love is You' (Byrne) 91–93
Myrbeck, S. 16

Namblard, M. 438
Namblard, O. 438
Nancy, J.-L. 141–143
Napolin, J.B. 244
National Sawdust *368*
natural frequencies 72–73
Navigations: Scoring the Moment 226
Negus-Fancey, C. *63*
Neimanis, A. 330
Nendel, C. 333, 334
Neuhaus, M. 224, 278, 279
Neumann 167
neumatic notational approach 132, 133, 135
neutral spaces 432
Newcomb Bay *442*
Newell, C. 18

Newman 443–445
next-generation audio (NGA) 38, 39, 46, 48
Nicolescu, B. 3; *Manifesto of Transdisciplinarity* 2
Nicholson, H. 363, 364
noise cancellation systems 118
noise-cancelling headphones 8
noise reduction coefficient (NRC) 100
noisy signals 305
non-binaural spatial cues 167–168
non-musical social spaces 417
Nono, L. 347
non-traditional research outputs (NTROs) 404
Noord, G. van 66
Normington, K. 363, 364
North Atlantic Treaty Organisation (NATO) 118
Novo, P. 344

Obadike, K. 240, 244
Obadike, M. 240
object-based audio 38–39, 43, 48
object-based signal representation 40
O'Callaghan, C. 141, 142
occlusion 173–174
ocularcentric paradigm 7
ocularcentrism 322
Odio-App *458, 459,* 464
The Office for Common Sound 225
O'Keeffe, C. 55, *56*
Oliveros, P. 280, 330
onomatopoeia 219–220
OpenAIR datase 44–45
open plan designs 320
open-plan layouts 30–31
operatic production 381–382
optimum space 182
Orbit Pavilion 25–26, *26*
Orchestra 430
Orchestrating Space by Icosahedral Loudspeaker (OSIL) 455
organism-environment ecosystem 219
organism physiology 309
'The Origin of Perspective' (Damisch) 130
ornamentation 206
Orpheus fragment 216
Orwell, G. 255
oscillation convergence *313*
O'Sullivan, S. 332
outdoor entertainment 428–431
outer sound 256–257
outside spaces 344
Ouzounian, G. 171, 210, 278
overtone system *140*

Paine, G. 330
paleohistoricity 308
panacusis loci 174–175

Panofsky, E.: 'Perspective as Symbolic Form' 130
Panourgia, E.I. 212
Parker, R. 344, 373
Parmegiani, B. 267
partimento techniques 287
Pasnau, R. 141
passive receptacle 367
Patel, R. 52–61, *56, 59, 61, 63,* 62–64, 66
Pätynen, J. 193
perceptions: auraltypical paradigm 170; ecological approach 29; impression of 302; physical and virtual realms 1; *vs.* sensation frequency *306*; sound 19; space 8; spectrum of 456–458; *see also* spatial hearing
percussion instruments 69
Perec, G. 412
Perez-Gomez, A. 133
performalism 367
performance events 363, 364
performance space 9, 360–361, 363
Performative and Public Spaces 109
peripheral auditory system 114
permutations 367
personal audio technology 8
'Perspective as Symbolic Form' (Panofsky) 130
Phantom Thread (Anderson) 271
Philip the Chancellor 291
philosophical-topological approach 211
philosophy 2, 10, 108, 141, 209–212, 246, 247, 457, 463; minor 292
photocopier test 44, 45
physicality of sense 306
physical scale models 53, 54
physiology 9, 107–109
Piano, R. 347
Pierrot Lunaire 381–382
piezoelectric mineral transductions 303
Piranesi, G.B. 131, *131,* 134
place of performance *see* performance space
platonic theory of knowledge 290
playback technologies 4
Pledger, D. 407
Plytas, N. *58*
poetic composition theory 218
The Poetics of Space (Bachelard) 88, 219
polar patterns 439–443
Poliakoff, J. 198
politics 8, 10, 140, 201, 209–212, 422, 437, 478; material sound transmission 243–244; posthuman approach 282; room tone 205
Pollard, J. 17, 52, 55–66
polychronicity 310
Ponting, H. 441
post-dualistic perspective 279
posthuman listening: decentring human listener 274–279; defined 282;

final proposition 281–282; First Nation deep listening practices 279–281; opening remark 274
posthuman principles 275–277
post-Young British Artists (YBA) 53
potentiality 366
Poulet, C. *56*
practice-based activities 404
practice-based learning 399
Praz, M. 365
'Prepared to Love' (James) 156
Prest, K.: Afternoon Delight 156
Pretty Valley *448*
Price, U. 134
privacy 30–31; experiment 33–34
private contemplative moment 22
private space 230, 380
Prokofiev, S. 135
Proof of Ohm's Law 149
proxemics 168
pseudorandom noise signals 78–79
psycho-acoustical studies 463
psychological spaces 63
psychology 9, 68, 87, 91, 107–109, 157, 168, 171, 209
Psycho-Politics (Han) 268
psycho-social disabilities 277
public address (PA) system 201; listening points 202–203
public concert 427–428
public spaces: settings 9; sound 8, 10
Purposeful Listening in Complex States of Time (1998) 142

quietness 8, 16
Quin, D. 438
Quinell, W. *59*

Raby, F. 331
Radio Mania (2009) 58, *58*
radio technology 154
radio transmission techniques 305
radio wavelengths *303*
Radique, E. 475
Rameau, J.-P. 287, 293
Rankin, S. 128
raytracing analysis 5, *189,* 349
readiness 270
reading aloud: anatomy 214; cultural prejudice 213; harmonious and rhythmical expression 216; history of 213–214; human cognition 214; mediation 217; moral thinking and behaviour 215; onomatopoeia 219–220; performance aptitude 216; poetic composition 218; principle of harmony 217; reduplication 220; self-talk 218; sound 218; vocalisation 220
Reaper™ techniques 401, 402

reasonable adjustments 160
recordings: ambisonic 353; audio and spatial improvements 4; binaural 17, 77, 112, 114; content 4; direct-to-disk systems 438; electromechanical 90; live setting projects 56–57; music 87–88; practice 93; production 93; studios 22–23
recreational tourism 449
reductionist approach 140
Reeves-Evison, T. 332
refraction-transduction: temporal shapes 310–311; waveforms *311,* 311–313
relationality 389–392
relational spatiality 300–301
religious worship 319
repertoire listening 401
res cogitans 285, 290
res extensa 285
resonances 261, 280, 471; acoustics 15; defined 2; disciplinary 2–4, 4; listening 424; low-frequency 80; symbolic 154; transdisciplinarity 2
reverberant spatial audio object (RSAO) 16, 39, 40, 42, 45–48
reverberation 22; accumulation effect 104; acoustic behaviours 7; chamber 22–25; electroacoustic 4; interior spaces 319; microphone system 48; mixing time 40; physical and virtual spatial contexts 9; sound energy 38; spatial effects 4; *see also* reverberation time (RT)
reverberation time (RT) 16, 44, 73–74, 82, *83,* 238, 240, 346–347
reverb simulation 88
rhythms 134
Riemann, B. 289
Riemann, H. 287
Risset, J.C. 462
Rivers, B.: Two Years At Sea 267
Roads, C. 266, 267, 462
Robinson, D. 175, 203, 280
rock-art sites 110
Rock petroglyphs of Valcamonica 110
Rode NT-SF1 376
Roederer, J. 290
Rohe, M. van der 287
Ronell, A. 174
room acoustics: analysis 79–80; defined 72; modal decomposition approach 72; modal density 73; modal theory 72–73; reverberation time 73–74; simulation 383n16
room impression: audio media 38; binaural reproduction 43–44; extracting parameters 41–42; flexibility 46; formal listening trials 45–46; hard *vs.* soft surfaces 38; loudspeakers 43; object-based signal representation 40; objective measures 44–45; outcomes 44; perspectives 48–49; pictures to sound space 45; production 43;

proposed methods 39; reflections 38; reverb effects 38; room impulse response *40*; RSAO parameter encoding process *42*; spatial audio's third dimension 46–47; SRIRs 40–41; users' assessment 47; VR experience *46*

room impulse response (RIR) 478; description *76, 76*–77; excitation signals 78–79; measurement system 77–78, *78*; room acoustics analysis 79–80

Rose, N. *56*

Rothko, M. 476

The Routledge Companion to the Sound of Space 1, 4

Royal Melbourne Institute of Technology (RMIT) 400

Runway, W. 439

rustic, idyllic and lyrical 141

Ruttmann, W. 437

Sabine equation 16, 74, 104, 105

Sabine, W.C. 16, 74, 99, 104, 240, 347

Sagrada Família Basílica: architectural implications 84; architectural standpoint 80–81; experimental setup 81–82; hyperboloid surfaces 32, 71; interior *75*; middle frequency measurement *84*; objective metrics 83–84; reverberation time 82, *83*; sound source and measurement locations *82*; virtual acoustic model 17, 74

Saldanha, S. 156

Salter, L. 387

Salzer, F. 140

Samartzis, P. 344

sanctuary 22, 224

San Francisco Museum of Modern Art (SFMOMA) 21, *21*

San Giorgio Maggiore 479

Sanguinetti, G. 287

Santa Maria della Salute 479

scattering coefficient 70–71

Schaeffer, P. 203, 437

Schafer, R.M. 145, 146, 165, 203, 277, 320, 322, 400, 401; Temples of Silence 278

Scheerbart, P. 244

Schenker, H. 139–140, *140,* 142, 146

schizophonia 203, 205

Schlieren photography 304, *304*

Schmidt, D. 20

Schroeder backward integration method 79

Schroeder, M.R. 79, 192

Schubart, C. 146

Schueller, H. 144

Schultz, F. 455

scientism 385

sculpturality 456, 466

Seashore, C. 142

selflessness 366

self-talk 218

sensation clusters: hearing long-distance 304–305; inorganic 304; noisy signals 305; physicality of sense 306

sensing technologies 439

sensory anthropology 109, 169, 211, 271

sensory response 266–268

sensory techniques 303

Severn, J. 214

Sgricci, T. 214–216

shaping sounds: Brandenburg region, drought scenarios 333–334; Potsdam, listening and field recording 332–333, *334*; recording underwater and surface vibrations *335*; water-drought patterns 334–336

Shapland, W. 56

shared perceptual space (SPS) 457, *463, 464*

Sharma, G.K. 6, 345, 454, 455

Shaw, J.K. 332

Shelley, P.B. 210, 213–218

shifting frequencies 330, 335

Shockz 121

Sibelius software 350

silences 104, 128, 146–150, 210, 220, 257–261, 264, 278, 325, 419–420, 424

silent disco systems 120

silent reading 213

Silent Sound (2006) 55–58, *56,* 60

simultaneity 18, 92–94, 204

situatedness 211, 317–319, 323, 325

six degrees of freedom (6DoF) 45, *46*

Slaby, A. *303*

Smalley, D. 90, 92, 462

Smith, N. 270

Smolicki, J. 330

Snider, A.-M. 270

Snow, M. 174

social morphology 362

social-space violation 169

socio-political space 9

solitude 258–259

sonic: agency 211; archaeology 474, 477; environmental states 336–337; environment constraints 22; fields 1–2; materialities 245–250; proxemics 168–169; qualities 301; space of appearance 206; spatiology 452

Sonic Urbanism 225–226

Sontag, S. 258

sound absorption average (SAA) 100

sound absorption coefficient 71

sound absorption rating 100

sound art 10, 246, 343–345; and acoustic ecology 437; compositions 393; developments 265; and political expression 210; practices 8; traditions 239, 241; *see also* music

sound cards 54, 56

sound column *20,* 20–21
Sound Craft (2019) 400
sound experiences 52, 167, 404
soundfield processing 116, 121
sound files 450
sound fixing and ranging channel (SOFAR) 117
Sound-Frameworks 225–228, 230–232, *232–234,* 234
sound generation 68–69; human speech 68–69; musical instruments 69
Sounding Vessels in the Theatre 7
sound-making practices 8
soundmapping technique 403
sound of nostalgia 141
the Sound of Silence 108, 139, 141, 142, 145, 146, 148, 149
Sound of Space Symposium 6
sound penetration 320, 325
sound perception 33–34
sound pressure level (SPL) 377
sound propagation 330; effects of boundaries 70–71; free-field 70; wave equation 69–70
sound recording technology 4
sound reflection 70–71
sound reinforcement system 27
soundscapes 26, 209–211, 274, 327n5, 459–460; acoustic shadows 31, 326; composition 401–403, 409; emotional change 186; immersion 241; liquid architectures 465; multitude 330; ochlophonic 168; positive design 148; posthumanist understanding 277–279; posthuman principles 275–277; spirituality 279; technologies 120; tuning 144–146; urban 139, 141, 144
sound's spatial-material circuitry: ambient selective attention 303; attention circuits 302–303; attention sediment 308–309; dynamics future-tensed 310; environmental frequency 309; interlinking 308; mechanism 298–299; operations 298; poly-temporal 310; refraction-transduction 310–313; sensation clusters 304–306; shared hearing 313–314; spaces of sound 299–302; vibrant subjectivity 306–307
space-making 8
Spaceman, J. 55, *56*
space-oriented curatorial strategy 429
spaces of sound: auditory event 299; context 300; definitions 299; fluctuant emplacement 301–302; ontological pluralism 300; relational spatiality 300–301; scale 300; spatial qualities 301
space synthesis 245
spaciousness 83–84, 89, 164
spanning, researchers and practitioners 1
spark-gap discharges *304*
spatial aliasing 117

spatial audio technologies 5, 344, 453, 461, 464–465; third dimension 46–47
spatial auditory cues **113**
spatial composition-design 402–403
spatial configuration 319
spatial diversity 302
spatial drift 461
spatial hearing: acoustic space 165; asymmetrical hearing 172–173; auraldiveristy 170; aurally diverse listening positionality 175; auraltypical perception 170; azimuth 166; binaural listener 171–172; cocktail party effect 168; dummy head 166–167; ears 173; ear wiggling 167; elevation 166; envelopment 167; field of 165; and freedom 175–176; free field and diffuse field 169; high-speed hand dryers, resonant and reverberant toilets 171; hyper-asymmetry 174; joy 169; lateralisation 167; leading ear 174; Lombard reflex 168; non-binaural spatial cues 167–168; occlusion 173–174; origin 165; otologically normal 171; panacusis loci 174–175; paracusis loci 172; sonic proxemics 168–169; speech perception and intelligibility 169–170; suitably equipped receptor 172; symmetrical hearing 166; topography 164
spatial imaging 394
spatial impression 375
Spatial Information Architecture Laboratory (SIAL) 400
spatiality: audio technologies 2, 5; auditory experiences 4; hearing 112; IKO 385–386; positionality 317–318; responsiveness 375; simultaneity 92–94; sonic fields 1–2; sound techniques 402
spatial room impulse responses (SRIR) 16–17, 39, 40–41, *42*
spatiosonic dialogues: acoustic response 348; activating space 354–355; architectural design practice 357; articulating space 351–354; context 347; defined 346; performer architecture 348–350; rehearsing and simulating space 350–351; relationships 355–357; reverberation time 346–347
spatiosonic practice 348
spatiotemporal materialities 300, 306, 310
speech production system 69
speech-singing 380
spiritualism 55, 279
Stacy, B. *61*
StarSound 402
SteadyHealth 266
Sterne, J. 170, 199, 277, 351
St. Gallen neumatic notation *129*
stillness 259–260
Stockhausen, K. 290–291, 422
Stoeckig, K. 142

Index

Stravinsky, I. 286
Street, S. 154
strings instruments 69
structure-borne sound 241
studio-based activities 402–403
Sueur, J. 330
suitably equipped receptor 172
surface materiality 319
surface sonic layer depth (SLD) 117
SurrRoom dataset 47
sweep sine signals 78–79
swept-sine technique 44
symbolism 160, 206
symmetrical hearing 166

tacit knowledge 354, 388
Tajadura-Jiménez, A. 91
Taj Mahal 264, 474, 480
Takemitsu, T. 291
Taun, Y.-F. 175
Taylor, M. 211
teaching 157, 161, 203, 277, 419, 423; acoustic design 400; activities **400**; class-based activities 401; composers 457; exhibitions 407–409, *409*; fieldwork 403–404; influence modes 399; intimate listening 404–405; intimate speaker orchestras 405–406, *406*; materiality 424; planting, embedding and configuring layers 406–407, *407, 408*; practice-based activities 404; research programs 466; and spaces 399–400; statements 409; studio-based activities 402–403
technical video collection 401
technology: alternative reality 48; ambisonics 114; assemblages 389; audio media 38; augmentation 108; avant-garde techniques 291; beam-forming 385; contemporary 55–56; customisation and personalisation 387; defence 117; in Europe 202; evolutionary programming and biometric sensing 191; of existence 279; innovation 181; military 290; music production 93; recording 88; revolution 295; sociocultural transformations 281; sound 4–5; spaces 4–5, 9, 57; ultrasonic scanning 29
telecoil ('T-coil') 197, 198, 201, 202, 205
Temple of Echo 132
Temples of Silence (Schafer) 278
Tempo Exercise techniques 403
temporalities 126, 131, 134–136, 300, 310, 317, 331, 335, 423
textural resources 401
texture 15, 21, 26, 31, 158, 181, 211, 245, 247, 266, 335, 336, 399, 421, 457
theatre performance 64
Theatrophone 202

Theatrum Mundi 225–226
theory of mind 215
theory of source bonding 90
Thibaud, J.-P. 432
thinking self 259, 260
30-dB-decay reverberation time (RT30) 45
Thomas, D. 154
Thompson, E. 145, 278
Thompson, M. 246
3D: acoustic computer model 382; audible space 4; audio branch technology 114; digital model 349; fully spatial audio 53, 59; immersive surround sound 172; soundfields 116; spatialisation systems 389; spatial modelling tools 5; stereoscopic video 58
time: accelerated 335; acoustic energy 80; amplitudes 77; continuity 256; cyclic 312; delay gap 40; distribution of neural pulses 290; geological 309; human interaction 15; limitations 136; minimal change 475; mixing 41; moving sound 381; musical 293; notation 134, 135; performance venues 430; radio transmissions 154; reverberation *see* reverberation time (RT); spatial design parameter 403; spatial locations 69; and transition 108; values 128; variance 310; warfare 260; zones 401
time-domain synthesis 79
Toepler, A. *304*
togethering 2–4
Tomatis, A. 174
tonal-harmonic function 287
tonal music 287, 290–295
Tonder, C. van 345, 479, 481
Tonkiss, F. 261
Toom, P.C. van den 141
Toop, D. 211
topology: communication and connection 326; connectedness 325; dreams 443–446; music *see* music; situatedness 317–319, 323, 325; vibrational 242
Toth, L. 256
tranquillity 258–260, 269, 480
transcription, screech and crack 418–420
transdisciplinarity 2, 3
transhumanism 279, 281
transmission loss (TL) 71
transportation systems 321
Transport for London (TFL): disability and deafness 200; integrated deafness training 197
transversal listening practices 279
transverse flows 307
Trevithick, J. 117
Tristan 291
Truax, B. 145–147, 401, 433
Tsunoda, T. 245

495

Index

Tuan, Y.-F. 361, 363, 365
Tudor, D. 143
tuning of soundscape 144–146
Turkle, S. 5
Turner, J.M.W. 108, 130, 131
Tvísöngur 22
Two Years At Sea (Rivers) 267
Tyburski, M. 108, 139, 141, 142, 145, 146, 149
Tyers, J. 428, 429
Tyrrell, J. 109, 240, 241

UK Holocaust Memorial International Design Competition *228*
Ultrasonic Acouspade 121
ultrasonic scanning technologies 29
ultrasonic sound 66n6
ultrasonography 29
universal constants 145
unpredictability 344, 394, 395
unvoiced speech 69
Urban Backstages 226
urban centres 8
urban development 227, 230–231
urban space 10, 29, 396, 438
Ursatz 139–141, 144

vagabond geometry 211, 288
Valcamonica astronauts 115
Valiant, D. 278
Vallee, M. 203
value chains: embedding sound 227–228; urban projects 210; *see also* affordances
Vandenbussche, L. 373
Varèse, E. 194, 239, 241, 290, 453
Västfjäll, D. 90
Vatalaro, P.A. 217
Vauxhall Gardens 428–431
vector base amplitude panning (VBAP) 43, 455
ventilation system 238, 376
verbalisation 457, 466
vibrational architectures: anti-monumental architectures 245; defined 238, 240; live sonifications 239; material transmission 241–244; reconfiguration 239; reverberant space 239–240; sonic materialities and architectural energies 245–250, *246–250*; sonic warfare 240
vibrational ecology 240, 242, 243, 251n16
vibrational space 199, 299
vibration-coupling transducers 299
video installation 55, *58*
violent protests 60–61
Virdi, J. 199
virtual acoustic model 17, 81, 91, 183, 186–187, 401, 474, 477, 478

virtual audio technologies 10
virtual loudspeakers 43–44, 117
virtual reality (VR) 4, 5, 58, 60, 114, 453, 454, 472
virtual spaces 4, 9, 17, 24, 45, 90, 91, 93, 268
Vismara, G. 479
visualisation 121, *232,* 323, 401
visuality 229, 323, 327n21
Visual Studio Team Services (VSTs) 402
Vitruvius 7, 15, 128, 166, 347
vocal movement positions diagram *65*
vocal sidetone 168
Voegelin, S. 243
voice box 68
voiced speech 69
voice performance 213
Vygotsky, L. 218

Walker, S. 62
Walsh, P. 62, *63*
Wand, A. 331
warmth concept 80
water-drought patterns 334–337, 339n39
Waterhouse, J.W.: Consulting the Oracle 205, *205*
wave physics 19
weathering 330–331
Weierstrass, C. 289
Wendt, F. 455
Werner, J.S. 210, 240, 245, *246–250*
Westerkamp, H. 145
White, Educated, Industrialised, Rich, Democratic (WEIRD) approach 171
Whitehead, G. 161
Wilkes Station *440*
Willaert, A. 347
Williams, N. 408
Woodger, N. 53
woodwind instruments 69
Woolf, V.: *Mrs Dalloway* 256
work identity 349
World Soundscape Project (WSP) 165, 277, 399

Xenakis, I. 239, 267, 291

Yavorsky, B. 292
Yeats, W.B. 214
Young, S. 418

Zak, A. 90
Zaunschirm, M. 455
Zhao, S. 17
Žižek, S. 417, 418, 424
zones of intensity 433, 453–454
Zotter, F. 454, 455, 456